# Significant Phased Array Papers

## MODERN FRONTIERS IN APPLIED SCIENCE

*Gallium Arsenide Microwave Bulk and Transit-Time Devices*
*by Lester F. Eastman*

*Laser Applications*
*by W. V. Smith*

*Microwave Engineers Handbook*
*by Ted Saad*

*Logarithmic Video Amplifiers*
*by R. S. Hughes*

*Parallel Coupled Lines and Directional Couplers*
*by Dr. Leo Young*

*Microwave Filters Using Parallel Coupled Lines*
*by Dr. Leo Young*

*Data Modem Selection and Evaluation Guide*
*by V. V. Vilips*

*Design, Performance and Application of Microwave Semiconductor Control Components*
*by Kenneth E. Mortenson and Jose M. Borrego*

*Adaptive Electronics*
*by Dr. Wolfgang Gaertner*

*Phased-Array Antennas*
*by Dr. Arthur A. Oliner and George H. Knittel*

*Noise Performance Factors in Communications Systems*
*by W. W. Mumford*

*Avalanche Transit-Time Devices*
*by Dr. George I. Haddad*

*Infrared to Millimeter Wavelength Detectors*
*by Frank R. Arams*

**ARTECH HOUSE, INC.**
**affiliated with Horizon House – Microwave, Inc.**
**610 Washington Street**
**Dedham, Massachusetts, 02026**

# Significant Phased Array Papers

R.C. Hansen

ARTECH HOUSE, INC.
Dedham, Massachusetts

affiliated with
Horizon House-Microwave, Inc.

ARTECH HOUSE, INC.

Printed and bound in the United States of America.



Library of Congress Catalog card number: 73-81240

# Significant Phased Array Papers

# TABLE OF CONTENTS

# Introduction

Phased array hardware and theory was at a primitive stage in the World War II era. Then came the rapid developments at Hughes Aircraft in the early fifties. Since that time the technology has reached a significant level of maturity.* Sophisticated theory has explained blind spot phenomena which were undiscovered and unexpected as late as 1955. Impedance matching techniques supported by analysis allow excellent performance over a wide range of scan angles. Near simultaneous search, track, discrimination, and designation of many high velocity targets is achievable. Perhaps the ultimate indicator of technical development is the serious consideration now being given to phased arrays for fire control radar in jet interceptor aircraft. Few applications are as demanding as this one.

The purpose of this book is to present in one volume the most important and classic papers that were published along the way as phased arrays developed. Only English language publications are represented; and, since most phased array work has been funded by the U.S. Department of Defense, all of the papers are from U.S. journals. As might be expected, the list of classic papers first compiled by the editor was much larger than the number that this book could economically include, and the editor's task was the thankless one of deciding which of the many excellent papers to omit. Since some readers may wish to take advantage of the extensive search and sorting process that produced this larger list, the papers omitted from this book are listed at the close of this introduction. It is a pleasure to note that this larger list contains other than U.S. contributions. Also, since several excellent books on phased arrays are in print, a list of these is also attached. Reports of government contractors have not been included, as essentially all the best material has subsequently been published in journal or book form. Before discussing the contents of this book, the other books will be mentioned. First, the only book extant covering the entire subject of phased arrays is the three-volume set entitled "Microwave Scanning Antennas," which contains contributions from many of the founders of this field. More specialized are two other books. One by Amitay et al treats mutual coupling in arrays of open ended waveguides, while a second by Mittra and Lee discusses Wiener-Hopf techniques as applied to mutual coupling calculations. Finally, the papers from a 1971 phased array conference at Brooklyn Polytechnic have been edited by Oliner and Knittel.

In this collection of significant papers, six groupings have been used, with no attempt to be either complete or chronological. Group 1 covers array analysis and synthesis. Elliott establishes general relationships for beamwidth and directivity of arrays, while Hansen evaluates various directivity approximations. Cheng surveys synthesis techniques in depth, including optimization of S/N. Matrix methods for constrained pattern synthesis are in the Strait and Hirasawa paper. The von Aulock paper was one of the first to use direction cosine coordinates and to discuss beam shape with scan in such coordinates. The Wheeler papers are classic in using simple models to obtain the basic behavior of impedance with E or H plane scan and in writing the array field in terms of a grating lobe series. The latter is a doubly infinite sum of element impedances written in impedance crater form.

The element-by-element sum using Carter mutual impedances (used in the late fifties) is useful for dipole arrays, but more complex elements demand correspondingly powerful methods. In the second group of papers on mutual coupling effects, Diamond computes complex power over the array aperture, including element multiple modes as necessary. Farrell and Kuhn perform a mode matching between modes external to the array and internal modes. Stark uses a different technique which is in principle that also used by Oliner, where a single cell in an infinite periodic array is analyzed using modes in this unit cell. This gives radiation resistance as a single term and yields important insight into parameter interrelationship. For geometries involving fences or baffles, DuFort relates element scattering parameters through the parallel plate region to external scattering parameters. Wasylkiwskyj and Kahn define a general array element in scattering terms and then derive the grating lobe series from this. Finally, integral equation methods are used by Amitay and Galindo for arrays of waveguide horns.

Several of these methods illuminate the perplexing problem of blind spots, where the overall pattern has a null at an angle near the grating lobe onset. As shown by Knittel, Oliner, and Hessel, these are caused by leaky waves over the aperture, supported by the element reactances. Surface waves propagating on dielectric covers also may exist and cause blind spots, as shown by these authors and by Mailloux. These papers, then, make up Group 3 on blind spots, although the demarcation between Groups 2 and 3 is not well defined.

Group 4 covers impedance matching and simulation. Contrary to an earlier hypothesis, it is shown by Varon that a perfect match for all scan angles is not possible, and that a good match may only be approached. Use of higher modes in the radiating elements is shown by Lee to allow excellent matching over a scan angle range; the higher modes are produced by irises or dielectric plugs. Waveguide simulators using a single mode are developed by Balfour and colleagues, and this technique is extended to multiple modes by Gustincic. This latter technique allows grating lobes to be included.

Group 5 papers represent somewhat of a departure in that conformal arrays are the subject. Sheleg describes a circular array for azimuth wide angle scan. The cylindrical array work of NELC is discussed in detail by Boyns, Provencher, et al. Finally, mutual coupling on a cylinder is studied by Sureau and Hessel, where the curvature allows the various leaky wave components to be identified and separated.

In the last group are papers more system oriented. Johnson describes the MOSAR beam forming system, which uses frequency multiplexed sampling. Sidelobe constraints on multiple beam Butler matrix arrays are investigated by Shelton. Beam position granularity when digital phase shift is used is treated by Hatcher. And Kmetzo looks at multiple face array tradeoffs for covering large scanning volumes.

Because of space limitations, it was not feasible to cover adaptive or data processing array, nor seismic or acoustic arrays. Radio astronomy arrays were likewise excluded. However, it is hoped that the classic papers on electronic scanning assembled here will be useful to array designers, to systems engineers, and to those who are increasing their knowledge of scanning arrays.

R. C. Hansen
September 1972
Encino, California

# References


N. Amitay, "Improvement of Planar Array Match by Compensation Through Contiguous Element Coupling," Trans. IEEE, Vol. AP-14, September 1966, pp. 580-586.

N. Amitay and V. Galindo, "The Analysis of Circular Waveguide Phased Arrays," Bell System Technical Journal, Vol. 47, November 1968, pp. 1903-1932.

N. Amitay et al, "Match Optimization of a Two-Part Phased Array Antenna Element," Trans. IEEE, Vol. AP-16, January 1968, pp. 47-57.

H. Bach, "Directivity of Basic Linear Arrays," Trans. IEEE, Vol. AP-18, January 1970, pp. 107-110.

C. V. Baker, "A Technique for the Analytical Synthesis of Shaped Beams for Arrays," Trans. IEEE, Vol. AP-17, November 1969, pp. 803-805.

G. V. Borgiotti, "A Novel Expression for the Mutual Admittance of Planar Radiating Elements," Trans. IEEE, Vol. AP-16, May 1968, pp. 329-333.

G. V. Borgiotti, "Modal Analysis of Periodic Planar Phased Arrays of Apertures," Proc. IEEE, Vol. 56, November 1968, pp. 1881-1892.

G. V. Borgiotti, "Edge Effects in Finite Arrays of Uniform Slits on a Ground Plane," Trans. IEEE, Vol. AP-19, September 1971, pp. 593-599.

G. V. Borgiotti and Q. Balzano, "Analysis of Periodic Phased Arrays of Circular Apertures," Trans. IEEE, Vol. AP-17, March 1969, pp. 224-226.

G. V. Borgiotti and Q. Balzano, "Mutual Coupling Analysis of a Conformal Array of Elements on a Cylindrical Surface," Trans. IEEE, Vol. AP-18, January 1970, pp. 55-63.

R. W. Breithaupt and G. C. McCormick, "Traveling Wave Array of Mismatched Elements," Trans. IEEE, Vol. AP-19, January 1971, pp. 4-11.

G. J. Buck, "Reflection-Lobe Dispersion in a Phased-Array Antenna," Trans. IEEE, Vol. AP-18, November 1970, pp. 802-804.

E. V. Byron and J. Frank, " 'Lost Beams' from a Dielectric Covered Phased-Array Aperture," Trans. IEEE, Vol. AP-16, July 1968, pp. 496-499.

E. V. Byron and J. Frank, "On the Correlation Between Wide-band Arrays and Array Simulators," Trans. IEEE, Vol. AP-16, September 1968, pp. 601-603.

D. K. Cheng, "Optimization Techniques for Antenna Arrays," Proc. IEEE, Vol. 59, December 1971, pp. 1664-1674.

R. M. Cox and W. E. Rupp, "Circularly Polarized Phased Array Antenna Element," Trans. IEEE, Vol. AP-18, November 1970, pp. 804-807.

R. B. Crane and C. B. Sharpe, "Limitations of Series-Fed Arrays in Broadband Communications," Trans. IEEE, Vol. AES-2, November 1966, pp. 659-664.

D. E. N. Davies and B. S. McCartney, "Cylindrical Arrays with Electronic Beam Scanning," Proc. IEE, Vol. 112, March 1965, pp. 497-505.

C. J. Drane, "Useful Approximations for the Directivity and Beamwidth of Large Scanning Dolph-Chebyshev Arrays," Proc. IEEE, Vol. 56, November 1968, pp. 1779-1787.

E. C. DuFort, "Finite Scattering Matrix for an Infinite Array," Radio Science, Vol. 2, January 1967, pp. 19-27.

E. C. DuFort, "Design of Corrugated Plates for Phased Array Matching," Trans. IEEE, Vol. AP-16, January 1968, pp. 37-46.

E. C. DuFort, "A Design Procedure for Matching Volumetrically Scanned Waveguide Arrays," Proc. IEEE, Vol. 56, November 1968, pp. 1851-1860.

A. G. Ehlenberger et al, "Design Criteria for Linearly Polarized Waveguide Arrays," Proc. IEEE, Vol. 56, November 1968, pp. 1861-1872.

G. H. Farrell and D. H. Kuhn, "Mutual Coupling Effects of Triangular-Grid Arrays by Modal Analysis," Trans. IEEE, Vol. AP-14, September 1966, pp. 652-654.

H. E. Foster and R. E. Hiatt, "Butler Network Extension to Any Number of Antenna Ports," Trans. IEEE, Vol. AP-18, November 1970, pp. 818-820.

J. Frank and J. Ruze, "Beam Steering Increments for a Phased Array," Trans. IEEE, Vol. AP-15, November 1967, pp. 820-821.

V. Galindo and C. P. Wu, "Asymptotic Behavior of the Coupling Coefficients for an Infinite Array of Thin-Walled Rectangular Waveguides," Trans. IEEE, Vol. AP-14, March 1966, pp. 248-249.

V. Galindo and C. P. Wu, "Dielectric Loaded and Covered Rectangular Waveguide Phased Arrays," Bell System Technical Journal, Vol. 47, January 1968, pp. 93-116.

V. Galindo and C. P. Wu, "On the Asymptotic Decay of Coupling for Infinite Phased Arrays," Proc. IEEE, Vol. 56, November 1968, pp. 1872-1880.

F. J. Goebels and R. K. Anderson, "A Dual-Band Slot Array Technique," Trans. IEEE, Vol. AP-18, March 1970, pp. 282-284.

N. Goto and D. K. Cheng, "Sidelobe-Reduction Techniques for Phased Arrays Using Digital Phase Shifters," Trans. IEEE, Vol. AP-18, November 1970, pp. 769-773.

P. W. Hannan, "The Ultimate Decay of Mutual Coupling in a Planar Array Antenna," Trans. IEEE, Vol. AP-14, March 1966, pp. 246-248.

P. W. Hannan and M. A. Balfour, "Simulation of a Phased-Array Antenna in Waveguide," Trans. IEEE, Vol. AP-13, May 1965, pp. 342-353.

P. W. Hannan et al, "Impedance Matching of a Phased-Array Antenna Over Wide Scan Angles by Connecting Circuits," Trans. IEEE, Vol. AP-13, pp. 28-34.

A. Hessel and G. H. Knittel, "On the Prediction of Phased-Array Resonances by Extrapolation from Simulator Measurements," Trans. IEEE, Vol. AP-18, pp. 121-123.

R. C. Hansen, "Directivity of Chebyshev Arrays," Trans. IEEE, Vol. AP-18, November 1970, pp. 815-818.

A. Hessel and J-C Sureau, "On the Realized Gain of Arrays," Trans. IEEE, Vol. AP-19, January 1971, pp. 122-124.

J. K. Hsiao, "Analysis of Interleaved Arrays of Waveguide Elements," Trans. IEEE, Vol. AP-19, Nov. 1971, pp. 729-735.

W. K. Kahn, "Ideal Efficiency of a Radiating Element in an Infinite Array," Trans. IEEE, Vol. AP-15, July 1967, pp. 534-538.

W. K. Kahn, "Active Reflection Coefficient and Element Efficiency in Arbitrary Antenna Arrays," Trans. IEEE, Vol. AP-17, September 1969, pp. 653-654.

G. H. Knittel, "Choosing the Number of Faces of a Phased-Array Antenna for Hemisphere Scan Coverage," Trans. IEEE, Vol. AP-13, November 1965, pp. 878-882.

L. W. Lechtreck, "Effects of Coupling Accumulation in Antenna Arrays," Trans. IEEE, Vol. AP-16, January 1968, pp. 31-37.

S. W. Lee, "Radiation From an Infinite Aperiodic Array of Parallel-Plate Waveguides," Trans. IEEE, Vol. AP-15, September 1967, pp. 598-606.

S. W. Lee and W. R. Jones, "On the Suppression of Radiation Nulls and Broadband Impedance Matching of Rectangular Waveguide Phased Arrays," Trans. IEEE, Vol. AP-19, January 1971, pp. 41-51.

S. W. Lee and W. R. Jones, "Surface Waves on Two-Dimensional Corrugated Structures," Radio Science, Vol. 6, August-September 1971, pp. 811-818.

D. J. Lewis and G. J. Buck, "Phase-Monopulse Tracking Errors Due to Mutual Scattering Effects in Phase-Scanned Arrays," Radio Science, Vol. 5, August-September 1970, pp. 1215-1220.

I. D. Longstaff et al, "Directional Properties of Circular Arrays," Proc. IEE, Vol. 114, June 1967, pp. 713-718.

A. R. Lopez, "Monopulse Networks for Series Feeding an Array Antenna," Trans. IEEE, Vol. AP-16, July 1968, pp. 436-440.

E. G. Magill and H. A. Wheeler, "Wide-Angle Impedance Matching of a Planar Array Antenna by a Dielectric Sheet," Trans. IEEE, Vol. AP-14, January 1966, pp. 49-53.

H. J. Moody, "The Systematic Design of the Butler Matrix," Trans. IEEE, Vol. AP-12, November 1964, pp. 786-788.

E. A. Nelson, "Quantization Sidelobes of a Phased Array with a Triangular Element Arrangement," Trans. IEEE, Vol. AP-17, May 1960, pp. 363-365.

R. W. Redlich, "Sampling Synthesis of Ring Arrays," Trans. IEEE, Vol. AP-18, January 1970, pp. 116-118.

W. H. Schaedla, "Equivalent Circuit Formulation for an Array of Phased Waveguide Apertures," Trans. IEEE, Vol. AP-18, January 1970, pp. 28-33.

L. Schwartzman and L. Topper, "Analysis of Phased Array Lenses," Trans. IEEE, Vol. AP-16, November 1968, pp. 628-632.

D. L. Sengupta et al, "Radiation Characteristics of a Spherical Array of Circularly Polarized Elements," Trans. IEEE, Vol. AP-16, January 1968, pp. 2-7.

C. B. Sharpe and R. B. Crane, "The Effects of Dispersion in Linear Arrays," Trans. IEEE, Vol. AP-15, March 1967, pp. 295-296.

J. P. Shelton, "Multibeam Planar Arrays," Proc. IEEE, Vol. 56, November 1968, pp. 1818-1821.

H. Steyskal, "On Antenna Power Pattern Synthesis," Trans. IEEE, Vol. AP-18, January 1970, pp. 123-124.

J-C Sureau and A. Hessel, "Element Pattern for Circular Arrays of Axial Slits on Large Conducting Cylinders," Trans. IEEE, Vol. AP-17, November 1969, pp. 799-803.

R. Tang and N. S. Wong, "Multimode Phased Array Element for Wide Scan Angle Impedance Matching," Proc. IEEE, Vol. 56, November 1968, pp. 1951-1959.

F. I. Tseng and D. K. Cheng, "Optimum Scannable Planar Arrays with an Invariant Sidelobe Level," Proc. IEEE, Vol. 56, November 1968, pp. 1771-1778.

D. Varon and G. I. Zysman, "On the Mismatch of Electronically Steerable Phased-Array Antennas," Radio Science, Vol. 3, May 1968, pp. 487-489.

W. Wasylkiwskyj and W. K. Kahn, "Mutual Coupling and Element Efficiency for Infinite Linear Arrays," Proc. IEEE, Vol. 56, November 1968, pp. 1901-1907.

W. Wasylkiwskyj and W. K. Kahn, "Scattering Properties and Mutual Coupling of Antennas with Prescribed Radiation Pattern," Trans. IEEE, Vol. AP-18, November 1970, pp. 741-752.

H. A. Wheeler, "A Systematic Approach to the Design of a Radiator Element for a Phased-Array Antenna," Proc. IEEE, Vol. 56, November 1968, pp. 1940-1951.

C. P. Wu, "Numerical Solutions for the Coupling Between Waveguides in Finite Arrays," Radio Science, Vol. 4, March 1969, pp. 245-254.

C. P. Wu, "Resonances in Waveguide Antennas with Dielectric Plugs," Bell System Technical Journal, Vol. 48, September 1969, pp. 2305-2324.

C. P. Wu, "Characteristics of Coupling Between Parallel-Plate Waveguides With and Without Dielectric Plugs," Trans. IEEE, Vol. AP-18, March 1970, pp. 188-194.

C. P. Wu, "Analysis of Finite Parallel-Plate Waveguide Arrays," Trans. IEEE, Vol. AP-18, May 1970, pp. 328-334.

C. P. Wu and V. Galindo, "Surface-Wave Effects on Dielectric Sheathed Phased Arrays of Rectangular Waveguides," Bell System Technical Journal, Vol. 47, January 1968, pp. 117-142.

A. I. Zaghloul and R. H. MacPhie, "Effects of Positioning Errors on Grating-Lobe Reduction in Uniform Linear Arrays," Radio Science, Vol. 6, March 1971, pp. 407-419.

## BOOKS ON PHASED ARRAYS

"Microwave Scanning Antennas," R. C. Hansen, editor, Vol. I: Apertures, 1964; Vol. II: Array Theory and Practice, 1966; Vol. III: Array Systems, 1966; Academic Press.

"Theory and Analysis of Phased Array Antennas," N. Amitay, V. Galindo, and C. P. Wu, Wiley-Interscience, 1972.

"Analytical Techniques in the Theory of Guided Waves," R. Mittra and S. W. Lee, Macmillan, 1971.

"Phased-Array Antennas," A. A. Oliner and G. H. Knittel, Artech House, 1972.

# ARRAY ANALYSIS AND SYNTHESIS I

# BEAMWIDTH AND DIRECTIVITY OF LARGE SCANNING ARRAYS*



R. S. ELLIOTT
PROFESSOR OF ENGINEERING
UNIVERSITY OF CALIFORNIA AT LOS ANGELES
Los Angeles • California

*First of Two Parts*

## INTRODUCTION

Many aspects of antenna array theory, particularly for uniform element spacing, are so widely known that they form a standard body of knowledge in textbooks.[1] In the case of linear arrays, general formulas for beamwidth and directivity have been established. However, in the case of planar arrays only the case of uniform excitation has been extensively inve tigated.[2,3] Surprisingly, no great advantage has been taken of the use of a Fourier series representation of the aperture distribution. Such a representation yields useful results of fairly general validity in the case of planar arrays, and even enhances the well known aspects of the linear theory. It is a purpose of the present paper to exhibit the advantages of such a representation.

## LINEAR ARRAY THEORY

### Pattern Shape

Consider first an array of $2N_z+1$ identical elements, symmetrically disposed along the Z axis, with common spacing $d_z$, as suggested by Figure 1. The conventional relation between spherical and rectangular coordinates is adopted and the far field array factor can then be written

$$A(\theta) = \sum_{n=-N_z}^{N_z} I_n e^{jn\,(kd_z \cos\theta - \alpha_z)} = \sum_{-N_z}^{N_z} f_n \qquad (1)$$

in which it has been assumed that the current distribution has a uniform progressive phase, governed by the parameter $\alpha_z$, and a symmetrical amplitude distribution. Thus, $I_n =$

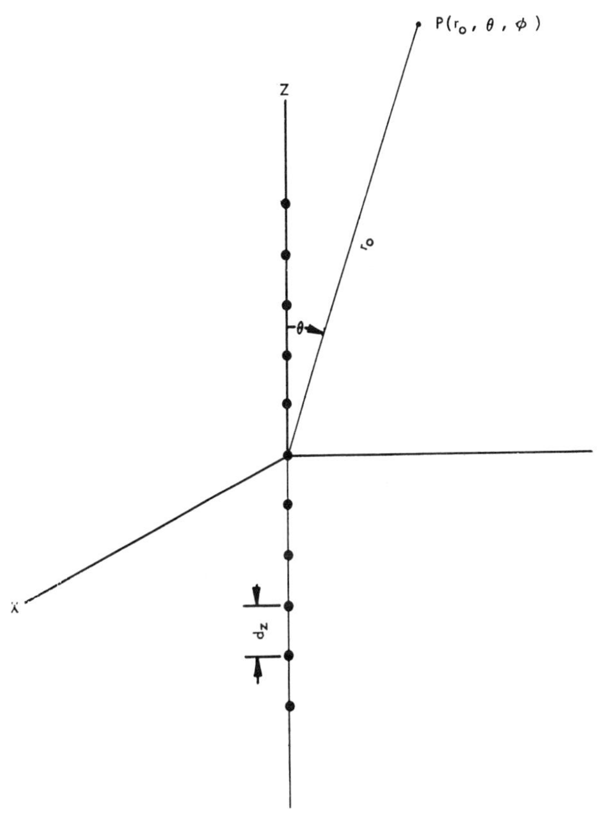


Figure 1 — Linear array geometry.

* *This study was supported under a consulting arrangement with Lockheed Aircraft Corp., Burbank, Cal.*

$I_{-n}$ is pure real for all values of n. $k=\frac{2\pi}{\lambda}$ is the wave number.

Equation (1) can be interpreted as a symmetrical family of phasors, fanned out in the complex plane as shown in Figure 2. The separation angle between adjacent phasors is $\psi=kd_z \cos\theta-\alpha_z$, and thus varies with $\theta$. At an angle $\theta_o$ defined by

$$\psi=kd_z \cos\theta_o-\alpha_z=0 \qquad (2)$$

all the phasors are aligned and thus

$$A(\theta_o)=\sum_{-N_z}^{N_z} I_n \qquad (3)$$

is the maximum value of the array factor. Therefore, $\theta_o$ defines the direction of the main beam.

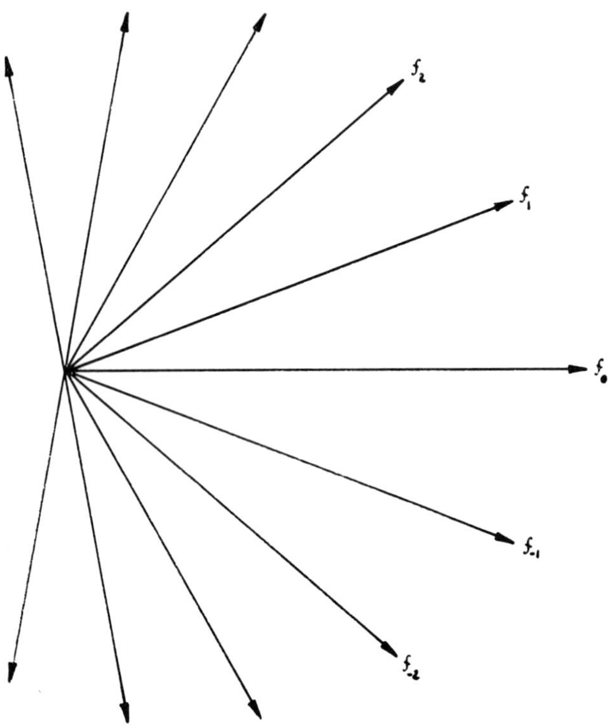


**Figure 2 — Phasor representation of linear array pattern.**

As $\theta$ departs from $\theta_o$, the phasors spread out in the complex plane. For large arrays, an angle $\theta_1<\theta_o$ is reached at which the sum of the phasors is zero, defining the first null on one side of the main beam. Continuing, an angle $\theta_3<\theta_1$ is reached at which the phasors are even further spread out, and their sum is once again zero. This defines the second null on one side of the main beam. Intermediate between $\theta_1$ and $\theta_3$ there is an angle at which the phasors sum to a secondary maximum, defining the height of the first sidelobe. As $\theta$ is decreased still further, subsequent sidelobes and nulls are traced out.

Similarly, if $\theta$ departs in the other direction from $\theta_o$, a sequence of angles $\theta_o<\theta_2<\theta_4<\ldots$ is reached at which array pattern nulls are obtained, with sidelobes in between. Because the array pattern is symmetrical in the variable $\psi=kd_z\cos\theta-\alpha_z$, the heights of the sidelobes on the two sides of the main beam are symmetrical also, but their positions are not necessarily symmetrical in $\theta$. Since $A(\theta)$ is independent of $\phi$, the three dimensional array pattern consists of conical lobes whose directions nest among the succession of angles $\ldots\theta_3,\theta_1,\theta_o,\theta_2,\theta_4,\ldots$

It is conceivable that the phasors of Equation (1) can spread out so far that they are once again aligned. This will occur if $kd_z\cos\theta-\alpha_z=\pm 2\pi$. If this is not to occur before the limits of real space are reached, then use of (2) establishes the conditions

$$|\cos\theta_o|+\frac{\lambda}{d_z}>1$$

$$|\cos\theta_o|-\frac{\lambda}{d_z}<-1$$

The second of these inequalities is more stringent and requires that

$$\frac{d_z}{\lambda}<\frac{1}{1+|\cos\theta_o|} \qquad (4)$$

Equation (4) is often referred to as the condition for avoidance of multiple main beams. It is a result which is independent of current distribution.* If the principal main beam is to be scanned close to endfire, so that $|\cos\theta_o|\rightarrow 1$, it is evident that the elements must be spaced only a half wavelength apart if a second main beam is to be avoided. In what follows, only linear arrays causing a single main beam will be considered.

**Beamwidth**

The half-power beamwidth of a linear array is defined as the angular separation between the two directions, one on each side of the main beam maximum, at which the power density is reduced by half. Let $\textcircled{H}=\theta'_2-\theta'_1$ be this beamwidth, in which $\theta'_2$ and $\theta'_1$ are the two values of $\theta'$ which satisfy the relation

$$A(\theta')=0.707\,A(\theta_o)=\sum_{-N_z}^{N_z} I_n e^{jn(kd_z\cos\theta'-\alpha_z)} \qquad (5)$$

The amplitude distribution $I_n$ can be represented by a Fourier series, viz.,

$$I_n=\sum_{p=-P}^{P} a_p e^{jp\frac{2\pi n}{2N_z+1}} \qquad (6)$$

in which P is the highest spatial harmonic needed to represent the distribution. $a_p=a_{-p}$ is pure real because the distribution is symmetric. Inserting (2) and (6) in (5) gives

* *Except, of course, that the current distribution is symmetrical, and the phase is uniform progressive.*

$$A(\theta') = \sum_{-P}^{P} a_p \sum_{-N_z}^{N_z} e^{jn\,kd_z\left[\cos\theta' - \cos\theta_0 + \frac{p\lambda}{L_z}\right]}$$

$$= \sum_{-P}^{P} a_p \frac{\sin\left[\frac{2N_z+1}{2} kd_z\left(\cos\theta' - \cos\theta_0 + \frac{p\lambda}{L_z}\right)\right]}{\sin\left[\frac{1}{2} kd_z\left(\cos\theta' - \cos\theta_0 + \frac{p\lambda}{L_z}\right)\right]} \tag{7}$$

in which $L_z = (2N_z+1)d_z$ is the length of the array. (The array length is assumed to include a distance $\frac{d_z}{2}$ beyond each end element). It is shown in Appendix A that for large arrays with conventional distributions, Equation (7) can be transformed to

$$\frac{\sin K\pi}{K\pi} = \frac{0.707}{\sum_{-P}^{P} \frac{a_p}{a_0}\left(-1\right)^{p+1} \frac{K^2}{p^2 - K^2}} \tag{8}$$

in which

$$K = \frac{L_z}{\lambda}(\cos\theta' - \cos\theta_0) \tag{9}$$

is a substitution variable from which the beamwidth can be deduced. For small P it is a simple matter to find K from (8). Several cases will now be considered.

*Case 1: Uniform Distribution*

This is the simplest case of all and extremely useful as a reference. Only $a_0$ has a value and Equation (8) yields two solutions for K which can be written

$$\cos\theta_1' - \cos\theta_0 = 0.443\frac{\lambda}{L_z} \tag{10}$$

$$\cos\theta_2' - \cos\theta_0 = -0.443\frac{\lambda}{L_z} \tag{11}$$

from which it follows that the half power beamwidth is given by

$$Ⓗ = \theta_2' - \theta_1' = \cos^{-1}\left[\cos\theta_0 - 0.443\frac{\lambda}{L_z}\right] - \cos^{-1}\left[\cos\theta_0 + 0.443\frac{\lambda}{L_z}\right] \tag{12}$$

$$\left(0 < \theta_0 \leqslant \frac{\pi}{2}\right) \qquad (\theta_1' \geq 0)$$

As the main beam is scanned around from broadside ($\theta_0 = \frac{\pi}{2}$) to endfire ($\theta_0 = 0$), a cross-section of the beam takes on a succession of positions as indicated by Figure 3. As the conical beam closes toward endfire, a position is reached at which $\theta'_1 = 0$ and from this position to endfire, there is no half-power point on one side of the beam. For this reason $\theta'_1 = 0$ is known as the scan limit. Equation (10) will not give a real value for $\theta'_1$ beyond this limit.

When the endfire position is reached, the concept of beamwidth once again takes on meaning. Equation (11) is still valid and one can write

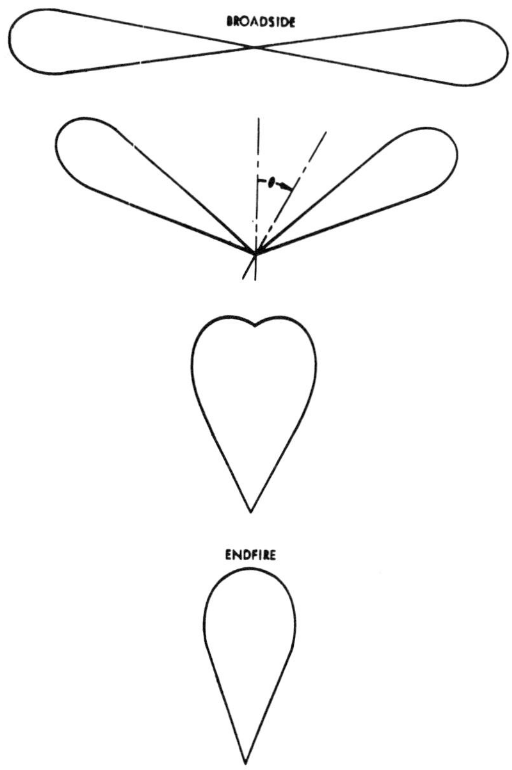


Figure 3 — Linear array pattern versus scan angle.

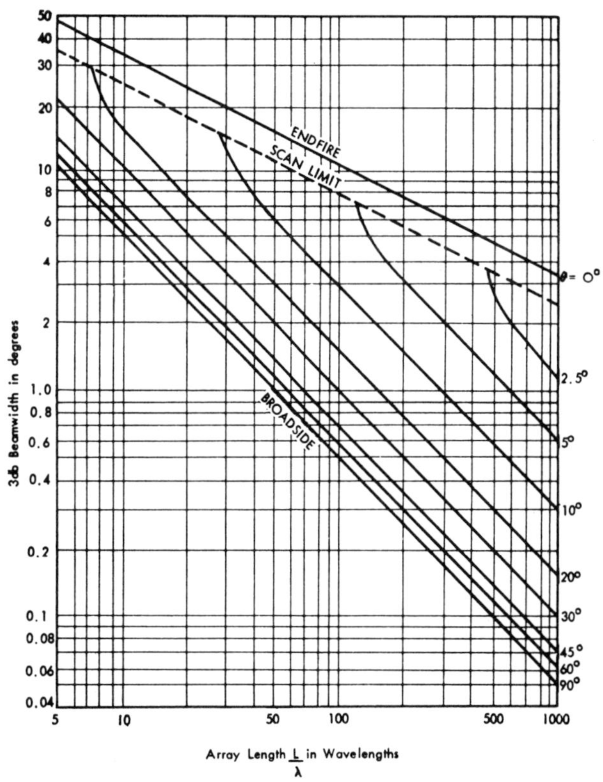


Figure 4 — Beamwidth versus linear array length and scan angle.

$$Ⓗ = 2\theta_2' = 2\cos^{-1}\left[1 - 0.443\frac{\lambda}{L_z}\right] \tag{13}$$

$$(\theta_0 = 0, \pi)$$

The beamwidths given by Equations (12) and (13) are plotted in Figure 4 as functions of array length and scan position. These curves will prove to be useful beyond the present case of a uniform linear array, since they can be used for all the linear and planar arrays we shall discuss.

Each of these equations for beamwidth — Equation (12), which is valid to within one beamwidth of endfire, and Equation (13), which is valid at endfire — has an approximate form when $L_z >> \lambda$. Using small angle expansions, one obtains

$$Ⓗ \cong 0.886 \frac{\lambda}{L_z} \csc \theta_o \quad (14)$$

(At or near broadside)

$$Ⓗ \cong 2 \sqrt{0.886 \frac{\lambda}{L_z}} \quad (15)$$

(At endfire)



For $L_z > 5\lambda$, Equation (14) is in error by less than 0.2 per cent at broadside and is in error by less than 4 per cent when the beam has been scanned to within two beamwidths of endfire. For $L_z > 5\lambda$, Equation (15) is in error by less than one per cent.

*Case 2: Cosine-on-a-Pedestal Distribution*

For this type of excitation, Equation (8) is simply

$$\frac{\sin K\pi}{K\pi} = \frac{0.707}{1 + \frac{2a_1}{a_0} \frac{K^2}{1-K^2}} \quad (16)$$

The range of interest is $0 \leqslant 2a_1 \leqslant a_0$ which covers the span from uniform excitation to a taper so severe that the excitation drops to zero at the ends of the array. In this span, K has an approximately parabolic dependence on taper, being given by

$$K = \pm \left[ 0.282 \left( \frac{2a_1}{a_0} \right)^2 + 0.443 \right] \quad (17)$$

For the extreme taper of zero excitation at the edges ($2a_1 = a_0$), the value $K = 0.725$ agrees with other published results.[4]

For large arrays scanned not too close to endfire, Equation (17) gives for the beamwidth

$$Ⓗ = 0.886 \frac{\lambda}{L_z} \csc \theta_o \left[ 1 + 0.636 \left( \frac{2a_1}{a_0} \right)^2 \right] \quad (18)$$

As one expects, increasing $\frac{2a_1}{a_0}$ to lower the sidelobe level has as a penalty an increase in the beamwidth. The quantity in brackets in (18) is called the beam broadening factor, f.

The radiation pattern from an equi-spaced linear array excited in a cosine-on-a-pedestal distribution consists of a main beam plus sidelobes which get successively lower as one departs in either direction from the main beam. Defining the sidelobe level as the height of the biggest sidelobe, relative to the height of the main beam, one finds by using (1) and (6), that for this case,

$$S \cong -\left[ 13 + 29 \left( \frac{2a_1}{a_0} \right) - 10 \left( \frac{2a_1}{a_0} \right)^2 \right] \quad (19)$$

in which S is the sidelobe level in db. Combining (18) and (19) permits us to plot the beam broadening factor as a function of sidelobe level, as shown in Figure 5.

*Case 3: Dolph-Chebyshev Distribution*

It is shown elsewhere[5] that if the current distribution is chosen so that the pattern can be represented by the Chebyshev polynomial $T_{2N_z}$ $(u_o \cos \frac{\psi}{2})$ with $\psi$ defined by (2), then the Fourier coefficients for such a distribution are given by

$$(2N_z + 1)\, a_p = T_{2N_z} \left( u_o \cos \frac{p\pi}{2N_z + 1} \right)$$

It follows that $(2N_z + 1)a_o = T_{2N_z}(u_o) = r$ is the main beam to sidelobe voltage ratio. For large arrays, and for sidelobe levels in the range from $-20$ db to $-60$ db, only $a_o$ and $a_1$ are significant in determining the beamwidth, and $a_1$ is given quite precisely by[5]

$$(2N_z + 1)\, a_1 = \cosh \left[ \sqrt{(\text{arc} \cosh r)^2 - \pi^2} \right] \quad (20)$$

Thus, for a Chebyshev distribution, the beamwidth can be determined as a function of sidelobe level from (18) by first finding $a_1$ from (20). The beam broadening factor is therefore approximately

$$f = 1 + 0.636 \left\{ \frac{2}{r} \cosh \left[ \sqrt{(\text{arc} \cosh r)^2 - \pi^2} \right] \right\}^2 \quad (21)$$

Equation (21) is also plotted in Figure 5. For a linear array of any length, with the main beam pointing in any general direction, the beamwidth can be determined for a Chebyshev pattern by reading the appropriate beamwidth off Figure 4 and multiplying by the appropriate value of f, as read from Figure 5. The Chebyshev beam broadening factor of Figure 5 is in substantial agreement with the results of Stegen,[6] who used a different technique to determine the beamwidth.

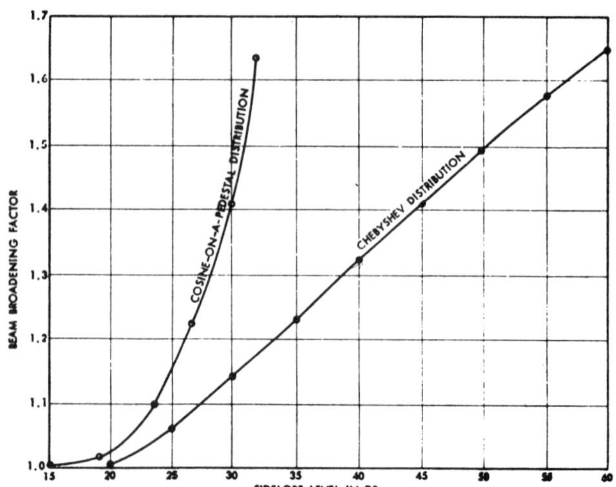


Figure 5 — Beam broadening versus side lobe level for linear arrays.

Figure 5 reveals the interesting and well-known fact that, for a given amount of beam broadening, a Chebyshev distribution is capable of yielding a lower sidelobe level than a cosine-on-a-pedestal distribution. Since only the first two Fourier terms representing the Chebyshev distribution were used in computing the Chebyshev curve in Figure 5, this result may appear paradoxical. The explanation lies in the fact that the higher Fourier terms are significant in determining sidelobe level, whereas their influence on beamwidth is slight.

### Directivity

If the element pattern is isotropic, the directivity is governed entirely by the array factor. It is defined as the power density in the direction of the main beam maximum divided by the average power density from the array. Thus,

$$D = \frac{A(\theta_0)\,A^*(\theta_0)}{\frac{1}{4\pi r_0^2}\int_0^{\pi}\int_0^{2\pi} A(\theta)\,A^*(\theta)\,r_0^2 \sin\theta \,d\theta\, d\phi} \qquad (22)$$

which simplifies to

$$D = \frac{2A(\theta_0)\,A^*(\theta_0)}{\int_0^{\pi} A(\theta)\,A^*(\theta)\,\sin\theta\, d\theta} \qquad (23)$$

Letting $\psi = kd_z \cos\theta - \alpha_z$ so that $d\psi = -kd_z \sin\theta d\theta$, and making use of Equations (1) and (3), this becomes

$$D = \frac{2\left(\sum_{-N_z}^{N_z} I_n\right)^2}{\frac{1}{kd_z}\int_{-kd_z-\alpha}^{kd_z-\alpha}\left(\sum_{-N_z}^{N_z} I_n e^{jn\psi}\right)\left(\sum_{-N_z}^{N_z} I_m e^{-jm\psi}\right) d\psi} \qquad (24)$$

If $d_z$ equals $\frac{\lambda}{2}$ (or any multiple thereof), this reduces very simply to

$$D = \frac{\left(\sum_{-N_z}^{N_z} I_n\right)^2}{\sum_{-N_z}^{N_z} I_n^2} \qquad (25)$$

which is a most interesting formula in several respects. The directivity, as given by Equation (25), turns out to be a measure of the coherence of radiation from the linear array. The numerator is proportional to the total coherent field squared whereas the denominator is proportional to the sums of the squares of the individual fields from each element.

Furthermore, the directivity, as given by Equation (25), is seen to be independent of scan angle. On the face of it, this seems surprising, since we have already observed that the main beam broadens as it is scanned away from broadside, a manifestation which usually signifies lowered directivity. However, for a linear array, as the conical beam is scanned toward endfire, the cone tends to occupy a smaller solid angle in space, an effect which just cancels the beam broadening. This compensation holds until the beam approaches endfire, when another compensation takes over — the appearance of a second main beam at reverse endfire.

Whereas Equation (25) is independent of scan angle, it is not independent of current distribution. Using the Fourier series description of the excitation embodied in Equation (6), one finds that

$$\left(\sum_{-N_z}^{N_z} I_n\right)^2 = (2N_z+1)^2\, a_0^2 \qquad (26)$$

$$\sum_{-N_z}^{N_z} I_n^2 = (2N_z+1)\sum_{-P}^{P} a_p^2 \qquad (27)$$

so that

$$D = \frac{2N_z+1}{\sum_{-P}^{P}\left(\frac{a_p}{a_0}\right)^2} \qquad (28)$$

For half-wave spacing, $L_z = (2N_z+1)\frac{\lambda}{2}$, so that Equation (28) can be rewritten

$$D = \frac{\frac{2L_z}{\lambda}}{1+2\sum_{p=1}^{P}\left(\frac{a_p}{a_0}\right)^2} \qquad (29)$$

For element spacings in the range $\frac{\lambda}{2} \leqslant d_z \leqslant \lambda$, if $L_z$ is held fixed, the directivity is found to be quite insensitive to element spacing. Since this is the range of element spacings which avoids either supergaining or multiple beams, we shall adopt Equation (29) as the expression for directivity of a linear array. Let us now interpret this equation for several cases.

*Case 1: Uniform Distribution*

Once again, this is the simplest case of all, and gives

$$D = \frac{2L_z}{\lambda} \qquad (30)$$

2 ⇒ unidirectional

which is sometimes referred to as the standard directivity (or gain). It is the maximum directivity which can be obtained from a linear array of length $L_z$, using an aperture distribution which has uniform progressive phase, and an element spacing of $\frac{\lambda}{2}$.

*Case 2: Cosine-on-a-Pedestal Distribution*

This is a special case of (29) in which only two terms appear in the denominator. It is more useful to display the normalized directivity, $D$, found by dividing (29) by (30). For this case,

$$D = \frac{1}{1+2\left(\frac{a_1}{a_0}\right)^2} \qquad (31)$$

Since the practical range of $\frac{a_1}{a_0}$ extends from zero to one half, the normalized directivity extends from unity to two-thirds, and is independent of the size of the array

*Case 3: Chebyshev Distribution*

Making use of the Fourier representation of a Chebyshev distribution, the directivity can be written

$$D = \frac{\frac{2L_z}{\lambda}}{1+\frac{2}{r^2}\sum_{p=1}^{N_z}\left[T_{2N_z}\left(u_0 \cos\frac{p\pi}{2N_z+1}\right)\right]^2} \qquad (32)$$

a result which agrees with Stegen.[7]

Unlike the computation for beamwidth of a Chebyshev array, in which only the first two Fourier coefficients were significant, it develops that all the Fourier coefficients which appear in the denominator of (32) must be considered. Indeed, if the array becomes large enough, the sum of the squares of these Fourier coefficients becomes proportional to $N_z$, and thus the directivity tends to a limit. (cf. Appendix B for a proof of this statement).

It is a tedious computation to determine all the Fourier coefficients in (32), particularly for large arrays. Fortunately, this is not necessary. In Appendix C, it is shown that (32) can be written alternatively as

$$D = \frac{2r^2}{1+(r^2-1)f\frac{\lambda}{L_z}} \qquad (33)$$

in which r is the main beam to sidelobe voltage ratio and f is the beam broadening factor. Equation (33) is quite accurate for large arrays and has the limit

$$D_{max} = 2r^2 \qquad (34)$$

which is reached when $L_z \rightarrow \infty$ and which agrees with the result of Appendix B. Thus the maximum gain for a Chebyshev array is three db more than the sidelobe level. This means, for example, that if one wishes to design a linear array to have uniform sidelobes and a gain of 43 db, it is necessary also to design it to have sidelobes that are down at least 40 db. Actually, this maximum gain is approached rather rapidly at first, as $L_z$ is increased, but then additional gain is bought very dearly in terms of increased array length. This point can be appreciated clearly by studying Figure 6, which is a plot of Equation (33) for various values of sidelobe level. An optimum gain (and thus array length) can be selected for a given sidelobe level by specifying a point on the appropriate curve of Figure 6 at which the curve has just barely begun to bend significantly. R. C. Hansen[8] has made an excellent study of such gain limitations for Taylor-type distributions and has come to similar conclusions.

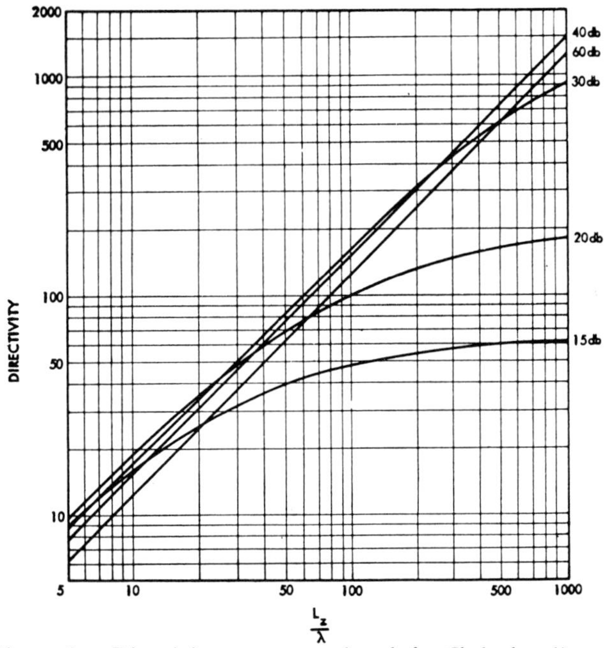


Figure 6 — Directivity versus array length for Chebyshev linear arrays.

This phenomenon of a gain limit was not observed for the cases of a uniform distribution and a cosine-on-a-pedestal distribution. Those distributions gave patterns with tapered sidelobes, and the Chebyshev feature of uniform sidelobes, while giving minimum beamwidth, is also responsible for the gain limitation. However, current antenna practice is to combine large arrays with low sidelobe levels, so this gain limitation is not so serious as to preclude the use of Chebyshev designs. For example, Figure 6 reveals that for arrays as long as one thousand wavelengths, very little bending has occurred in the curve for a 40 db sidelobe level.

### Beamwidth and Directivity

Equations (9) and (29) reveal that for a linear array, both the beamwidth and the directivity depend linearly on the array length. Upon eliminating $\frac{L_z}{\lambda}$ from these two expressions, one obtains

$$D = \frac{1.77}{\Theta_0}\,\frac{f}{\left[\sum_{p=-P}^{P}\left(\frac{a_p}{a_0}\right)^2\right]} \qquad (35)$$

in which $\Theta_0$ is the broadside beamwidth.

If the beamwidth is expressed in degrees instead of radians, and if the distribution is uniform, Equation (35) reduces to the simple relation

$$D = \frac{101.5}{\textcircled{H}_o} \qquad (36)$$

If a cosine-on-a-pedestal distribution is employed, the factor in brackets in (35) is a slowly-varying function of taper, only suffering a three per cent rise as $a_1$ increases to 0.25 $a_o$ and only taking on a maximum increase of nine per cent at the extreme taper of $a_1=0.5\ a_o$. For a Chebyshev distribution, until an array length is reached at which the directivity begins to limit, the factor in brackets in (35) is unity. Thus Equation (36) is a good working relation between beamwidth and directivity for most useful excitations, and this can be rounded off by saying that the product of broadside beamwidth and directivity for a linear array is about one-hundred.

## Appendix A

Let

$$u_p = \frac{\pi d_z}{\lambda}\left(\cos\theta' - \cos\theta_o + \frac{p\lambda}{L_z}\right) \qquad (A\text{-}1)$$

so that Equation (7) can be written

$$A(\theta') = \sum_{-P}^{P} a_p \frac{\sin(2N_z+1)\,u_p}{\sin u_p} = \sum_{-P}^{P} a_p \frac{\sin[(2N_z+1)\,u_o + p\pi]}{\sin u_p}$$

$$= \frac{\sin[(2N_z+1)\,u_o]}{\sin u_o}\sum_{-P}^{P} a_p \cos p\pi \left[\frac{\sin u_o}{\sin u_p} - 1 + 1\right] \qquad (A\text{-}2)$$

For large arrays with conventional distributions, $u_p$ and $u_o$ are small angles and P is a small integer. Then (A-2) can be written

$$A(\theta') = (2N_z+1)\frac{\sin[(2N_z+1)\,u_o]}{(2N_z+1)\,u_o}\left\{a_o + \sum_{p=1}^{P} 2a_p \cos p\pi\right\}$$

$$-(2N_z+1)\frac{\sin[(2N_z+1)\,u_o]}{(2N_z+1)\,u_o}\sum_{p=1}^{P} 2a_p \cos p\pi \frac{p^2}{p^2-K^2} \qquad (A\text{-}3)$$

in which

$$K = \frac{L_z}{\lambda}(\cos\theta' - \cos\theta_o) \qquad (A\text{-}4)$$

But

$$A(\theta') = 0.707\ A(\theta_o) = 0.707 \sum_{-N_z}^{N_z} I_n = 0.707 \sum_{-P}^{P}\sum_{-N_z}^{N_z} a_p\, e^{jn\frac{2\pi p}{2N_z+1}}$$

$$A(\theta') = 0.707\ (2N_z+1)\ a_o \qquad (A\text{-}5)$$

Combining (A-3) and (A-5) gives

$$\frac{\sin K\pi}{K\pi}\left\{a_o + \sum_{p=1}^{P} 2a_p \cos p\pi\left[1 - \frac{p^2}{p^2-K^2}\right]\right\} = 0.707\ a_o$$

from which

$$\frac{\sin K\pi}{K\pi} = \frac{0.707}{1 - \sum_{p=1}^{P} \frac{2a_p}{a_o}(-1)^p \frac{K^2}{p^2-K^2}}$$

which can also be written

$$\frac{\sin K\pi}{K\pi} = \frac{0.707}{\sum_{-P}^{P} \frac{a_p}{a_o}(-1)^{p+1} \frac{K^2}{p^2-K^2}} \qquad (A\text{-}6)$$

## Appendix B

Making use of the properties of the Chebyshev polynomials, we have

$$(2N_z+1)\ a_p = T_{2N_z}\left((u_o \cos\frac{p\pi}{2N_z+1}\right) = \cos 2N_z v_p \qquad (B\text{-}1)$$

in which $v_p$ is defined by

$$\cos v_p = u_o \cos\frac{p\pi}{2N_z+1} \qquad (B\text{-}2)$$

and p is a large enough integer that $v_p$ is real. (For practical distributions, $p \geq 3$ suffices.) Let

$$S = \sum_{p=3}^{N_z} (2N_z+1)^2 a_p^2 = \sum_{3}^{N_z} \cos^2 2N_z v_p$$

$$S = \sum_{3}^{N_z} \left(\frac{1}{2} + \frac{1}{2} \cos 4 N_z v_p\right) \qquad \text{(B-3)}$$

Since $u_0 \cong 1$,

$$v_p = \frac{p\pi}{2N_z+1} + \delta_p \qquad \text{(B-4)}$$

in which $\delta_p$ is small, and tends to zero as $u_0 \rightarrow 1$. Thus for large arrays, as p ranges from 3 to $N_z$, $v_p$ ranges from a value close to zero to a value close to $\frac{\pi}{2}$, with the values of $v_p$ being regularly spaced. Therefore the term

$$\sum_{3}^{N_z} \frac{1}{2} \cos 4 N_z v_p$$

is quite small, and, in (B-3),

$$S \rightarrow \frac{N_z}{2} \qquad \text{(B-5)}$$

Since, for a fixed sidelobe level, $a_0$, $a_1$, and $a_2$ remain constant as $N_z$ is increased,[5] and since $(2N_z+1)a_0 = r$, the main beam to sidelobe voltage ratio, Equation (28) yields the limit

$$D \rightarrow \frac{2N_z+1}{\frac{1}{r^2}(2)\left(\frac{N_z}{2}\right)} = 2r^2 \qquad \text{(B-6)}$$

Thus the limiting directivity of a Chebyshev linear array is governed by the sidelobe level.

## Appendix C

For a large Chebyshev array, the denominator of Equation (23) can be approximated by assuming the main beam and each of the sidelobes to be a half-sinusoid in field distribution. The main beam has its maximum at $\theta_0$ and nulls at $\theta_2$ and $\theta_1$. In this range we assume

$$A(\theta) = r \cos \left[\pi \frac{\theta - \frac{\theta_2+\theta_1}{2}}{\theta_2 - \theta_1}\right] \qquad \text{(C-1)}$$

and determine a contribution to the denominator of (23) as

$$F_0 = r^2 \sin \theta_0 \int_{\theta_1}^{\theta_2} \cos^2 \left[\pi \frac{\theta - \frac{\theta_2+\theta_1}{2}}{\theta_2 - \theta_1}\right] d\theta \qquad \text{(C-2)}$$

in which $\sin\theta$ has been brought out in front of the integrand because a large array is assumed so that $\theta_2 - \theta_1$ is small and the variation in $\sin\theta$ is small. Equation (C-2) can be integrated easily, giving

$$F_0 = r^2 \sin \theta_0 \left(\frac{\Delta \theta_0}{2}\right) \qquad \text{(C-3)}$$

in which $\Delta\theta_0 = \theta_2 - \theta_1$ is the angular width between nulls.

When a similar calculation is performed for one of the sidelobes, the only difference is that a unit amplitude replaces the amplitude r in (C-1). Thus Equation (23) can be written

$$D = \frac{2r^2}{r^2 \sin \theta_0 \cdot \left(\frac{\Delta \theta_0}{2}\right) + \sum_m{}' \sin \theta_m \cdot \left(\frac{\Delta \theta_m}{2}\right)} \qquad \text{(C-4)}$$

in which $\Delta\theta_m$ is the angular width of the m[th] sidelobe and the sum $\Sigma'$ extends over the entire angular region except for the portion occupied by the main beam. This can be rephrased as

$$D = \frac{4r^2}{(r^2-1)\sin \theta_0 \cdot \Delta \theta_0 + \sum_m \sin \theta_m \cdot \Delta \theta_m} \qquad \text{(C-5)}$$

in which the sum $\Sigma$ is over all space. For large arrays,

$$\sum_m \sin \theta_m \cdot \Delta \theta_m \cong \int_0^{\pi} \sin \theta \cdot d\theta = 2 \qquad \text{(C-6)}$$

Furthermore, the angular extent of the main beam is related to the half-power beamwidth by the expression

$$\Delta \theta_0 \cong \frac{\textcircled{H}}{0.443} \qquad \text{(C-7)}$$

whereas the half-power beamwidth, using (14), is given by

$$\textcircled{H} = 0.886\, f \frac{\lambda}{L_z} \csc \theta_0 \qquad \text{(C-8)}$$

in which f is the beam broadening factor. Thus

$$\sin \theta_0 \cdot \Delta \theta_0 = 2f \frac{\lambda}{L_z} \qquad \text{(C-9)}$$

and therefore

$$D = \frac{2r^2}{(r^2-1)\, f \frac{\lambda}{L_z} + 1} \qquad \text{(C-10)}$$

## References


1. Jordan, E. C., *Electromagnetic Waves and Radiating Systems,* Prentice-Hall, 1950, Chapter 12; and J. E. Eaton, L. J. Eyges, and G. G. MacFarlane, *Microwave Antenna Theory and Design,* McGraw-Hill, ed. S. Silver, 1949, Chapter 9.
2. Von Aulock, W. H., "Properties of Phased Arrays," *Proc. IRE,* Vol. 48, No. 10, Oct. 1960, pp. 1715-28.
3. King, M. J. and R. K. Thomas, "Gain of Large Scanned Arrays," *Trans. IRE,* Vol. AP-8, No. 6, Nov. 1960, pp. 635-6.
4. Silver, S., *Microwave Antenna Theory and Design,* McGraw Hill, 1949, p. 187.
5. Elliott, R. S., "An Approximation to Chebyshev Distributions," to appear in *AP Trans. of IEEE.*
6. Stegen, R. J., "Excitation Coefficients and Beamwidths of Tchebycheff Arrays," *Proc. IRE,* Vol. 41, Nov. 1953, pp. 1671-4.
7. Stegen, R. J., "Gain of Tchebycheff Arrays," *Trans. IRE,* Vol. AP-8, Nov. 1960, pp. 629-31.
8. Hansen, R. C., "Gain Limitations of Large Antennas," *Trans. IRE,* Vol. AP-8, Sept. 1960, pp. 490-5.


**Part II of Dr. Elliott's paper, which will deal with planar array theory, will appear in the January issue of *the microwave journal.***

# Beamwidth and Directivity of Large Scanning Arrays

**LAST OF TWO PARTS**



## PLANAR ARRAY THEORY

*Pattern Shape*

Next consider the radiating elements to be arranged in a regular rectangular array in the x-y plane, as shown in Figure 7. Let there be $2N_x+1$ rows of elements, each row parallel to the y axis, with common spacing $d_x$ between rows. Let each row contain $2N_y+1$ elements, whose common spacing is $d_y$. By the $mn^{th}$ element, we shall mean the element whose positional coordinates are

$$x_m = md_x \qquad y_n = nd_y$$

in which $-N_x \leq m \leq N_x$, and $-N_y \leq n \leq N_y$. Similarly, the current at the terminals of the $mn^{th}$ element will be designated $I_{mn}$. With this notation, the array factor can be written

$$A(\theta,\phi) = \sum_{m=-N_x}^{N_x} \sum_{n=-N_y}^{N_y} \frac{I_{mn}}{I_{oo}} e^{jk \sin\theta [md_x \cos\phi + nd_y \sin\phi]}$$

If each row has the same current distribution — albeit the current levels are different in different rows — that is, if

$$\frac{I_{mn}}{I_{mo}} = \frac{I_{on}}{I_{oo}}$$

then the array factor can be written

$$A(\theta,\phi) = A_x(\theta,\phi) \cdot A_y(\theta,\phi) \qquad (37)$$

in which

$$A_x(\theta,\phi) = \sum_{-N_x}^{N_x} I_m e^{jm\, kd_x \sin\theta \cos\phi} \qquad (38)$$

$$A_y(\theta,\phi) = \sum_{-N_y}^{N_y} I_n e^{jn\, kd_y \sin\theta \cos\phi}$$

and

$$I_m = \frac{I_{mo}}{I_{oo}} \qquad I_n = \frac{I_{on}}{I_{oo}}$$

are the normalized current distributions in a row of elements parallel to the x axis and y axis respectively.

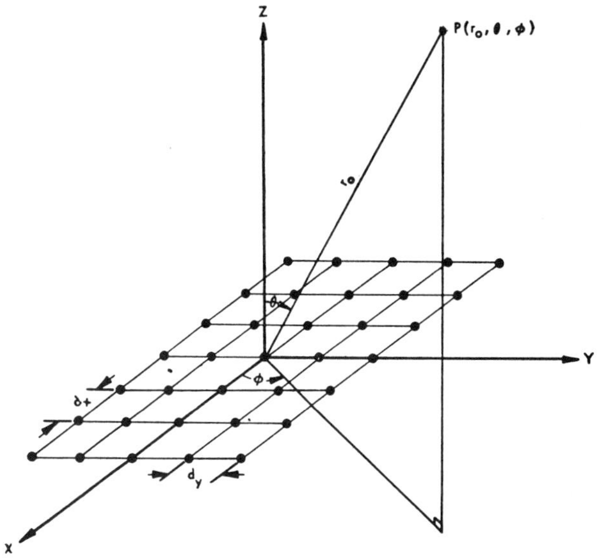


Figure 7 — Planar array geometry.

**R. S. ELLIOTT**
UNIVERSITY OF CALIFORNIA AT LOS ANGELES

A professor of engineering at UCLA, Dr. Elliott was one of the founders of Rantec Corp. and served as its first vice president and technical director. Prior to this he worked for Hughes in research, and also served for the Naval Ordnance Laboratory during the Korean War. Mr. Elliott holds a BA degree in English Literature and a BS in Electrical Engineering from Columbia, and an MS and PhD in Electrical Engineering from the University of Illinois. He is a member of Sigma Xi, Tau Beta Pi, and a Fellow of the IEEE.

Equation (37) is an example of the principle of pattern multiplication. It reveals that, under the stated restrictions, the array factor for a planar array is the product of the array factors for two linear arrays — one laid out along the x axis, and the other laid out along the y axis. All that we have discussed about linear arrays in Part I can thus be brought to bear on the problem of understanding such planar arrays.

If, additionally, $I_{mn}$ differs in phase from $I_{oo}$ by the factor

$$e^{-m\alpha_x \, n\alpha_y}$$

then

$$A(\theta,\phi)=\left(\sum_{-N_x}^{N_x} I_m\, e^{jm[kd_x \sin\theta \cos\phi - \alpha_x]}\right)$$

$$\left(\sum_{-N_y}^{N_y} I_n\, e^{jn[kd_y \sin\theta \sin\phi - \alpha_y]}\right) \quad (39)$$

and the array distribution has a uniform phase progression $\alpha_x$ in the x direction and a uniform phase progression $\alpha_y$ in the y direction. The amplitude distributions $I_m$ and $I_n$ in (39) are now pure real. If they are also symmetric, the factor $A_x$ gives a pattern which consists of a conical main beam and sidelobes, symmetric about the x axis. The main beam of $A_x$ makes an angle $\theta_x$ with the positive x axis which satisfies the relation

$$kd_x \cos\theta_x - \alpha_x = kd_x \sin\theta \cos\phi - \alpha_x = o$$

$$\cos\theta_x = \frac{\alpha_x}{kd_x} = \sin\theta \cos\phi \quad (40)$$

Similarly, the factor $A_y$ gives a pattern which consists of a conical main beam and sidelobes, symmetric about the y axis. The main beam of $A_y$ makes an angle $\theta_y$ with the positive y axis which satisfies the relation

$$\cos\theta_y = \frac{\alpha_y}{kd_y} = \sin\theta \sin\phi \quad (41)$$

The criterion developed in Part I applies to these two patterns with respect to the avoidance of multiple conical main beams in $A_x$ and $A_y$. Both $d_x$ and $d_y$ must not exceed a half wavelength if the two conical patterns are to be scanned to the vicinity of endfire.

The planar array factor $A=A_xA_y$, being the product of these two linear array factors, is then principally the intersection of the two conical main beams, plus those sidelobes of each conical pattern which coincide with the conical main beam of the other conical pattern. Of course, it is possible to scan one conical beam so close to the x axis and the other conical beam so close to the y axis that the two conical beams do not intersect. This is an impractical situation, and later in this section a criterion will be established for avoiding it.

Assuming that the two conical beams do intersect, their product gives two pencil beams, one pointing in the half-space $z>0$ and the other pointing in the half-space $z>0$. In most practical applications, the element pattern will be selected so as to give negligible radiation in the half-space $z>0$ (through use of a ground plane, for example). There is then left the single main pencil beam, pointing in the direction $(\theta_o, \phi_o)$ which satisfies the equations

$$\tan\phi_o = \frac{\alpha_y\, d_x}{\alpha_x\, d_y} \quad (42)$$

$$\sin^2\theta_o = \frac{\alpha_x^2}{k^2 d_x^2} + \frac{\alpha_y^2}{k^2 d_y^2} \quad (43)$$

These relations are obtained by solving Equations (40) and (41) simultaneously. For given spacings $d_x$ and $d_y$, and for given inter-element phase shifts $\alpha_x$ and $\alpha_y$, Equations (42) and (43) give a unique direction $(\theta_o, \phi_o)$ in the half-space $z>0$.

Equation (43) can be used as the criterion for avoiding the situation that $A_x$ and $A_y$ contain conical main beams which do not intersect. This situation would just be reached if $\sin^2\theta_o = 1$. Then the elliptical relation

$$\frac{\alpha_x^2}{k^2 d_x^2} + \frac{\alpha_y^2}{k^2 d_y^2} = 1 \quad (44)$$

limits the range of $\alpha_x$ (or $\alpha_y$) for specified values of $kd_x$, $kd_y$, and $\alpha_y$ (or $\alpha_x$). In all that is to follow, we shall assume the existence of a single main pencil beam.

*Beamwidth*

Since the significant sidelobes are in the two cones defined by (40) and (41), these cones constitute the pattern cuts which should be taken to determine the sidelobe level. However, the profiles of the main pencil beam which are obtained in these two cuts are not, in general, due to two orthogonal slices through the pattern. (For example, if the pencil beam lies close to the x-y plane at $\phi_o = \frac{\pi}{4}$, these two cuts are almost coincident). Thus it is desirable to define beamwidth in another fashion, one which will reveal more information about the structure of the pencil beam.

A suitable definition, and one which we shall adopt, is to use the pattern cuts in the half-plane $\phi=\phi_o$, and in the orthogonal plane which contains the directional line $(\theta_o, \phi_o)$. Referring to Figure 8, in which the main pencil beam is shown pointing in an arbitrary direction $(\theta_o, \phi_o)$, the two half-power beamwidths are designated as Ⓗ and $\widetilde{\text{Ⓗ}}$.

In the half-plane $\phi=\phi_o$, the beamwidth is simply

$$\text{Ⓗ} = \theta'_2 - \theta'_1 \quad (45)$$

in which $\theta'_2$ and $\theta'_1$ are the two values of $\theta'$ which satisfy

$$A(\theta', \phi_o) = 0.707\, A(\theta_o, \phi_o) \quad (46)$$

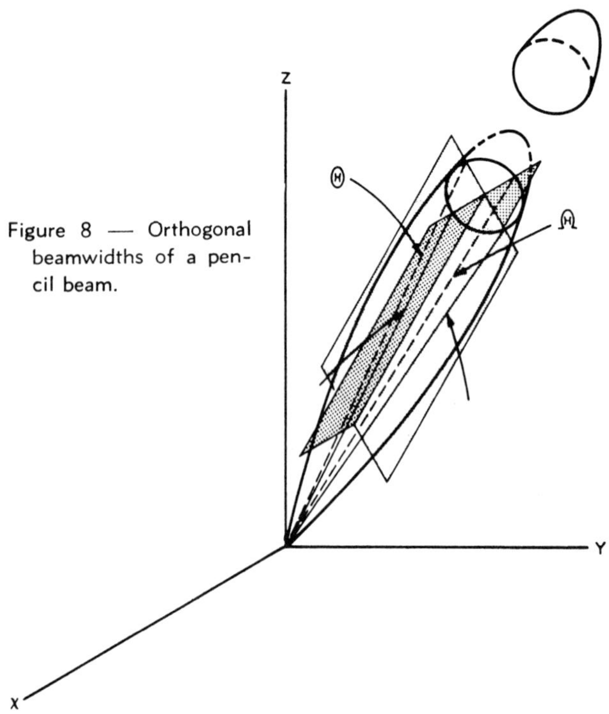


Figure 8 — Orthogonal beamwidths of a pencil beam.

Using the fact that $\alpha_x = kd_x \sin\theta_o \cos\phi_o$ and $\alpha_y = kdy \sin\theta_o \sin\phi_o$, and using (39) gives

$$\sum_{-N_x}^{N_x} I_m e^{jmkd_x \cos\phi_o(\sin\theta' - \sin\theta_o)} \sum_{-N_y}^{N_y} I_n e^{jnkd_y \sin\phi_o(\sin\theta' - \sin\theta_o)} = 0.707 \sum_{-N_x}^{N_x} \sum_{-N_y}^{N_y} I_m I_n \qquad (47)$$

For large arrays, with the main beam scanned not too far off broadside, one can write $\theta'_2 = \theta_o + \triangle\theta_2$ and $\theta'_1 = \theta_o - \triangle\theta_1$, in which $\triangle\theta_2$ and $\triangle\theta_1$ are small. Substituting in (47) yields

$$\sum_{-N_x}^{N_x} \sum_{-N_y}^{N_y} I_m I_n e^{jk \cos\theta_o[md_x \cos\phi_o + nd_y \sin\phi_o]\triangle\theta_2} = 0.707 \sum_{-N_x}^{N_x} \sum_{-N_y}^{N_y} I_m I_n \qquad (48)$$

$$\sum_{-N_x}^{N_x} \sum_{-N_y}^{N_y} I_m I_n e^{-jk \cos\theta_o[md_x \cos\phi_o + nd_y \sin\phi_o]\triangle\theta_1} = 0.707 \sum_{-N_x}^{N_x} \sum_{-N_y}^{N_y} I_m I_n \qquad (49)$$

Since $I_m$ and $I_n$ are symmetrical distributions, it follows that in this approximation

$$\triangle\theta_1 = \triangle\theta_2 = \frac{Ⓗ}{2} \qquad (50)$$

The left side of (48) consists of a family of phasors, symmetrically spread out in the complex plane, much as in the case of the linear array described in Figure 2 of Part I. This sum is to be 0.707 times the sum of the phasors when they are all aligned. For normal distributions, the phasors do not need to fan out too far to reduce the sum by the factor 0.707. The outermost phasor normally is not fanned out more than about $\frac{\pi}{2}$ radians. Thus the phase factor

$$e^{jk \cos\theta_o[md_x \cos\phi_o + nd_y \sin\phi_o]\triangle\theta_2}$$

can be expanded in a power series which will converge reasonably rapidly for even the largest values of m and n. When this is done, (48) becomes

$$0.707 \sum_{-N_x}^{N_x} \sum_{-N_y}^{N_y} I_m I_n = \sum_{-N_x}^{N_x} \sum_{-N_y}^{N_y} I_m I_n$$

$$+ jk\triangle\theta_2 \cos\theta_o \sum_{-N_x}^{N_x} \sum_{-N_y}^{N_y} (md_x \cos\phi_o + nd_y \sin\phi_o) I_m I_n$$

$$+ \frac{(jk \triangle\theta_2 \cos\theta_o)^2}{2} \sum_{-N_x}^{N_x} \sum_{-N_y}^{N_y} (md_x \cos\phi_o + nd_y \sin\phi_o)^2 I_m I_n$$

$$+ \; . \; . \; . \; . \; . \; . \; . \qquad (51)$$

Since the distributions $I_m$ and $I_n$ are symmetrical, all summations which contain m and/or n to an odd power are zero. Thus, to *third order,*

$$2.344 \sum_{-N_x}^{N_x} \sum_{-N_y}^{N_y} I_m I_n = k^2 Ⓗ^2 \cdot \cos^2\theta_o \sum_{-N_x}^{N_x} \cdot \sum_{-N_y}^{N_y} (m^2 d_x^2 \cos^2\phi_o + n^2 d_y^2 \sin^2\phi_o) I_m I_n \qquad (52)$$

But, if the beam is scanned in the x-z plane, Ⓗ is precisely the beamwidth of the *linear* array laid out along the x axis. Let us call this beamwidth $Ⓗ_x$. Similarly, if the beam is scanned in the y-z plane, Ⓗ is precisely the beamwidth of the *linear* array laid out along the y axis. Calling this beamwidth $Ⓗ_y$, one can write for these two special cases of (52),

$$2.344 \sum_{-N_x}^{N_x} \sum_{-N_y}^{N_y} I_m I_n = k^2 Ⓗ_x^2 \cdot \cos^2\theta_o \sum_{-N_x}^{N_x} \sum_{-N_y}^{N_y} m^2 d_x^2 I_m I_n \qquad (53)$$

$$2.344 \sum_{-N_x}^{N_x} \sum_{-N_y}^{N_y} I_m I_n = k^2 Ⓗ_y^2 \cdot \cos^2\theta_o \sum_{-N_x}^{N_x} \sum_{-N_y}^{N_y} n^2 d_y^2 I_m I_n \qquad (54)$$

13

Let $Ⓗ_{xo}$ represent the beamwidth in (53) for the particular case that $\theta_o=0$, i.e., for the linear array of $2N_x+1$ elements phased to give a broadside beam. Similarly, let $Ⓗ_{yo}$ represent the beamwidth in (54) for the particular case that $\theta_o=0$, i.e., for the linear array of $2N_y+1$ elements phased to give a broadside beam. Inserting these particular cases of (53) and (54) in (52) yields a simple expression for Ⓗ for the general case of an arbitrary scan direction, namely,

$$Ⓗ^{-2}=\cos^2\theta_o\,[Ⓗ^{-2}_{xo}\cos^2\phi_o+Ⓗ^{-2}_{yo}\sin^2\phi_o] \quad (55)$$

To use Equation (55), one needs to determine $Ⓗ_{xo}$, $Ⓗ_{yo}$ and the pointing direction $(\theta_o,\phi_o)$. For uniform excitation $Ⓗ_{xo}$ or $Ⓗ_{yo}$ can be determined by using $\frac{L_x}{\lambda}$ or $\frac{L_y}{\lambda}$ in place of $\frac{L_z}{\lambda}$ and reading the appropriate beamwidth off the broadside curve of Figure 4. For cosine-on-a-pedestal or Chebyshev distributions, the beam broadening factors of Figure 5 must be incorporated. It is then a simple calculation to determine Ⓗ.

Equation (55) indicates that for scan in a $\phi=$constant plane, $Ⓗ\sim\sec\theta_o$, which is consistent with the idea that the projected aperture in the pointing direction is reduced by the factor $\cos\theta_o$, suggesting a broadened beam.

Finally, for square arrays, $Ⓗ_{xo}=Ⓗ_{yo}$ and (55) reduces to

$$Ⓗ=Ⓗ_{xo}\cdot\sec\theta_o \quad (56)$$

indicating that the beamwidth Ⓗ is independent of $\phi_o$ and that it broadens with scan in accordance with the projected aperture.

Turning to the half-power beamwidth $\mathcal{H}$ in the perpendicular plane, it is evident from Figure 8 that

$$\mathcal{H}=(\phi'_2-\phi'_1)\cdot\sin\theta_o \quad (57)$$

in which $\phi'_2$ and $\phi'_1$ are the two values of $\phi'$ which satisfy

$$A(\theta_o,\phi')=0.707\,A(\theta_o,\phi_o) \quad (58)$$

Proceeding similarly, it follows that to third order

$$2.344\sum_{-N_x}^{N_x}\sum_{-N_y}^{N_y}I_m I_n=k^2\mathcal{H}^2\sum_{-N_x}^{N_x}\sum_{-N_y}^{N_y}(m^2 d_x^2\sin^2\phi_o+n^2 d_y^2\cos^2\phi_o)\,I_m I_n \quad (59)$$

Once again, if the beam is scanned in the x-z plane, but not too far off broadside, $\mathcal{H}$ is the beamwidth of the linear array laid out along the y axis, and is thus $Ⓗ_{yo}$. In like manner, if the beam is scanned in the y-z plane, $\mathcal{H}$ is the beamwidth of the linear array laid out along the x axis, and is thus $Ⓗ_{xo}$. It follows that (59) can be rewritten in the simple form

$$\mathcal{H}^{-2}=Ⓗ^{-2}_{xo}\cdot\sin^2\phi_o+Ⓗ^{-2}_{yo}\cdot\cos^2\phi_o \quad (60)$$

Unlike Ⓗ, $\mathcal{H}$ does not depend on $\theta_o$. For square arrays it also does not depend on $\phi_o$. For any scan direction, $\mathcal{H}$ can be determined from (60) by using Figure 4 (plus Figure 5, if needed) to obtain $Ⓗ_{xo}$ and $Ⓗ_{yo}$.

It is useful to define an areal beamwidth B by the relation

$$B=Ⓗ\,\mathcal{H} \quad (61)$$

Upon substitution of expressions (55) and (60), this becomes

$$B=\frac{Ⓗ_{xo}\,Ⓗ_{yo}\cdot\sec\theta_o}{\left[\sin^2\phi_o+\frac{Ⓗ^2_{yo}}{Ⓗ^2_{xo}}\cos^2\phi_o\right]^{1/2}\left[\sin^2\phi_o+\frac{Ⓗ^2_{xo}}{Ⓗ^2_{yo}}\cos^2\phi_o\right]^{1/2}} \quad (62)$$

For square arrays, the areal beamwidth is independent of $\phi_o$. For rectangular arrays, for a given $\theta_o$, it has the same value in the x-z plane that it does in the y-z plane. However, for intermediate planes, B is smaller; for example, if $L_x=10\,L_y$ so that $Ⓗ_{yo}\cong 10\,Ⓗ_{xo}$, one finds that in the half plane $\phi=\frac{\pi}{4}$, B is only 20 per cent of the value it assumes in the half-planes $\phi=0,\frac{\pi}{2}$.

Using the concept of areal beamwidth, the general effect of scanning a pencil beam can be constructed as suggested in exaggeration by Figure 9. At broadside-broadside, the cross-section of the beam is approximately elliptical (position $P_1$) with dimensions proportional to $L_x^{-1}$ and $L_y^{-1}$. As the beam scans in the x-z plane, the beam cross-section elongates in that direction (position $P_2$). Scanning in the y-z plane causes elongation in the other beam dimension. For a constant angle from the zenith, $\theta_o$, as the beam is moved from $\phi_o=0$ to $\phi_o=\frac{\pi}{2}$, the two half-power beam-

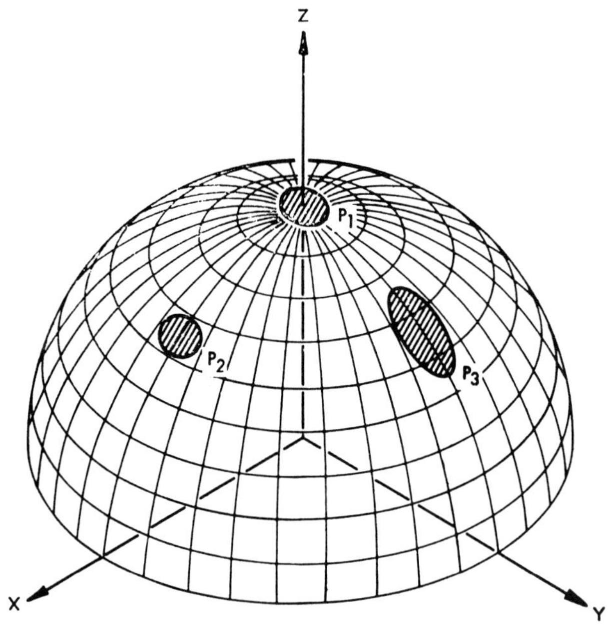

Figure 9 — Beam shape versus scan position for a pencil beam.

widths smoothly change according to formulas (55) and (60) and the beam area smoothly changes according to formula (62).

Thus, for narrow pencil beams from planar arrays, if the beam is not scanned too far from broadside, the entire subject of beamwidth can be based on the results previously obtained for linear arrays. The foregoing results are quite good to within several beamwidths of the limiting condition of no main beam at all, defined by (44).

The special case of a narrow pencil beam from a planar array, with the beam scanned to endfire, is principally of academic interest, but for completeness it is treated in Appendix E.

## Directivity

The directivity of the array factor of a planar array is given by

$$D = \frac{4\pi\, A(\theta_o,\phi_o)\, A^*(\theta_o,\phi_o)}{\int_0^{\frac{\pi}{2}}\int_0^{2\pi} A(\theta,\phi)\, A^*(\theta,\phi)\cdot \sin\theta \cdot d\theta \cdot d\phi} \qquad (63)$$

in which it is assumed that the element pattern is such as to eliminate the total pattern in the half-space $\theta > \frac{\pi}{2}$. It is shown in Appendix D that for large arrays which are not scanned closer than several beamwidths to endfire, an approximate reduction of Equation (63) is

$$D = \pi\cos\theta \cdot \frac{\frac{2L_x}{\lambda}}{\sum\limits_p \left(\frac{a_p}{a_o}\right)^2} \cdot \frac{\frac{2L_y}{\lambda}}{\sum\limits_q \left(\frac{b_q}{b_o}\right)^2} \qquad (64)$$

in which the Fourier coefficients $a_p$ and $b_q$ describe the aperture distribution in the x and y directions and $L_x = (2N_x+1)d_x$ $L_y = (2N_y+1)d_y$ are the dimensions of the array. This equation has the simple interpretation that the directivity of a planar array is

$$D = \pi\, D_x\, D_y \cos\theta_o \qquad (65)$$

in which $D_x$ and $D_y$ are the directivities of the two linear arrays. Thus D is $\pi\cos\theta_o$ times the product of the directivities of the two linear arrays. The factor $\cos\theta_o$ accounts for the decrease in projected aperture with scan. Unlike the directivity of a linear array, which was found to be independent of scan angle, the directivity of a planar array is dependent on $\theta_o$. However, it is independent of the azimuthal coordinate $\phi_o$.

Because of the form of (64), many of the remarks which have been made about the directivity of linear arrays can be carried over to planar arrays. For aperture distributions which have a uniform progressive phase and a symmetric amplitude, (64) indicates that maximum directivity results from a choice of uniform excitation. Cosine-on-a-pedestal distributions give lowered sidelobes at the penalty of some loss in directivity. However, this penalty is not very great. The most extreme aperture taper corresponds to $2a_1 = a_o$ and $2b_1 = b_o$. For this extreme taper, (64) indicates a normalized directivity $D = 4/9$, or a loss of 3.5 db, but this normalized gain is independent of aperture size.

Chebyshev distributions, however, do suffer a gain limit and the curves of Figure 6 are applicable to planar arrays. They can be used to determine $D_x$ and $D_y$ with $\frac{L_x}{\lambda}$ or $\frac{L_y}{\lambda}$ used as abscissas in place of $\frac{L_z}{\lambda}$.

The directivity of a planar array with the beam scanned to endfire is discussed in Appendix E.

## Areal Beamwidth and Directivity

Just as in the case of linear arrays, we find from Equations (62) and (64) that for planar arrays, both areal beamwidth and directivity depend linearly on the area of the planar aperture. Upon eliminating $\frac{L_x L_y}{\lambda^2}$ from these two expressions, there results

$$D = \frac{9.87}{B_o}\left[\frac{f_x}{\sum\limits_{-P}^{P}\left(\frac{a_p}{a_o}\right)^2}\;\frac{f_y}{\sum\limits_{-Q}^{Q}\left(\frac{b_q}{b_o}\right)^2}\right] \qquad (65)$$

in which $B_o$ is the areal beamwidth when the beam is in either principal plane, and $f_x$ and $f_y$ are the beam broadening factors for the linear arrays of $2N_x+1$ and $2N_y+1$ elements which comprise the two dimension of the array. The quantity in brackets is unity for a uniform distribution, is almost unity for cosine-on-a-pedestal distributions, and is unity for Chebyshev distributions until gain limiting sets in. Thus, for most practical aperture distributions,

$$D = \frac{32{,}400}{B_o} \qquad (66)$$

in which the areal beamwidth is now expressed in square degrees rather than in square radians.

When discussing linear arrays, we noted that beamwidth depended on the scan angle $\theta_o$ but that directivity did not. Thus, in establishing the relation (35), the beamwidth at a specific scan angle (broadside) was used. Similarly, in the case of planar arrays, we observe that the areal beamwidth depends on the scan angle $\phi_o$ but that the directivity does not. Therefore, a specific scan angle ($\phi_o = 0$ or $\frac{\pi}{2}$) was selected for the specification of the areal beamwidth $B_o$. However, for planar arrays, both areal beamwidth and directivity depend on the other scan angle $\theta_o$, but in such a way that their product is independent of $\theta_o$.

Equation (65) can be rounded off by saying that the product of areal beamwidth and directivity for a planar array is about 32,000, with the proviso that both quantities are measured at the same tilt angle $\theta_o$ and the beamwidth is referred to a principal plane.

* Radiation is confined to a half-space. Beam is no longer disk-like or conical.

## CONCLUSIONS

Formulas have been deduced for the beamwidth and directivity of linear and planar arrays as functions of aperture size, scan angle, and sidelobe level, for various practical aperture distributions. Inter-relations between these formulas have also been noted, and curves have been computed from which one can determine beamwidth and directivity for most practical arrays by means of simple calculations.

## APPENDIX D

Using Equation (39), one can write

$$AA^* = \sum_{-N_x}^{N_x} I_m e^{jm\,kd_x(\sin\theta\cos\phi - \sin\theta_o\cos\phi_o)}$$

$$\cdot \sum_{-N_x}^{N_x} I^*_{m'} e^{jm'\,kd_x(\sin\theta\cos\phi - \sin\theta_o\cos\phi_o)}$$

$$\cdot \sum_{-N_y}^{N_y} I_n e^{jn\,kd_y(\sin\theta\sin\phi - \sin\theta_o\sin\phi)}$$

$$\cdot \sum_{-N_y}^{N_y} I^*_{n'} e^{-jn'\,kd_y(\sin\theta\sin\phi - \sin\theta_o\sin\phi_o)} \qquad \text{(D-1)}$$

The current distributions $I_m$ and $I_n$ can be represented by Fourier series, viz.,

$$I_m = \sum_{p=-P}^{P} a_p e^{jp\frac{2\pi md_x}{L_x}} \qquad \text{(D-2)}$$

$$I_n = \sum_{q=-Q}^{Q} b_q e^{jq\frac{2\pi nd_y}{L_y}} \qquad \text{(D-3)}$$

For symmetric distributions, $a_p$ and $b_q$ are pure real and $a_p = a_{-p}$, $b_q = b_{-q}$. Then

$$AA^* = \left|\sum_{-P}^{P}\sum_{-N_x}^{N_x} a_p e^{jm\,kd_x\left(\sin\theta\cos\phi - \sin\theta_o\cos\phi_o + \frac{p\lambda}{L_x}\right)}\right|^2$$

$$\cdot \left|\sum_{-Q}^{Q}\sum_{-N_y}^{N_y} b_q e^{jn\,kd_y\left(\sin\theta\sin\phi - \sin\theta_o\sin\phi_o + \frac{q\lambda}{L_y}\right)}\right|^2 \qquad \text{(D-4)}$$

$$AA^* = \left|\sum_{-P}^{P} a_p \frac{\sin[(2N_x+1)u_p]}{\sin u_p}\right|^2$$

$$\left|\sum_{-Q}^{Q} b_q \frac{\sin[(2N_y+1)v_q]}{\sin v_q}\right|^2 \qquad \text{(D-5)}$$

in which

$$u_p = \frac{\pi d_x}{\lambda}\left(\sin\theta\cos\phi - \sin\theta_o\cos\phi_o + \frac{p\lambda}{L_x}\right) \qquad \text{(D-6)}$$

$$v_q = \frac{\pi d_y}{\lambda}\left(\sin\theta\sin\phi - \sin\theta_o\sin\phi_o + \frac{q\lambda}{L_y}\right) \qquad \text{(D-7)}$$

For the close spacings normally encountered in large scanning arrays ($d_x \cong \frac{\lambda}{2}$, $d_y \cong \frac{\lambda}{2}$), the denominators in (D-5) are slowly varying functions of $\theta$ and $\phi$, and one can write with negligible error

$$AA^* = \left|\sum_{-P}^{P} a_p \frac{\sin[(2N_x+1)u_p]}{u_p}\right|^2$$

$$\left|\sum_{-Q}^{Q} b_q \frac{\sin[(2N_y+1)v_q]}{v_q}\right|^2 \qquad \text{(D-8)}$$

It is convenient to rephrase this expression in terms of spherical coordinates for which the latitudinal angle $\theta_y$ is measured from the y axis and the azimuthal coordinate $\phi_y$ is measured from the z axis in the x-z plane. Then

$$x = r_o \sin\theta\cos\phi = r_o \sin\theta_y \sin\phi_y$$

$$y = r_o \sin\theta\sin\phi = r_o \cos\theta_y$$

$$z = r_o \cos\theta = r_o \sin\theta_y \cos\phi_y$$

so that

$$u_p = \frac{\pi d_x}{\lambda}\left(\sin\theta_y\sin\phi_y - \sin\theta_{yo}\sin\phi_{yo} + \frac{p\lambda}{L_x}\right)$$

$$v_q = \frac{\pi d_y}{\lambda}\left(\cos\theta_y - \cos\theta_{yo} + \frac{q\lambda}{L_y}\right)$$

Thus, the formula for directivity becomes

$$D = \frac{4\pi\left|\sum_{-N_x}^{N_x} I_m\right|^2 \left|\sum_{-N_y}^{N_y} I_n\right|^2}{\int_0^{\pi}\left|\sum_{-Q}^{Q} b_q \frac{\sin[(2N_y+1)v_q]}{v_q}\right|^2 \sin\theta_y\left[\int_{-\frac{\pi}{2}}^{\frac{\pi}{2}}\left|\sum_{-P}^{P} a_p \frac{\sin[(2N_x+1)u_p]}{u_p}\right|^2 d\phi_y\right]d\theta_y} \qquad \text{(D-9)}$$

The denominator of (D-9) consists of a sum of integrals. The integrand of each integral involves the product of four functions of the form sinU/U, with the origins for these four functions not, in general, coinciding. The displacement of origin in each of these four functions is governed by $p\lambda/L_x$ or $q\lambda/L_y$. For large arrays with realistic sidelobe levels, the aperture distributions are such that it only takes a few terms in (D-2) and (D-3) to approximate the distributions adequately. Then the maximum significant values of q and p are small, and the maximum displacement of origin which we need to consider is small. For such arrays, it is useful to let

$$\theta_y = \theta_{yo} + \theta' \quad \text{(D-10)}$$

$$\phi_y = \phi_{yo} + \phi' \quad \text{(D-11)}$$

in which the range of $\theta'$ and $\phi'$ need only be small in order to cover the principal part of sin U/U for each of the four functions in the integrand of each of the integrals in the denominator of (D-9).

With these substitutions, one finds

$$u_p \cong \frac{\pi d_x}{\lambda}\left(\cos\theta_{yo}\cdot\sin\phi_{yo}\cdot\theta' + \sin\theta_{yo}\cdot\cos\phi_{yo}\cdot\phi' + \frac{p\lambda}{L_x}\right)$$

$$v_q \cong \frac{\pi d_y}{\lambda}\left(-\sin\theta_{yo}\cdot\theta' + \frac{q\lambda}{L_y}\right)$$

If the main beam does not point within several beamwidths of the extreme values of $\theta_y$ and $\phi_y$, the finite limits of integration in (D-9) can be replaced by infinite limits with negligible error. Then a typical first integral from (D-9) can be written

$$L_{rs} = \int_{-\infty}^{\infty} a_r a_s \frac{\sin\left[\left(2N_x+1\right)\frac{\pi d_x}{\lambda}\left(\cos\theta_{yo}\cdot\sin\phi_{yo}\cdot\theta' + \sin\theta_{yo}\cdot\cos\phi_{yo}\cdot\phi' + \frac{r\lambda}{L_x}\right)\right]}{\frac{\pi d_x}{\lambda}\left(\cos\theta_{yo}\cdot\sin\phi_{yo}\cdot\theta' + \sin\theta_{yo}\cdot\cos\phi_{yo}\cdot\phi' + \frac{r\lambda}{L_x}\right)}$$

$$\frac{\sin\left[\left(2N_x+1\right)\frac{\pi d_x}{\lambda}\left(\cos\theta_{yo}\cdot\sin\phi_{yo}\cdot\theta' + \sin\theta_{yo}\cdot\cos\phi_{yo}\cdot\phi' + \frac{s\lambda}{L_x}\right)\right]}{\frac{\pi d_x}{\lambda}\left(\cos\theta_{yo}\cdot\sin\phi_{yo}\cdot\theta' + \sin\theta_{yo}\cdot\cos\phi_{yo}\cdot\phi' + \frac{s\lambda}{L_x}\right)} d\phi'$$

For $2N_x >> 1$, this integral has a significant value only when r=s. Thus,

$$L_{rs} = \delta_s^r \left\{\frac{a_r a_s (2N_x+1)}{\frac{d_x}{\lambda}\sin\theta_{yo}\cdot\cos\phi_{yo}}\right\} \quad \text{(D-12)}$$

Since (D-12) is independent of $\theta'$, the integral

$$J_{rs} = \int_{-\infty}^{\infty} b_r b_s \frac{\sin[(2N_y+1)v_r]}{v_r}\cdot\frac{\sin[(2N_y+1)v_s]}{v_s}\sin\theta_y\cdot d\theta \quad \text{(D-13)}$$

can be evaluated in the same fashion. The factor $\sin\theta_y$ can be replaced by $\sin\theta_{yo}$, its value in the region of principal importance of the integrand. Then

$$J_{rs} = \delta_s^r \left\{\frac{b_r b_s (2N_y+1)}{\frac{d_y}{\lambda}}\right\} \quad \text{(D-14)}$$

Therefore, the directivity can be written

$$D = \frac{4\pi\left|\sum_{-N_x}^{N_x}\sum_{-N_y}^{N_y} I_m I_n\right|^2 \sin\theta_{yo}\cos\phi_{yo}}{\sum_{-P}^{P}\sum_{-Q}^{Q} a_p^2 b_q^2 \frac{2N_x+1}{\frac{d_x}{\lambda}}\cdot\frac{2N_y+1}{\frac{d_y}{\lambda}}} \quad \text{(D-15)}$$

From (D-2),

$$I_m^2 = \sum_r\sum_s a_r a_s e^{j(r-s)\frac{2\pi m d_x}{L_x}}$$

$$\sum_{-N_x}^{N_x} I_m^2 = \sum_r\sum_s a_r a_s \frac{\sin\left[(r-s)\frac{\pi d_x}{L_x}(2N_x+1)\right]}{\frac{\pi d_x}{L_x}(r-s)}$$

$$\sum_{-N_x}^{N_x} I_m^2 = \sum_r\sum_s a_r a_s (2N_x+1)\,\delta_s^r$$

$$\sum_{-N_x}^{N_x} I_m^2 = \sum_r a_r^2 (2N_x+1) \quad \text{(D-16)}$$

Similarly,

$$\sum_{-N_y}^{N_y} I_n^2 = \sum_s b_s^2 (2N_y+1) \quad \text{(D-17)}$$

Inserting (D-16) and (D-17) in (D-15), and making use of the fact that $\sin\theta_{yo}\cdot\cos\phi_{yo} = \cos\theta_o$, one obtains

$$D = 4\pi\frac{d_x}{\lambda}\frac{d_y}{\lambda}\cos\theta_o\frac{\left|\sum_{-N_x}^{N_x}\sum_{-N_y}^{N_y} I_m I_n\right|^2}{\sum_{-N_x}^{N_x}\sum_{-N_y}^{N_y} I_m^2 I_n^2} \quad \text{(D-18)}$$

For $d_x = d_y = \frac{\lambda}{2}$, this formula can be compared with (25), which gave the directivity for a linear array. Under such an assumption about element spacing, (D-18) is $\pi$

times the product of the directivities of the two linear arrays when the pencil beam is broadside-broadside. The factor $\cos\theta_o$ accounts for the shrinkage of the projected aperture in the direction of the main beam.

For the general distribution given by (D-2) and (D-3),

$$\left\{\sum_{-N_x}^{N_x} I_m\right\}^2 = a_o^2\,(2N_x+1)^2 \qquad \text{(D-19)}$$

$$\left\{\sum_{-N_y}^{N_y} I_n\right\}^2 = b_o^2\,(2N_y+1)^2 \qquad \text{(D-20)}$$

Using these results together with (D-16) and (D-17) in (D-18), one obtains

$$D=4\pi\frac{L_x L_y}{\lambda^2}\cos\theta_o\frac{1}{\displaystyle\sum_{-P}^{P}\left(\frac{a_p}{a_o}\right)^2\sum_{-Q}^{Q}\left(\frac{b_q}{b}\right)^2} \qquad \text{(D-21)}$$

This formula gives the classical result for uniform excitation and indicates that the directivity is also maximum for uniform excitation.

## APPENDIX E

*Beamwidth of a Planar Array Scanned to Endfire*

If the beam of a planar array is scanned to endfire ($\theta_o=\frac{\pi}{2}$), the expansions leading to Equations (48) and (49) are not valid, and one must return to (47). Its proper expansion gives

$$\sum_{-N_x}^{N_x}\sum_{-N_y}^{N_y} I_m I_n\, e^{-jk\frac{(\Delta\theta_2)^2}{2}[md_x\cos\phi_o+nd_y\sin\phi_o]} = 0.707\sum_{-N_x}^{N_x}\sum_{-N_y}^{N_y} I_m I_n \qquad \text{(E-1)}$$

$$\sum_{-N_x}^{N_x}\sum_{-N_y}^{N_y} I_m I_n\, e^{-jk\frac{(\Delta\theta_1)^2}{2}[md_x\cos\phi_o+nd_y\sin\phi_o]} = 0.707\sum_{-N_x}^{N_x}\sum_{-N_y}^{N_y} I_m I_n \qquad \text{(E-2)}$$

so that once again, to this approximation, $\Delta\theta_1=\Delta\theta_2=\frac{\textcircled{H}}{2}$.

Proceeding as before, the exponentials in (E-1) can be expanded in a power series, giving to third order

$$2.344\sum_{-N_x}^{N_x}\sum_{-N_y}^{N_y} I_m I_n = k^2\left(\frac{\textcircled{H}}{2}\right)^4\sum_{-N_x}^{N_y}\sum_{-N_x}^{N_y} I_m I_n\,(m^2 d_x^2\cos\phi_o+n^2 d_y^2\sin^2\phi_o) \qquad \text{(E-3)}$$

If $\phi_o=0$, $\textcircled{H}$ is the vertical beamwidth of an endfire array of $2N_x+1$ elements. We shall designate this $\textcircled{H}_{xe}$. Similarly, if $\phi_o=\frac{\pi}{2}$, $\textcircled{H}=\textcircled{H}_{ye}$ is the vertical beamwidth of an endfire array of $2N_y+1$ elements. Using these two special cases of (E-3), the general case can be written

$$\textcircled{H}^{-4}=\textcircled{H}^{-4}_{xe}\cos^2\phi_o+\textcircled{H}^{-4}_{ye}\sin^2\phi_o \qquad \text{(E-4)}$$

However, if the beam of a planar array is scanned to endfire ($\theta_o=\frac{\pi}{2}$), the expansions leading to Equation (59) are still valid, and all that is needed is a re-interpretation of the horizontal beamwidth $\mathcal{H}$. If $\phi_o=0$, one has a broadside array of $2N_y+1$ elements in which each element is an endfire array of $2N_x+1$ sub-elements. Calling the horizontal beamwidth for this case $\mathcal{H}_{xe}$, we find approximately that

$$\mathcal{H}^{-2}_{xe}=\textcircled{H}^{-2}_{xe}+\textcircled{H}^{-2}_{yo} \qquad \text{(E-5)}$$

Similarly, if $\phi_o=\frac{\pi}{2}$, one has a broadside array of $2N_x+1$ elements, in which each element is an endfire array of $2N_y+1$ sub-elements. Calling the horizontal beamwidth for this case $\mathcal{H}_{ye}$, one can write approximately

$$\mathcal{H}^{-2}_{ye}=\textcircled{H}^{-2}_{xo}+\textcircled{H}^{-2}_{ye} \qquad \text{(E-6)}$$

Using these two special cases of (59) enables one to express the general case in the form

$$\mathcal{H}^{-2}=(\textcircled{H}^{-2}_{xo}+\textcircled{H}^{-2}_{ye})\sin^2\phi_o+(\textcircled{H}^{-2}_{xe}+\textcircled{H}^{-2}_{yo})\cos^2\phi_o \qquad \text{(E-7)}$$

The areal beamwidth for this endfire case is therefore

$$B_e=\frac{[(\textcircled{H}^{-2}_{xo}+\textcircled{H}^{-2}_{ye})\sin^2\phi_o+(\textcircled{H}^{-2}_{xe}+\textcircled{H}^{-2}_{yo})\cos^2\phi_o]^{-\frac{1}{2}}}{[\textcircled{H}^{-4}_{xe}\cos^2\phi_o+\textcircled{H}^{-4}_{ye}\sin^2\phi_o]^{1/4}} \qquad \text{(E-8)}$$

For the special cases that the beam lies along the x axis or along the y axis, (E-8) reduces to

$$B_{xe}=\frac{\textcircled{H}_{xe}}{[\textcircled{H}^{-2}_{xe}+\textcircled{H}^{-2}_{yo}]^{1/2}} \qquad \text{(E-9)}$$

$$B_{ye}=\frac{\textcircled{H}_{ye}}{[\textcircled{H}^{-2}_{xo}+\textcircled{H}^{-2}_{ye}]^{1/2}} \qquad \text{(E-10)}$$

In discussing Equation (62), we found that the areal beamwidth in the x-z plane was the same as in the y-z plane, if the beam was scanned not too far from broadside. Equations (E-9) and (E-10) indicate that when the beam is scanned to endfire this is no longer true.

On the other hand, for a square array, the areal beamwidth at endfire is independent of $\phi_o$ and given by

$$B_e=\frac{\textcircled{H}_{xe}}{[\textcircled{H}^{-2}_{xo}+\textcircled{H}^{-2}_{xe}]^{1/2}} \qquad \text{(E-11)}$$

*Directivity of a Planar Array Scanned to Endfire*

If the beam of a planar array is scanned to endfire ($\theta_o=\frac{\pi}{2}$), the analysis of Appendix D is not valid, and one must

18

return to the equations for $u_p$ and $v_q$ which immediately precede Equation (D-9) and include the second order terms. When this is done, one obtains

$$u_p = \frac{\pi d_x}{\lambda}\left[ -\frac{(\theta')^2}{2}\cdot\sin\theta_{yo} + \theta'\cdot\cos\theta_{yo} - \frac{(\phi')^2}{2}\cdot\sin\theta_{yo} + \frac{p\lambda}{L_x}\right] \quad \text{(E-12)}$$

$$v_q = \frac{\pi d_y}{\lambda}\left[ -\frac{(\theta')^2}{2}\cdot\cos\theta_{yo} - \theta'\cdot\sin\theta_{yo} + \frac{q\lambda}{L_y}\right] \quad \text{(E-13)}$$

Equation (D-9) is still a valid representation for the directivity if the $\phi_y$ limits are changed to $\pm\pi$ to accommodate the fact that half the pattern is in the lower hemisphere. As before, the only significant terms in the summation of integrals in (D-9) are the self-product terms. If the main beam is assumed to be endfire, but in either principal plane ($\theta_{yo}=0, \frac{\pi}{2}$) the integration of (D-9) is relatively simple. For example, if $\theta_{yo}=\frac{\pi}{2}$, the significant $\phi_y$ integrals are of the form

$$Ka_p^2 \int_{-\infty}^{\infty} \frac{\sin^2[w^2-K_p]}{[w^2-K_p]^2}dw \quad \text{(E-14)}$$

in which

$$K = \frac{(2N_x+1)^2}{\sqrt{\frac{\pi L_x}{2\lambda}}}$$

$$K_p = \frac{\pi L_x}{2\lambda}(\theta')^2 - p\pi$$

and w is a substantial variable given by

$$w = \sqrt{\frac{\pi L_x}{2\lambda}}\cdot\phi'$$

As before, the finite limits have been replaced by infinite limits, an adequate approximation for large arrays.

For practical aperture distributions, the maximum value of p is small and (E-14) can be further approximated by[9]

$$Ka_p^2 \int_{-\infty}^{\infty} \frac{\sin^2 w^2}{w^4}dw = \frac{4}{3}\sqrt{\pi}\,Ka_p^2 \quad \text{(E-15)}$$

so that the directivity can be written

$$D = \frac{4\pi\left\{\sum_{-N_x}^{N_x} I_m\right\}^2\left\{\sum_{-N_y}^{N_y} I_n\right\}^2}{\frac{4}{3}\sqrt{\pi}\,K\sum_{-P}^{P} a_p^2 \int_{-\infty}^{\infty}\left\{\sum_{-Q}^{Q} b_q(2N_y+1)\frac{\sin\frac{\pi L_y}{\lambda}\left(\theta'-\frac{q\lambda}{L_y}\right)}{\frac{\pi L_y}{\lambda}\left(\theta'-\frac{q\lambda}{L_y}\right)}\right\}^2 \sin\theta_y\cdot d\theta'} \quad \text{(E-16)}$$

The remaining integration parallels the development of Appendix D, and one obtains

$$D = \frac{3\pi}{\sqrt{2}}\frac{L_y}{\lambda}\sqrt{\frac{L_x}{\lambda}}\;\frac{\left\{\sum_{-N_x}^{N_x} I_m\right\}^2\left\{\sum_{-N_y}^{N_y} I_n\right\}^2}{\sum_{-P}^{P}[(2N_x+1)a_p]^2\sum_{-Q}^{Q}[(2N_y+1)b_q]^2} \quad \text{(E-17)}$$

Use of (D-19) and (D-20) reduces this to

$$D = \frac{\frac{3\pi}{\sqrt{2}}\frac{L_y}{\lambda}\sqrt{\frac{L_x}{\lambda}}}{\sum_{-P}^{P}\left(\frac{a_p}{a_o}\right)^2\sum_{-Q}^{Q}\left(\frac{b_q}{b_o}\right)^2} \quad \text{(E-18)}$$

Similarly, if the endfire beam is in the other principal plane ($\theta_{yo}=0$), there results

$$D = \frac{\frac{3\pi}{\sqrt{2}}\frac{L_x}{\lambda}\sqrt{\frac{L_y}{\lambda}}}{\sum_{-P}^{P}\left(\frac{a_p}{a_o}\right)^2\sum_{-Q}^{Q}\left(\frac{b_q}{b_o}\right)^2} \quad \text{(E-19)}$$

Presumably, as the endfire beam is scanned in the x-y plane ($\theta_{yo}$ variable) the directivity smoothly changes from (E-18) to (E-19), but this more general case has not been investigated. It seems reasonable to predict that the general endfire expression should be

$$D = \frac{\frac{3\pi}{\sqrt{2}}\sqrt{\frac{L_x}{\lambda}}\sqrt{\frac{L_y}{\lambda}}}{\sum_{-P}^{P}\left(\frac{a_p}{a_o}\right)^2\sum_{-Q}^{Q}\left(\frac{b_q}{b_o}\right)^2}\cdot\sqrt{\frac{L_x}{\lambda}\cos^2\theta_{yo}+\frac{L_y}{\lambda}\sin^2\theta_{yo}} \quad \text{(E-20)}$$

For the case of a uniform aperture distribution, (E-18) is lower by a factor of 2 than the expression derived by King and Thomas[10] but this is because they confined their integration to a hemisphere. Even the expression (E-18) can be too high; if there is more than one main beam, such as would occur at endfire for half-wavelength spacing.

## ACKNOWLEDGMENTS

The author is grateful to his friends at the Lockheed Antenna Laboratory who were kind enough to review and criticize this work, and special thanks are due to Dale Milligan and Fred Zboril.

## REFERENCES

9. Groebner und Hofreiter, *Bestimmte Integraltafel*, Springer-Verlag, Table 333, No. 19e, Wien, 1961, p. 120.
10. King, M. J., and R. K. Thomas, "Gain of Large Scanned Arrays," *Trans. IRE*, Vol. AP-8, No. 6, Nov. 1960, pp. 635-6.

*Reprinted by permission from* IEEE TRANSACTIONS ON ANTENNAS AND PROPAGATION
Vol. AP-20, No. 1, January 1972, pp. 100-102


20

## Comparison of Square Array Directivity Formulas

ROBERT C. HANSEN

***Abstract*—Directivity of rectangular planar arrays is often approximated using the directivity of two constituent linear arrays, either of dipoles or isotropic elements. A result has been derived for directivity of a rectangular planar array of short dipoles (in the plane of the array) using a method from Forman. The several approximate directivity calculations are compared with this more accurate result. Curves of exact directivity of square arrays of from 4 to 20 elements on a side using short dipoles are presented.**

### I. Introduction

Directivity of planar arrays is readily calculated when the spacing between elements is $\lambda/2$. Few arrays of course are built with this spacing, as larger spacings (with a fixed number of elements) reduce mutual coupling and produce higher directivity; element spacing is a tradeoff between mutual coupling, grating lobe effects, and directivity. Beam scanning often produces worsened mutual coupling for spacings near the grating lobe spacing, but in spite of this scanning arrays usually have spacings greater than $\lambda/2$. Calculation of directivity for variable spacings is then a useful task.

For linear arrays extensive data are available. Tai [1], in a most useful paper, presented calculations and results for directivity of uniformly spaced, uniformly excited broadside linear arrays with several orientations of short dipoles. Endfire arrays were covered in two reports by Tai [2]. Interesting plots of directivity versus spacing and phasing for a few elements have been published by Bach and Hansen [3]. Tables of Chebyshev linear array parameters, including directivity, have been developed by Brown and Sharp [4]. Optimum Chebyshev array directivity, optimum in the sense that the grating lobe at 90° is of a level equal to the sidelobe level, has been published by Hansen [5]. Linear arrays with other nonuniform excitations can also be handled by the techniques previously referred to.

Until recently the only results available for planar arrays were two approximations, one developed by Elliott [6] and another by Kraus [7]. Elliott showed that pencil beam planar array directivity is approximately $\pi G_x G_y$, where $G_x$ and $G_y$ are the directivities of the $x$ and $y$ constituent linear arrays. This result is for an array over a ground plane, (i.e., for an array with a beam each side the coefficient is $\pi/2$). Kraus used the mutual resistance, where the directivity is given by

$$D = \frac{120N^2}{\sum_{i=1}^{N} R_i}. \tag{1}$$

Unfortunately, $R_i$ is the input (self + mutual) resistance of the $i$th element with all others present. Using superposition this can be calculated from individual mutual impedances between pairs of elements, but with all intervening elements present. This last condition is difficult to fulfill and seldom is, thereby giving an approximate result. Forman [8] derived a result for planar arrays with general spacing and excitation, both for isotropic and for $\cos^{n+0.5}\theta$ conical element patterns. The latter means a pattern symmetric in $\theta$. An extension of this to short dipole patterns is part of this communication and will be discussed later. All of the discussion here overlooks gain loss due to impedance mismatch. This is especially important for scanning arrays [9].

Two objectives of this communication are to compare several directivity approximations for square arrays, and to present directivity versus spacing curves for square arrays of short dipoles. Although radiation patterns multiply, i.e., the overall pattern is equal to the array factor times the element pattern, it is known that directivities do not multiply. This is particularly true when isotropic array directivity is multiplied by a short dipole pattern. The difference in beamwidths is sufficiently large that the far-out grating lobes contribute differently; this is because the pattern integration is a nonlinear process. For example, consider the beamwidths and directivities of two elements. The directivities were calculated by direct pattern integration as follows.

| | Directivity (dB) | Beamwidth (deg) |
|---|---|---|
| Short dipole | 1.76 | 90 × isotropic |
| Conical cos $\theta$ | 4.77 | 90 × 90 |

### II. Exact Calculation for Short Dipole Array

Forman [8] devised an ingenious method for integrating the terms of the directivity expression for iostropic and conical patterns, and this method will be used here.[1] For the sake of brevity, only the key steps will be included herein. The directivity is

$$D = \frac{4\pi E(0,0)E^*(0,0)}{\iint EE^* \sin\theta \, d\theta \, d\varphi}. \tag{2}$$

A conventional spherical coordinate system is used, with the polar axis also the array normal, and the array in the $x$–$y$ plane. The dipoles are in the $x$ direction. For brevity, let

$$\alpha = kd_x \sin\theta \cos\varphi$$

$$\beta = kd_y \sin\theta \sin\varphi \tag{3}$$

with $k = 2\pi/\lambda$ and the $d$'s the interelement spacings. The pattern of a short dipole is

$$E = (1 - \sin^2\theta \cos^2\varphi)^{1/2} \tag{4}$$

and the field is simply

$$E(\theta,\varphi) = \sum_{n=0}^{N-1}\sum_{m=0}^{M-1} \exp(-j\alpha n - j\beta m)(1 - \sin^2\theta\cos^2\varphi)^{1/2}. \tag{5}$$

Here the array has $N$ by $M$ elements. The intensity is

$$EE^* = \sum_{n}\sum_{n'}\sum_{m}\sum_{m'} \exp(-jr\alpha - jq\beta)(1 - \sin^2\theta\cos^2\varphi) \tag{6}$$

with $r = n - n'$ and $q = m - m'$. Taking first the $n,n'$ sums, removing the diagonal, and rearranging allows one sum to be

Manuscript received December 14, 1970; revised August 6, 1971.
The author was with KMS Technology Center, Van Nuys, Calif. He is now with R. C. Hansen, Inc., Encino, Calif. 91316.

[1] Numerical integration is of course possible, but double integration, even of a single term is expensive.

performed:

$$\sum_n \sum_{n'} \exp(-jr\alpha) = N + 2\sum_{n=1}^{N-1}\sum_{r=1}^{n} \cos r\alpha$$
$$= N + 2\sum_{r=1}^{N-1}(N-r)\cos r\alpha. \tag{7}$$

With a similar result for the $m, m'$ series, the intensity becomes

$$\frac{EE^*}{NM} = \left[1 + 2\sum_{r=1}^{N-1}\left(1-\frac{r}{N}\right)\cos r\alpha + 2\sum_{q=1}^{M-1}\left(1-\frac{q}{M}\right)\cos q\beta\right.$$
$$\left. + 4\sum_{r=1}^{N-1}\sum_{q=1}^{M-1}\left(1-\frac{r}{N}\right)\left(1-\frac{q}{M}\right)\cos r\alpha\cos q\beta\right]$$
$$\cdot(1-\sin^2\theta\cos^2\varphi). \tag{8}$$

These then are inserted into the directivity expression (2). From the four major terms of (8), three types of integrals occur. The one and the element factor integrate immediately to $8\pi/3$. The second type, which includes the $r$ and the $q$ sums, is integrated in $\varphi$ using Grobner and Hofreiter [10, formula 334]. The argument is either $\gamma\cos\varphi$ or $\gamma\sin\varphi$ where $\gamma$ includes appropriate constants and functions of $\theta$; the upper signs belong with $\gamma\cos\varphi$ as follows:

$$\int_0^{\pi/2}\cos\begin{bmatrix}\gamma\cos\varphi\\ \gamma\sin\varphi\end{bmatrix}\cos^2\varphi\,d\varphi = \pi(1\pm1)J_0(\gamma) \pm \frac{2\pi J_1(\gamma)}{\gamma}. \tag{9}$$

The $\theta$ integrals are now integrals of Bessel functions and are immediately done by Grobner and Hofreiter [10, formula 541.5]. The results are

$$\iint \cos r\alpha(1-\sin^2\theta\cos^2\varphi)\,d\Omega = 8\pi\left[\frac{\sin A}{A^3} - \frac{\cos A}{A^2}\right],$$
$$A = rkd_x \tag{10}$$

$$\iint \cos q\beta(1-\sin^2\theta\cos^2\varphi)\,d\Omega = 4\pi\left[\frac{\sin B}{B} - \frac{\sin B}{B^3} + \frac{\cos B}{B^2}\right],$$
$$B = qkd_y. \tag{11}$$

The last type of integral is the most difficult. The cosine product of $r\alpha$ and $q\beta$ is expanded into sums of $(r\alpha + q\beta)$ and $(r\alpha - q\beta)$. Later it will appear that these two give identical results, so they are here combined as follows:

$$\iint \cos r\alpha\cos q\beta(1-\sin^2\theta\cos^2\varphi)\,d\Omega$$
$$= \iint \cos(r\alpha+q\beta)(1-\sin^2\theta\cos^2\varphi)\,d\Omega. \tag{12}$$

Using Forman's technique, let $r\alpha + q\beta = C\sin(\varphi + \mathrm{AT})$, where $\mathrm{AT} = \arctan rd_x/qd_y$ and $C = \sin\theta(r^2k^2d_x^2 + q^2k^2d_y^2)^{1/2}$. When this is substituted into (12), the $\varphi$ integral becomes, after some manipulation,

$$\int_0^{2\pi}\cos(C\sin(\varphi+\mathrm{AT}))\cos^2\varphi\,d\varphi$$
$$= \int_0^{2\pi}\cos(C\sin x)\cos^2(x-\mathrm{AT})\,dx. \tag{13}$$

These are integrated again using [10], with the result

$$-2\pi\sin^2\mathrm{AT}\,J_0(C) - 2\pi\cos 2\,\mathrm{AT}\,\frac{J_1(C)}{C}. \tag{14}$$

Now the earlier assertion that $(r\alpha - q\beta)$ gave the same result as $(r\alpha + q\beta)$ can be proved; the sign changes the sign of AT, but no change results in (14). Integration in $\theta$ now proceeds directly with the result:

$$\iint \cos r\alpha\cos q\beta(1-\sin^2\theta\cos^2\varphi)\,d\Omega$$
$$= 4\pi\left[\cos^2\mathrm{AT}\frac{\sin E}{E} + (3\sin^2\mathrm{AT}-1)\left(\frac{\sin E}{E^3} - \frac{\cos E}{E^2}\right)\right] \tag{15}$$

where $E$ is $C$ with the $\sin\theta$ removed. Combining all these results gives the directivity expression $D = NM/\text{denominator}$, where the denominator is

$$\text{denom} = \frac{2}{3} + 4\sum_{r=1}^{N-1}\left(1-\frac{r}{N}\right)\left(\frac{\sin A}{A^3} - \frac{\cos A}{A^2}\right)$$
$$+2\sum_{q=1}^{M-1}\left(1-\frac{q}{M}\right)\left(\frac{\sin B}{B} - \frac{\sin B}{B^3} + \frac{\cos B}{B^2}\right)$$
$$+4\sum_{r=1}^{N-1}\sum_{q=1}^{M-1}\left(1-\frac{r}{M}\right)\left(1-\frac{q}{M}\right)\left[\cos^2\mathrm{AT}\frac{\sin E}{E}\right.$$
$$\left.+(3\sin^2\mathrm{AT}-1)\left(\frac{\sin E}{E^3} - \frac{\cos E}{E^2}\right)\right] \tag{16}$$

with $A = rkd_x$, $B = qkd_y$, $E = r^2k^2d_x + q^2k^2d_y^2$.

## III. Comparison of Results

Square arrays of short dipoles, with equal spacing and uniform excitation, of size 4 by 4 and 20 by 20 have been calculated several ways[2]. Perhaps the poorest approximation assumes isotropic arrays along $x$ and $y$, with the directivity for short dipoles given by

$$G \simeq \frac{\pi}{2} G_{x\text{ iso}} G_{y\text{ iso}} G_{\text{short dipole}}. \tag{17}$$

Here the linear array values were obtained using Tai's formulas. A better approximation uses the $x$ and $y$ values for broadside parallel and collinear linear arrays, respectively. Again using Tai's short dipole array directivities, the approximation is

$$G \simeq \frac{2\pi}{3\times 2} G_{x\text{ short}} G_{y\text{ short}}. \tag{18}$$

The factor of 2/3 is needed as both the $x$ and $y$ values include the 3/2 short dipole directivity. Another approximation uses the Forman isotropic square array results

$$G \simeq G_{\text{sq iso}} G_{\text{short dipole}}.$$

Finally, the most accurate result, uses the exact short dipole results described in the previous section:

$$G = G_{\text{sq short}}. \tag{19}$$

Fig. 1 shows these three approximations and exact results for a 4 by 4 array versus spacing, and Fig. 2 for a 20 by 20 array. These two sizes should be typical of the relative accuracy for both small and large square arrays. In Fig. 2, the broadside × collinear approximation has not been plotted as it is very close to the square isotropic case. From these figures several interesting items appear. In general, up to the first grating lobe ($d/\lambda < 0.9$), the two linear array approximations and the square isotropic are close together, but the exact result is 1 to 2 dB lower. Above this point the two linear array approximations are still 2 dB high, but the square isotropic is quite close to the best result. This is perhaps surprising; one might have expected the linear broadside and collinear result to be better.

[2] A Com-Share code was written for formula (16).

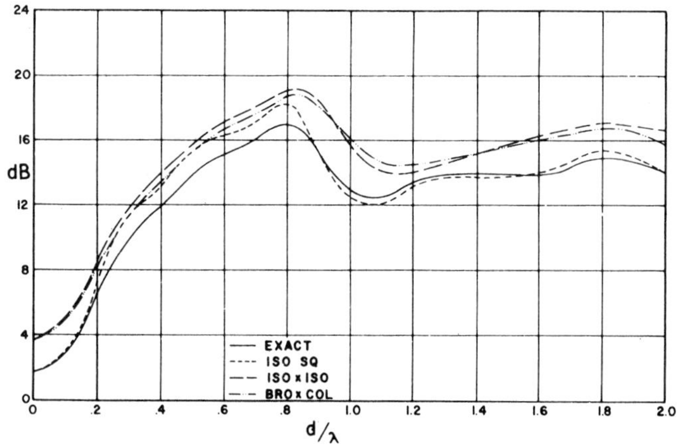


**Fig. 1.** Approximate and exact directivities of 4 by 4 array.

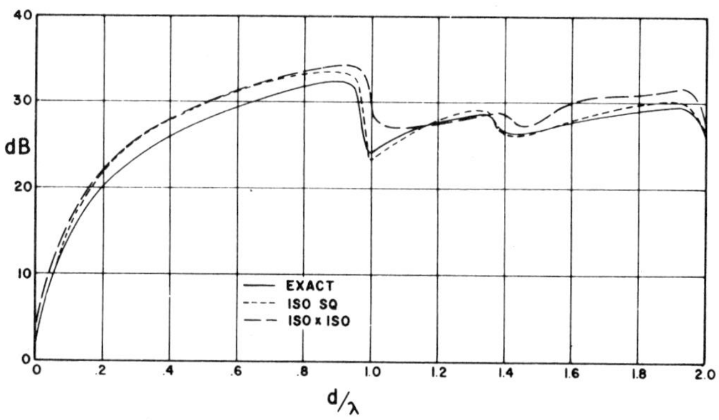


**Fig. 2.** Approximate and exact directivities of 20 by 20 array.

It must be emphasized that the purpose here is to evaluate the approximate techniques in comparison with the exact results, for short dipoles. Fig. 3 gives square array directivity versus spacing for short dipoles.

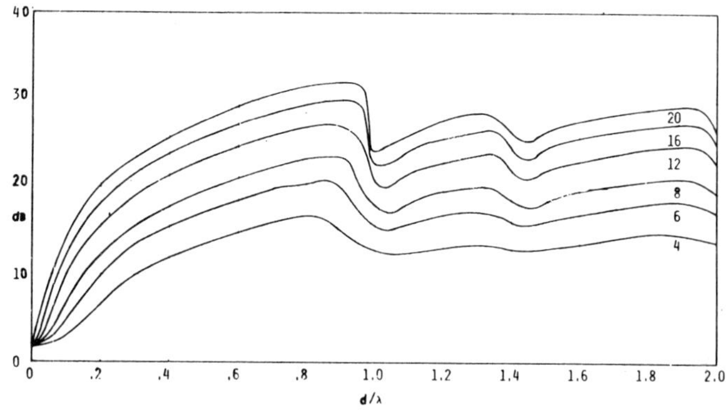


**Fig. 3.** Directivities of square array of short dipoles. (Number on curve is number of elements per side).

## IV. Conclusions

Several approximate directivity formulas for square arrays have been evaluated. Directivity computed from orthogonal linear arrays is roughly 2 dB high; accurate formulas are developed for planar arrays of short dipoles. Directivity curves are given for various numbers of elements.

## References


[1] C-T Tai, "The optimum directivity of uniformly spaced broadside arrays of dipoles," *IEEE Trans. Antennas Propagat.*, vol. AP-12, pp. 447–454, July 1964.

[2] ——, "On the optimum gain of uniformly spaced arrays of isotropic sources or dipoles," Antenna Lab., Ohio State Univ., Columbus, Tech. Rep. 1522-1, Mar. 31, 1963.

——, "The gain of uniform arrays of isotropic sources and dipoles," Antenna Lab., Ohio State Univ., Columbus, Tech. Rep. 1522-2, Mar. 31, 1963.

[3] H. Bach and J. E. Hansen, "Uniformly spaced arrays," in *Antenna Theory*, pt. 1, R. E. Collin and F. J. Zucker, Eds. New York: McGraw-Hill, 1969, ch. 5.

[4] L. B. Brown and G. A. Scharp, "Tschebyscheff antenna distribution, beamwidth, and gain tables," Naval Ordnance Lab., Corona, Calif., Tech. Rep. 383, Feb. 28, 1958.

[5] R. C. Hansen, "Directivity of Chebyshev arrays," *IEEE Trans. Antennas Propagat.* (Commun.), vol. AP-18, pp. 815–818, Nov. 1970.

[6] R. S. Elliott, "The theory of antenna arrays," in *Microwave Scanning Antennas*, vol. 2, R. C. Hansen, Ed. New York: Academic Press, 1966, ch. 1.

[7] J. D. Kraus, *Antennas*. New York: McGraw-Hill, 1950, ch. 11.

[8] B. J. Forman, "A novel directivity expression for planar antenna arrays," *Radio Sci.*, vol. 5, pp. 1077–1083, July 1970.

[9] A. A. Oliner and R. G. Malech, "Mutual coupling in infinite scanning arrays," in *Microwave Scanning Antennas*, vol. 2, R. C. Hansen, Ed., New York: Academic Press, 1966, ch. 3, p. 305.

[10] W. Grobner and N. Hofreiter, *Integraltafel*, pt. 2. New York: Springer, 1958.


22

# Optimization Techniques for Antenna Arrays

DAVID K. CHENG, FELLOW, IEEE

*Invited Paper*

**Abstract**—Various techniques for optimizing the performance indices of antenna arrays are discussed. In particular, methods for maximizing array directivity and signal-to-noise power ratio are reviewed. These performance indices are expressible as a ratio of two Hermitian forms, which has a certain special property to enable the facile determination of both its maximum value and the conditions under which this maximum is attained. Special maximization procedures by excitation amplitude and phase adjustments, by spacing perturbation, by phase adjustments only, and by a coordinate transformation with constraints on the array pattern are examined. For wire antennas the method of moments using a subsectioning technique can be applied to obtain numerical answers which include mutual-coupling effects. Methods for considering large arrays and for the maximization of power gain are indicated.

## I. Introduction

An antenna is an essential part of any electronic system which transmits or receives electromagnetic energy in a wireless fashion. Without an antenna, electromagnetic energy will be localized, and interaction at a distance between unconnected points in space does not occur. An antenna can be considered as a transducer which converts electromagnetic waves in space to current or voltage variations in a circuit, or vice versa. It must be an efficient radiator (or collector) of electromagnetic energy, and it should direct the energy to certain desired directions and suppress it in other specified directions. Thus one must be concerned not only with the conversion efficiency of an antenna, but also with its spatial response, or *radiation pattern*. In an environment which does not involve nonlinear media, a reciprocity relation holds such that the properties (pattern, gain, impedance) of an antenna used for receiving are identical with those when it is transmitting. For simplicity we shall then refer all discussions to radiation properties.

A radiating source of electromagnetic energy may take many different forms. It may be a piece of conducting wire, a dielectric rod, a metallic horn, or a slot on the side of a waveguide. The radiation pattern of a single element is fixed for a given frequency of excitation and contains, in general, a main beam and a number of smaller sidelobes. In practical applications there is quite often a need for either improving the directive properties or controlling the sidelobe structure of the radiation pattern. Two methods are available for this purpose: one method is to use an appropriately shaped reflector or lens fed by a radiating element, and the other is to employ a number of radiating elements properly arranged in space to form an antenna array. When it is necessary to steer (scan) the main beam of the radiation pattern, the requisite motion of heavy reflector- or lens-type antennas entails both mechanical and structural problems. Moreover, the possible rate of scan is severely limited. On the other hand, beam-steering for an antenna array can be accomplished electronically by adjusting the relative phase of excitation in the array elements with no need for mechanical motion, resulting in a *phased array*. We shall concern ourselves only with phased array antennas in this paper.

The radiation pattern of an array antenna obviously depends on both the array geometry and the pattern of the individual array elements. Aside from such circuit properties as impedance, bandwidth, and efficiency, the parameters which characterize antenna performance are all based on the shape of the radiation pattern. Performance optimization then is a procedure for the maximization or minimization of certain measures on the radiation pattern. One well-known result in this regard is the Chebyshev array which makes use of the properties of Chebyshev-Akhiezer polynomials [1], [2]. A Chebyshev array is optimum in the sense that for a specified sidelobe level the width of the main beam of its radiation pattern is a minimum. Conversely, for a specified beamwidth all the sidelobes of a Chebyshev array are of equal height and are at a lowest level. The original Chebyshev design considered by Dolph [1] was limited to linear broadside arrays of isotropic elements uniformly spaced at a distance equal to or larger than a half-wavelength. It has since been extended in various ways (see, for instance, [3]). The Chebyshev distribution of currents for a planar array radiating in a broadside direction was first considered by Baklanov [4]. Recently methods for determining the current distribution, the minimum required number of controlled elements, and other properties of optimum rectangular arrays with a steerable main beam and constant sidelobes have been formulated [5].

Besides the beamwidth–sidelobe relationship, an important performance index for any antenna is its directive gain, or *directivity*. Directivity is defined as the ratio of the radiation intensity (radiated power per unit solid angle) in the direction of the main beam to the average radiation intensity. To put it another way, the directivity of an antenna or an array is the ratio of its maximum radiation intensity to the radiation intensity of an isotropic (omnidirectional) source[1] radiating the same total power. It measures the ability of concentrating the radiated energy in the main-beam direction. Our attention in this paper will be directed toward the various techniques for maximizing the directivity of antenna arrays.

The performance of an antenna system as a receiving device is often constrained by the presence of a spatially distributed background noise as well as by the noise generated in the receiving system. A useful performance index of a receiving array is the *signal-to-noise power ratio* (*SNR*) at the system output. The problem of finding the complex weighting factors of the individual array elements such that the SNR is maximized for a signal coming from a given direction and a noise of a given power

Manuscript received June 21, 1971; revised July 23, 1971. *This invited paper is one of a series planned on topics of general interest—The Editor.* This work was supported in part by the National Science Foundation and in part by the Advanced Research Projects Agency.

The author is with the Electrical and Computer Engineering Department, Syracuse University, Syracuse, N. Y. 13210.

[1] It should be noted that an isotropic source radiating uniformly in all directions is physically unrealizable for vector fields.

spectral density and a given spatial distribution is of considerable importance. It can be shown [6], [7] that techniques similar to those for the maximization of directivity are applicable for SNR maximization. The formulation becomes more involved when the element space-frequency response, the signal power spectral density, the internal noise power spectral density, the spatial noise cross-power spectral density, and the receiver or filter frequency responses are to be considered [8]. We shall not attempt to discuss the more general case in this paper.

The problem of maximizing the directivity of a linear array with equally spaced isotropic elements was first studied by Uzkov in 1946 [9]. He demonstrated that the maximum obtainable directivity for an array with $N$ elements spaced at a half-wavelength ($\lambda/2$) apart is $N$ and that it tends to $N^2$ as the spacings approach zero. Bloch, Medhurst, and Pool [10] examined the maximum directivity of a linear array of half-wave dipoles from the point of view of self and mutual resistances of the elements. Gain optimization under a specified constraint was investigated by Uzsoky and Solymar [11], and Lo, Lee, and Lee [6]. In 1964 Tai [12] published many interesting curves showing the optimum directivity of various types of uniformly spaced broadside arrays, linear arrays with maximum radiation in the direction of the array normal. The optimization problem was generalized by Cheng and Tseng [13], [14] to include arrays of nonisotopic elements arranged in an arbitrary configuration with a main beam pointing at an arbitrary direction. By making use of a theorem on the properties of a ratio of two Hermitian forms in matrix algebra, the optimization procedure was formalized in a concise manner. It turned out that Krupitskii in USSR [15] had used the same theorem to prove the existence and uniqueness of a solution for exciting an array of discrete radiators for maximum directivity.

In the following we shall first express, in Section II, the directivity and SNR of an array of discrete elements as a ratio of two Hermitian forms. The optimization principle for a ratio of Hermitian forms is then reviewed and applied to antenna arrays in Section III. With a given array configuration where the element positions are not to be changed, the excitation amplitudes and phases in the array elements can be adjusted for the optimization of a performance index. If the array has a total of $N$ elements, the optimization procedure involves the determination of $2N$ parameters. Typical results for linear and circular arrays will be presented. For linear arrays the spacings between the array elements represent another convenient set of parameters that can be adjusted to improve the performance index further. A spacing-perturbation technique which can be used in conjunction with the adjustments in excitation amplitudes and phases is discussed in Section IV. This process of excitation and spacing adjustments may be repeated until the improvement obtained by further adjustments is no longer significant.

In practice it is perhaps inconvenient to adjust element spacings. Even adjustments in excitation amplitudes are difficult and expensive to make. Techniques for maximizing the directivity of a fixed array by keeping the amplitudes equal and adjusting the phases only are therefore of interest. These techniques are explored in Section V.

The adjustment of the excitation amplitudes, phases, or spacings of an array in order to achieve a maximum directivity changes the array radiation pattern. In practice it is often desirable to control some aspects of the array pattern. For example, one may wish to have a maximum directivity in one direction while requiring a null in certain other directions in order to minimize interference. The method of optimization under constraints is discussed in Section VI.

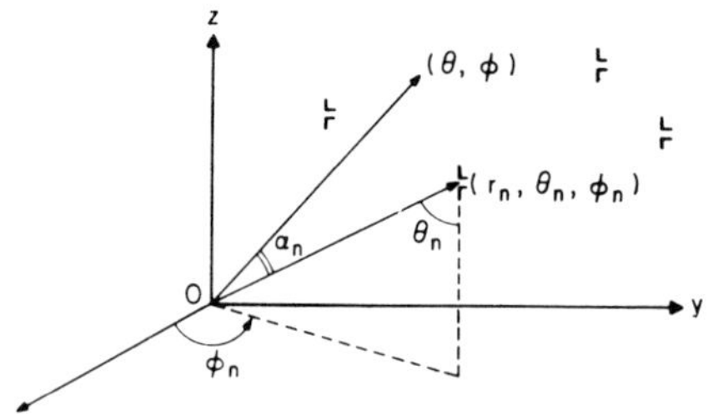


Fig. 1. Reference coordinates of an arbitrary array.

Initially, as we develop the optimization procedure, we shall assume that all the elements in an array are identically polarized and have the same radiation pattern (element pattern). Once the element pattern is specified, the physical structure of the array elements is no longer important in our problem, and it is immaterial whether the elements are dipoles, horns, slots, or other apertures. This assumption neglects the implications of mutual coupling. For large arrays with many elements this assumption gives acceptable results, although the element pattern must first be found. However, for arrays with a small number of closely spaced elements or in cases where more accurate results are desired, mutual-coupling effects must be taken into account. Section VII reviews the moment method for optimizing the directivity of arrays of wire antennas without neglecting mutual coupling. The moment method provides numerical solutions by first converting the governing integro-differential equations into matrix equations.

Finally, in Section VIII, we discuss various factors which are relevant in the optimization of discrete antenna arrays.

## II. General Formulation

Consider an array of $N$ discrete similarly oriented identical elements arranged arbitrarily in a 3-dimensional space, as shown in Fig. 1. Denoting the excitation in the $n$th element located at $(r_n, \theta_n, \phi_n)$ by $I_n \exp(j\psi_n)$, we may write the array factor for electric field intensity as

$$E(\theta, \phi) = \sum_{n=1}^{N} I_n \exp\left[j(\psi_n + kr_n \cos\alpha_n)\right] \tag{1}$$

where

$$\cos\alpha_n = \sin\theta \sin\theta_n \cos(\phi - \phi_n) + \cos\theta \cos\theta_n \tag{2}$$

$k = 2\pi/\lambda$ is the wavenumber, and $\lambda$ is the operating wavelength. Let $g(\theta, \phi)$ denote the element power-pattern function which is normalized such that

$$g(\theta_0, \phi_0) = 1 \tag{3}$$

in the direction $(\theta_0, \phi_0)$ of the main beam. The directivity of the array is then

$$D = \frac{\text{radiation intensity in direction } (\theta_0, \phi_0)}{\text{average radiation intensity}} = \frac{|E(\theta_0, \phi_0)|^2}{\frac{1}{4\pi}\int_0^{2\pi} d\phi \int_0^{\pi} |E(\theta, \phi)|^2 g(\theta, \phi) \sin\theta d\theta}. \tag{4}$$

We now define two $N$-element column vectors $J$ and $F_0$, with $J$ representing the set of complex element excitation functions:

$$J = [J_n] = \begin{bmatrix} I_1 \exp(j\psi_1) \\ I_2 \exp(j\psi_2) \\ \cdot \\ \cdot \\ \cdot \\ I_N \exp(j\Psi_N) \end{bmatrix} \qquad (5)$$

and $F_0$ representing the set of phase factors due to differences in distance

$$F_0 = [F_{0n}] = \begin{bmatrix} \exp(-jkr_1 \cos \alpha_{01}) \\ \exp(-jkr_2 \cos \alpha_{02}) \\ \cdot \\ \cdot \\ \exp(-jkr_N \cos \alpha_{0N}) \end{bmatrix} \qquad (6)$$

where $\cos \alpha_{0n}$ $(n=1, 2, \cdots, N)$ is obtained from (2) by setting $\theta=\theta_0$ and $\phi=\phi_0$. From (1) and (4)–(6) it is readily verified that the directivity can be written as

$$D = \frac{J^\dagger AJ}{J^\dagger BJ} \qquad (7)$$

where † on a matrix indicates the adjoint, or the conjugate transpose, of the matrix. $A$ and $B$ are $N \times N$ square matrices defined as

$$A = [a_{mn}] = F_0 F_0^\dagger \qquad (8)$$

and

$$B = [b_{mn}] \qquad (9)$$

with

$$b_{mn} = \frac{1}{4\pi} \int_0^{2\pi} d\phi \int_0^{\pi} g(\theta, \phi) \cdot \exp[-jk(r_m \cos \alpha_m - r_n \cos \alpha_n)] \sin \theta d\theta. \qquad (10)$$

It is obvious that matrices $A$ and $B$ are both Hermitian, i.e., $A^\dagger = A(a_{nm}{}^* = a_{mn})$, and $B^\dagger = B(b_{nm}{}^* = b_{nm})$. Hence $D$ in (7) is a ratio of two Hermitian forms.[2] In addition, $B$ is positive-definite, which implies that for any $J \neq \mathbf{0}$, $J^\dagger BJ > 0$. This is proved in [14]. Since the elements of matrices $A$ and $B$ are known when the array geometry, the operating wavelength, and the scan angle are given, the optimization problem reduces to the determination of the excitation matrix $J$ such that $D$ in (7) is maximized.

In receiving systems the output SNR, instead of the directivity, is of interest. The *output SNR* may be defined as the ratio of the power received per unit solid angle in the direction of the signal to the average noise power received per unit solid angle. It is only necessary to replace the power-pattern function $g(\theta, \phi)$ in (4) by a more general weighting function $w(\theta, \phi)$ which includes the spatial distribution of noise power.[3] We write

$$w(\theta, \phi) = g(\theta, \phi) T(\theta, \phi) \qquad (11)$$

where $T(\theta, \phi)$ is the spatial distribution function of noise power. Here we understand noise to be a combination of interference, clutter, atmospherics, and random noise. It is clear that replacing $g(\theta, \phi)$ by $w(\theta, \phi)$ does not change the nature of the optimization problem. In fact, the expression for SNR reduces to that for $D$ when $T(\theta, \phi) = 1$. We expect a suppression of the sidelobes in the directions of high noise power for SNR improvement [7]. In the next section we state the theorem for maximizing $D$ by adjusting $J$.

## III. Optimization by Excitation Adjustments

A theorem in matrix algebra on the peoperties of a ratio of two Hermitian forms [16] is useful for the maximization of directivity by excitation adjustments. It may be stated as follows:

*Theorem 1*

If a quantity $D$ is expressible as a ratio of two Hermitian forms as in (7) and if $B$ is nonsingular and positive definite, then the largest eigenvalue $\lambda_M$ of the "regular pencil" of matrices $A$-$\lambda B$ is the maximum obtainable $D$ when $J$ is the eigenvector satisfying the homogeneous equation



$$AJ = \lambda_M BJ. \qquad (12)$$

For our case, the following corollary, proved in [14], makes the optimization procedure particularly straightforward and simple.

*Corollary:* If $A$ in (7) is expressible in the form of (8), then

1) the largest and only nonzero eigenvalue of the regular pencil $A$-$\lambda B$ is

$$\lambda_M = D_M = F_0^\dagger B^{-1} F_0 \qquad (13)$$

2) the eigenvector corresponding to $\lambda_M$ is

$$J_M = B^{-1} F_0. \qquad (14)$$

Equations (13) and (14) solve the problem of optimization by excitation adjustments for an arbitrary array. For a given array configuration, it is only necessary to determine the elements of the matrices $F_0$ and $B$, in accordance with (6) and (10), respectively.

For a linear array with elements located arbitrarily at distances $d_n$ from a reference point

$$a_{mn} = \exp[-jk(d_n - d_m) \sin \theta_0] \qquad (15)$$

where $\theta_0$ denotes the main-beam direction measured from the normal to the array, and

$$b_{mn} = \frac{1}{4\pi} \int_0^{2\pi} d\phi \int_0^{\pi} g(\theta, \phi) \exp[-jk(d_n - d_m) \sin \theta] d\theta. \qquad (16)$$

If the array elements are isotropic, $g(\theta, \phi) = 1$, and are equally spaced, $d_n - d_m = (n-m)d$, (15) and (16) reduce to

$$a_{mn} = \exp[jk(m-n)d \sin \theta_0] \qquad (17)$$

and

$$b_{mn} = \frac{\sin k(m-n)d}{(m-n)d}. \qquad (18)$$

As an example, it has been shown [13] that an endfire array with 8 isotropic elements equally spaced at 0.425 λ apart has a directivity of 12.5 with a uniform amplitude and cophasal excitation.[4] Optimization by the above procedure results in a directivity

---

[2] Sometimes Hermitian forms are written as general inner products; for instance,

$$J^\dagger AJ = \sum_{m=1}^{N} \sum_{n=1}^{N} J_m{}^* a_{mn} J_n = \langle J, AJ \rangle.$$

They are quadratic forms in Hilbert space in the variables $J_n$.

[3] Monochromatic operation is assumed. The consideration of signal and noise power spectral densities, antenna space-time response function, and system impulse response makes the problem slightly more involved [8].

[4] An excitation is said to be cophasal when the relative phases in the elements are adjusted such that the contributions of all the elements add in phase in the main-beam direction. Thus for an endfire array with 0.425 λ spacing, the progressive phases in the neighboring elements differ by 0.425×360° or 153°.

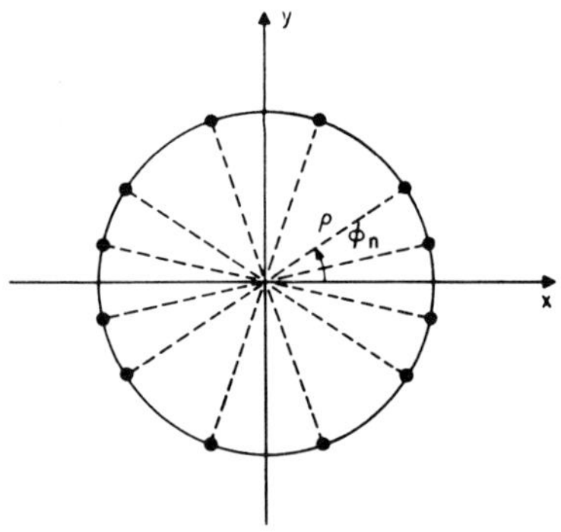


Fig. 2. A circular array.



of 22.0. The amplitude is tapered (center-to-edge ratio: 1.69) and the phase shift between adjacent elements is roughly 170°. Other examples of linear-array optimization by excitation adjustments will be given in Section IV when spacing-perturbation techniques are discussed.

Besides the linear array, the circular array represents another class of arrays with a simple geometry and important practical applications. Fig. 2 shows a general circular array with non-uniformly spaced elements in the $x$–$y$ plane. Substituting $\theta_n = \pi/2$ in (3), we obtain

$$\cos \alpha_n = \sin \theta \cos (\phi - \phi_n). \tag{19}$$

Since $r_n = \rho_n = \rho$ for all $n$, $b_{mn}$ in (10) simplifies to

$$b_{mn} = \frac{1}{4\pi} \int_0^{2\pi} d\phi \int_0^{\pi} g(\theta, \phi) \cdot \exp \{ -jk\rho_{mn} \cos (\phi - \phi_{mn}) \sin \theta \} \sin \theta d\theta \tag{20}$$

where

$$\rho_{mn} = 2\rho \left| \sin (\phi_m - \phi_n)/2 \right| \tag{21}$$

and

$$\phi_{mn} = \tan^{-1} \left( \frac{\sin \phi_m - \sin \phi_n}{\cos \phi_m - \cos \phi_n} \right). \tag{22}$$

For a circular array of $N$ uniformly spaced isotropic elements, $g(\theta, \phi) = 1$, $\phi_n = 2n\pi/N$, and

$$\rho_{mn} = 2\rho \left| \sin (m - n)\pi/N \right|. \tag{23}$$

We have, from (20),

$$b_{mn} = \frac{\sin (k\rho_{mn})}{k\rho_{mn}}. \tag{24}$$

The expressions of $b_{mn}$ for some directive elements with typical power-pattern functions have also been given [14].

Fig. 3 compares the maximum directivity $D_M$ with the directivity $D_u$ under a uniform amplitude and cophasal excitation for a 12-element circular array as a function of the array diameter. We note that $D_M$ is everywhere higher than $D_u$ and that $D_M$ increases very rapidly when the array diameter is less than $2\lambda$ (a superdirective situation). Since superdirective arrays require very large currents of opposite signs in neighboring elements, resulting in excessive heat loss and very low radiation intensity in the direction of the main beam, it is appropriate to define a *main-beam radiation efficiency* $\eta$. A practical optimum design then would be a suitable compromise between $D_M$ and $\eta$. We define

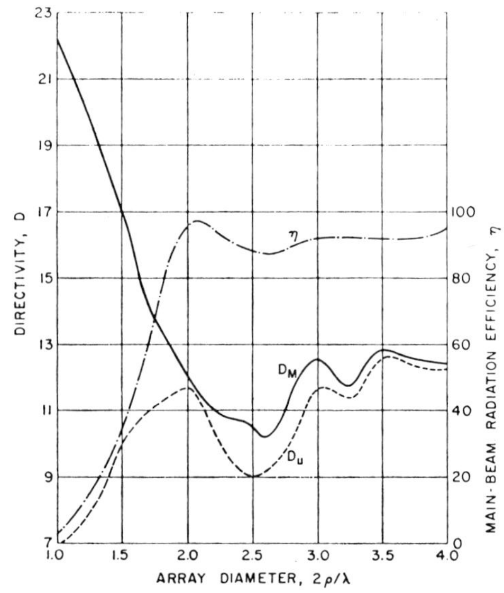


Fig. 3. Directivity and main-beam radiation efficiency for a circular array with isotropic elements ($N = 12$, $\theta_0 = 90°$, $\phi_0 = 0°$).

$$\eta = \frac{\left| E(\theta_0, \phi_0) \right|^2}{N \sum_{n=1}^{N} I_n^2} \times 100 \text{ percent}. \tag{25}$$

By the use of Schwarz's inequality, it is easy to show that [14] $\eta$ equals 100 percent only for uniformly excited cophasal arrays and becomes very small under superdirective situations. The main-beam radiation efficiency turns out to be the reciprocal of the tolerance sensitivity used by Uzsoky and Solymar [11] to measure the mean-square variation of the maximum field with respect to the mean-square deviation of the excitation. The values of $\eta$ under the conditions for $D_M$ are also plotted in Fig. 3.

It has been pointed out [17] that the optimization problem becomes easier by applying an orthogonalization process for arrays possessing a cyclic symmetry. For large circular arrays with many uniformly spaced elements, the tedium of inverting a large $B$ matrix [in (14)] can be circumvented through the introduction of a rotational operator [18].

## IV. Optimization by Spacing Adjustments

In the preceding section optimization was achieved by adjusting the excitation amplitudes and phases in the fixed elements of a given array. If the element positions are also allowed to vary, we acquire an additional dimension of freedom (which represents an additional $N-1$ degrees of freedom for a linear array with $N$ elements), and we expect to be able to improve on the results obtained by excitation adjustments only. A spacing-perturbation technique exists which is useful for optimizing the directivity in a given direction or the SNR in a given noise environment [7]. This technique will be developed in this section.

Consider a linear array of $N$ identical elements symmetrically located about the origin along the $x$ axis. Let $(\theta_0, \phi_0)$ be the direction of the main beam, and $I_n \exp(j\Phi_n)$ and $I_n \exp(-j\Phi_n)$ be the excitations in the $n$th and $-n$th elements, respectively, where

$$\Phi_n = kd_n \sin \theta_0 \cos \phi_0 + \psi_n. \tag{26}$$

In (26), $d_n$ is the distance of the $n$th element from the origin, and $\psi_n$ is the phase shift from cophasal operation. Note that, except for the assumed symmetry about the origin, there is no restriction on element spacings which may be nonuniform and the elements themselves need not be omnidirectional. If $N$ is odd[5] and equals $2M-1$, the array factor is

$$E(u) = I_0 + 2 \sum_{n=1}^{M} I_n \cos(\delta_n u + \psi_n) \tag{27}$$

where

$$u = kd(\sin\theta\cos\phi - \sin\theta_0\cos\phi_0) \tag{28}$$

$$\delta_n = d_n/d \tag{29}$$

and $d$, a normalizing distance, may be any choice of convenience. For example, if one starts with an equally spaced array, it would be natural to make $d$ the uniform spacing between neighboring elements. We write the output SNR in the direction of the signal ($\theta=\theta_0$, $\phi=\phi_0$, $u=0$) as

$$\mathrm{SNR} = \frac{|E(0)|^2}{\frac{1}{4\pi}\int_0^{2\pi} d\phi \int_0^{\pi} |E(u)|^2 w(\theta,\phi)\sin\theta d\theta} \tag{30}$$

where $w(\theta, \phi)$ is a weighting function defined in (11). Our problem is to find the set of normalized element positions $\{\delta_n\}$ such that SNR is maximized for a given set of excitation amplitudes and phases. Note that this becomes a directivity maximization problem when $w(\theta,\phi)=g(\theta,\phi)$.

Let $\{\delta_n^0\}$ denote the original normalized element positions, and

$$\delta_n = \delta_n^0 + x_n \tag{31}$$

where $x_n$ represents the spacing perturbation for the $n$th element and $x_n \ll 1$. Substitution of (31) in (27) yields, approximately,

$$E(u) = E^0(u) - 2\sum_{n=1}^{M} I_n x_n u \sin(\delta_n^0 u + \psi_n) \tag{32}$$

where $E^0(u)$ is the original unperturbed array factor with $\delta_n^0$ substituted for $\delta_n$ in (27). Using (32), we can write (30) in the following form:

$$\mathrm{SNR} = \frac{|E^0(0)|^2}{\alpha - 2\mathbf{x}'\boldsymbol{\beta} + \mathbf{x}'C\mathbf{x}} \tag{33}$$

where

$$\alpha = \frac{1}{4\pi}\int_0^{2\pi} d\phi \int_0^{\pi} |E^0(u)|^2 w(\theta,\phi)\sin\theta d\theta \tag{34}$$

$$\mathbf{x}' = [x_1, x_2, \cdots, x_n, \cdots, x_N] \tag{35}$$

is the transpose of the column matrix of spacing perturbations $\mathbf{x}$; $\boldsymbol{\beta}$ is a column matrix of typical element

$$\beta_n = \frac{1}{2\pi}\int_0^{2\pi} d\phi \int_0^{\pi} I_n u E^0(u) w(\theta,\phi) \cdot \sin(\delta_n^0 u + \psi_n)\sin\theta d\theta \tag{36}$$

and $C=[C_{mn}]$ is an $N\times N$ square matrix with

$$C_{mn} = \frac{1}{\pi}\int_0^{2\pi} d\phi \int_0^{\pi} I_m I_n u^2 w(\theta,\phi) \cdot \sin(\delta_m^0 u + \psi_m)\sin(\delta_n^0 u + \psi_n)\sin\theta d\theta. \tag{37}$$

[5] Only very minor modifications are needed when $N$ is even. The formulation is entirely parallel.

It can readily be shown that $C$ is symmetric and positive definite. Use can then be made of the following theorem which is proved in the Appendix.

*Theorem 2*

If a quantity SNR can be expressed in terms of an $N\times 1$ real column vector $\mathbf{x}$ as in (33), where $\alpha$ is a constant, $\boldsymbol{\beta}$ is another $N\times 1$ real column vector, and $C$ is an $N\times N$ positive definite symmetric square matrix, then

$$\max_{(\mathbf{x}=\mathbf{x}_M)} \mathrm{SNR} = \frac{|E^0(0)|^2}{\alpha - \boldsymbol{\beta}C^{-1}\boldsymbol{\beta}} \tag{38}$$

and

$$\mathbf{x}_M = C^{-1}\boldsymbol{\beta}. \tag{39}$$

Equations (38) and (39) give the results of a first-order perturbation. After the components of $\mathbf{x}_M$ have been determined from (39), one can then use $(\boldsymbol{\delta}^0+\mathbf{x}_M)$ as the new normalized element-position column matrix and perform a second-order perturbation to obtain further improvement in the performance index. This process can be repeated until it becomes evident that further iteration yields a negligible improvement. The final values of $\{\delta_n\}$ determine the element positions for a maximum SNR for the given excitation. Now this perturbed nonuniformly spaced array can be further optimized by proper amplifications and phase shifts in the array elements using the method developed in the preceding section [7]. A second local maximum will be reached, which may possibly be further improved by holding the excitation unchanged and again perturbing the spacings. The cycle may be repeated until further adjustments are no longer worthwhile.

We illustrate the application of the above technique with a boradside array of 7 isotropic elements. It is desired to optimize the array for maximum SNR in a noise environment, $T(u)$ specified by Fig. 4 with $a=15$, $u_1=(1/4)(2\pi d/\lambda)$, and $u_2=(1/12)(2\pi d/\lambda)$. The normalizing distance $d$ is chosen to be 0.885 $\lambda$ which corresponds to the spacing for maximum directivity in a uniformly spaced linear array with 7 isotropic elements. The results for the 1) space-perturbed, 2) excitation-adjusted (by amplification), and 3) optimized arrays are listed in Table I. We note that large improvements in SNR are possible by optimization through either spacing perturbation or excitation adjustments, and that the SNR of the optimized array is about 11.4 times that of the original uniform array. Of course, the array optimized for a maximum SNR is not the same as the one for a maximum directivity. The directivity of the SNR-optimized array in Table I (sketched in Fig. 5(a)) is 10.6, whereas a directivity-optimized broadside array of 7 isotropic elements has a directivity of 10.63 [7]. The radiation patterns of the SNR-optimized array are plotted in Fig. 5(b), where the spatial distribution of the noise or interference power is also shown. It is interesting to see that, at the expense of a slightly wider main beam, the sidelobes of the optimized array are everywhere lower than those of the uniform array. In particular, the first sidelobe, which normally occurs in a region where the noise power is high, is much suppressed and its position slightly shifted.

## V. Optimization by Phase Adjustments

In Section III we discussed the method for maximizing array directivity by adjusting the excitation amplitudes and phases in the array elements. Accurate amplitude adjustments require the use of precision amplifiers or attenuators, or specially designed directional couplers. It is hence of both theoretical and practical interest to develop a technique for optimizing the directivity of

27

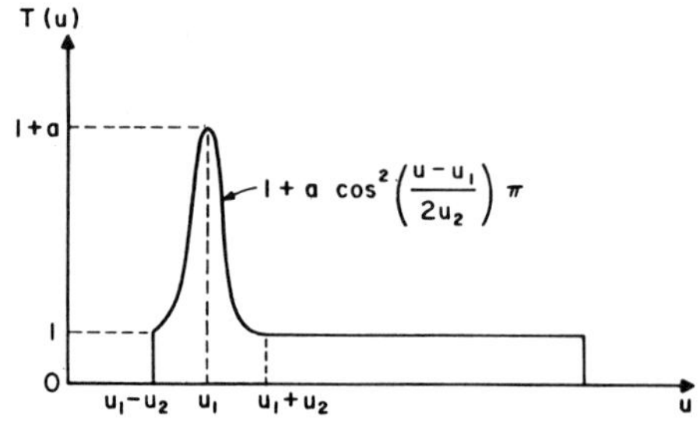


Fig. 4. Spatial distribution of noise power; $u=(2\pi d/\lambda)\sin\theta$.

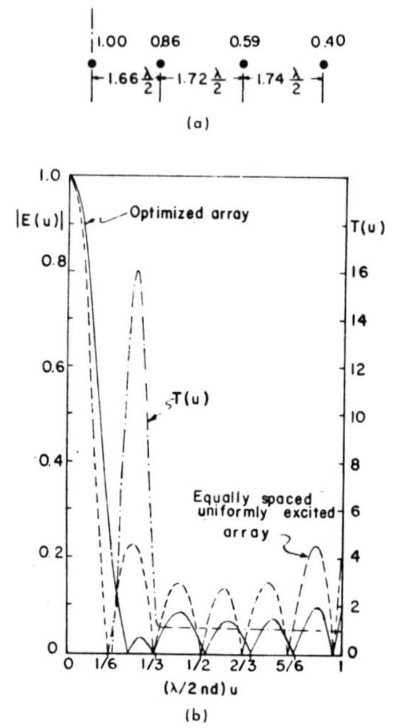


Fig. 5. SNR optimization for 7-element broadside array. (a) SNR optimized array. (b) Radiation patterns.

TABLE I

SNR Optimization for 7-Element Broadside Array

$$u_1 = \frac{1}{4}\left(\frac{2\pi d}{\lambda}\right); \quad u_2 = \frac{1}{12}\left(\frac{2\pi d}{\lambda}\right); \quad d_0 = 0.$$

| | $I_0$ | $I_1$ | $I_2$ | $I_3$ | $(d_1-d_0)\frac{2}{\lambda}$ | $(d_2-d_1)\frac{2}{\lambda}$ | $(d_3-d_2)\frac{2}{\lambda}$ | SNR |
|---|---|---|---|---|---|---|---|---|
| Uniform array | 1.00 | 1.00 | 1.00 | 1.00 | 1.77 | 1.77 | 1.77 | 16.0 |
| Space-perturbed array | same as above | | | | 1.65 | 1.76 | 2.10 | 19.8 |
| Exc.-adjusted perturbed array | 1.00 | 0.89 | 0.67 | 0.39 | same as above | | | 78.1 |
| Optimized array | 1.00 | 0.86 | 0.59 | 0.40 | 1.66 | 1.72 | 1.74 | 181.9 |

uniformly excited arrays requiring only phase adjustments. The elimination of the need for amplitude adjustments would result in a simplified feed structure and a reduced cost.

Directivity maximization under the constraint of a uniform amplitude in all the array elements can be formulated by Lagrange multiplier methods in several ways. However, it was found that the resulting equations were not amenable to a stable solution even by iterative methods, because of convergence problems. On the other hand, a perturbation procedure similar to that employed in Section IV for spacing adjustments can be used for the phase-adjustment problem [19]. The essential steps of this procedure will be developed in the following.

Consider a linear array of $2M+1$ symmetrically located equally spaced identical elements, all with the same excitation amplitude $I$. The array factor is then, from (27),

$$E(u) = I\left[1 + 2\sum_{n=1}^{M}\cos(nu+\psi_n)\right] \tag{40}$$

where $u$ is defined in (28) and $\psi_n$ is the phase shift from cophasal operation. We may write

$$\psi_n = \psi_n{}^0 + x_n \tag{41}$$

where $\psi_n{}^0$ is an assumed initial value and $x_n \ll 1$ is a small perturbation. Ordinarily it is convenient to start from a cophasal excitation, i.e., $\psi_n{}^0=0$. Substituting (41) in (40), we obtain approximately

$$E(u) = E^0(u) - 2I\sum_{n=1}^{M}x_n\sin(nu+\psi_n{}^0). \tag{42}$$

Using (42) in (4), we can express the directivity of the array in the following form

$$D = \frac{|E(\theta_0,\phi_0)|^2}{\alpha_1 - \mathbf{x}'\boldsymbol{\beta}_1 + \mathbf{x}'\mathbf{C}_1\mathbf{x}} \tag{43}$$

where

$$\mathbf{x}' = [x_{-M},\cdots,x_{-1},x_0,x_1,\cdots,x_M] \tag{44}$$

and $\alpha_1$, $\boldsymbol{\beta}_1$, and $\mathbf{C}_1$ are similar to $\alpha$, $\boldsymbol{\beta}$, and $\mathbf{C}$ in (33). $\alpha_1$ is identical to $\alpha$ in (34), and the elements of $\boldsymbol{\beta}_1$ and $\mathbf{C}_1$ can be obtained from (36) and (37), respectively, with $I_m=I_n=I$ and the argument $(\delta_n{}^0u+\psi_n)$ of the sine functions replaced by $(nu+\psi_n{}^0)$. Element power-pattern function $g(\theta,\phi)$ replaces weighting function $w(\theta,\phi)$ in computing directivity.

It is now obvious that the same Theorem 2, which is proved in the Appendix and has been found useful for optimization by spacing adjustments, can be used for maximizing $D$ in (43) by phase adjustments. The required phase changes in the array elements are determined from

$$\mathbf{x}^1 = \mathbf{C}_1^{-1}\boldsymbol{\beta}_1 \tag{45}$$

which represents a first-order perturbation from the initial values $\{\psi_n{}^0\}$. One may then treat $\{\psi_n{}^0+x_n{}^1\}$ as the new initial values and perform a second-order perturbation. This process may be repeated until it is apparent that a maximum directivity has been obtained.

Although the preceding formulation for optimization by phase adjustments starts with a linear array, it is clear that the procedure can be applied to an arbitrary 3-dimensional array. In particular, for a circular array of $N$ elements with radius $\rho$ in the $x$–$y$ plane, a phase perturbation as indicated in (41) results in an array factor

$$E(\theta,\phi) = E^0(\theta,\phi) - 2I\sum_{n=1}^{N}x_n\sin(\Delta_n+\psi_n{}^0) \tag{46}$$

with

$$\Delta_n = k\rho[\sin\theta\cos(\phi-\phi_n) - \sin\theta_0\cos(\phi_0-\phi_n)] \tag{47}$$

where $\phi_n$ denotes the location of the $n$th element and $(\theta_0,\phi_0)$ is the direction of the main beam. We note that (46) is entirely similar to (42). Substitution of (46) in (4) will yield a directivity expression in the form of (43), and hence the same optimization procedure follows. The optimum directivity $D_0$ obtained by phase adjustments only for a circular array with 12 uniformly spaced

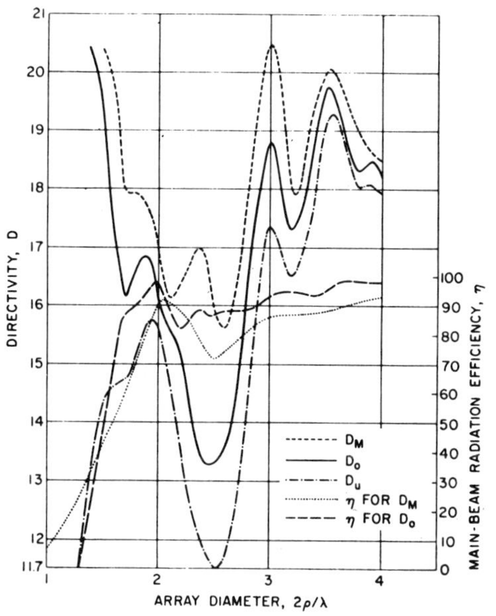


Fig. 6. Optimum directivity and main-beam radiation efficiency for a circular array with dipole elements ($N=12$, $\theta_0=90°$, $\phi_0=0°$).

short dipoles is plotted in Fig. 6 as a function of array diameter. In the same figure are also plotted $D_M$, the maximum directivity when both amplitudes and phases are adjusted, and $D_u$, the directivity under a uniform amplitude and cophasal excitation. We note that the $D_0$ curve lies everywhere between the $D_M$ and $D_u$ curves. For most array diameters less than 3λ (element spacing less than 3λ/4) an improvement of about 2 dB in directivity is possible by phase adjustments alone. When the array diameter is very small, the directivity of the phase-adjusted array increases rapidly, indicating a superdirective situation which is absent in a uniform cophasal array. The main-beam radiation efficiency of a superdirective array tends to be very low, as has been pointed out in Section III.

## VI. Optimization with Constraints

In previous sections we discussed techniques for maximizing the directivity or the SNR of an antenna array without constraints; that is, without imposing at the same time a requirement on any other performance index of the array. We have already seen that a maximum directivity for an array with closely spaced elements is accompanied by a low main-beam radiation efficiency. In practice, we may desire that a maximum directivity be obtained together with a prescribed value for main-beam radiation efficiency. The existence of constraints or auxiliary conditions effectively reduces the total number of independent variables which can be adjusted for optimization. In such cases, a procedure using Lagrange multipliers can be applied to determine the stationary value of directivity [20]. This approach has been employed [6] to maximize SNR under a constraint on a $Q$ factor. The $Q$ factor is a quantity which is proportional to the ratio of the directivity and the main-beam radiation efficiency. It turns out that this approach results in a rather involved numerical procedure. We shall not go into it further here.

A more useful class of optimization problems with constraints pertains to the maximization of some performance index while controlling the array pattern in certain definite ways. For instance, one may wish to have a maximum directivity in the direction of some distant transmitter or receiver and, at the same time, to minimize the interference from some other directions. This is equivalent to the problem of directivity maximization with controlled locations of certain pattern nulls, and can be reduced to that of maximizing the ratio of two Hermitian forms, as studied in Section III, through the introduction of a constraint matrix.

The array factor in (1) can be written as the inner product of the space vector $F=[\exp(-jkr_n\cos\alpha_n)]$ and the excitation vector $J$ defined in (5); i.e.,

$$E(\theta,\phi) = \langle F, J\rangle = F^{\dagger}J. \tag{48}$$

Pattern nulls in the directions $(\theta_i, \phi_i)$ are specified by homogeneous equations

$$F_i^{\dagger}J = \sum_{n=1}^{N} I_n \exp\left[j(\psi_n + kr_n\cos\alpha_{in})\right] = 0,$$
$$i = 1, 2, \cdots, M < (N-1) \tag{49}$$

where $\alpha_{in}$ are obtained from (2) by setting $\theta=\theta_i$ and $\phi=\phi_i$. A geometrical interpretation of (49) is that the excitation vector $J$ which we seek to maximize the directivity in the direction $(\theta_0, \phi_0)$ is now required to be simultaneously orthogonal to $M$ independent constraint vectors $F_i$. The $N$-dimensional space is divided into two mutually orthogonal subspaces: an $M$-dimensional subspace containing the constraint vectors and an $(N-M)$-dimensional subspace where the excitation vector $J$ must lie. The mathematical procedure [21] for maximization under constraints consists of 1) finding an appropriate set of $M$ mutually orthogonal vectors that occupy the same subspace as the constraint vectors $F_1, F_2, \cdots, F_M$; 2) obtaining an additional orthogonal $(N-M)$ vectors which at the same time are orthogonal to the first $M$ vectors; 3) forming from steps 1) and 2) an $N\times N$ normalized constraint matrix which is the unitary matrix for coordinate transformation; 4) transforming the directivity expression as a ratio of two Hermitian forms to the new orthogonal coordinate system; and 5) maximizing the directivity in the same manner as outlined in Section III.

The Gram-Schmidt procedure [20] provides a method for determining an orthonormal basis for a vector space in which any set of spanning vectors is known. This method can be used to find the constraint matrix in steps 1) and 2). Assuming $P$ to be the normalized constraint matrix which is also the transformation matrix, we write

$$PJ = J_c. \tag{50}$$

The directivity in (7) becomes, after the coordinate transformation,

$$D = \frac{J_c^{\dagger}(PAP^{\dagger})J_c}{J_c^{\dagger}(PBP^{\dagger})J_c} = \frac{J_c^{\dagger}A_cJ_c}{J_c^{\dagger}B_cJ_c}. \tag{51}$$

Inasmuch as each of the first $M$ mutually orthogonal vectors is a linear combination of the constraint vectors, the first $M$ rows of $J_c$ in (50) are linear combinations of the homogeneous constraint equations (49) and are therefore zeros. Hence the first $M$ entries in $J_c$ and the first $M$ rows and $M$ columns of $A_c$ and $B_c$ can be discarded, resulting in an abridged form for directivity [22]

$$D_a = \frac{J_a^{\dagger}A_aJ_a}{J_a^{\dagger}B_aJ_a} \tag{52}$$

where $A_a$ and $B_a$ are the abridged $(N-M)\times(N-M)$ matrices and $J_a$ is the abridged $(N-M)$-element column vector. The remainder of the optimization process then follows in exactly the same manner as that pertaining to the unconstrained $D$ in (7), Section III, except that the numerical problem is now simpler because the matrices involved are of a lower rank. It is obvious that the maximum obtainable directivity with constraints will be less than that without constraints on account of the reduced freedom. Typical results on maximum directivity with null placements and with reduced radiation level in an angular sector have been published [22].

## VII. Consideration of Mutual Coupling

One tacit assumption implied in the optimization techniques which we have considered thus far is that all the elements in an array have the same radiation pattern. This is tantamount to assuming that mutual-coupling effects are negligible. For large arrays with many elements this assumption is acceptable if our interest lies in array directivity and not in the current distribution or the exact radiation pattern of each element. The consideration of mutual-coupling effects in array optimization greatly complicates the problem. To the author's knowledge, no work has been published on array synthesis or optimization for mutually coupled aperture-type radiators. With wire antennas the method of moments can be used to obtain numerical answers [23]–[26]. In this section we will outline the optimization procedure for array of wire antennas when mutual coupling is not neglected.



The moment method for solving electromagnetic problems consists mainly of three steps; namely, the formulation of the governing integro-differential equations, the expansion of the unknown functions in terms of a set of linearly independent basis functions, and the testing of the expanded equation by forming inner products with a set of linearly independent set of weighting functions [27]. The result is a set of simultaneous equations which can be solved by matrix methods. The choice of the basis and weighting functions depends on the desired accuracy and the ease in evaluating the coefficients in the simultaneous equations. One convenient technique in the choice of basis functions is to divide the domain of the unknown function into small intervals or subsections. Simple basis functions such as pulses or triangles are defined to exist over one or a few such subsections and to be zero elsewhere. The simplest weighting functions are Dirac delta functions defined at discrete points at which the expanded governing equations are to be satisfied. This is the point-matching or sampling method.

For thin wire antennas, the currents and charges on the wires can be approximated by current and charge filaments along the antenna axes. We consider an array of thin linear antennas parallel to the $z$ axis. From the Maxwell's equation for time-harmonic fields in a homogeneous medium

$$\bar{\nabla} \times \bar{H} = j\omega\epsilon\bar{E} + \bar{J} \tag{53}$$

and $\bar{H} = 1/\mu \bar{\nabla} \times \bar{A}$, we have

$$\bar{E} = \frac{1}{j\omega\epsilon}\left(\frac{1}{\mu}\bar{\nabla} \times \bar{\nabla} \times \bar{A} - \bar{J}\right). \tag{54}$$

Now

$$\bar{\nabla} \times \bar{\nabla} \times \bar{A} = \bar{\nabla}\bar{\nabla}\cdot\bar{A} - \nabla^2\bar{A} \tag{55}$$

and

$$(\nabla^2 + k^2)\bar{A} = -\mu\bar{J} \tag{56}$$

where $k^2 = \omega^2\mu\epsilon$. Combination of (55) and (56) with (54) yields

$$\bar{E} = \frac{1}{j\omega\mu\epsilon}(\bar{\nabla}\bar{\nabla}\cdot\bar{A} + k^2\bar{A}). \tag{57}$$

Since $\bar{J}$ has only a $z$ component, we can rewrite (57) as a scalar equation

$$E_z = \frac{1}{j\omega\mu\epsilon}\left(\frac{\partial^2}{\partial z^2} + k^2\right)A_z. \tag{58}$$

For thin linear antennas, the tangential electric field $E_{zp}$ at the center of a typical $p$th subsection is then [26]

$$E_{zp} = \frac{1}{j\omega\epsilon}\left(\frac{\partial^2}{\partial z^2} + k^2\right) \cdot \int_{\substack{\text{all}\\\text{antennas}}} \frac{I(z')\exp\left[-jk\,|\bar{r}_p - \bar{r}'|\right]}{4\pi\,|\bar{r}_p - \bar{r}'|}\,dz' \tag{59}$$

where $\bar{r}_p$ is the position vector to the center of the $p$th subsection under consideration, and $\bar{r}'$ is a vector from the origin to a point on an antenna at which the current is $I(z')$. If each of the $N$ antennas in the array is divided into $S$ subsections of length $\Delta l_q$ which carries a constant current $I_q$, (59) can be approximated by a summation

$$E_{zp} = \sum_{q=1}^{M} I_q \left\{\frac{1}{j\omega\epsilon}\left(\frac{\partial^2}{\partial z^2} + k^2\right) \cdot \int_{\Delta l_q} \frac{\exp\left[-jk\,|\bar{r}_p - r'|\right]}{4\pi\,|\bar{r} - r'|}\,dz'\right\}. \tag{60}$$

In (60), $M = NS$. The quantity in the wavy brackets is the electric field at the center of subsection $p$ due to a unit current in subsection $q$ and can be written as $Z_{pq}/\Delta l$, where equal subsections ($\Delta l_q = \Delta l$) are assumed for simplicity. Thus

$$E_{zp}(\Delta l) = V_p = \sum_{q=1}^{M} Z_{pq} I_q \tag{61}$$

or, in matrix form,

$$\boldsymbol{V} = \boldsymbol{ZI} \tag{62}$$

where $\boldsymbol{V} = [V_p]$ and $\boldsymbol{I} = [I_q]$ are $M \times 1$ column matrices and $\boldsymbol{Z} = [Z_{pq}]$ is an $M \times M$ square matrix. $\boldsymbol{Z}$ may be called a generalized impedance matrix and depends only on the geometrical configuration of the array.

In the terminology of the method of moments, (62) is the result of 1) using pulse basis functions each of which exists over only one subsection, 2) using an integral-type inner product, and 3) using Dirac delta functions as weighting functions. The differential operator in (60) may be approximated by a second-order difference operator and computer subroutines for the calculation of $Z_{pq}$ are available [26]. In a radiation problem $\boldsymbol{V}$ is a column matrix of known voltages which are all zero except for the excitation voltages at feed points. Column matrix $\boldsymbol{I}$ determines the current distributions on the wire antennas

$$\boldsymbol{I} = \boldsymbol{Z}^{-1}\boldsymbol{V} = \boldsymbol{YV}. \tag{63}$$

With $\boldsymbol{I}$ known, all field quantities of interest can be determined.

The array directivity defined in (4) can be expressed in terms of the total power input to the array $P_{in}$ if the antenna wires are assumed to be perfectly conducting. The electric field in the far zone of an array consisting of $z$-directed wire antennas is

$$E_\theta = -j\omega A_\theta = j\omega A_z \sin\theta. \tag{64}$$

Thus

$$D = \frac{4\pi r^2\,|E(\theta_0)|^2/2}{\zeta P_{in}} = \frac{2\pi r^2\,|\omega A_z \sin\theta_0|^2}{\zeta P_{in}} \tag{65}$$

where $\zeta$ is the intrinsic impedance of the medium. The time-average input power to the array is

$$P_{\text{in}} = \frac{1}{2} \text{Re} \sum (V_{\text{in}} I_{\text{in}}^*). \tag{66}$$

Since nonzero voltages exist only at the feed points, (66) can be written as

$$\begin{aligned} P_{\text{in}} &= \frac{1}{2} \text{Re} (I^\dagger V) \\ &= \frac{1}{2} \text{Re} (V^\dagger Y^\dagger V) \\ &= \frac{1}{2} V_1^\dagger \left[ \frac{Y_1 + Y_1^*}{2} \right] V_1 \end{aligned} \tag{67}$$

where $V_1$ is a column matrix reduced from $V$ by retaining only the $N$ (number of wire antennas in the array) nonzero feed-point voltages, and $Y_1$ is an $N \times N$ admittance matrix reduced from $Y$ by deleting all elements that do not correspond to the feed subsections.

In the far zone, the magnetic potential can be approximated by

$$A_z = \frac{\mu(\Delta l)}{4\pi r} \exp(-jkr) \sum_{q=1}^{M} I_q \exp[jkr_q \cos \alpha_q] \tag{68}$$

where $\alpha_q$ is the angle between the position vector to the $q$th subsection carrying current $I_q$ and that to the field point. Using the matrix representations of $F$ as in (6) and of $I$ as in (62), we can write (68) as

$$A_z = \frac{\mu(\Delta l)}{4\pi r} \exp(-jkr) F^\dagger I \tag{69}$$

or in reduced matrices as

$$A_z = \frac{\mu(\Delta l)}{2\pi r} \exp(-jkr) F_1^\dagger Y_1 V_1. \tag{70}$$

For simplicity we write

$$G_1^\dagger = F_1^\dagger Y_1. \tag{71}$$

With (67), (70), and (71), $D$ in (65) becomes

$$D = \frac{(\omega\mu\Delta l \sin\theta_0)^2}{4\pi\zeta} \frac{V_1^\dagger G_1 G_1^\dagger V_1}{V_1^\dagger \left[ \frac{Y_1 + Y_1^*}{2} \right] V_1}. \tag{72}$$

It can be shown [24] that the excitation voltage matrix required for making $D$ in (72) a maximum is

$$V_1 = \left[ \frac{Y_1 + Y_1^*}{2} \right]^{-1} G_1 \tag{73}$$

and

$$D_M = \frac{(\omega\mu\Delta l)^2}{4\pi\eta} G_1^\dagger \left[ \frac{Y_1 + Y_1^*}{2} \right] G_1. \tag{74}$$

The directivity optimization problem with voltage excitation is now completely solved, and the effect of mutual coupling has been taken into consideration. As can be seen, the main task in obtaining numerical solutions lies in the determination of the admittance matrix $Y_1$.

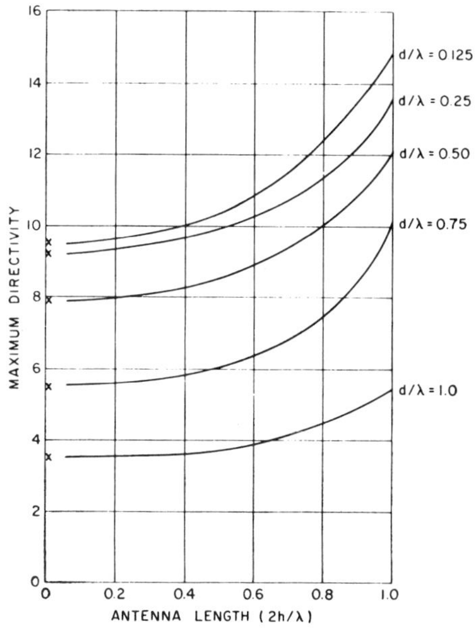


Fig. 7. Maximum directivity of a 4-element circular array in direction $\theta = 90°$ and $\phi = 45°$.

Using the above procedure, Cummins [24] determined the maximum directivity in the principal $H$ plane of a circular array of 4 uniformly spaced center-fed wire antennas. Fig. 7 shows the variation of maximum directivity ($D_M$) versus antenna length ($2h/\lambda$) with array diameter ($d/\lambda$) as the parameter. The points marked by crosses correspond to the maximum directivity of a circular array of 4 isotropic sources as computed by the method outlined in Section III.

The solution of the dual problem of an array excited by a set of current sources follows an entirely similar procedure [24].

## VIII. Other Considerations

We discuss here several related aspects of the array optimization problem which either can be treated by an extension of some of the preceding techniques or need special attention.

### *A. Maximization of Power Gain*

When the antenna wires are not perfectly conducting, resistive losses occur and power gain is no longer the same as directivity or directive gain [33]. Electromagnetically speaking, $E_{zp}$ in (61) on the surface of subsection $p$ is no longer zero, but is equal to the product of $I_p$ and $Z_i$, the internal impedance per unit length of the wire conductor.

$$V_p = E_{zp}(\Delta l) = Z_{pp}' I_p \tag{75}$$

where

$$Z_{pp}' = Z_i(\Delta l) \tag{76}$$

and

$$Z_i = \left( \frac{j\omega\mu}{\sigma} \right)^{1/2} = \frac{1+j}{\sigma\delta}. \tag{77}$$

In (77), $\sigma$ and $\delta$ are, respectively, the conductivity and the skin depth of the wire conductor at the operating frequency. Hence if the generalized impedance matrix $Z$ for perfectly conducting wires has been found, the only modification needed for a moment solution with finitely conducting wires is the addition of $Z_{pp}' = Z_i(\Delta l)$ to the diagonal elements of $Z$.



31

It is obvious that the achievable maximum power gain for a given array configuration is lower than the maximum directivity because of the resistive losses. Numerical results have also shown [24] that for arrays with closely spaced elements the required excitations for maximum power gain are not the same as those for maximum directivity.

*B. Arrays with Random Errors*

Under practical applications random errors exist in excitation amplitudes and phases as well as in element positions. It is then of interest to examine the effect of random errors in these design parameters on the optimization procedure. This has been done for arrays with an arbitrary geometrical configuration [28]–[31]. Correlations are allowed to exist between the errors in the array parameters and no restrictions are necessary either on the magnitude or on the probability distribution of the random errors. The dependence of the expected directivity or SNR, the main-beam radiation efficiency, the optimum excitation amplitudes and phases and the radiation pattern on the variance and correlation distance of parameter errors has been studied [30], [31]. It was found that the excitations calculated on the basis of no random errors do not yield a maximum expected directivity when parameter errors exist.

*C. Techniques for Large Arrays*

In the numerical solution of the array optimization problem by the method of moments in Section VII, it is necessary to invert the generalized impedance matrix $Z$, as defined in (62). The order of the matrix $Z$ is $M=NS$, which is the product of the number of wire antennas in the array and the number of subsections for each antenna (assuming equal antenna lengths). Hence $M$ can be very large for large arrays with many elements, especially when the elements are of a length which is an appreciable fraction of the operating wavelength. The feasibility of inverting large matrices is constrained by computer memory capacity and the cost. Alternative techniques which will relax these constraints are therefore of great importance.

For circular arrays with many uniformly spaced elements, the inversion of the relevant large matrix is facilitated through diagonalization by a change of basis with a unitary transformation matrix [18]. The columns of the transformation matrix are the eigenvectors of a rotational operator. With this technique the inversion of the large matrix can be evaluated by straightforward matrix multiplication.

For array configurations with no rotational symmetry other methods must be sought. The far-zone electric field due to an array of $N$ parallel $z$-directed dipoles is

$$E(\theta, \phi) = \frac{j\omega\mu}{4\pi r} \exp(-j\beta r) \sum_{n=1}^{N} \exp(j\beta \bar{r}_n \cdot \hat{u}) \int_{-h}^{h} I_n(z') \cdot \exp(j\beta z' \cos\theta) \sin\theta \, dz' \quad (78)$$

where $\beta$ is the phase constant, $\bar{r}_n$ is the vector from the origin to the center of the $n$th dipole, and $\hat{u}$ is the unit vector from the origin to the observation point. The first step in the optimization problem is the insertion of (78) in the expression for the performance index of interest. In order to maximize the performance index, it is necessary to use an appropriate expansion for the currents $I_n(z')$. The moment method discussed in Section VII used expansion functions defined over subsections, which resulted in $M\times M$ matrices. An alternative is to use the three-term theory for cylindrical antennas developed by King and his associates [32]. This has been done [34], and it has been found that, by using some special properties of the integrals involved, accurate numerical solutions of array optimization problems can obtained by working with matrices of order $N$ (not $3N$ as first suspected, where $N=M/S$). Details of this technique together with numerical results will be reported on at a later date.

## Appendix

### Proof of Theorem 2

Let

$$P = \alpha - 2\mathbf{x}'\boldsymbol{\beta} + \mathbf{x}'C\mathbf{x} \quad (79)$$

which is the denominator of the quantity in (32). It is necessary to prove that if $C$ is positive definite $P$ will be minimum at

$$\mathbf{x}_M = C^{-1}\boldsymbol{\beta} \quad (80)$$

and

$$\min_{(\mathbf{x}=\mathbf{x}_M)} P = \alpha - \boldsymbol{\beta}'C^{-1}\boldsymbol{\beta}. \quad (81)$$

*Proof:* If $C$ is positive definite, it is known that [16]

$$(\boldsymbol{\beta}'C^{-1}\boldsymbol{\beta})(\mathbf{x}'C\mathbf{x}) \geq (\mathbf{x}'\boldsymbol{\beta})^2 \quad (82)$$

or

$$\mathbf{x}'C\mathbf{x} = \frac{1}{\boldsymbol{\beta}'C^{-1}\boldsymbol{\beta}} (\mathbf{x}'\boldsymbol{\beta})^2 \quad (83)$$

where the equality sign applies when

$$\mathbf{x} = \mathbf{x}_M = C^{-1}\boldsymbol{\beta}. \quad (80)$$

Let $c=\boldsymbol{\beta}'C^{-1}\boldsymbol{\beta}>0$, and $b=\mathbf{x}'\boldsymbol{\beta}$. We have, from (79) and (83)

$$P = A - 2b + \mathbf{x}'C\mathbf{x} \geq A - 2b + \frac{b^2}{c}. \quad (84)$$

But

$$A - 2b + \frac{b^2}{c} = A - c + \frac{1}{c}(c-b)^2 \geq A - c. \quad (85)$$

Combining (84) and (85), we obtain

$$P \geq A - \boldsymbol{\beta}'C^{-1}\boldsymbol{\beta} \quad (86)$$

where the equality sign holds with (80); hence Theorem 2 is proved.

## References


[1] C. L. Dolph, "A current distribution for broadside arrays which optimizes the relationship between beam width and side lobe level," *Proc. IRE*, vol. 34, June 1946, pp. 335–348.

[2] V. L. Pokrovskii, "On optimum linear antennas," *Radio Eng. and Electronic Phys.* (in Russian), vol. 1, 1956, pp. 593–600 (translated by M. D. Friedman, ASTIA Doc. AD-110229, Feb. 1957).

[3] M. T. Ma and D. K. Cheng, "A critical study of linear arrays with equal side lobes," *IRE Int. Conv. Rec.*, pt 1, 1961, pp. 110–122.

[4] Y. V. Baklanov, "Chebyshev distribution of currents for a plane array of radiators," *Radio Engrg. Electron. Phys.* (USSR), vol. 11, Apr. 1966, pp. 640–642.

[5] F. I. Tseng and D. K. Cheng, "Optimum scannable planar arrays with an invariant sidelobe level," *Proc. IEEE*, vol. 56, Nov. 1968, pp. 1771–1778.

[6] Y. T. Lo, S. W. Lee, and Q. H. Lee, "Optimization of directivity and signal-to-noise ratio of an arbitrary antenna array," *Proc. IEEE*, vol. 54, Aug. 1966, pp. 1033–1045.

[7] F. I. Tseng and D. K. Cheng, "Spacing perturbation techniques for array optimization," *Radio Sci.*, vol. 3, May 1968, pp. 451–457.

[8] D. K. Cheng and F. I. Tseng, "Signal-to-noise ratio maximization

for receiving arrays," *IEEE Trans. Antennas Propagat.* (Commun.), vol. AP-14, Nov. 1966, pp. 792–794.
[9] A. I. Uzkov, "An approach to the problem of optimum directive antenna design," *C. R. Acad. Sci. USSR.*, vol. 35, 1946, p. 35.
[10] A. Bloch, R. G. Medhurst, and S. D. Pool, "A new approach to the design of superdirective aerial arrays," *Proc. Inst. Elec. Eng.*, vol. 100, Sept. 1953, pp. 303–314.
[11] M. Uzsoky and L. Solymar, "Theory of superdirective linear arrays," *Acta Phys.* (Budapest), vol. 6, 1956, pp. 185–204.
[12] C. T. Tai, "The optimum directivity of uniformly spaced broadside arrays of dipoles," *IEEE Trans. Antennas Propagat.*, vol. AP-12, July 1964, pp. 447–454.
[13] D. K. Cheng and F. I. Tseng, "Gain optimization for arbitrary antenna arrays," *IEEE Trans. Antennas Propagat.* (Commun.), vol. AP-13, Nov. 1965, pp. 973–974.
[14] ——, "Maximisation of directive gain for circular and elliptical arrays," *Proc. Inst. Elec. Eng.*, vol. 114, May 1967, pp. 589–594.
[15] E. I. Krupitskii, "On the maximum directivity of antennas consisting of discrete radiators," *Sov. Phys.—Dokl.*, vol. 7, Sept. 1962, pp. 257–259.
[16] F. R. Gantmacher, *The Theory of Matrices*, vol. I. New York: Chelsea Publishing Co., 1960.
[17] B. Das, "On maximum directivity of symmetrical systems of radiators," *Radio Eng. Electron. Phys.* (USSR), vol. 10, June 1965, pp. 853–859.
[18] D. K. Cheng and F. I. Tseng, "Pencil-beam synthesis for large circular arrays," *Proc. Inst. Elec. Eng.*, vol. 117, July 1970, pp. 1232–1234.
[19] P. D. Raymond, Jr., "Optimization of array directivity by phase adjustment," M.Sc. thesis, Elec. Eng. Dept., Syracuse University, Syracuse, N. Y., June 1970.
[20] F. B. Hildebrand, *Methods of Applied Mathematics*. Englewood Cliffs, N. J.: Prentice-Hall, 1965.
[21] E. A. Guillemin, *The Mathematics of Circuit Analysis*. New York: Wiley, 1951.
[22] C. J. Drane, Jr., and J. F. McIlvenna, "Gain maximization and controlled null placement simultaneously achieved in aerial array patterns," Air Force Cambridge Res. Labs., Bedford, Mass., Rep. AFCRL-69-0257, June 1969.
[23] R. F. Harrington, "Matrix methods for field problems," *Proc. IEEE*, vol. 55, Feb. 1967, pp. 136–149.
[24] J. A. Cummins, "Analysis of a circular array of antennas by matrix methods," Ph.D. dissertation, Elec. Eng. Dept., Syracuse University, Syracuse, N. Y., Dec. 1968.
[25] B. J. Strait and K. Hirasawa, "On radiation and scattering from arrays of wire antennas," *Proc. Nat. Elec. Conf.*, vol. 25, 1969.
[26] A. T. Adams and B. J. Strait, "Modern analysis methods for EMC," *IEEE/EMC Symp. Rec.*, July 1970, pp. 383–393.
[27] R. F. Harrington, *Field Computation by Moment Methods*. New York: Macmillan, 1968.
[28] E. N. Gilbert and S. P. Morgan, "Optimum design of directive antenna arrays subject to random variations," *Bell Syst. Tech. J.*, vol. 34, May 1955, pp. 637–663.
[29] M. D. Smaryshev, "Maximizing the directive gain of an antenna array," *Radio Eng. Electron. Phys.* (USSR), vol. 9, Sept. 1964, pp. 1399–1400.
[30] F. I. Tseng and D. K. Cheng, "Gain optimization for arbitrary antenna arrays subject to random fluctuations," *IEEE Trans. Antennas Propagat.*, vol. AP-15, May 1967, pp. 356–366.
[31] D. K. Cheng and F. I. Tseng, "Optimum spatial processing in a noisy environment for arbitrary antenna arrays subject to random errors," *IEEE Trans. Antennas Propagat.*, vol. AP-16, Mar. 1968, pp. 164–171.
[32] R. W. P. King, R. B. Mack, and S. S. Sandler, *Arrays of Cylindrical Dipoles*. Cambridge, England: Cambridge University Press, 1967.
[33] R. F. Harrington, "Antenna excitation for maximum gain," *IEEE Trans. Antennas Propagat.*, vol. AP-13, Nov. 1965, pp. 896–903.
[34] D. K. Cheng, "Antenna optimization criteria," Rome Air Develop. Cent., Rome, N.Y., Rep. RADC-TR-71-156, Apr. 1971.

*Reprinted from* IEEE TRANSACTIONS ON ANTENNAS AND PROPAGATION
Vol. AP-17, No. 2, March 1969
pp. 237–239


# ARRAY DESIGN FOR A SPECIFIED PATTERN BY MATRIX METHODS

34

B. J. Strait and K. Hirasawa

## Array Design for a Specified Pattern by Matrix Methods

*Abstract*—**A new method is presented for designing arrays of parallel wire antennas. The method relates a desired pattern with required feed voltages such that mutual coupling is taken into account in the design procedure. The technique uses matrix methods together with results of the "Method of Moments" to determine the self- and mutual impedances and the required excitations. The design method can be applied to circular, elliptical, and other planar arrays although attention is restricted here to linear arrays. The feed points and the array geometry must be specified.**

### Introduction

Harrington [1], [2] has shown that the "Method of Moments" can be applied for analyzing the electromagnetic behavior of thin-wire objects. A general procedure was pointed out for treating thin wires of arbitrary shape either as antennas or as obstacles or scatterers in a known incident field. Cummins [3] has extended the analysis technique for circular arrays of wire antennas and has found that for a given array the current distribution, input and mutual impedances, and radiation field can be determined very accurately given the set of voltages applied at the feed points of the array elements.[1] The method presented here for array design is an extension of those of Harrington and Cummins and is for the purpose of relating desired array patterns directly with required feed voltages. No unrealistic assumptions are necessary with regard to current distributions on the array elements, and mutual coupling is taken into account [4].

The basic method calls for 1) an approximation of the exact integral equation for conducting bodies by an approximate equation valid for thin wires, 2) replacement of derivatives by finite difference approximations, 3) use of pulse functions for expansion functions resulting in step approximations for the current and charge (Cummins makes use of triangle expansion functions), and 4) use of point-matching or impulse-testing functions. There are a total of $N$ wire elements making up the array, and each wire is considered as $S$ short segments connected together. The end points of each segment define a pair of terminals in space, and these $M = N \times S$ pairs are thought of as the ports of an $M$-port network. The array specialized for the example presented later is shown in Fig. 1, where it is noted that the $n$th segment is identified by its starting point $n^-$, its midpoint $n$, and its termination $n^+$. The length of each segment used in this work is $\Delta l$, where $\Delta l_n^-$ and $\Delta l_n^+$ denote increments shifted ½ segment minus or plus along the wire.

The wires are specified to be thin, and it is assumed that 1) current flows only in the axial direction, and 2) current and charge densities are approximated by filaments of

Manuscript received October 28, 1968. This research was sponsored by the U.S. Air Force Cambridge Research Laboratories, Office of Aerospace Research, under Contract AF19628-68-C-0180.

[1] Cummins has succeeded in determining the feed voltages required for optimum gain in a circular array of parallel wire elements in a specified geometry and also has presented a perturbation technique for modifying the feed voltages to achieve a more desirable sidelobe level than that yielded in the maximum gain pattern.

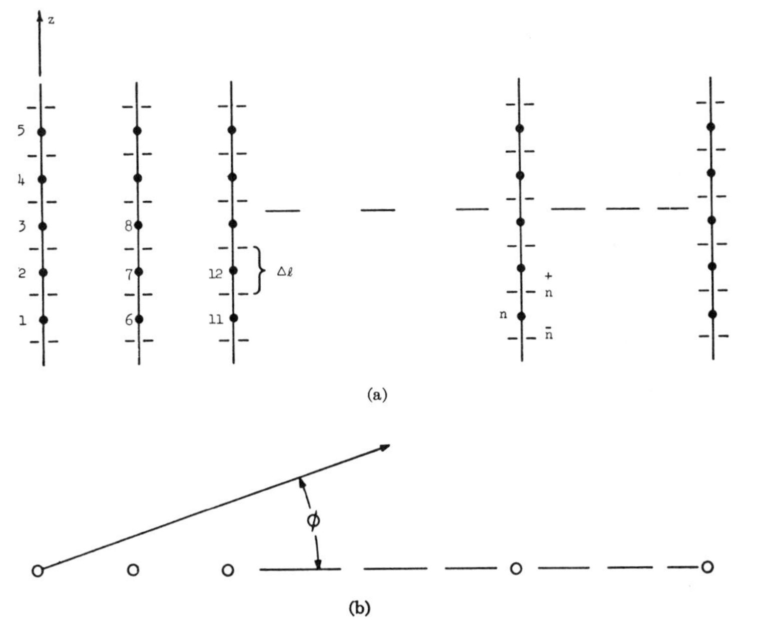


Fig. 1. Equally spaced linear array. (a) Rear view. (b) Top view. There are five segments shown per element as in the example of this communication.

current $I$ and charge $\sigma$ on the wire axes. The boundary condition $I = 0$ at the ends of the wires is provided by using an extra half-segment at each end of each wire. The boundary condition regarding the tangential component of the electric field $\mathbf{E}$ at the wire surfaces is accounted for approximately by requiring that the axial component of $\mathbf{E}$ vanish at the surface of each wire. Harrington defines

$$[I] = \begin{bmatrix} I(1) \\ I(2) \\ \cdot \\ \cdot \\ \cdot \\ I(M) \end{bmatrix}, \quad [V] = \begin{bmatrix} V(1) \\ V(2) \\ \cdot \\ \cdot \\ \cdot \\ V(M) \end{bmatrix} \tag{1}$$

such that

$$[V] = [Z][I] \tag{2}$$

and provides formulas for the self- and mutual impedances in $[Z]$.

Once the current distribution given by

$$[I] = [Z]^{-1}[V] \tag{3}$$

is known, the magnetic vector potential in the far-zone field can be found using

$$A_z = \frac{e^{-jkr}}{4\pi r} \Delta l \sum_{n=1}^{M} I(n) e^{jkr_n \cos \xi_n} \tag{4}$$

where $\mathbf{r}$ and $\mathbf{r}_n$ are radius vectors to the field point and the source point, respectively, and $\xi_n$ is the angle between $\mathbf{r}$ and $\mathbf{r}_n$.

## Array Design for a Specified Pattern

The design problem calls for determination of excitation voltages necessary to provide a desired pattern. Specializing to the principal $H$ plane of a linear array

$$r_n \cos \xi_n = \rho_n \cos \phi \tag{5}$$

where $\rho_n$ is simply the distance from the origin to the wire containing the $n$th segment and $\phi$ is measured from the array axis. If values of the vector potential are specified for a number $q$ of points in the far-zone field, and if these complex numbers are arranged appropriately in matrix form as $[A]$, then using (4)

$$[A] = [F][I] \tag{6}$$

where $[A]$ is given as a column vector with each element corresponding to a different value of $\phi$. Of course, the typical element of $[F]$ is $e^{jk\rho_n \cos \phi}$; as indicated by (4) and (5). The matrix $[F]$ is completely specified once the array geometry is given.

Using (3) in (6),

$$[A] = [F][Z]^{-1}[V]. \tag{7}$$

Note that $[F]$ is a $q \times M$ matrix while $[Z]^{-1}$ is a $M \times M$ matrix. Thus the matrix product $[F][Z]^{-1}$ is of dimension $q \times M$, where it is assumed that $q \ll M$. For lossless antenna elements $[V]$ is a column vector with all $M$ elements equal to zero except those corresponding to feed points. Assuming use of the same number of feed points as there are specified field points, $[V]$ contains $M - q$ elements that are identically zero and $q$ nonzero elements. Thus the $(M - q)$ columns in the matrix product $[F][Z]^{-1}$ corresponding to the zero elements in $[V]$ can be deleted. With these columns deleted the result is a matrix $[B]$ which is of dimension $q \times q$ suitable for inversion. Then

$$[A] = [B][V'] \tag{8}$$

where $[V']$ contains only the nonzero elements of $[V]$. Then

$$[V'] = [B]^{-1}[A] \tag{9}$$

which is the desired result. Thus the real and imaginary parts of the vector potential at $q$ points in the far-zone field are related directly to the real and imaginary parts of $q$ excitation voltages, provided the array geometry and the positions of the feed points are specified.

## Design Example

Consider the design of an 8-element linear broadside array of half-wavelength parallel vertical wires to provide a Dolph–Chebyshev pattern with sidelobes at the $-20$ dB level. The wires are to be center fed and equally (half-wavelength) spaced. The desired pattern is shown in Fig. 2(a).

To design the array using the method presented in this communication it is easiest to specify a feed voltage distribution that is symmetrical about the array center so that the pattern will be symmetrical about the main beam and then specify four

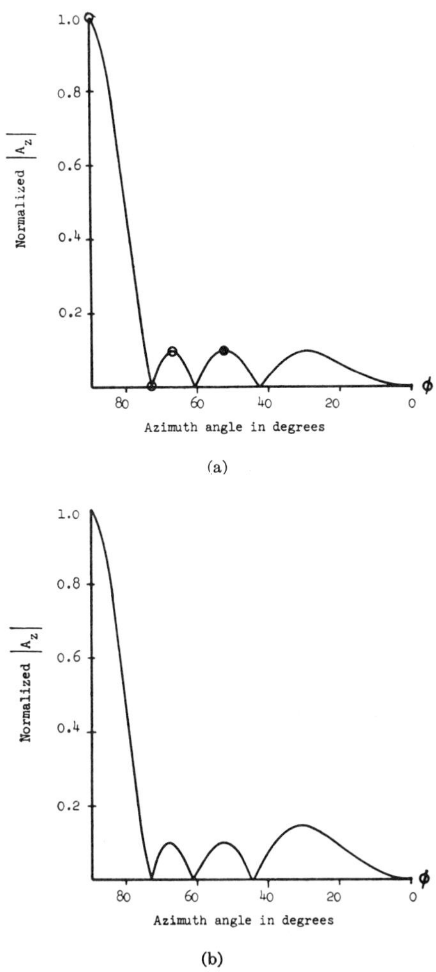


Fig. 2. (a) Desired Dolph–Chebyshev pattern with $-20$-dB sidelobes. (b) Pattern resulting from design procedure of this communication. Patterns are symmetrical about $\phi = \pi/2$.

values (the array is center fed and symmetrically excited; hence, there are only four independent excitation voltages) on one side of the pattern, as the $[A]$ matrix. The values specified in this example are marked with circles on the desired pattern of Fig. 2(a). These values are used to specify Re $[A]$. In this case it was convenient to set Im $[A] = 0$.

In this example five segments are taken per element (per wire) so that $M = 40$, and the wires are taken to be $\lambda/50$ in diameter. Straightforward inversion of the impedance matrix requires (at worst) inversion of a $40 \times 40$ matrix although in many cases symmetry allows reduction of the problem and inversion of much smaller matrices. Input impedances, gain, and other array characteristics in general are most accurately determined when the number of segments taken per wire is large. On the other hand, design for a specified field pattern can be accomplished by using a relatively small number of segments. A limiting factor regarding choice of the number of segments used per element is the dimension of the impedance matrix to be inverted, since this dimension is proportional to the number of elements in the array and to the number of segments per element.

The matrix $[A]$ resulting from the use of the values indicated in Fig. 2(a) is inserted in (9) to obtain the required voltage matrix $[V']$ from which $[V]$ can be constructed by adding zeros in the appropriate places. Then using (3) and (4) the actual pattern can be found, and the result is shown in Fig. 2(b). It is noted that the agreement between desired and actual patterns is quite good near values of $\phi$ corresponding to values specified in $[A]$. However, more values need to be specified if it is desired to improve the result near $\phi = 0, \pi$. This can be done by allowing use of more feed points or possibly by proper loading of the antennas. The latter subject is presently under study.

## Conclusion

In this communication a new method was presented for designing arrays of wire elements to provide specified patterns. The technique uses matrix methods to relate a desired pattern directly with voltages required at the feed points of the array elements. Mutual coupling is included in the design procedure, and no unreasonable assumptions are necessary regarding current distributions on the array elements. The method requires only that the array geometry and the feed points be specified.

It should be pointed out that, although this example involves equally spaced, center fed elements of equal length, no such restrictions are necessary in general. It is only necessary that the array geometry and the feed points be specified. The array elements do not even have to be straight and parallel although, if they are not, the computer program becomes considerably more complicated. Treatments of circular, elliptical, and other planar arrays of parallel wire elements are handled just as easily as the linear array by this method regardless of the lengths of the individual elements and the positions of the specified feed points.

A total of about three minutes was required to complete the example presented here using an IBM 360/50. It is felt that this time can be reduced with presently planned modifications of the program.

B. J. Strait
K. Hirasawa
Dept. of Elec. Engrg.
Syracuse University
Syracuse, N.Y. 13210

## Acknowledgment

The authors wish to thank Dr. R. F. Harrington and Dr. J. Mautz of the Department of Electrical Engineering, Syracuse University, for their helpful suggestions.

## References

[1] R. F. Harrington, "Matrix methods for field problems," *Proc. IEEE*, vol. 55, pp. 136–149, February 1967.
[2] ——, *Field Computation by Moment Methods*. New York: Macmillan, 1968.
[3] J. Cummins, Ph.D. dissertation, Syracuse University, Syracuse, N. Y. (to be published).
[4] B. Strait and K. Hirasawa, "Array design by matrix methods," Contract F19628-68-C-0180, AFCRL-68-0542, Sci. Rept. 1, October 1968.

# Properties of Phased Arrays*

WILHELM H. VON AULOCK†, ASSOCIATE MEMBER, IRE

*Summary*—**This paper discusses in some detail the scanning properties of planar and linear phased arrays consisting of a large number of equispaced radiators, either omnidirectional or directive. Assuming that a suitable beam shape and sidelobe level in the broadside direction have been attained, it is shown how scanning of the array through introduction of a phase delay into the wave front distorts this pattern and introduces new sidelobes. The study uses contour maps of the antenna pattern in "sin θ-space," which is a projection of the unit sphere on the plane of the array. In this space the antenna pattern remains invariant with scan angle and merely undergoes a translation proportional to the phase delay between adjacent radiators. Using a parallel projection of the unit sphere onto the plane of the array, the actual antenna patterns are then readily obtained in conventional spherical coordinates. This method also permits ready evaluation of the influence of the directivity of the radiating elements, including a slight shift of the beam maximum with respect to that of an array of isotropic radiators. Furthermore, a pictorial representation of the coverage of a tilted planar array of given scanning properties can be obtained in terms of an earth-fixed coordinate system. Finally, it is shown how the beam-pointing error in a phased array is related to systematic errors in the phase delay.**

## I. Introduction

MECHANICALLY scanned antennas were employed almost exclusively in the radar systems of World War II and are still being used in the majority of all operational radar systems. However, new requirements for radar systems, including rigidity and shock resistance of the antenna structure, volumetric rapid scanning of large antennas, and avoidance of any mechanical motion in the radar system, have led to an ever-increasing interest in electrically scanned antennas in which the antenna stays fixed in space and the radar beam is moved by introducing a phase delay into the radiated wave front. Such an antenna is called, briefly, a "phased array." Its properties are well understood and readily amenable to numerical and graphic analysis.

Although the purpose of a two-dimensional phased array is the same as that of a parabolic reflector in a two-gimbal mount, namely, to scan a radar beam, the characteristics of these two antennas have some fundamental differences, thus making it impossible simply to substitute one for the other in an existing radar system. Phased arrays can provide scanning patterns and scanning rates which are impossible to attain with mechanically scanned antennas. Conversely, a parabolic reflector can be turned and tilted to radiate an identical antenna pattern toward every point in a hemisphere, a task which is inherently impossible with a planar phased array. Furthermore, the scanning properties of a mechanically scanned antenna are readily described in a spherical coordinate system with its origin at the phase center of the antenna, whereas scanning of a phased array can be discussed more conveniently in a different coordinate system.

This paper presents a brief review of the scanning properties of phased arrays. The discussion is restricted to arrays of equispaced radiators located in a plane. Further, it is assumed that the variation of scan angle $\theta_s$ of these arrays,[1] hereinafter referred to as "scan sector," is large compared to the beamwidth. This assumption implies that the phased arrays consist of many radiating elements, possibly 50 to 100 radiators for a linear array and 2500 to 10,000 elements for two-dimensional arrays.

The introduction of a phase delay across the wave front radiating from an array has many consequences besides the obvious shift of the beam maximum.

1) The beamwidth increases as one scans away from broadside.
2) The beam shape changes slightly as one scans from broadside to moderate scan angles and is modified drastically for extreme scan angles close to 90° from broadside.
3) New sidelobes may appear at moderate to large scan angles.
4) The beam direction is slightly different from that computed by standard formulas.

All these effects are well understood. They have to be taken into account if one attempts to approximate the precise performance of a mechanically scanned antenna with a phased array. This discussion therefore deals with beam direction, beamwidth, and beam shape as functions of phase delay. Furthermore, the sensitivity of array performance to systematic excitation errors is reviewed.

As a result of this study, it is possible to state accuracy requirements for the phase delay of practical arrays. The capabilities and limitations of this type of scanning antenna will be apparent, and techniques for the design of such an array will emerge.

* Received by the IRE, October 30, 1959; revised manuscript received, March 4, 1960.
† Bell Telephone Labs., Inc., Whippany, N. J.

[1] The scan angle $\theta_s$ is defined as the angle between the array normal and the beam maximum.

The deterioration of array characteristics with random excitation errors has received considerable attention in the literature.[2–5] It is not to be discussed here because it is believed that systematic errors, rather than random errors, will limit the performance of multielement phased arrays.

Whereas mechanically scanned antennas are adequately described by giving two antenna patterns in orthogonal planes, this method of presentation is entirely unsatisfactory for a phased array whose pattern changes appreciably as the beam is scanned away from the array normal. To appreciate fully the characteristics of both linear and two-dimensional phased arrays, their patterns must be studied in a hemisphere rather than in two planes, and a pattern representation which permits the study of pattern changes as a function of scan angle must be found. Thus, the hemisphere, rather than two orthogonal planes, becomes the preferred space to be considered, and the planar, rather than the linear, phased array emerges as the principal object of this discussion. In particular, the two-dimensional array of isotropic radiators is used to show the variation of beam position, beam shape, beamwidth, and secondary beams with phase delay. Directive radiating elements will be introduced later to demonstrate their contribution to beam distortion, sidelobe level, and beam-pointing errors.

## II. Two-Dimensional Phased Arrays of Isotropic Radiators

Consider a two-dimensional array of $M \cdot N$ equispaced isotropic radiators in the $xy$ plane (Fig. 1). The radiators are excited by individual currents $I_{mn}$ and provisions are made to introduce independent progressive phase delays in the $x$ and $y$ directions, so that the delay between adjacent radiators is $\psi_x$ and $\psi_x$ (radians), respectively. We want to examine the motion of the radar beam as a function of the phase delays and to study the variations of beam shape with the scan angle. Hence, we are not concerned with the classic problem of array theory which attempts to correlate the antenna pattern with the complex excitation coefficients $I_{mn}$, but we simply assume that this problem has been solved and that means exist to produce an acceptable amplitude pattern in the $z$ direction normal to the array when all array elements are excited in phase ($\psi_x = \psi_y = 0$).

[2] L. A. Rondinelly, "Effects of random errors on the performance of antenna arrays of many elements," 1959 IRE National Convention Record, pt. 1, pp. 174–189.

[3] L. A. Kurtz and R. S. Elliot, "Systematic errors caused by the scanning of antenna arrays: phase shifters in the branch lines," IRE Trans. on Antennas and Propagation, vol. AP-4, pp. 619–627; October, 1956.

[4] D. K. Cheng, "Effect of arbitrary phase errors on the gain and beamwidth characteristics of radiation pattern," IRE Trans. on Antennas and Propagation, vol. AP-3, pp. 145–147; July, 1955.

[5] R. S. Elliot, "Mechanical and electrical tolerances for two-dimensional scanning antenna arrays," IRE Trans. on Antennas and Propagation, vol. AP-6, pp. 114–120; January, 1958.

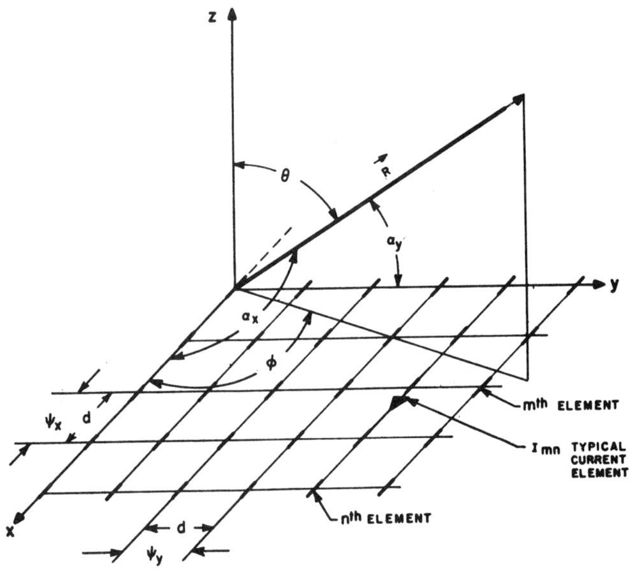


Fig. 1—Planar array of equispaced current elements.

### A. The Antenna Pattern

The amplitude pattern of an array of isotropic radiators is given by[6]

$$S_a = \sum_{m=0}^{M-1} \sum_{n=0}^{N-1} |I_{mn}| \, e^{jd_r(m\tau_x + n\tau_y)}, \qquad (1)$$

where

$M, N$ = number of elements in $x$ and $y$ directions,

$|I_{mn}|$ = amplitude of excitation coefficient,

$d_r = 2\pi d/\lambda_0$ = spacing between radiating elements in radians,

$\lambda_0$ = wavelength in free space,

$\tau_x = \cos\alpha_x - \cos\alpha_{xs}$,

$\tau_y = \cos\alpha_y - \cos\alpha_{ys}$,

$\cos\alpha_x, \cos\alpha_y$ = direction cosines of radius vector specifying the point of observation,

$\cos\alpha_{xs}, \cos\alpha_{ys}$ = direction cosines of radius vector specifying beam maximum (scan direction).

The conventional pattern representation of this array uses a spherical coordinate system with azimuth and elevation angles $\phi$ and $\theta$ (Fig. 1). These angles are related to the direction cosines as follows:

$$\sin^2\theta = \cos^2\alpha_x + \cos^2\alpha_y,$$

$$\tan\phi = \frac{\cos\alpha_y}{\cos\alpha_x}. \qquad (2)$$

[6] S. Silver, "Microwave Antenna Theory and Design," M.I.T. Rad. Lab. Ser., McGraw-Hill Book Co., Inc., New York, N. Y., vol. 12, pp. 104–106; 1949.

After fixing the direction of the beam maximum by specifying $\alpha_{xs}$ and $\alpha_{ys}$ and computing $\theta_s$ and $\phi_s$ from (2), one selects a plane $\phi=$const and computes the pattern

$$S_a = f(\theta) \qquad (\theta_s, \phi_s, \phi = \text{const}).$$

Similarly, one may compute the pattern

$$S_a = f(\phi) \qquad (\theta_s, \phi_s, \theta = \text{const}).$$

Clearly, this pattern representation has two disadvantages.

1) The pattern depends on the position of the beam maximum $\theta_s$, $\phi_s$.
2) The pattern representation is valid only for one particular plane ($\phi=$const) or conical surface ($\theta=$const). Hence, a complete representation of the antenna pattern in the hemisphere requires many curves for many different planes.

Both these difficulties can be overcome by using the differences of direction cosines $\tau_x$ and $\tau_y$ as variables and plotting a contour map of the amplitude pattern,

$$f(\tau_x, \tau_y) - S_a = 0 \qquad (S_a = \text{const}) \qquad (3)$$

in direction cosine space.

Here we will assume that this particular contour is a circle for $M=N$ and an ellipse for $M\neq N$. It is interesting to note that this assumption is justified in the case of uniform illumination ($I_{mn}=I_0$). The amplitude pattern of an $M\cdot N$ array with uniform illumination[7] is given by

$$S_a = \frac{\sin \frac{1}{2}Md_r\tau_x}{M\sin \frac{1}{2}d_r\tau_x}\cdot\frac{\sin \frac{1}{2}Nd_r\tau_y}{N\sin \frac{1}{2}d_r\tau_y}. \qquad (4)$$

The contour map of the main beam can be approximated by

$$S_a = \frac{\sin x}{x}\cdot\frac{\sin y}{y},$$
$$x = \tfrac{1}{2}Md_r\tau_x,$$
$$y = \tfrac{1}{2}Nd_r\tau_y. \qquad (5)$$

This map is shown in Fig. 2. It is apparent that the contour is an almost perfect circle at the half-power point but approaches a square as $S_a$ approaches zero. Hence, one may talk of a beam broadening in the diagonal. This beam broadening amounts to less than 2 per cent at the half-power point and may be neglected (Fig. 3). Thus, the half-power beamwidth of a uniformly illuminated square array ($M=N$) is given in $\tau$ space by

$$2\Delta\tau = \frac{C_0}{A/\lambda_0}, \qquad (6)$$

where $C_0=0.888$ and $A=Nd$ (aperture).

[7] J. D. Kraus, "Antennas," McGraw-Hill Book Co., Inc., New York, N. Y., ch. 4; 1950.

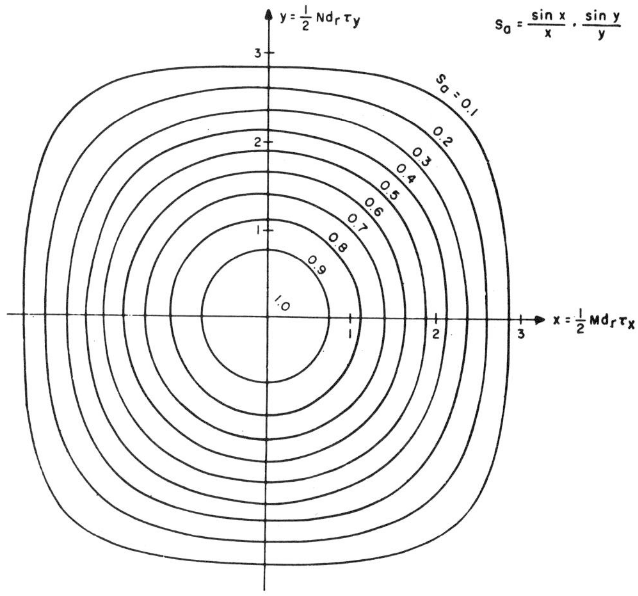


Fig. 2—Constant amplitude contours of the beam from a uniformly illuminated $M\cdot N$ planar array.

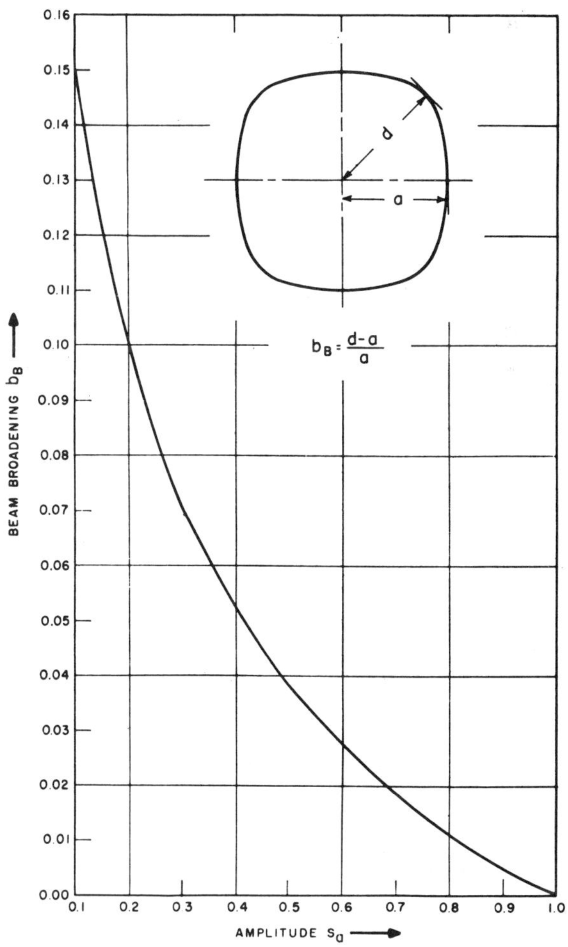


Fig. 3—Beam broadening of a uniformly illuminated planar array.

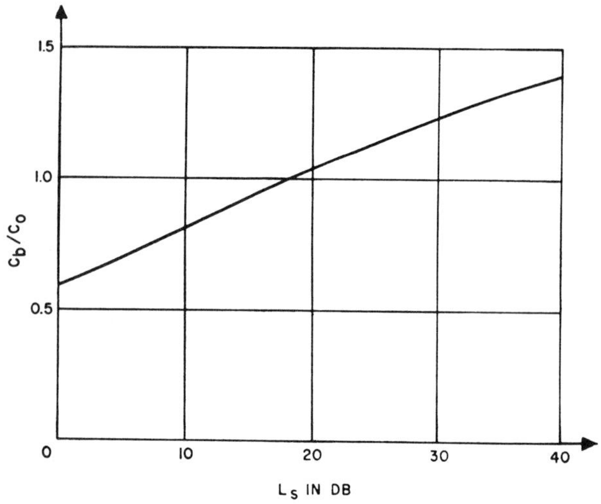


Fig. 4—Beamwidth of a Dolph-Tchebycheff array vs sidelobe level $L_S$ relative to a uniformly illuminated array of equal aperture.



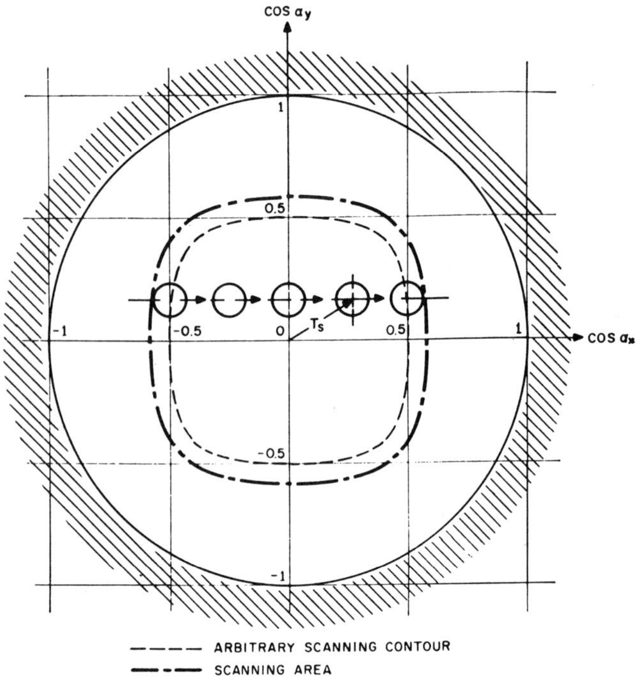


Fig. 5—Scanning in the complex $T$ plane.

For a narrow beam array whose beam is oriented normal to the plane of the array, $2\Delta\tau$ can be interpreted directly as half-power beamwidth in radians. The beamwidth is related to the sidelobe level of the array in such a manner that the beam broadens when the excitation coefficients $I_{mn}$ are tapered to satisfy specific sidelobe requirements. For a Dolph-Tchebycheff excitation function[8] (Fig. 4), the beamwidth can be expressed by

$$2\Delta\tau = \left(\frac{C_b}{C_0}\right)\left(\frac{C_0}{A/\lambda_0}\right), \qquad (7)$$

where $C_b/C_0$ expresses the beam broadening with respect to uniform excitation and as a function of design sidelobe level. Note that a design sidelobe level of 40 db introduces a beam broadening of 41 per cent for a given aperture.

### B. *Scanning of the Beam*

The direction of the beam maximum is a function of the two phase delays $\psi_x$ and $\psi_y$ so that the direction cosines are proportional to the respective phase delays:

$$\cos\alpha_{xs} = \frac{\psi_x}{d_r} \qquad \cos\alpha_{ys} = \frac{\psi_y}{d_r}. \qquad (8)$$

This fundamental property of phased arrays makes it desirable to discuss pattern shape and beam motion in a coordinate system which has $\cos\alpha_x$ and $\cos\alpha_y$ for the two axes.

For ease of manipulation, let us then define a complex $T$ plane with coordinates $\cos\alpha_x$ and $\cos\alpha_y$. This permits description of the point of observation by a single complex number $T$ and locates the beam maximum in the $T$ plane by another complex number, $T_s$.

$$T = \cos\alpha_x + i\cos\alpha_y,$$

$$T_s = \cos\alpha_{xs} + i\cos\alpha_{ys} = \frac{\psi}{d_r}, \qquad (9)$$

where $\psi = \psi_x + i\psi_y$ (phase delay).

Similarly, we can define a complex $\tau$ plane, which is suitable for contour maps of the amplitude pattern, by

$$\tau = T - T_s = \tau_x + i\tau_y. \qquad (10)$$

The antenna pattern, which is a function of the complex variable $\tau$, is invariant in the $T$ plane. Scanning the antenna by introducing the complex phase delay $\psi = \psi_x + i\psi_y$ simply translates the pattern in space so that the pattern center moves to $T_s = \psi/d_r$ (Fig. 5). If the phase delay is chosen so that $|T_s| > 1$, the half-power contour moves outside of the unit circle and becomes imaginary, *i.e.*, unobservable.

Two definitions are helpful in discussing array properties:

*The scanning contour*—a closed curve in the $T$ plane, traced by the beam maximum when the beam is scanned through the largest required deviations from the array normal (Fig. 5);

*The scanning area*—the area whose boundary is the outer envelope of all the half-power contours of the main beam when the beam is scanned along the scanning contour.

[8] L. B. Brown and G. A. Sharp, "Tschebyscheff Antenna Distribution, Beamwidth, and Gain Tables," Naval Ord. Lab., Corona, Calif., NAV-ORD Rept. No. 4629; 1958.

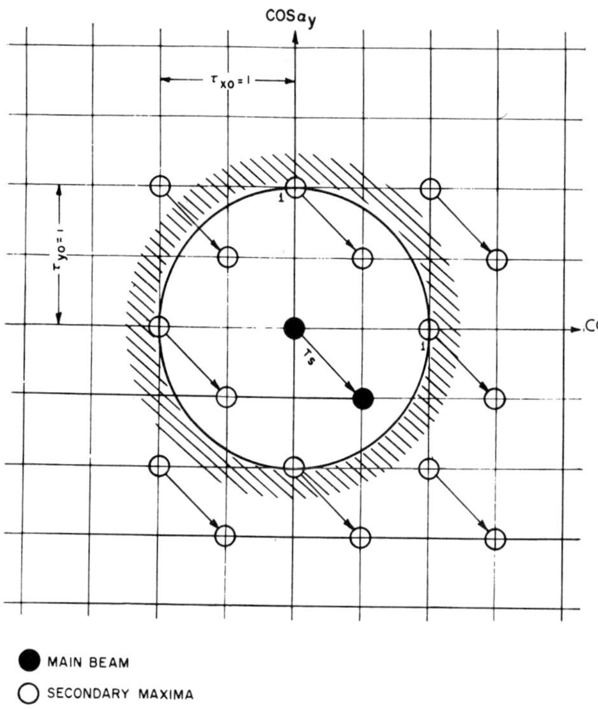


Fig. 6—Secondary maxima of a scanned array with one-wavelength element spacing.

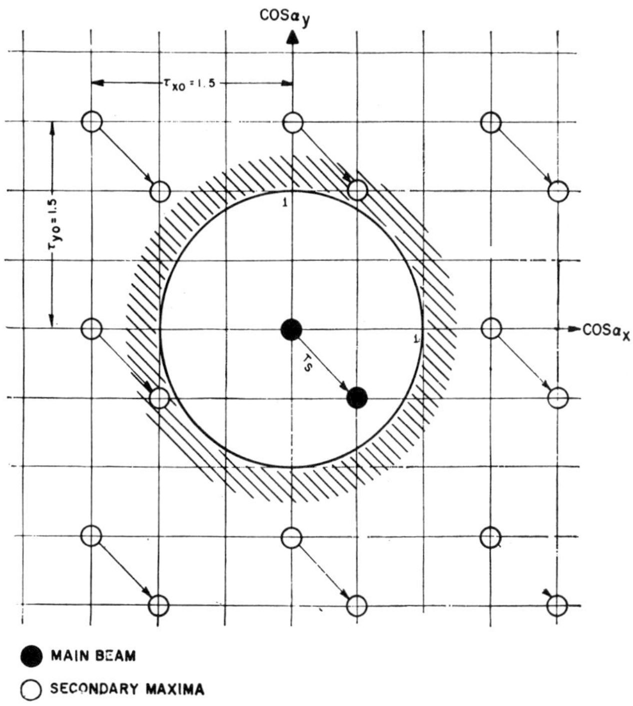


Fig. 7—Secondary maxima of a scanned array with two-thirds-wavelength element spacing.

*C. Secondary Maxima*

Let us now consider the effects of scanning on the beam pattern. The first is the occurrence of secondary maxima. An array of isotropic radiators has an infinite grid of these maxima spaced $\tau_{x0}=\tau_{y0}=\lambda_0/d$. The larger the element spacing, the closer the spacing of the maxima. Whenever the spacing is greater than $\lambda_0/2$, these maxima may become visible as they wander into the unit circle with each scanning of the array. If the elements in the array are spaced one free-space wavelength apart, the main beam and four endfire beams will be visible at zero scan angle (Fig. 6), and any scanning away from broadside will move a multiplicity of beams over the hemisphere. For example, if such an array is scanned by a phase delay $\psi=(1+i)\pi$, four beams will be seen on the circle $\theta=45°$. However, narrower spacing of the radiating elements (*e.g.*, $d=2\lambda_0/3$) will essentially suppress all secondary beams as long as $|T_s|\leqq\frac{1}{2}$ (Fig. 7).

Thus, the proper spacing of radiating elements is the first step in eliminating secondary beams in the array pattern. The second step is concerned with the proper choice of radiating elements (see Section IV).

*D. Scanning of a Phased Array in a Spherical Coordinate System*

All previous computations have been made in a coordinate system which was fixed with respect to the array, and it has been found that the antenna pattern remained invariant in direction cosine space ($T$ plane). To examine the scanning characteristics of a phased array in its associated spherical coordinate system (Fig. 1), we have to project the orthogonal coordinates $\theta=$constant and $\phi=$constant onto the $T$ plane. This projection permits immediate interpretation of the antenna pattern for any desired scan angle in terms of the spherical $\theta$, $\phi$ system.

Expressing the complex direction cosine $T$ in polar coordinates

$$T = \cos\alpha_x + i\cos\alpha_y = \sin\theta(\cos\phi + i\sin\phi),$$

$$T = \sin\theta e^{i\phi}, \tag{11}$$

immediately yields the desired projection which can be described as the parallel projection of the unit sphere on the equatorial plane (Fig. 8). Each circle $\sin\theta=$constant corresponds to two values of $\theta$. Let us therefore define $0\leqq\theta\leqq\pi/2$ as the positive hemisphere and restrict our discussion to it.

*E. Scanning Distortions in a Spherical Coordinate System*

The resolution of a radar system is limited by the width of the radar beam. If this beamwidth depends on beam direction, then the resolution becomes a function of scan angle. To illustrate the magnitude of beam distortion in a phased array consider a beam which is circular at the half-power points in $T$ space. If this beam is scanned away from the array normal, it is then necessary to differentiate between two beamwidths in the $\theta$ and $\phi$ directions, respectively (Fig. 9). The reference

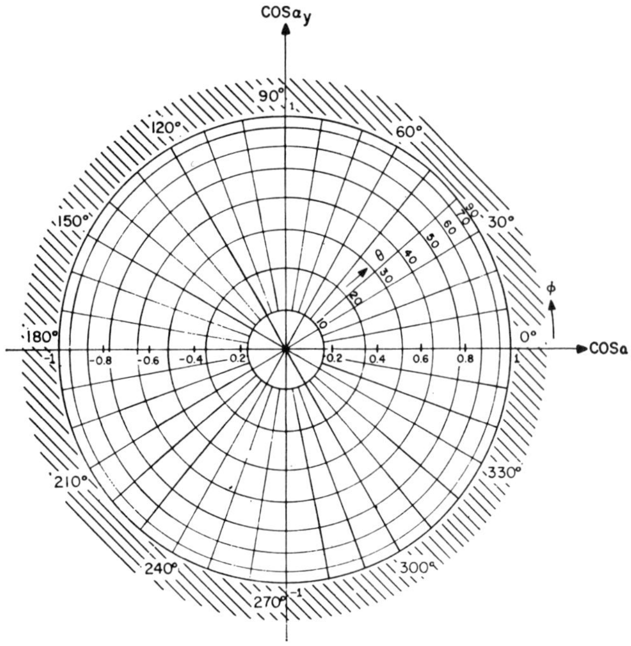


Fig. 8—Projection of unit sphere on $T$ plane.

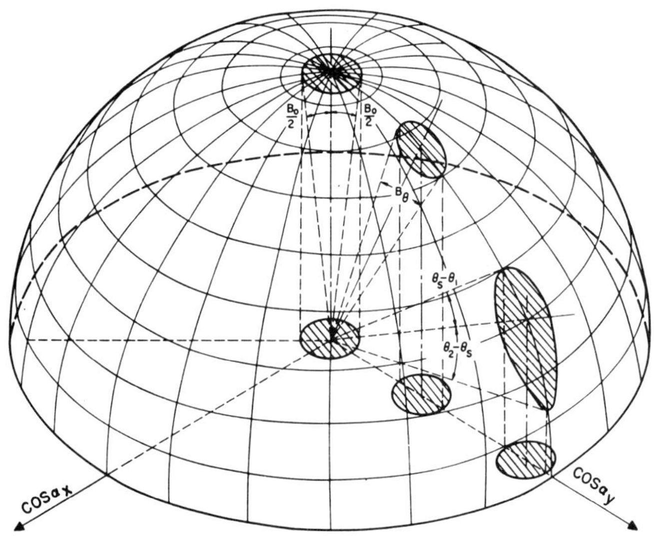


Fig. 9—Beamwidth and eccentricity of the scanned beam.

beamwidth is defined as [*cf.*, (7)]:

$$B_0 = 2 \sin^{-1} (\Delta\tau) \text{ (radians)}, \tag{12}$$

where $\Delta\tau = C_b\lambda_0/2A$. The beamwidth in the $\phi$ direction is constant and equal to $B_0$.

When the beam is directed away from the array normal, then the beamwidth $B_\theta$ in the $\theta$ direction is obtained from

$$2\Delta\tau = |T_2| - |T_1| = \sin \theta_2 - \sin \theta_1. \tag{13}$$

Thus

$$B_\theta = \theta_2 - \theta_1 = 2 \sin^{-1}\left(\frac{\Delta\tau}{\cos \frac{1}{2}(\theta_2 + \theta_1)}\right) \text{ (radians)}. \tag{14}$$

Eqs. (12) and (14) can be simplified for narrow-beam antennas in the usual manner, provided the scan angle $\theta_s \simeq (\theta_1+\theta_2)/2$ is not larger than, say, 45°:[9]

$$B_0 = 2\Delta\tau \text{ (radians)},$$

$$B_\theta = \frac{2\Delta\tau}{\cos \frac{1}{2}(\theta_2 + \theta_1)} \simeq \frac{B_0}{\cos \theta_s} \text{ (radians)}. \tag{15}$$

As the beam is scanned away from the array normal, it broadens in the $\theta$ direction. Hence, its contour at the half-power points changes gradually from a circle to an ellipse.

### *F. Beam Eccentricity*

The assumption in (15) that $(\theta_2+\theta_1)/2 = \theta_s$ is inaccurate because the beam maximum is not centered between the two half-power points at $\theta_2$ and $\theta_1$. Thus, a beam eccentricity (Fig. 9) can be defined as

$$2e = \frac{(\theta_2 - \theta_s) - (\theta_s - \theta_1)}{(\theta_2 - \theta_s) + (\theta_s - \theta_1)}. \tag{16}$$

Use of a trigonometric identity and (15) as applied to $\theta_2 - \theta_s$ and $\theta_s - \theta_1$, the two "half" beams, in (16) yields

$$2e = \tan \tfrac{1}{4}(\theta_1 + \theta_2 + 2\theta_s) \tan \tfrac{1}{4}(\theta_2 - \theta_1). \tag{17}$$

As we are dealing with a second-order effect, (17) may be approximated by

$$e = \frac{B_\theta}{8} \tan \theta_s. \tag{18}$$

Note that the center of the beam moves relative to the contour by $(B_\theta^2/8) \tan \theta_s$ radians. For a narrow beam, this eccentricity is very small and may be neglected; however, for a broad beam, it can be appreciable.

Additional beam eccentricity is observed when the isotropic radiators are replaced by actual radiating elements which exhibit a particular element pattern. This effect is discussed in Section IV-A.

### *G. Endfire Beam*

It has been shown that two phenomena limit the size of the practical scan sector,

1) the appearance of secondary maxima,
2) the broadening of the beam.

Both effects suggest maximum scan sectors of the order of ±30° ($|T_s|_{max} = \frac{1}{2}$). If an array with $2\lambda_0/3$ spacing is

[9] R. W. Bickmore, "A note on the effective aperture of electrically scanned arrays," IRE Trans. on Antennas and Propagation, vol. AP-6, pp. 194–196; April, 1958.

scanned this far from its normal, maximum beam broadening in the $\theta$ direction is about 13 per cent and secondary maxima begin to appear at $\theta = 90°$. It is this appearance of secondary maxima which requires a brief discussion of the beam shape close to $\theta = 90°$.

The secondary maximum of an array of isotropic radiators appearing at $\theta = 90°$ is, in effect, an endfire beam directed parallel to the array. Its beamwidth in the $\phi$ direction is equal to $B_0$, whereas its beamwidth $B_{\theta e}$ in the $\theta$ direction[10] is obtained from (13) by noting that $\theta_2 = \pi/2$ and $\theta_1 = (\pi/2) - B_{\theta e}$ and that $\Delta\tau$, rather than $2\Delta\tau$, applies (Fig. 10). We obtain the following approximation for a narrow-beam array:

$$B_{\theta e} = \sqrt{2\Delta\tau} = \sqrt{B_0} \text{ (radians)}. \tag{19}$$

Thus, an array with a reference beamwidth of 1° (0.017 radian) has an endfire beamwidth of $B_{\theta e} = 7.5°$. Should the secondary maximum move just one-half beamwidth farther into the positive hemisphere, then its beamwidth increases by 41 per cent and is given by $\sqrt{2B_0}$. This substantial broadening of the endfire beam shows the importance of suppressing secondary maxima at $\theta = 90°$ in a practical phased array through selection of appropriate radiating elements.

## III. Special Cases of Two-Dimensional Phased Arrays

### A. The Linear Array

The discussion of two-dimensional phased arrays (see Section II) includes linear arrays as a special case. Putting $N = 1$ in (1), we obtain the amplitude pattern of a linear array of isotropic radiators,

$$S_a = \sum_{m=0}^{M-1} |I_m| e^{jd_r m\tau_x}. \tag{20}$$

The contour map of the pattern of a linear array in direction cosine space is just a grid of lines parallel to the $y$ axis (Fig. 11). In particular, the beamwidth between half-power points is again given by

$$2\Delta\tau_x = \frac{C_b}{A/\lambda_0}, \tag{7}$$

where $A = Md$ (aperture). Scanning of a linear array means moving a strip of width $2\Delta\tau_x$ across the $T$ plane so that

$$T_{xs} = \sin\theta_s \cos\phi = \frac{\psi_x}{d_r}. \tag{21}$$

[10] Consistent with our statement that we restrict our attention to the positive hemisphere, the endfire beamwidth is defined as $B_{\theta e} = (\pi/2) - \theta_1$.

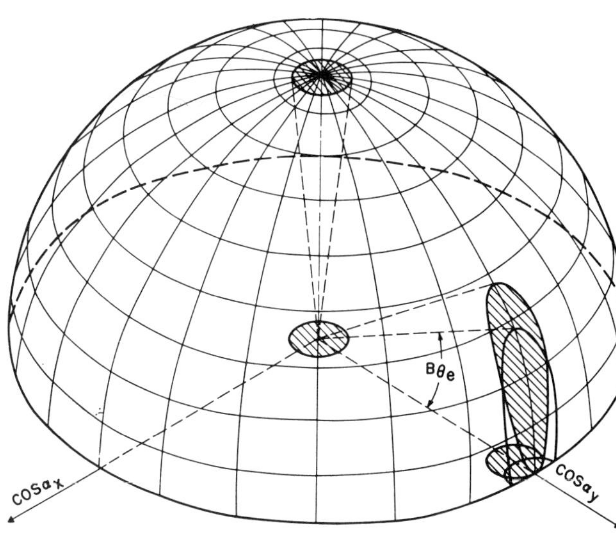


Fig. 10—Beam approaching endfire position.

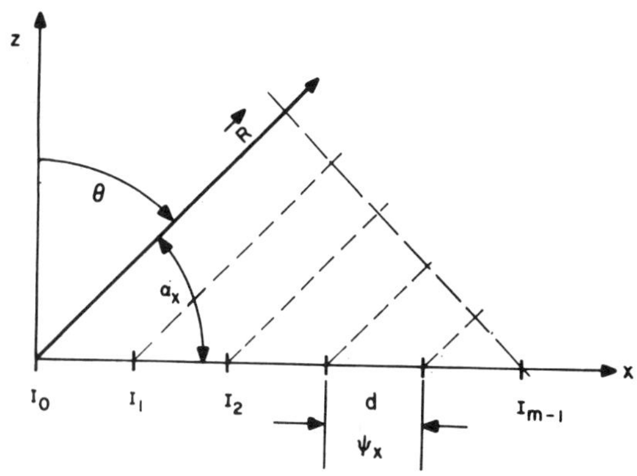


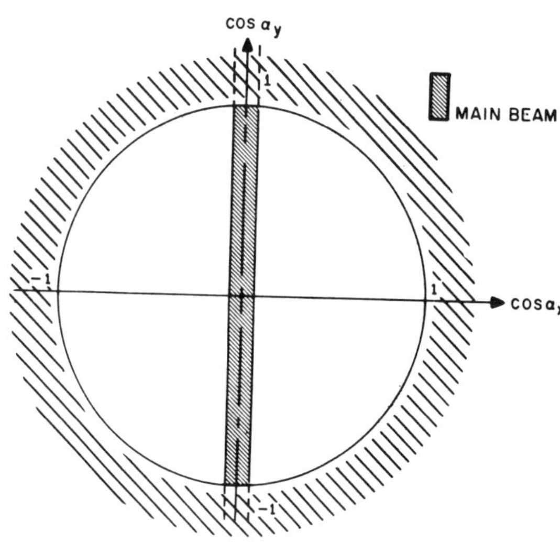


Fig. 11—Linear phased array.

If we again confine our attention to the positive hemisphere, then the actual beam of a linear array of isotropic radiators is a conical half shell. As the beam is scanned away from the array normal, the cone angle decreases and the cone contracts, much like an umbrella being folded, until at $\theta = 90°$ it becomes a single endfire beam. The discussion of the endfire beam in Section II-G is readily adapted to the linear array and secondary maxima have the same spacing as before,

$$\tau_{x0} = \frac{\lambda_0}{d},$$

but have the appearance of strips in the $T$ plane (Fig. 12). Their suppression through proper element spacing and suitably chosen radiating elements is guided by the considerations outlined in Section II-C and Section IV.



*B. The Stacked Beam*

A two-dimensional array of $M \cdot N$ radiators permits formation of a stacked beam (Fig. 13) by applying a multiplicity of fixed phase delays in the $y$ direction. The stacked beam is then scanned by applying a variable phase delay in the $x$ direction. There is a close similarity between the deformation of the stacked beam and of the beam of a linear array as a function of phase delay $\psi_x$. Here again, the aggregate of beams can be compared to a folding umbrella.

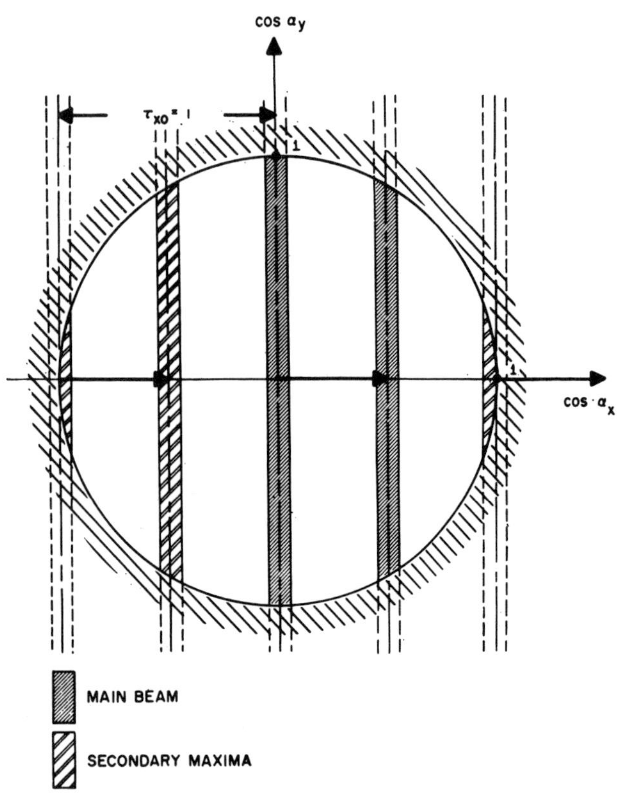


Fig. 12—Scanning of a linear array with one-wavelength element spacing.

*C. The Four-Beam Cluster*

In conventional radar systems, a four-beam cluster is employed to obtain monopulse information.[11] Such a four-beam cluster can readily be formed with a phased array by appropriate antenna feed structures with multiple outputs. In a phased array, the interpretation of the signal obtained from beams $A$, $B$, $C$, $D$ (Fig. 14) must take into account the fact that the beam cluster cannot be rotated in the spherical coordinate system. Hence, the signals derived from $(A-C)$ and $(B-D)$ provide information relative to the $\theta$ direction when the cluster is scanned along the $x$ axis ($\phi = 0$). The same beam combination supplies information pertaining to the $\phi$ direction when the cluster is scanned along the $y$ direction ($\phi = 90°$). Similar reasoning applies to the difference signals $(A-B)$ and $(C-D)$.

[11] D. R. Rhodes, "Introduction to Monopulse," McGraw-Hill Book Co., Inc., New York, N. Y.; 1959.

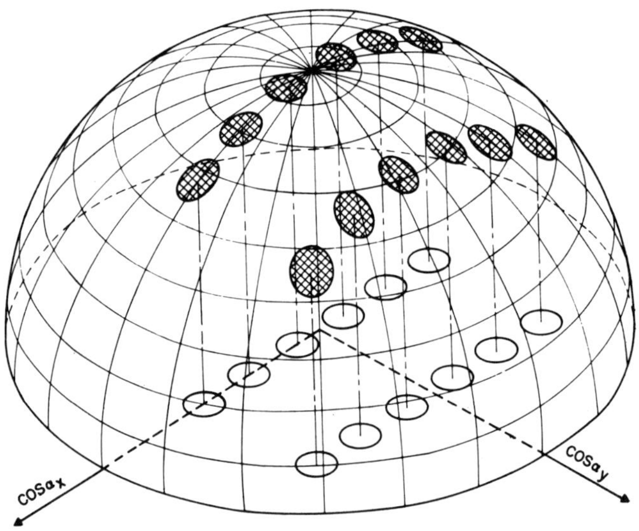


Fig. 13—Scanning of a stacked beam.

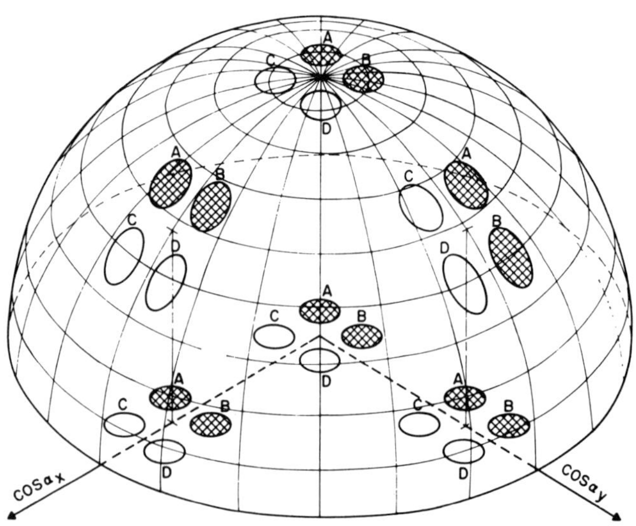


Fig. 14—Scanning of a 4-beam cluster.

## IV. The Radiating Elements

In general, a phased array consists of identical directive radiators adjacent to a large reflecting ground plane. The pattern of such an array can be obtained by multiplying the pattern of the equivalent array of isotropic radiators $S_a$ [see (1)] with the representative pattern of a single radiating element $S_e$, taking into account the effects of the ground plane and of mutual coupling between radiating elements. For the purpose of this discussion, let us assume that it is possible to find a single representative element pattern $S_e$. This implies that the array has many elements, thus providing a uniform environment for all elements except for a few at the very edge of the array. As the latter usually receive less illumination than the elements closer to the array center, their contribution to the over-all pattern is small.

Radiating elements assume many forms and shapes, such as dipoles, slots, open-end and tapered waveguides, polyrods, and helices. The free-space patterns of these radiators are reported in the literature and will not be discussed in detail. Rather, the requirements of an ideal radiating element will be stated, and the limitations of phased arrays will be discussed when the actual radiating elements fail to approximate this ideal.

First, the representative pattern $S_e$ of the radiating element remains fixed in $T$ space. It is, in effect, a window through which the scanned beam radiates into space. The ideal element factor $S_e$ is therefore given by $S_e=1$ over the entire scanning area (Fig. 15) and $S_e \leqq L_s$ everywhere else ($L_s$ is the highest acceptable sidelobe level).

Such an ideal pattern would permit scanning without introducing any additional distortion and would suppress all secondary maxima outside of the scanning area. In practical cases, the element pattern will drop from $S_e=1$ at $\theta=0$ to a prescribed value, such as $S_e=0.707$ (3 db), on the scanning contour. It is highly desirable to provide for a rapid decrease of $S_e$ to 0 outside of the scanning area and for $S_e=0$ on the unit circle ($\theta=90°$). As an example, the pattern of a half-wave dipole at a distance of $\lambda_0/8$ from the ground plane is shown in Fig. 16.

Other radiating elements, such as horns or polyrods, can be designed to provide a better approximation to the ideal pattern. However, these radiators may not meet the fundamental requirement, namely, that their physical size permits close spacing of the radiating elements to suppress secondary maxima.

A detailed contour map of the representative pattern $S_e$ of the array elements is required to determine the largest possible scanning contour (generally, the half-power contour) and the necessary element spacing. For example, the dipole pattern (Fig. 16) will permit scanning to 40.5° and 48.5° in $x$ and $y$ directions, respectively. This pattern has a sharp zero at $\theta=90°$. Hence, element spacing should be close to $\lambda_0/2$ to insure that secondary maxima stay outside the unit circle even at the largest practical scan angles.

If better radiating elements with steeper slopes of the pattern outside of the scanning area are available, the secondary maxima can be spaced closer to the main beam, thus permitting wider spacing of the radiating elements with attendant savings in hardware and circuitry. The theoretical maximum for element spacing is obtained from

$$\frac{\lambda_0}{d} = 2(\sin \theta_s + \Delta\tau). \tag{22}$$

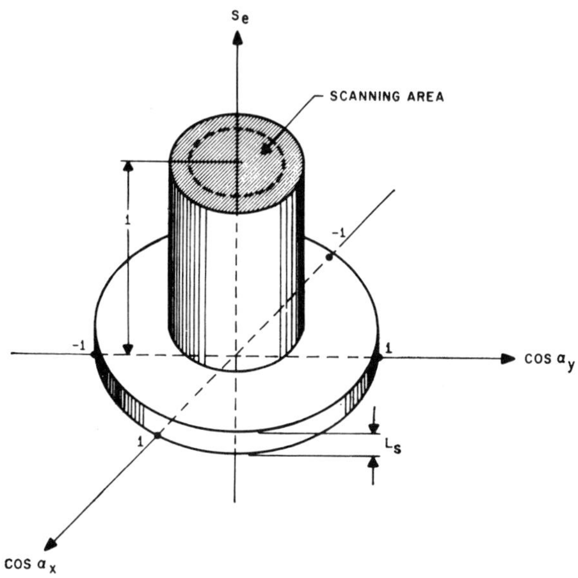


Fig. 15—Ideal element factor.

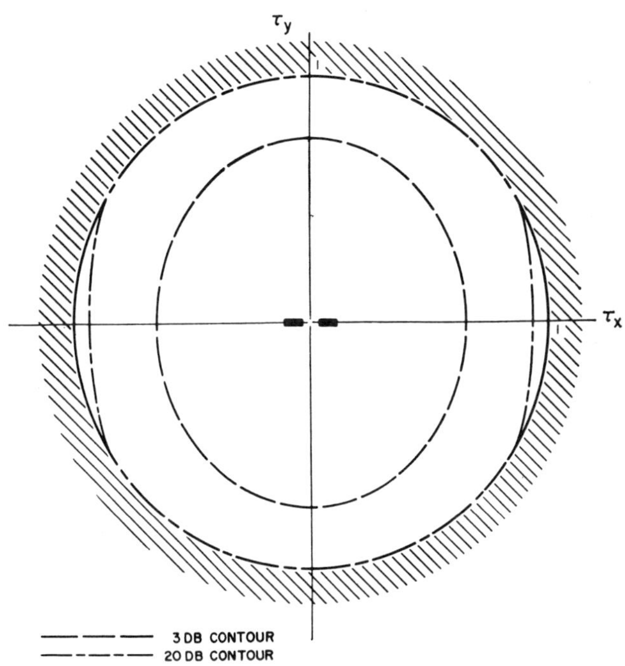


Fig. 16—Contour map of the pattern of a half-wave dipole located one-eighth wavelength above an infinite ground panel.

Hence, in special cases where only a small scan angle is desired, considerable savings can be realized by designing radiating elements which closely approximate the ideal pattern. If, for example, a $\frac{1}{2}$-degree beam were scanned $\pm 15°$, the radiating elements could be spaced $2\lambda_0$ apart, provided that the element factor $S_e$ closely approximated Fig. 15.

For linear arrays, which are scanned in the $x$ direction, the radiating elements determine the pattern shape in the $y$ direction. Here, special requirements may have to be met, such as a squared cosecant pattern.

*A. Correction for Beam Direction*

The beam direction of an array of isotropic radiators was previously given [see (9)] as

$$\sin \theta_s e^{i\phi_s} = T_s = \frac{\psi}{d_r} \cdot$$

It can be shown that the scan angle $T_s$ must be corrected by a small quantity $\Delta T$ to account for the beam distortion (Fig. 17) which is introduced by the pattern $S_e$ of the radiating elements. If $S_e$ does not exhibit circular symmetry (*e.g.*, if $S_e$ is a dipole pattern of the type shown in Fig. 16), then $\Delta T$ is a complex quantity: *i.e.*, it contains corrections for the $\theta$ and $\phi$ directions. If $S_e$ is independent of $\phi$ or is circularly symmetric, the correction applies only to the $\theta$ direction. Restricting this study to the latter case, let us assume scanning along the real axis, where $T_s = \sin \theta_s$. The radiated amplitude pattern is given by

$$S = S_a S_e. \tag{23}$$

The correction for the beam maximum can be found by solving

$$\left.\frac{dS}{dT}\right|_{T=T_s-\Delta T} = 0 \tag{24}$$

for $\Delta T$.[12] To simplify computation, the logarithmic patterns $F = \log S$, $F_a = \log S_a$, and $F_e = \log S_e$ are introduced. Then the first derivative of (23) is

$$F'(T) = F_a'(T - T_s) + F_e'(T), \tag{25}$$

where $T_s$ is assumed constant. Expanding the right-hand side of (25) into a power series about $T_s$, we can solve (24) for $\Delta T$,

$$\Delta T = \frac{F_e'(T_s)}{F_a''(0) + F_e''(T_s)} \simeq \frac{F_e'(T_s)}{F_a''(0)} \cdot \tag{26}$$

In general, the curvature of the element pattern $F_e''(T_s)$ is much smaller than that of the array factor $F_a''(0)$ and may be neglected.

[12] The minus sign is chosen because the expected shift of the beam maximum is toward the array normal, *i.e.*, in the direction of decreasing $T$.

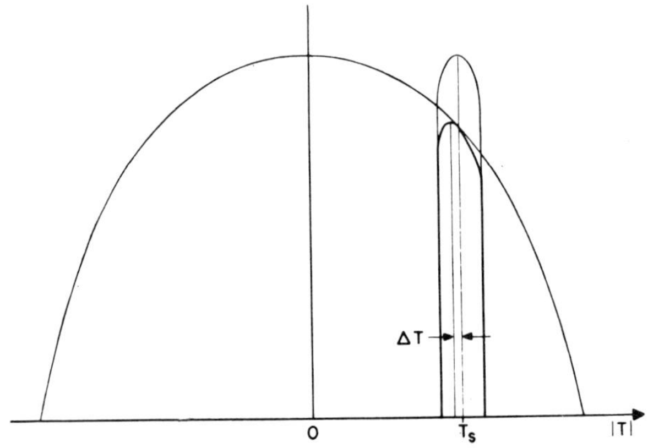


Fig. 17—Correction for beam direction of arrays consisting of directive radiators.

Thus, (26) permits ready evaluation of the correction factor $\Delta T$ from measured or computed antenna patterns. As an example, consider an array where both the array factor and the element pattern can be approximated by $S = (\sin x)/x$. Using the logarithmic pattern

$$F = \log \sin x - \log x, \tag{27}$$

where $x = Md_rT/2$, we find

$$\frac{dF}{dT} = -\frac{1}{T}(1 - x \cot x). \tag{28}$$

$$\frac{d^2F}{dT^2} = -\frac{1}{T^2}\left[\left(\frac{x}{\sin x}\right)^2 - 1\right]. \tag{29}$$

Using (29) to obtain the curvature of the array factor at $T = 0$ and introducing the beamwidth $2\Delta\tau_a$ from (6), we find that

$$F_a''(0) = -\frac{x^2}{3T^2} = -\frac{2.6}{(2\Delta\tau_a)^2} \cdot \tag{30}$$

The quantity $F_a''(0)$ is inversely proportional to the square of the beamwidth. Using (28) to obtain the slope of the element factor at the half-power point and introducing the beamwidth $2\Delta\tau_e$, then

$$F_e'(T_s) = -\frac{1.5}{2\Delta\tau_e} \cdot \tag{31}$$

The slope $F_e'$ is inversely proportional to the beamwidth. Thus the maximum correction term $\Delta T_{max}$, which applies to the maximum scan angle, can be expressed in terms of array and element beamwidth,

$$\Delta T_{max} = 0.57 \frac{(2\Delta\tau_a)^2}{2\Delta\tau_e} \cdot \tag{32}$$

For a narrow-beam array, $\Delta T_{max}$ is quite small and of the order of $10^{-3}$ to $10^{-4}$. However, $\Delta T_{max}$ increases with the square of the beamwidth and cannot be ignored for phased arrays which have a relatively broad beam.

Thus, the radiating elements may introduce beam distortion, sidelobes, and a small shift in beam direction. It may be concluded that proper design of radiating elements for phased arrays is of crucial importance to insure optimum performance. Despite substantial information on free-space patterns of a large variety of radiators, there is not yet enough information on the representative patterns of such elements as part of multielement phased arrays, and further investigations are needed.

## V. Accuracy of Phased Arrays

In this study, the accuracy of a phased array is defined as the accuracy with which the direction of the beam maximum can be determined and reproduced by measuring and adjusting the phase delay between radiating elements. Using the conventional method of expressing beam position and its accuracy, it can be stated that the radiated beam maximum is determined by the spherical coordinates $\theta_s \pm \Delta\theta$ and $\phi_s \pm \Delta\phi$, with $\Delta\theta$ and $\Delta\phi$ designating the $2\sigma$-values of a normal distribution around $\theta_s$ and $\phi_s$. If the antenna were mechanically scanned around two or more orthogonal shafts, it would be reasonable to give values for $\Delta\theta$ and $\Delta\phi$ which would apply to scanning throughout the entire hemisphere. It is readily seen, however, that no such numbers $\Delta\theta$ and $\Delta\phi$ can exist for phased arrays. If it can be assumed that all necessary phase delays between zero and $\psi_{max}$ can be produced with equal accuracy, then the two direction cosines $\cos \alpha_{xs}$ and $\cos \alpha_{ys}$ will have constant tolerances but $\Delta\theta$ and $\Delta\phi$ will depend upon the scan angle $\theta_s$. From Section II-B, we obtain

$$d(\cos \alpha_{xs}) = \frac{1}{d_r} d\psi_x \qquad d(\cos \alpha_{ys}) = \frac{1}{d_r} d\psi_y.$$

These statements can be combined into a single equation in the complex $T$ plane,

$$dT = \frac{1}{d_r} d\psi \tag{33}$$

A small change in complex phase shift is given by

$$d\psi = |d\psi| e^{i\epsilon}. \tag{34}$$

The differential $dT$ can be expressed as an arc $d\gamma$ on unit sphere (Fig. 18) by rewriting it in spherical coordinates

$$dT = d(\sin \theta_s e^{i\phi_s}) = (\cos \theta_s d\theta + i \sin \theta_s d\phi)e^{i\phi_s}, \tag{35}$$

and computing $d\theta$ and $d\phi$

$$d\theta = \frac{\cos(\epsilon - \phi_s)}{d_r \cos \theta_s} |d\psi| \qquad d\phi = \frac{\sin(\epsilon - \phi_s)}{d_r \sin \theta_s} |d\psi|. \tag{36}$$

Then the tolerance for a phased array can be stated in terms of the angle $\Delta\gamma$:

$$\Delta\gamma \equiv \sqrt{(\Delta\theta)^2 + \sin^2 \theta_s (\Delta\phi)^2}$$
$$= \frac{|\Delta\psi|}{d_r} \sqrt{1 + \tan^2 \theta_s \cos^2(\epsilon - \phi_s)}. \tag{37}$$

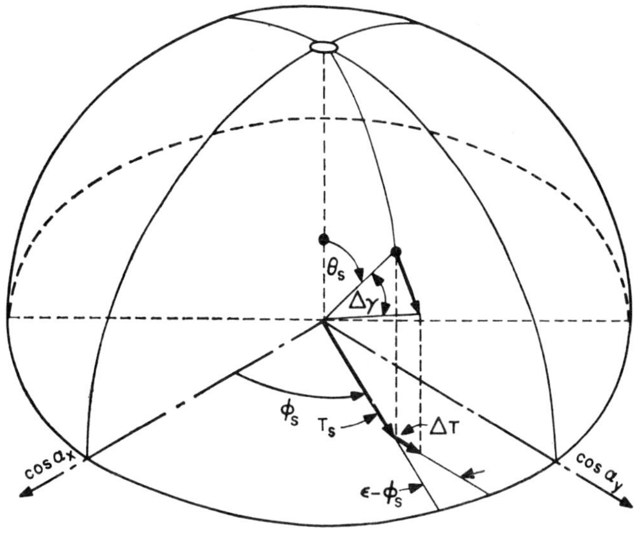


Fig. 18—Accuracy of phased arrays.

This result has the following significance:

1) Given a change in phase delay $\Delta\psi$ in the $\theta$ direction ($\epsilon = \phi_s$), the beam motion is proportional to $\Delta\psi$ in $T$ space. Hence, for relatively small $\theta_s$, the tolerance $\Delta\gamma$ is proportional to $\Delta\psi$. However, $\Delta\gamma$ increases with $\theta_s$ as $1/\cos \theta_s$ in exactly the same way as the beam broadens (Section II-E).
2) Given a change in phase delay $\Delta\psi$ in the $\phi$ direction ($\epsilon = \pi/2 + \phi_s$), the beam motion along the arc $\Delta\gamma$ is proportional to $\Delta\psi$ and hence independent of the scan angle $\theta_s$.
3) The tolerance $\Delta\gamma$ is generally smaller than the tolerance $|\Delta\psi|$ because the constant of proportionality $1/d_r = \lambda_0/2\pi d$ is smaller than unity. For practical cases, $\frac{1}{2} > 1/d_r > \frac{1}{10}$; typically, $1/d_r = \frac{1}{4}$. This implies that a systematic error in phase delay of, say, $|\Delta\psi| = 1°$ produces a beam-direction accuracy of the order of $\frac{1}{4}$ degree, or less than 0.005 radian. This accuracy is not so good as that of modern, mechanically scanned antennas.

It is beyond the scope of this paper to discuss the design of accurate phased arrays, but we can point out here that the main problem appears to be the development of accurate methods to measure phase delay across the aperture of an array and of feedback and control circuitry which permits precise adjustment of phase shifters to the required values. Mechanical and electrical phase shifters are available for introducing the required phase delay.[13–15] Both appear to be capable of

[13] G. C. Southworth, "Principles and Applications of Waveguide Transmission," D. Van Nostrand Co., Inc., New York, N. Y., pp. 325–335; 1950.

[14] F. Reggia and E. G. Spencer, "A new technique in ferrite phase shifting for beam scanning of microwave antennas," Proc. IRE, vol. 45, pp. 1510–1517; November, 1957.

[15] F. E. Goodwin and H. R. Senf, "Volumetric scanning of a radar with ferrite phase shifters," Proc. IRE, vol. 47, pp. 453–454; April, 1959.

generating phase delays with an accuracy of $|\Delta\psi| = 1°$. These phase shifters may be used in parallel and series feed structures as well as in combinations of both (Fig. 19). In a series feed, all phase shifts $\psi_i$ are equal to $\psi$; hence, programming is simple but errors in $\psi$ show up directly as beam-direction errors. In a parallel feed, $\psi_{i+1} - \psi_i = \psi$; this implies that all $\psi_i$ are different, requiring elaborate programming, but systematic errors in $\psi_i$ have a tendency to compensate each other, at least partially. Furthermore, phase shifters in a parallel feed can be designed with considerably lower peak-power rating than phase shifters in a series feed. Thus, each of these feeds has its own peculiar problem with respect to systematic errors, frequency dispersion, complexity, and power-handling capability.

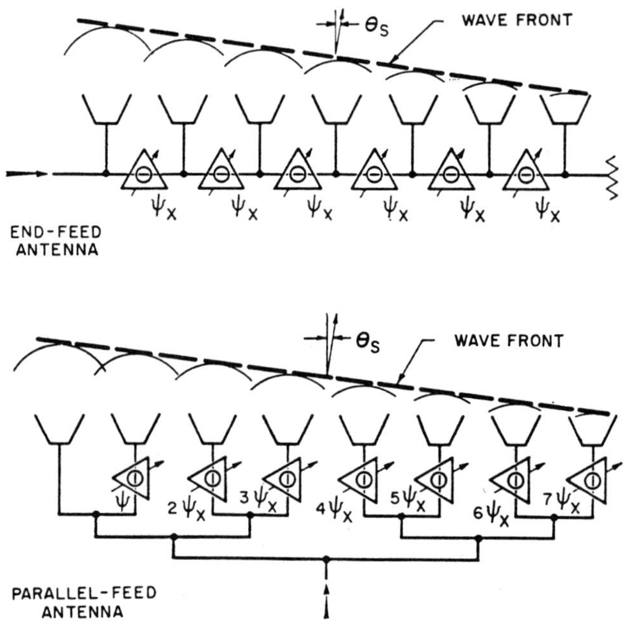


Fig. 19—Parallel-feed and series-feed antenna systems.

## VI. Scanning of Phased Arrays in a Ground-Based Coordinate System

So far, the scanning characteristics of a phased array have been described in a spherical coordinate system which is fixed with respect to the array (Fig. 1). If the array were horizontal and oriented so that its $x$ axis pointed north and its $y$ axis pointed west, then the same coordinate system could be used as a ground-based system and the beam position and pattern shape could be interpreted immediately in terms of the angles $\theta$ and $\phi$ of a ground-based coordinate system. In general, this will not be the case; the array normal will be inclined by an angle $\theta_a$ from the vertical, and the plane of the array will not be horizontal. In fact, if the array were mounted on a ship or airplane, neither its $x$ nor its $y$ axis would normally be horizontal. Then the interpretation of beam direction in terms of a ground-based system would require a rotation of the array coordinate system through the Eulerian angles and a suitable coordinate transformation to find the beam direction and describe scanning distortions in a ground-based system.

This general case,[16] which is perfectly straightforward but somewhat involved and cumbersome, will not be treated here. But a simpler case, that of the ground-based tilted array, is given in some detail to show how its scanning performance can readily be studied by a parallel projection of a unit sphere belonging to the ground-based system on the $T$ plane, which is fixed with respect to the array.

As an example, let the $y$ axis of the array remain horizontal and point west. The array normal is then tilted forward through an angle $\theta_a$ so that the normal points north (Fig. 20). The array coordinates are primed to distinguish them from the coordinates $x$, $y$, $z$ of the ground-based system. Rotating the array in the $xy$ plane to obtain a principal direction (that of the $z'$ axis) different from north does not introduce any complications and therefore will not be considered.

[16] H. Goldstein, "Classical Mechanics," Addison-Wesley Publishing Co., Inc., Reading, Mass., ch. 4; 1950.

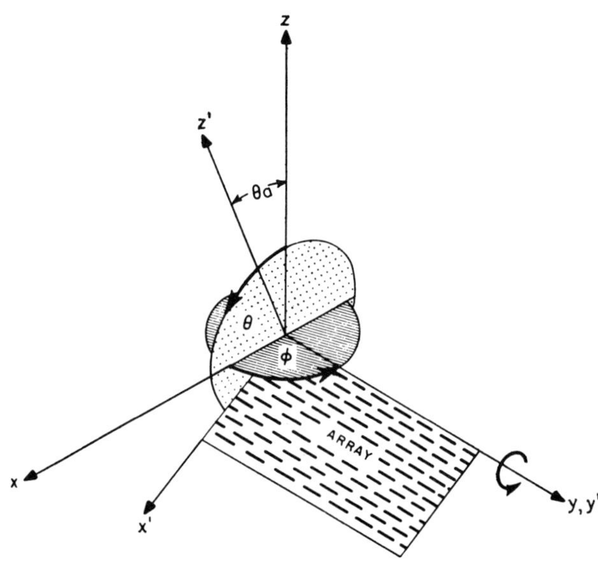


Fig. 20—Coordinate system of a tilted planar array.

Let $\vec{x}$ and $\vec{x}'$ be radius vectors in the ground-based and array coordinate systems, respectively:

$$\vec{x} = \begin{pmatrix} x \\ y \\ z \end{pmatrix} \qquad \vec{x}' = \begin{pmatrix} x' \\ y' \\ z' \end{pmatrix}.$$

Then $\vec{x}$ and $\vec{x}'$ are connected by the orthogonal transformations

$$\vec{x}' = \overleftrightarrow{A}\vec{x}, \tag{38}$$

$$\vec{x} = \overleftrightarrow{\tilde{A}}\vec{x}', \tag{39}$$

where the rotation matrix $\overleftrightarrow{A}$ is given by

$$\overleftrightarrow{A} = \begin{pmatrix} \cos\theta_a & 0 & -\sin\theta_a \\ 0 & 1 & 0 \\ \sin\theta_a & 0 & \cos\theta_a \end{pmatrix}, \tag{40}$$

and $\overleftrightarrow{\tilde{A}}$ is the transpose of $\overleftrightarrow{A}$.

Two computations can be carried out, noting that on unit sphere

$$\begin{aligned} x &= \sin\theta\cos\phi &&= \cos\alpha_x, \\ y &= \sin\theta\sin\phi &&= \cos\alpha_y, \\ z &= \cos\theta &&= \cos\alpha_z, \end{aligned}$$

$$x^2 + y^2 + z^2 = x'^2 + y'^2 + z'^2 = 1. \tag{41}$$

First, for a desired beam direction $\theta_s$, $\phi_s$ in the ground-based coordinate system, the required phase shift is derived immediately from (38), (41), and (8),

$$\frac{\psi_x}{d_r} = \cos\theta_a \sin\theta_s \cos\phi_s - \sin\theta_a \cos\theta_s,$$

$$\frac{\psi_y}{d_r} = \sin\theta_s \sin\phi_s. \tag{42}$$

Second, a mapping of the ground-based $\theta$, $\phi$ system on the $T$ plane is obtained by expanding (39),

$$\begin{aligned} \sin\theta\cos\phi &= x'\cos\theta_a + z'\sin\theta_a, \\ \sin\theta\sin\phi &= y', \\ \cos\theta &= -x'\sin\theta_a + z'\cos\theta_a. \end{aligned} \tag{43}$$

Using (41) and rearranging terms, the equations for lines of constant longitude $\phi$ and latitude $\theta$ are, respectively,

$$(y\cot\phi - x'\cos\theta_a)^2 = (1 - x'^2 - y'^2)\sin^2\theta_a, \tag{44}$$

$$(\cos\theta + x'\sin\theta_a)^2 = (1 - x'^2 - y'^2)\cos^2\theta_a. \tag{45}$$

The latter can be rewritten to show that the lines of constant $\theta$ are ellipses with centers on the $x'$ axis (see Fig. 21),

$$\frac{(x' + \sin\theta_a\cos\theta)^2}{(\cos\theta_a\sin\theta)^2} + \frac{y'^2}{\sin^2\theta} = 1. \tag{46}$$

The north pole ($\theta = 0$) is located at $x' = -\sin\theta_a$. We are interested only in that part of these ellipses which corresponds to the projection on the positive hemisphere. Hence, the intersection of these ellipses with the unit circle is located at $x' = -\cos\theta/\sin\theta_a$.

The lines $\phi = \text{const}$ appear as ellipses whose main axes are rotated with respect to the $x'$ and $y'$ directions. By rearranging terms in (44) and applying additional algebra, the equation of an ellipse in polar form ($x' = \rho\cos\epsilon$; $y' = \rho\sin\epsilon$) is obtained,

$$\frac{1}{\rho^2} = \frac{\cos^2(\epsilon - \epsilon_0)}{\sin^2\theta_a \sin^2\phi} + \sin^2(\epsilon - \epsilon_0), \tag{47}$$

in which $\cot\epsilon_0 = -\tan\phi\cos\theta_a$.

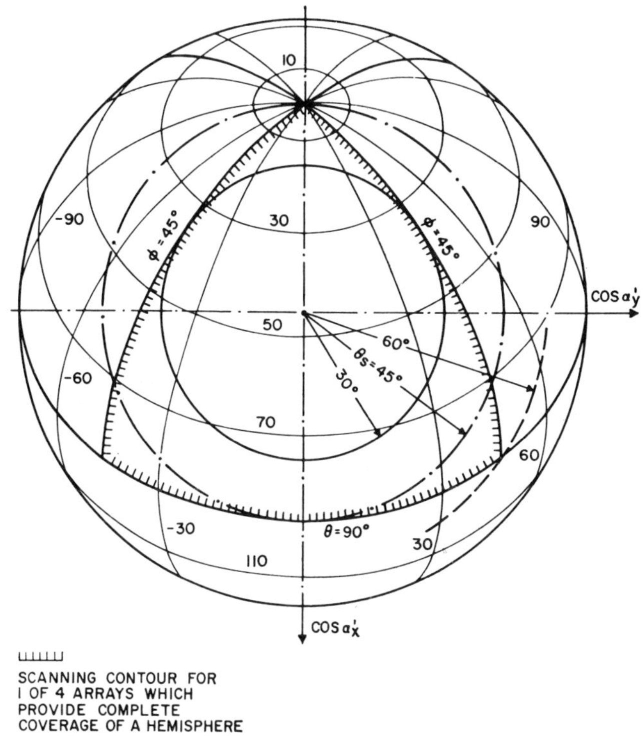


Fig. 21—Mapping of a ground-based spherical coordinate system onto the plane of an array whose normal is tilted 45° from the vertical.

The mapping given by (46) and (47) is very useful in determining the optimum tilt angle and scanning contour for a given scanning task. In general, a compromise will be needed between the scanning contour which gives the least beam distortion (see Sections II-E through II-G) and the scanning contour which provides the best coverage of the desired solid angle of the hemisphere. In the case of a phased array whose normal points straight up (Fig. 8), a circular scanning contour provides the best coverage. However, if complete coverage of a hemisphere is desired, then the scanning contours of several tilted arrays must be combined. If relatively large scanning distortions are permissible, then it is possible to cover the hemisphere with four phased arrays tilted 45° (Fig. 21). In this instance, a maximum scan angle of 45° would permit complete coverage to $\theta = 70°$, down to 20° elevation, whereas a maximum scan angle of 60° would be required to cover all points down to zero elevation. A circular scanning contour would provide considerable overlap at the north pole. This could be avoided by suitable programming of the phased arrays to obtain triangular scanning contours.

Acknowledgment

The author is indebted to W. E. Danielson, J. N. Hines, and I. Jacobs for many stimulating discussions and to R. S. McCarter for his critical reading of the manuscript. Miss J. Peterson and W. Pelish were most helpful in the preparation of the figures.

*Reprinted from* IEEE TRANSACTIONS
ON *ANTENNAS AND PROPAGATION*
Volume AP-13, Number 4, July, 1965
pp. 506-514


# Simple Relations Derived from a Phased-Array Antenna Made of an Infinite Current Sheet

50

HAROLD A. WHEELER, FELLOW, IEEE

*Abstract*—The simplest concept of a phased array is an infinite planar current sheet backed by a reflecting boundary. The electric current sheet, or resistance sheet, is the limiting case of many small electric dipoles, closely spaced, and backed by an open-circuit boundary. If this array is viewed as a receiver, a plane wave incident on the array at some angle ($\theta$) meets a boundary resistance varying in proportion to cos $\theta$ for angles in the $H$ plane, and 1/cos $\theta$ for angles in the $E$ plane. If the array is matched at broadside ($\theta=0$), the corresponding reflection coefficient has the magnitude $(\tan \frac{1}{2}\theta)^2$.

While the electric current sheet is realizable, the open-circuit boundary is not. However, a magnetic current sheet can be simulated by a conductive sheet with holes utilized as magnetic dipoles, such a sheet providing the backing equivalent to a short-circuit boundary. The latter case is related to the former by electromagnetic inversion or duality. Therefore, an incident plane wave meets a boundary conductance varying in proportion to cos $\theta$ for angles in the $E$ plane, and 1/cos $\theta$ for angles in the $H$ plane. The predicted behavior is verified qualitatively by tests of such a model with elements of a practical size.

The derivation is based on the principle of dividing the space in front of the array into parallel tubes or waveguides, one for each element cell in the sheet or array. This is one of the principles published by the author in 1948. A related principle enables the simulation of an infinite array by imaging a few elements in the walls of a waveguide. This latter principle is utilized for making tests of the array.

Manuscript received September 28, 1964; revised February 4, 1965. This paper was originally presented at the IEEE Symposium on Antennas and Propagation, September, 1964.
The author is President of Wheeler Labs., Inc., Great Neck, N. Y.

## I. Introduction

A PLANAR ARRAY of many radiating elements is used to develop a narrow pencil beam. The connections between the elements and the sending or receiving circuit are phased to control the beam angle of deflection from broadside (the "scan angle"). This is the type of antenna commonly called a "phased array," implying beam steering by electrical phasing without mechanical motion of the array.

This type of antenna introduces an unusual problem of design: the variation of element impedance with scan angle. This variation is caused by the inherent coupling of the elements, which contributes to the apparent impedance of each element in a manner dependent on the phasing required for the scan angle. With phasing for wide-angle scanning, the impedance variation from this cause becomes substantial and should be taken into account, along with the variation from other causes such as the frequency bandwidth.

There have been various studies of the element impedance variation with scan angle. Some are based on laborious computation of an array of many elements [5], [7], [12], [13]. Others are based on the concept of an infinite array [2], [6], [8], [14], [17], [18], [21]. This latter concept has been found by the author to offer the best basis for the design of an element for use at all locations in a finite array of many elements. Moreover, it offers the only evaluation that is independent of the number of elements. Therefore, there is a particular interest in the behavior of an element in an infinite array.

In general, the variation of element impedance with angle depends on the "fine structure" of the element, and on the arrangement of elements in the array. However, there is a hypothetical array which avoids any question of fine structure. This is the array that is formed of an infinite current sheet, representing elements so small and so closely spaced that the fine structure is not resolved. On the basis of such an array, there is found a simple derivation for impedance variation with angle, which is inherent in arrays fitting into a certain pattern. Furthermore, it is found that some practical types do fall into this pattern, so the conclusions from this hypothesis are useful in practice.

It is the purpose of this paper to describe the hypothetical array formed of an infinite current sheet, to derive its impedance variation with angle, and to introduce some corollary concepts, such as the "ideal element pattern," which theoretically would avoid the implement variation.

The scope of this treatment is limited to the infinite planar array, having a well-defined planar face (ground plane), which separates the radiation region (free space, in front) from the associated circuits (shielded, behind). A practical example is an array of radiating holes in a sheet of conducting material (metal).

It is noted that the apparent impedance of the element, as formulated here, is the "active impedance" that would be seen looking out of one element with all the other elements excited in the phase required for the scan angle.

## II. Symbols

MKS rationalized units.

$l$ = effective length of small electric dipole.
$a$ = element spacing in square array.
$A = a^2$ = area of element cell in square array.
$\theta$ = scan angle from broadside.
$\lambda$ = free-space wavelength.
$\epsilon$ = electrivity (electric permittivity)
$\mu$ = magnetivity (magnetic permeability)
$H$ = magnetic field intensity (A/m).
$E$ = electric field intensity (V/m).
$I$ = current in electric dipole.
$H_s$ = magnetic field intensity adjacent to electric current sheet (A/m).
$I_s$ = electric current density in sheet (A/m).
$F(\theta)$ = relative field intensity in radiation pattern of isolated element.
$R$ = radiation resistance of electric dipole in active array phased for some angle.
$R_c$ = 377 ohms = wave resistance across square area of wavefront in free space.
$R_s$ = resistance across square area of sheet.
$G$ = radiation conductance of magnetic dipole (across a slot) in active array phased for some angle.
$\rho$ = voltage-reflection coefficient at face of array.
sub-$c$ = free space
sub-$o$ = broadside ($\theta = 0$).
sub-$s$ = sheet.
sub-$m$ = plane of magnetic field ($H$).
sub-$e$ = plane of electric field ($E$).
sub-$i$ = ideal.
sub-$h$ = Huygens source.

## III. Theoretical Background

The concept of the infinite array was introduced by the author in 1948 [2], long before the recent intensive work on phased arrays with beam steering. That publication showed the opportunity for simple computation of some properties of a radiating element in an infinite array. This approach had been suggested by the theory of such an element in a waveguide. Now it turns out that the same principle is applicable to phased arrays. Moreover, it is found that the experimental design of an element for a phased array is facilitated by imaging a few elements in waveguide walls to simulate an infinite array [17], [20], [23], [24].

As an introduction to the present discussion, Figs. 1 and 2 are reproduced from the 1948 publication. Figure 1 shows an infinite planar array of electric dipoles in a square arrangement in a plane with free space in front and "open circuit behind." This is the same basis to be used for the infinite current sheet. Fig. 2 shows a modification of the same to form an array phased for an oblique angle of radiation. This is a special case of a phased array, in which adjacent elements are phased in opposition; a condition that is most simply simulated by the metal walls of a rectangular waveguide. (The same publication also describes the simulation of an infinite array in a rectangular waveguide, with oblique angles of radiation.)

Among the formulas in that publication, there are two that form the background for the present discussion. Referring to Fig. 1, the radiation resistance of each element is given as

$$R_0 = R_c l^2/a^2 = R_c l^2/A. \tag{1}$$

Referring to Fig. 2, for oblique radiation at an angle $\theta$ from broadside, in the $H$ plane, the radiation resistance happens to be given for a pair of directions ($\pm\theta$). Taking one-half of this for one direction,

$$R = \frac{R_c}{\cos\theta}\,\frac{l^2}{a^2}\,. \tag{2}$$

This forms an introduction to the angle variation of mpedance that is to be derived. In this case, the radiation resistance increases with angle from broadside.

In the 1948 publication, some reliance is placed on

51

earlier derivations of the radiation resistance of an electric dipole in a rectangular waveguide. The complete derivation can be described briefly with reference to the tubes of radiation in front of the array, as shown in Figs. 1 and 2. In the broadside direction ($\theta=0$), shown in Fig. 1, the tubes are square and are bounded by hypothetical image planes. The wave resistance between top and bottom of each square cell ($a^2$) is that of free space ($R_c$). The electric dipole (of effective length $=l$) is coupled to this resistance by an effective transformation ratio ($l/a$), so its radiation resistance is proportional to the square of this ratio, as appears in formula (1). In an oblique direction ($\theta$), as shown in Fig. 2, each tube is decreased in width by the projection cosine, so the resistance of the tube and of the dipole is increased by the factor $1/\cos\theta$.



It is noted that this simple computation for an infinite array is based on real power, and, therefore, only the resistance component is given. In general, there is also a variation of the reactance component, so that the impedance variation is the resultant of both components.

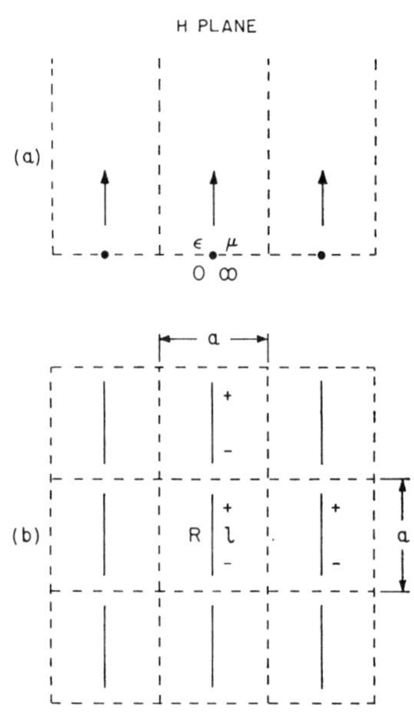


Fig. 1. An infinite planar array at the boundary of half-infinite space. (a) Plan. (b) Elevation.

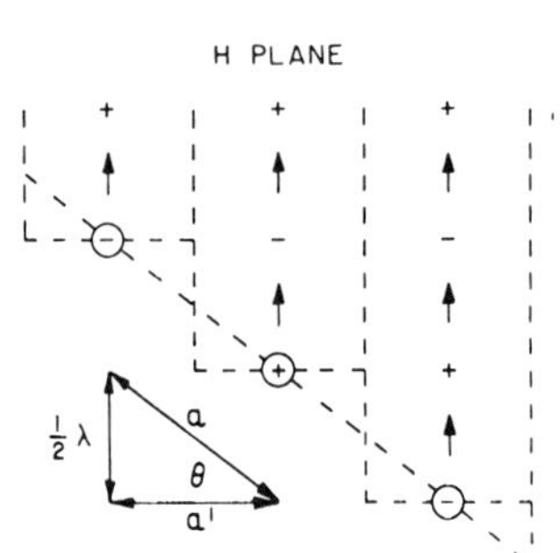


Fig. 2. The same array phased for an oblique angle of radiation.

Further theoretical background is found in more recent publications, which will be reviewed in a later section of this paper.

## IV. The Infinite Current Sheet

The present discussion is to be based on an infinite current sheet. There are two kinds of current sheet that may be assumed, electric and magnetic. The former is conceptually the easier, because an electric conductor is within our experience. Therefore it will be utilized for the derivations. On the other hand, the magnetic current sheet (and not the electric) is found to be realizable in the required environment of the array, so it will be relied on for experimental verification.

Figure 3 shows the basis for a planar phased array made of an electric current sheet. The baseline in Fig. 3(a) is the scan-plane projection of the plane of the array. Above the line is represented the radiation region (free space) in front of the array. Below the line is assumed to be an "open-circuit" reflector behind the array ($\epsilon=0$, $\mu=\infty$). The latter region is not physically realizable, but is conceptually simple; it is carried over from Figs. 1 and 2.

The electric current sheet is in the plane of the array. It has a current density $I_s$ of uniform amplitude, but phased for radiating a plane wave in a direction at an angle $\theta$ from broadside. The current density is associated with a magnetic intensity $H_s$ of equal amplitude and

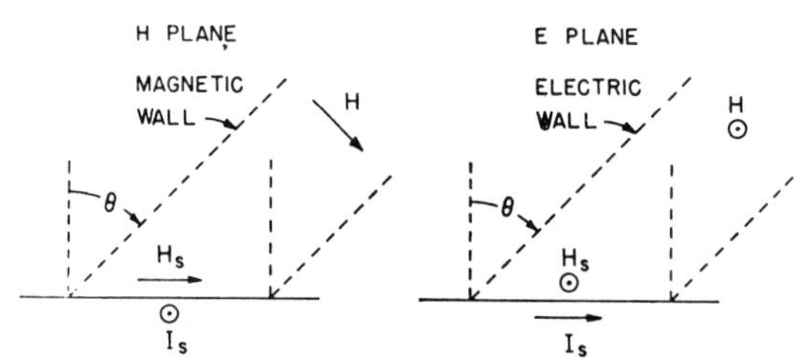


(a) Plane of scan, showing current sheet and magnetic fields.

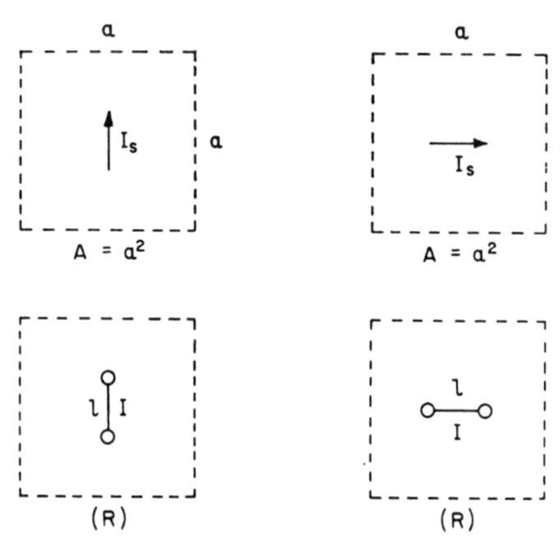


(b) Plane of array, showing element cell of current sheet and equivalent current element.

Fig. 3. An idealized model of a planar array, made of an electric current sheet.

phase, but crossed in space. In the radiation region, the $H$ plane and $E$ plane are determined to include, respectively, the $H_s$ and $I_s$ vectors. The scan angle $\theta$ may be in either of these planes, as shown.

Figure 3(b) shows the face of the array, the infinite current sheet being exemplified by a square cell of area $A = a^2$. For some purposes, there will be used also the illustrated concept of a small electric current element, or electric dipole, equivalent to a small cell. These are said to be equivalent if they have equal moment. If the moment of the latter is $lI$, as indicated, the condition for equivalence is

$$lI = a(aI_s) = AI_s. \tag{3}$$

For the broadside direction of radiation ($\theta = 0$), obtained by uniform phase over the array, the radiation resistance $R_0$ of this element is given by formula (1).

The current sheet may have any value of uniform surface resistance $R_s$ (the resistance across any square of the surface). For present purposes, it is assumed to be "matched" to free space for broadside radiation, so $R_s = R_c$ (377 ohms, the wave resistance across a square area of wavefront in free space).

## V. The Variation of Impedance with Scan Angle

Since the receiving viewpoint may be conceptually simpler, it will be taken first for deriving the variation of impedance with scan angle. This derivation will be based on Fig. 3(a), showing the two principal planes of scan angle, the $H$ plane and the $E$ plane. Any reflection from the face of the array, as will be derived, appears in the form of a beam radiated at the opposite angle from broadside.

Referring to an oblique angle of scanning in the $H$ plane, we have a situation similar to Fig. 2. The incoming wave sees a section of current sheet which is wider han the section of wavefront, so the wave appears to be reflected by a resistance that is lesser in the ratio, $\cos\theta$. The resulting voltage-reflection coefficient is negative:

$$\rho_m = \frac{\cos\theta - 1}{\cos\theta + 1} = -(\tan\tfrac{1}{2}\theta)^2. \tag{4}$$

Referring instead to an oblique angle of scanning in the $E$ plane, we find an opposite situation. The incoming wave sees a section of current sheet which is longer than the section of wavefront, so the wave appears to be reflected by a resistance that is greater in the ratio, $1/\cos\theta$. The resulting voltage-reflection coefficient is positive:

$$\rho_e = \frac{1/\cos\theta - 1}{1/\cos\theta + 1} = \frac{1 - \cos\theta}{1 + \cos\theta} = (\tan\tfrac{1}{2}\theta)^2. \tag{5}$$

Changing over to the sending viewpoint, we may derive the same relations, but naturally the reflection coefficient will be reversed. Here we refer also to the equivalent current elements in Fig. 3(b).

The method is given in the author's early paper on this subject [2], here represented by Figs. 1 and 2. The space in front of the array is partitioned by hypothetical waveguide walls and the radiation resistance of each element is formulated. Referring to Fig. 3(a), these walls are located by projecting the element cell in the beam direction determined by the array phasing. The nature of the walls is determined by the requirement for imaging the fields of a uniform plane wave: the walls parallel to the $H$ field are "electric" walls, or "short-circuit" boundaries ($\epsilon = \infty$, $\mu = 0$); and the walls parallel to the $E$ field are "magnetic" walls, or "open-circuit" boundaries ($\epsilon = 0$, $\mu = \infty$).



From (1), the radiation resistance of each element is known for broadside phasing ($R_0$ for $\theta = 0$). Changing to the phasing for an oblique angle modifies the situation in two respects. First, the area of wavefront is decreased by the factor $\cos\theta$, increasing the resistance by the factor $1/\cos\theta$, as in (2). Second, the radiation pattern of the element may affect the relative amount of radiation in the beam direction. The latter effect is to be formulated here.

The radiation pattern of a small electric dipole has the following dependence on the angle $\theta$ in Fig. 3.

$$H \text{ plane: } E/E_0 \equiv F_m(\theta) = 1 \tag{6}$$

$$E \text{ plane: } E/E_0 \equiv F_e(\theta) = \cos\theta. \tag{7}$$

This is the familiar "doughnut" pattern, or specifically the half of this pattern that is located in front of the array. These factors may be regarded as foreshortening of the element length by projection on the wavefront. The effect on the radiation resistance is naturally proportional to the square of this factor.

The resulting variation of element resistance with angle is given by the following formulas, which are essentially similar to some derived by Stark [6].

$$H \text{ plane: } R/R_0 = \frac{1}{\cos\theta}[F_m(\theta)]^2 = 1/\cos\theta \tag{8}$$

$$E \text{ plane: } R/R_0 = \frac{1}{\cos\theta}[F_e(\theta)]^2 = \cos\theta. \tag{9}$$

As would be expected from the principle of reciprocity, these ratios are the inverse of those derived from the receiving viewpoint. Each of the corresponding reflection coefficients is merely reversed.

The factor $1/\cos\theta$, which appears in (8) and (9), may be inferred from another viewpoint in transmission. Let us consider a finite planar array of many elements, so large that the beam width is a small fraction of one radian. In any particular plane of scan, the aperture width projected in the beam direction is proportional to $\cos\theta$, so the beam width is inversely proportional to this ratio. The power in the beam is proportional to the factor $F^2$ in the beam direction, and also to the beam width. The latter introduces the factor $1/\cos\theta$ in the radiation resistance.

From the sending viewpoint, the reflected power is returned to the generator, or may be absorbed in an intervening "isolator." In the case of a generator matched to the element broadside resistance $R_0$, the net result of the reflection is to decrease the radiated power below the "available" power by the factor $(1-|\rho|^2)$.

Figure 4 shows the variation of element resistance and reflection coefficient with scan angle in the $H$ and $E$ planes. The polarity of $\rho$, in Fig. 4(a), is given for the electric dipole, as in Fig. 3(b).

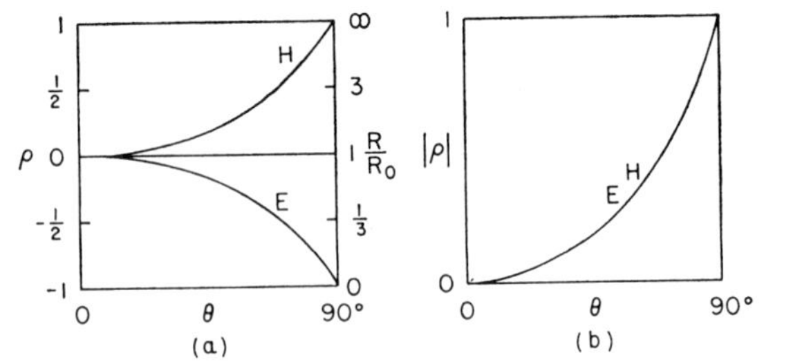


Fig. 4. The variation of reflection with scan angle, in the idealized model. (a) For the case of electric dipoles, from the sending viewpoint. (b) For either kind of dipoles, from either viewpoint.

In general terms, the same derivation is applicable to the inverse case of a magnetic current sheet, with transformation by the principles of duality. This means interchanging $H$ and $E$, interchanging open-circuit and short-circuit reflector, interchanging resistance $R$ and conductance $G$ (or inverting $R$), and interchanging voltage and current-reflection coefficients (or reversing the sign of the voltage-reflection coefficient). The radiation pattern of a small magnetic dipole (or slot in a metal sheet) is as follows,

$$E \text{ plane: } E/E_0 \equiv F_e(\theta) = 1 \tag{10}$$

$$H \text{ plane: } E/E_0 \equiv F_m(\theta) = \cos\theta. \tag{11}$$

The resulting variation of element conductance is as follows,

$$E \text{ plane: } G/G_0 = \frac{1}{\cos\theta}[F_e(\theta)]^2 = 1/\cos\theta \tag{12}$$

$$H \text{ plane: } G/G_0 = \frac{1}{\cos\theta}[F_m(\theta)]^2 = \cos\theta. \tag{13}$$

Of greatest importance is the fact that the magnetic current sheet can be simulated in a physical structure. A metal sheet can be used as the required short-circuit reflector, and holes in the sheet can be used to simulate the required magnetic dipoles as radiating elements. Such a practical structure will be described further on.

The current sheet has only resistance and no reactance, so the simple theory gives the entire variation of impedance with angle. When approximated by an array of elements of appreciable size and spacing, there is a departure from this simple rule. Usually, the resistance variation is somewhat different, and there is also appreciable variation of reactance.

In general, the angle from broadside either increases or decreases the radiation loading of the element, from the sending viewpoint. The radiation loading of an electric dipole is proportional to resistance, while that of a magnetic dipole is proportional to conductance. From the formulas given, the following conclusions can be drawn.

*Variation of radiation loading by increasing angle ($\theta$)*

| *Plane of angle ($\theta$)* | *Radiating element* | |
|---|---|---|
| | *E dipole* | *M dipole* |
| $H$ plane | increasing | decreasing |
| $E$ plane | decreasing | increasing |

## VI. The Ideal Element Pattern

The formulas (8) and (9) give a basis for describing an "ideal" element pattern that would yield an element resistance invariant with scan angle, likewise in any scan plane (or in any linear polarization relative to the scan plane). This pattern is $F_i(\theta)$, defined as follows.

$$R/R_0 = \frac{1}{\cos\theta}[F_i(\theta)]^2 = 1; \qquad F_i(\theta) = \sqrt{\cos\theta}. \tag{14}$$

Figure 5 shows the three patterns mentioned here. In either plane, the ideal pattern ($E'H'$) is a change from the dipole pattern ($H$ to $H'$ or $E$ to $E'$).

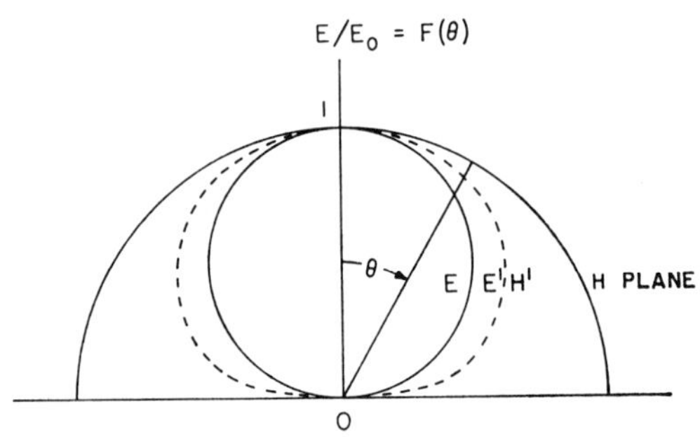


Fig. 5. Isolated element patterns of a small electric dipole.

The element pattern, as used here, is that of an isolated element, and the ideal pattern is derived on the same basis. This same concept might have been inferred from previous articles, though not explicitly stated [5], [6]. An ideal element pattern has been implied or described also on a different basis, the element being located in the environment of an infinite array of passive elements [18]. Both viewpoints happen to yield the same shape for the ideal element pattern, but with different significance.

It is noted in passing that there is a kind of simple radiator that offers a theoretical attraction for use as an element in an array. It is the "electromagnetic dipole" or "Huygens source" [1], [3], [4], [9], [11]. It is formed by crossed superposition of an electric dipole (or current element) and a magnetic dipole (or current loop) radiating equal (maximum) fields in one direction (the axis). This gives the same amplitude of field in all planes through the axis, namely,

$$|F_h(\theta)| = \tfrac{1}{2}(1 + \cos\theta) = (\cos \tfrac{1}{2}\theta)^2. \qquad (15)$$

This has a cardioid shape which is a close approximation to the ideal shape (14). Aside from its favorable pattern, it has some peculiarities that require further study before it can be used as an element in an array. 1) It is difficult to realize in a physical structure. 2) It is a two-port for each linear polarization, as distinguished from the usual one-port. 3) Its pattern would be destroyed by any reflector behind the element. In the meantime, this type of radiator offers some intriguing opportunities for theoretical studies.

## VII. A Practical Structure

As previously mentioned, the relations derived here are applicable to a practical structure simulating a magnetic current sheet backed by a short-circuit reflector. Figure 6 shows such a structure. The array is formed of a metal sheet with the elements radiating through holes in the sheet. It has some properties in common with the arrays of slots, as exemplified in the studies by Rabinowitz, Edelberg, and Oliner [5], [8].

There is a periodic array of holes in either of the two forms that are useful. Figure 6(a) shows a square array, which is the simplest form. Figure 6(b) shows a triangular array, which is still simple and offers some practical advantages. Also it gives the closer approximation to a current sheet, for the same separation of adjacent elements. Figure 6(c) shows a cross section of either form through a "cardinal" plane (*C-C*).

The resistance sheet in this presentation is isotropic. Following this concept here, each element hole is made circular, and is backed by a circular waveguide. Simulation of the current sheet requires that the hole diameter and center-to-center spacing be sufficiently small in terms of the wavelength, so that this "fine structure" cannot be "resolved" by the wave. This condition is nearly satisfied if these dimensions are much less than one-half wavelength, whereas dimensions of this order are required in practice. In the circular waveguide, the TE-11 mode is to be utilized. Its propagation requires dielectric loading in this diameter. Such loading may be provided by spaced circular disks of high dielectric constant, to avoid proportionate loading of the undesired TM-01 mode [19].

This structure will handle any polarization. In practice, it requires impedance matching over the required frequency band, which can be approximated by inserting obstacles in the circular waveguide. There remains the impedance variation with angle, which is the subject of this monograph.

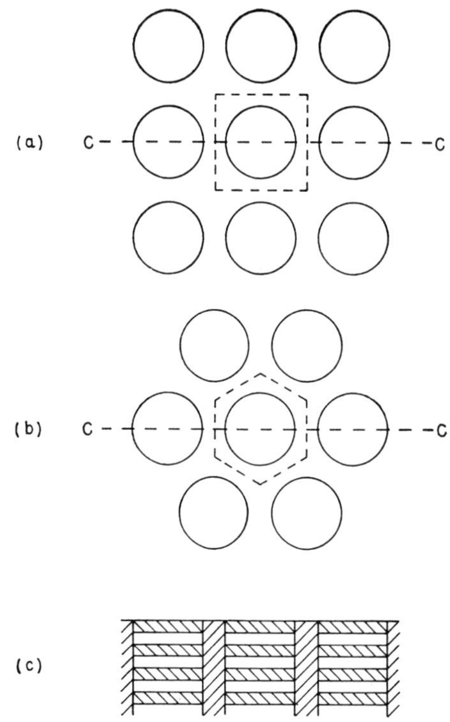


Fig. 6. An array of circular holes in a metal sheet. (a) Square array. (b) Triangular array. (c) Cross section in cardinal plane *C-C*, showing dielectric-disk loading in circular waveguide.

## VIII. Experimental Verification

The author's laboratory has made tests of a practical array utilizing such a structure. An infinite array has been simulated by imaging a few elements in the plane walls of a waveguide [17], [20], [23], [24]. Then the reflection coefficient has been measured from the viewpoint of reception.

The experimental structure is a triangular array, as shown in Fig. 6(b). The hole diameter is 0.45 wavelength, and the center-to-center spacing is 0.60 wavelength. The dielectric loading is sufficient to lower the TE-11 cutoff frequency down to about $1/\sqrt{2}$ or 0.7 of the operating frequency. Figure 7 shows on the reflection chart the voltage-reflection coefficients for several scan angles. Each point is observed in an individual waveguide to simulate the particular angle in an infinite array.

In this experiment, there is no attempt to provide impedance matching at the array face for broadside phasing, which is assumed in the theoretical derivation. Each waveguide is provided with a matched termination to assure a pure traveling wave in reception.

On the reflection chart, Fig. 7, the two extreme points represent a large scan angle of 56° in the cardinal plane, *C-C* in Fig. 6(b), with *E* or *H* vector in this plane as indicated. The intermediate point represents a small scan angle of 29°, so it is not much different from 0°

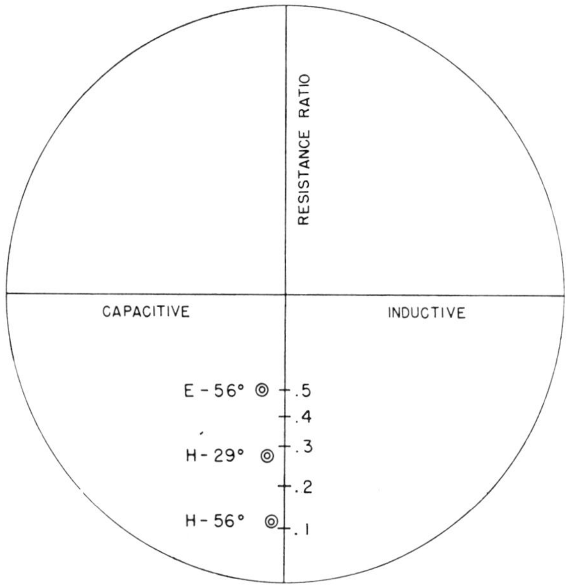


Fig. 7. The coefficient of reflection from the face of an array, at various angles, from the receiving viewpoint.

(broadside); this angle happens to be in the intercardinal plane (30° from *C-C*) with *H* in this plane. These angles are well within the limits out to which the array would develop a single beam free of grating lobes.

On this chart, the scale of normalized resistance is marked in the region of interest.

The relation of the three points on the reflection chart is, roughly, what would be expected from the simple theory based on a magnetic current sheet, off-center, though as not matched for broadside. 1) The points are near the real axis of impedance and reflection coefficient. 2) The resistance is higher for *E* in the scan plane. 3) The extreme points are separated slightly more than the predicted closest resistance ratio, the lowest value being roughly 0.22 times the highest value, as compared with $(\cos 56°)^2 = 0.31$. 4) The intermediate point is not far from the ratio mean of the extreme value of resistance.

Regarding the experimental points more critically, it is reasonable to say that all the discrepancies may be caused by the size of the element, specifically the hole diameter and the spacing. The resistance variation is close to the predicted amount and there is little variation of reactance. At least, there is qualitative agreement and a fair approximation to quantitative agreement.[1] It is remarkable that it has been possible to measure the behavior of a practical element in an environment simulating an infinite array.

[1] More recent theoretical and experimental studies have evaluated the reflection coefficient from the transmitting viewpoint. From this viewpoint, the phase of reflection varies with scan angle and polarization, as affected mainly by element spacing.

## IX. Historical Evolution of the Concepts

Having developed a simple picture of the array reflection, as it varies with scan angle, we shall review the evolution of ideas that has led to this result. We shall pass over the historic studies of the coupling between elementary antennas, which enabled the detailed computation of finite arrays, since this method becomes unduly laborious for a large number of elements. The present objective is to evaluate the typical or average behavior of an element in a very large planar array, the number of elements being in the range from 10 by 10 = 100, up to 100 by 100 = 10,000.

The viewpoint of the infinite array was introduced by the author's publication of 1948 [2]. In a unified presentation, it showed the identity of behavior of a radiating element in a waveguide with plane walls and in an infinite array. It then applied the waveguide viewpoint for computation of the element radiation resistance in closed form, avoiding the summation of a doubly infinite series. Inherent in this presentation were some more specific principles whose significance was not appreciated until later.

The infinite array which received most attention in that monograph was the planar array phased for a broadside pencil beam, Fig. 1 here. For theoretical purposes, this was identified with a nonphysical rectangular waveguide, so the element behavior in the array was found to be different from what had previously been computed for practical waveguides. On the other hand, the infinite array which was shown phased for oblique radiation, Fig. 2, was intended only for its theoretical interest in relation to the well-known wave behavior in physical rectangular waveguides. At that time, the author did not envision the future opportunities for an array with electrical phasing for beam steering without mechanical motion. However, the behavior of such a phased array was inherent in the theoretical derivation for waveguides, as seen here (2).

About ten years later, the growing interest in phased arrays began to be reflected in the published literature. There were three principal sources of publications related to the present subject, namely, S. Edelberg and A. A. Oliner [8], P. S. Carter, Jr. [7], and J. L. Allen [13]. Of special significance, also, were the reports of S. J. Rabinowitz [5], J. L. Allen [12], and L. Stark [6]. All of these sources were concerned with the variation of element impedance with scan angle.

Among the recent publications, the one most closely related to the present subject is that of Edelberg and Oliner [8]. They extend the waveguide analogy to a rectangular array of holes in a metal sheet, and to any scan angle. While their presentation is complicated by the finite dimensions of the element cell and the hole, one can glean from it the simple relations that are presented here, for the limiting case of small elements and close spacing, simulating the magnetic current sheet.

The concept of an "ideal" element pattern has not yet been published in clear and explicit terms. In the formulas of Rabinowitz [5], and in those of Stark [6], the radiation resistance of an element is seen to depend on the radiation pattern of an isolated element (including its reflector). From these formulas, one might infer that an element having a particular pattern (14) would cause the radiation resistance to be invariant with scan angle. An attempt is made here to present this topic clearly. (This concept of isolated element pattern should not be confused with the different concept of array element pattern, the latter being effective in the array environment [18]. It happens that the ideal pattern has the same shape in both cases.)

While beyond the scope of this report, it is natural to speculate on the significance of the impedance variation with a current sheet, and how it might be decreased in practice. The current sheet is a thin array, having no depth in the transition face between the circuits and the radiation region. One approach for decreasing the impedance variation involves structural depth with dimensional controls, such as projections in front, or the spacing between dipoles and reflector, or the addition of interconnecting circuits behind the array [15], [22]. We may infer a principle that the simple relations are associated with a thin array of small elements, and that element size or structural depth offers some opportunity for departure from these relations.

The problems of evaluating an array are those of computation, on one hand, and of measurement, on the other. Since the computations were greatly simplified by the analogy between an infinite array and a waveguide, it should not be surprising that a similar approach has been found invaluable in measurements. This application is naturally restricted to the cases where the hypothetical waveguide is physically reliable. Since 1960, the author and his associates have been exploring the discrete cases where a physical waveguide is available for simulating the behavior of an infinite array [17], [20], [23], [24]. Rectangular waveguides have been utilized for square and triangular arrays, and there are some cases of the latter type where a triangular waveguide is particularly useful.

The concept of an element in an infinite array has been acquiring stature also as the sound basis for designing a practical element for use in a large array of many elements. The element impedance in an infinite array appears to be centered in the range of element impedance at all locations in the finite array. This practical utility excites further interest in the concept.

It is common experience in science, that the ultimate simplification in concepts comes as a culmination of more complicated studies. The present topic appears to be an example of this principle, and an attempt has been made to reduce it to the simplest terms.

## X. Conclusion

The current sheet is a highly simplified conceptual model of a phased array. It is used as the basis for deriving some simple relations. One is the variation of element impedance and reflection coefficient with scan angle. Another is the ideal isolated-element pattern that would make the element radiation resistance invariant with angle.

The derivation relies on the principle that an element in an infinite array behaves like one in a waveguide with properly defined walls, as presented in the author's monograph of 1948.

In this model, the reflection coefficient varies with the scan angle, oppositely for the $H$ or $E$ vector in the scan plane. The predicted variation has been verified qualitatively by tests of an array approximately simulating a magnetic current sheet.



## Acknowledgment

The author's organization has been engaged in active work on the design of phased arrays under the direction of Bell Telephone Laboratories. That work has provided the stimulus for this study, and the author is particularly grateful to R. L. Mattingly and P. H. Smith for their encouragement.

The staff of Wheeler Laboratories has participated in this study in various ways, particularly in helpful discussions and the experimental verification, for which the author wishes to express his appreciation.

## References

[1] S. A. Schelkunoff, *Electromagnetic Waves*. Princeton, N. J.: Van Nostrand, 1943, pp. 354–355. (Huygens source, cardioid patterns.)

[2] H. A. Wheeler, "The radiation resistance of an antenna in an infinite array or waveguide," *Proc. IRE*, vol. 36, pp. 478–488, April 1948.

[3] E. C. Jordan, *Electromagnetic Waves and Radiating Systems*. Englewood Cliffs, N. J.: Prentice-Hall, 1950, pp. 521–523. (Huygens source, cardioid pattern.)

[4] S. A. Schelkunoff and H. T. Friis, *Antennas—Theory and Practice*. New York: Wiley, 1952, pp. 531–523. (Huygens source, cardioid pattern.)

[5] S. J. Rabinowitz, "The conductance of a slot in an array antenna," M.I.T. Lincoln Lab., Lexington, Mass., Tech. Rept. 192, December 31, 1958. (Rectangular array of slots in metal sheet. Computation by summation of series. Relation between element pattern and variation of radiation conductance.)

[6] L. Stark, "Radiation impedance of a dipole in an infinite array," Hughes Aircraft Co., Fullerton, Calif., Tech. Rept. FL 60-230, May 1, 1960. (Impedance matrix, resistance given by first term of summation. Relationship between element pattern and variation of element resistance.) Submitted to *IEEE Trans. on Antennas and Propagation.*

[7] P. S. Carter, Jr., "Mutual impedance effects in large beam scanning arrays," *IRE Trans. on Antennas and Propagation*, vol. AP-8, pp. 276–285, May 1960. (Comparison between finite and infinite width of array.)

[8] S. Edelberg and A. A. Oliner, "Mutual coupling effects in large antenna arrays: Part I—Slot arrays," *IRE Trans. on Antennas and Propagation*, vol. AP-8, pp. 286–297, May 1960. (Waveguide viewpoint, starting with all modes and simplifying to a single mode.)

[9] J. D. Hanfling, "Mapping of the far field polarization of an-

tennas by the stereographic projection," M.S. thesis, Polytechnic Institute of Brooklyn, New York, N. Y., June 1960. (Crossed Huygens sources for unidirectional radiation with pure circular polarization, realization by electric dipole and current loop.)
[10] J. L. Allen, et al., "Phased array antenna studies, 1 July 1959 to 1 July 1960," M.I.T. Lincoln Lab., Lexington, Mass., Tech. Rept. 228, ASTIA Doc. 249, 470, August 1960.
[11] H. Jasik, "Fundamentals of antennas," in *Antenna Engineering Handbook*. New York: Wiley, 1961, ch. 2, pp. 7–10. (EM dipole cardioid pattern, realization by electric dipole and current loop.)
[12] J. L. Allen, et al., "Phased array radar studies, 1 July 1960 to 1 July 1961," M.I.T. Lincoln Lab., Lexington, Mass., Tech. Rept. 236, November 1961. (Computed variation of reflection with angle for several cases of dipoles with reflector.)
[13] ——, "Gain and impedance variation in scanned dipole arrays," *IRE Trans. on Antennas and Propagation*, vol. AP-10, pp. 566–572, September 1962. (Many references.)
[14] H. A. Wheeler, "Principles of phased arrays," Wheeler Labs., Great Neck, L. I., N. Y., P-1116, May 1963. (Notes for three seminars, diagrams, many references.)
[15] P. W. Hannan, D. S. Lerner, and G. H. Knittel, "Wide-angle impedance matching calculated for a phased array antenna," *1963 G-AP Symp. Digest*, pp. 228–233. (Achieved by interconnecting circuits behind the array.)
[16] J. L. Allen, "The theory of array antennas," M.I.T. Lincoln Lab., Lexington, Mass., Tech. Rept. 323, July 25, 1963. (Excellent survey, many references.)
[17] P. W. Hannan, P. J. Meier, and M. A. Balfour, "Simulation of phased array antenna impedance in waveguide," *IEEE Trans. on Antennas and Propagation* (*Communications*), vol. AP-11, pp. 715–716, November 1963.
[18] P. W. Hannan, "The element-gain paradox for a phased-array antenna," *IRE Trans. on Antennas and Propagation*, vol. AP-12, pp. 423–433, July 1964. (The ideal element pattern is stated on the basis of an element located in the array environment, which is not the same as the isolated element pattern used in the present paper, though the shape of the ideal pattern happens to be the same in both cases.)
[19] P. J. Meier, M. A. Balfour, and H. A. Wheeler, "Circular waveguide loaded with dielectric discs for increased usable bandwidth," *1964 G-MTT Internat'l Symp. Program and Digest*, pp. 33–36.
[20] P. W. Hannan and M. A. Balfour, "Phased-array element radiation determined by simulation in waveguide," *1964 G-AP Internat'l Symp. Program and Digest*, pp. 143–149.
[21] H. A. Wheeler, "Simple relations derived from a phased array made of an infinite current sheet," *1964 G-AP Internat'l Symp. Program and Digest*, pp. 157–160.
[22] P. W. Hannan, D. S. Lerner, and G. H. Knittel, "Impedance matching a phased-array antenna over wide scan angles by connecting circuits," *IEEE Trans. on Antennas and Propagation*, vol. AP-13, pp. 28–34, January 1965.
[23] P. W. Hannan and M. A. Balfour, "Simulation of a phased-array antenna in waveguide," *IEEE Trans. on Antennas and Propagation*, vol. AP-13, pp. 342–353, May 1965.
[24] M. A. Balfour, "Phased-array simulators in waveguide for a triangular arrangement of elements," *IEEE Trans. on Antennas and Propagation* (*Communications*), vol. AP-13, pp. 475–476, May 1965.

*Reprinted from* IEEE TRANSACTIONS
ON *ANTENNAS AND PROPAGATION*
Volume AP-14, Number 6, November, 1966
pp. 707-714




# The Grating-Lobe Series for the Impedance Variation in a Planar Phased-Array Antenna

HAROLD A. WHEELER, FELLOW, IEEE

***Abstract*—In an infinite planar array of elements with periodic spacing, the element active impedance varies with phasing for beam steering. This impedance variation may be expressed as the sum of a double Fourier series. This series is identified with the periodic grating-lobe pattern on the "sin $\theta$ plane" which is also the plane of two-dimensional phasing coordinates. An "impedance crater," with contours peculiar to the kind of element, is placed on every grating-lobe center. The inside of the central crater, which coincides with the unit circle of real space on this plane, determines the resistance variation with scan angle of the main lobe. The central crater and the skirts of the surrounding craters overlap in this circle; their sum determines the accompanying reactance variation. All craters together form the "grating-lobe series," which gives a picture of the entire impedance variation with scan angle. In a simple example, the reactance variation associated with half-wave spacing of the elements is found to be nearly equal to the resistance variation associated with the kind of element.**

## I. INTRODUCTION

A PLANAR broadside-array antenna of many radiating elements is used to develop a pencil beam. In the "phased array," the beam is steered in angle by phasing of the element excitation.

A design problem peculiar to the phased array is presented by the element-impedance variation with phasing for beam steering. This variation is caused by the intercoupling of the elements in the array [5], [8], [17]. While it can be computed for some cases from this viewpoint, there are other approaches that are more helpful in predicting the impedance variation.

This monograph is directed to a pictorial presentation of the element impedance variation with phasing. It is based on two previous presentations that are essentially the same. One is the modal series of Edelberg and Oliner [6]. The other is the double Fourier series of Stark [4], [18]. Both are made possible by the infinite-array concept introduced earlier by the author [1].

The pictorial presentation is to be described as a "grating-lobe series" of "impedance craters" deployed on the "sin $\theta$ plane" in terms of the phasing coordinates described by von Aulock [7]. In this form, the peculiarities of Stark's formula are displayed in a graphical form that is most helpful in understanding and predicting the impedance variation with phasing. It is to be introduced in terms of the simplest radiating element, which is a small dipole (electric or magnetic).

In the simple case of a single main lobe free of grating lobes in real space, the principal term gives all of the resistance variation, while all terms may contribute to the reactance variation. The former is determined by the element radiation pattern and the element spacing in the array. In a practical case, the reactance variation is determined mainly by the first few terms in the series; these are identified with nearby grating lobes in imaginary space.

This presentation is restricted to the kind of array and the kind of element for which the grating-lobe series is valid and most useful, outlined in Tables I and II. The theory is based on the infinite array and therefore gives an approximation for elements far from the edge in a finite array of many elements.

Referring to Table II, the theory presented here is based on the concept of a one-port element. Ideally, it is a very small element, so that only one mode of excitation is effective. It may be generalized to include an ele-

Manuscript received March 7, 1966. This work was presented at the 1965 IEEE International Antenna and Propagation Symposium, Washington, D. C.
The author is with Wheeler Laboratories, Great Neck, N. Y.

TABLE I

DESCRIPTION OF ARRAY

| |
|---|
| Planar array<br>Pencil beam. |
| Periodic array<br>Square, triangular, etc., grid.<br>Infinite array.<br>Like elements.<br>Equal magnitude of excitation. |
| One-port elements<br>Small. |
| Beam steering (scanning)<br>Progressive linear phasing. |
| Single lobe (main lobe, beam)<br>Extra lobes (grating lobes). |



TABLE II

PECULIARITIES OF ONE-PORT RADIATING ELEMENTS

| | Electric Dipole | Magnetic Dipole |
|---|---|---|
| Radiation in single mode: | $RI^2$ | $GV^2$ |
| Invisible if: | $I=0$ (*OC*) | $V=0$ (*SC*) |
| Conductor form: | wire | slot |
| Small form: | short wire | short slot |
| Half-wave form: | thin wire | narrow slot |
| Resonance by: | length or end-loading or circuit | |

ment of any number of ports with coherent excitation. This means that the relative excitation of all ports is forced to be the same, invariant with phasing. This restriction is implicit in most of the studies of phased arrays. It is approximately satisfied in typical practical elements of moderate size. A small element must be resonated for impedance matching with a line. A somewhat larger element may experience coherency of excitation by virtue of resonance, as in the case of a half-wave dipole.

It is noted in passing that a magnetic dipole may be a loop of wire. Also, a short slot may be a square or circular hole in a conducting sheet.

The element impedance, for present purposes, is the active impedance, from the sending viewpoint. It is the apparent impedance of any element in the environment of an infinite array. All other elements are excited with the same amplitude, and with progressive linear phasing for any angle of beam steering.

The real part of the element impedance is the radiation resistance representing the radiated power. The imaginary part includes a constant term representing the stored energy in the near field; this term may be the largest component of the element impedance, but it can be tuned out for one frequency. The element impedance includes the total mutual impedance with all other elements, whose effect is subject to variation with the phasing. It happens that the actual value of the resistance is significant, but only the variation of the reactance. The term "radiation impedance," as used herein, includes the entire resistance and the entire variation of reactance with phasing; it may or may not include any part of the broadside reactance to be taken as a reference.

After reviewing the phasing coordinates, the impedance crater will be described. Then the grating-lobe series will be presented on these coordinates, followed by a simple example having comparable variation of resistance and reactance. Finally, there will be a discussion of the background and utility of the grating-lobe series.

## II. SYMBOLS

MKS units (meters, ohms); radians.

---

$x+jy$ = array coordinates

$w=\sin\theta$ = radial coordinate on $\sin\theta$ plane

$w\exp j\phi=u+jv$ = phasing coordinates on $\sin\theta$ plane

$\phi$ = angle of scan plane, perpendicular to array plane, from $x$ axis toward $y$ axis; or from $u$ axis toward $v$ axis

$\theta$ = scan angle from broadside, in scan plane

$\lambda$ = free-space wavelength at operating frequency

$a, b$ = spacing in $x, y$ directions, between consecutive rows in array

$\alpha, \beta$ = phase angle in $x, y$ directions, between consecutive rows in array

$\alpha_1, \beta_1$ = phase angle for scan to grazing incidence

$u_1, v_1$ = spacing of rows of grating lobes

$m, n$ = serial numbers in grating-lobe pattern on phasing coordinates.

---

$Z=R+jX$ = active radiation impedance of element in array, with all elements excited

$R_0$ = active radiation resistance associated with any one lobe at broadside; the reference for the impedance crater

$Z_c$ = active radiation impedance associated with any one lobe at any scan angle: the impedance crater

$\Delta R, \Delta X$ = change of $R$ or $X$ by scan from broadside

$Y=G+jB$ = active radiation admittance of element in array, with all elements excited

$I$ = current in $Z$

$V$ = voltage on $Y$

$F(\phi, \theta)$ = radiation pattern factor of isolated element; ratio of electric or magnetic field intensity relative to its value in broadside directon

$\rho$ = reflection coefficient at array element, matched at broadside

$S$ = voltage standing-wave ratio (VSWR > 1) associated with reflection coefficient

$\rho, \rho'$ = reflection coefficient for current sheet or spaced array

$S, S'$ = VSWR for current sheet or spaced array

$E, H$ = electric or magnetic-field intensity

$C, IC$ = cardinal or intercardinal plane of symmetry, perpendicular to plane of array ($C$ in lines of least spacing of elements).

## III. The Array Phasing Coordinates

A particular set of coordinates is chosen for present purposes. It is based on the "sin $\theta$ plane" which has been found useful in presenting the hemispherical directive patterns of antennas, especially arrays [2], [7]. In hemispherical beam scanning, it has been found to represent also the array phasing in two dimensions, hence the designation, "array phasing coordinates."

Figure 1 shows these coordinates in relation to scan angle and phasing. The beam steering is described by the angle ($\phi$) of the scan plane, from one coordinate axis ($u$) toward the other ($v$), and the angle ($\theta$) of the beam axis from broadside. The unit circle is the orthographic projection of the scan hemisphere onto the coordinate plane. The inside of the circle represents "real space" and the outside "imaginary space."

The same coordinates represent the two-dimensional phasing ($\alpha$ or $\beta$) relative to their values ($\alpha_1$ or $\beta_1$) that would give deflection to the "horizon" in the direction of either axis. The phasing has reality over the entire plane, but the main beam is developed only if the phasing is within the unit circle.

The mathematical relations of these coordinates are summarized as follows.

$$w \exp j\phi = u + jv \tag{1}$$

$$w = \sqrt{\mu^2 + v^2} = \sin\theta \tag{2}$$

$$u = \alpha/\alpha_1 = w\cos\phi = \sin\theta\cos\phi \tag{3}$$

$$v = \beta/\beta_1 = w\sin\phi = \sin\theta\sin\phi. \tag{4}$$

The "$2\pi$ ambiguity" in the phase angle along either axis causes a doubly periodic "grating-lobe" pattern, which will receive attention further on.

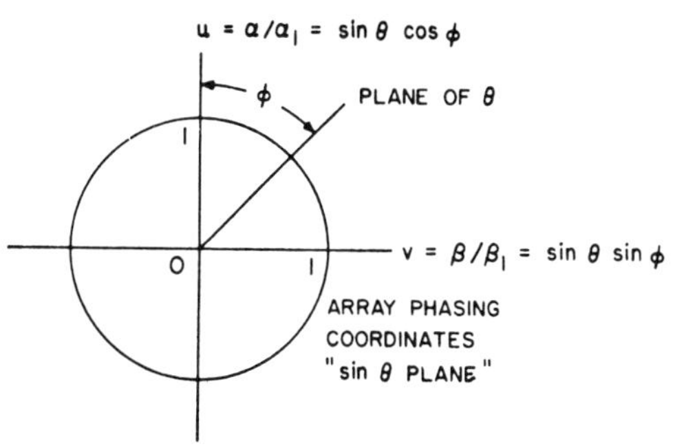


Fig. 1. The array phasing coordinates on the sin $\theta$ plane.

## IV. The Impedance Crater

The element of the grating-lobe series is the "impedance crater." This concept is expressed in terms of the ratio of the element radiation impedance ($Z$) over its radiation resistance at broadside ($R_0$). This ratio embodies the properties of a one-port radiating element in the limit of close spacing for all conditions of phasing, but without the effects of its spacing in the array. The resulting separation of these two effects is the basic feature of the grating-lobe series.

The concept of the impedance crater will be introduced with reference to the limiting case of a small dipole as the radiating element. The impedance crater represents the radiation impedance associated with the power radiated in the main beam, as distinguished from any other lobes (grating lobes). In the limit of close spacing, there are no other lobes, so this gives the entire impedance variation with scanning. Therefore, the elementary impedance crater may be based on such a model.

This model is the infinite current sheet of small dipoles with close spacing. The author previously presented the varying radiation resistance associated with beam scanning in such a model [9], [14]. Now it appears that the same formulas give also the varying reactance with phasing beyond the limits of beam scanning. Both will be required in the grating-lobe series.

Figure 2 shows, diagrammatically, the impedance variation with scanning in either of the principal planes. 61
For the case of electric dipoles, beam deflection from broadside in the $H$ plane (taken on the $u$ axis) increases the resistance. Beam deflection in the $E$ plane ($v$ axis) decreases the resistance. These cases are shown in Figs. 2(a) and 2(b).

The varying radiation reactance is mathematically a continuation of the radiation resistance. The following formulas give the entire radiation impedance ($Z$), comprising the resistance and the varying component of reactance, in the principal planes of scanning.

$$H \text{ plane } (\phi = 0): \quad \frac{Z}{R_0} = \frac{1}{\cos\theta} = \frac{1}{\sqrt{1 - u^2}} \tag{5}$$

$$E \text{ plane } (\phi = \tfrac{1}{2}\pi): \quad \frac{Z}{R_0} = \cos\theta = \sqrt{1 - v^2}. \tag{6}$$

These are expressed relative to the resistance at broadside ($R_0$).

It is seen that there is no variation of reactance in real space. Outside of real space, there is no resistance and the formulas give the varying component of reactance. However, the sign of the reactance is indeterminate. It has been determined separately by computation of the impedance of the current sheet as an infinite sum (integral). The reactance is found to be, respectively, positive and negative, as indicated in Figs. 2(a) and (b).

For angles in all scan planes, the radiation impedance may be expressed in various forms. They include the directional radiaton pattern ($F$) of a small dipole, which is the ratio of electric or magnetic field intensity over its maximum value (at broadside).

$$[F(\phi, \theta)]^2 = 1 - (\sin\phi\sin\theta)^2 = (\cos\phi)^2 + (\sin\phi\cos\theta)^2 \tag{7}$$

$$\frac{Z}{R_0} = \frac{1}{\cos\theta}[F(\phi, \theta)]^2 = \frac{1}{\cos\theta}(\cos\phi)^2 + \cos\theta(\sin\phi)^2 = \frac{1 - v^2}{\sqrt{1 - u^2 - v^2}}. \tag{8}$$

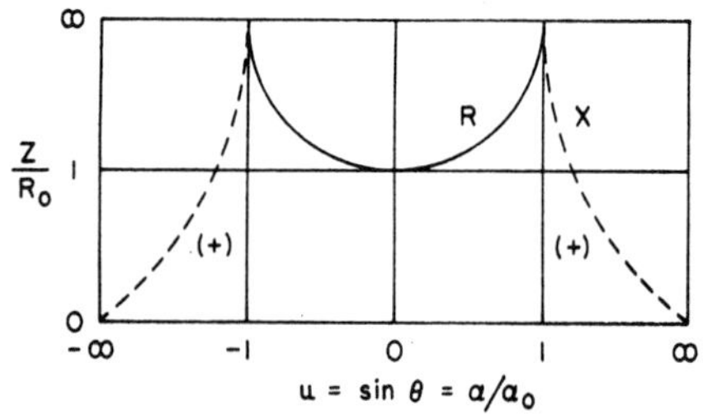


(a) $H$ plane.

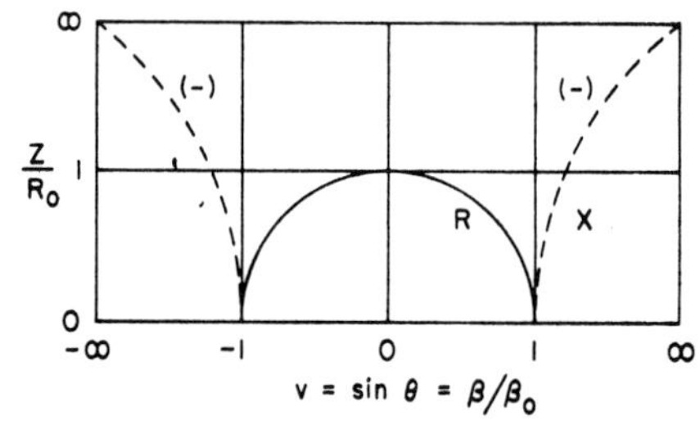


(b) $E$ plane.

Fig. 2. The impedance variation with phasing, based on the current sheet of electric dipoles.

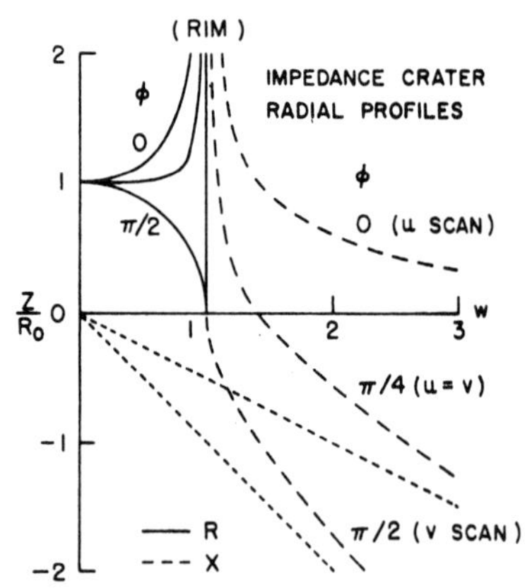


Fig. 3. The impedance crater for a small electric dipole, described by radial profiles.

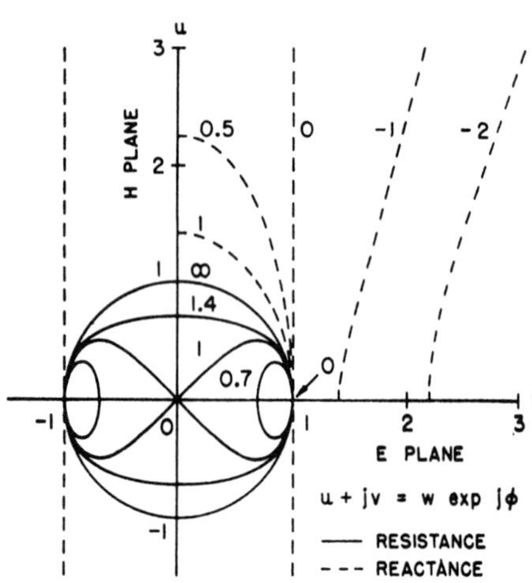


Fig. 4. The impedance crater for a small electric dipole, described by a contour map.

This formula has been verified by the infinite sum of mutual impedance between lines of elements in the limiting case where it becomes a Fourier integral.

It is seen that the radiation impedance is pure resistance within the unit circle ($u^2+v^2=1$) which includes real space, and is pure reactance outside of this circle. The variation in certain scan planes is graphed as radial profiles in Fig. 3. (The straight dotted lines are the asymptotes of the curved dashed lines.) The variation over all conditions of phasing is presented as a contour map in Fig. 4. These graphical forms suggest the topography of a crater, hence the term "impedance crater." There is a high rim of varying thickness, which disappears in the direction of the dipole axis, where there is no radiation.

In general, the impedance crater is determined by the form of the radiating element in the array. The one shown is the simplest and most fundamental. By duality, it also represents the small magnetic dipole, which can be approximately realized by a small slot or circular hole in a metal sheet. For the more complicated case of half-wave radiators, a contour map of an impedance crater has previously been presented [10]. There are available a fascinating variety of impedance craters related to some ordinary and extraordinary element configurations.

## V. The Grating-Lobe Series

In a periodic array, the phasing may cause extra lobes, termed "grating lobes" in recognition of the same phenomenon observed with optical gratings. The relation between the main lobe and the grating lobes depends on the element spacing relative to the wavelength. If any pattern of repeating rows of elements has a spacing exceeding one half-wavelength, one or more grating lobes may be radiated with some conditions of the scanning of the main lobe over the hemisphere. Otherwise, the grating lobes are radiated only by phasing beyond the hemispherical scanning of the main lobe. On the phasing coordinates, the grating lobes are brought closer by increasing the element spacing. They are avoided in the current sheet, because the spacing approaches zero.

There are two common periodic arrangements of elements in a planar array. Figure 5(a) shows the square array and Fig. 5(b) the triangular array. In each, the element spacing is identified (a) and is given a nominal value (relative to the free-space wavelength, λ) by way of example. There are indicated also the array coordinates ($x$, $y$), and some of the planes of symmetry ($C$ = cardinal, $IC$ = intercardinal).

Beside each array in Fig. 5 is shown its pattern of grating lobes on the phasing coordinates ($u$, $v$). Each grating lobe is represented by a unit circle enclosing the phasing range over which this lobe would exist in real space. The lobe circles are contiguous by virtue of the choice of spacing in each array; therefore, the main lobe can be scanned over a complete hemisphere, free of grating lobes, but the extremes of scan are approaching the condition of real grating lobes. The extra lobes appear in imaginary space, meaning that they would radiate only if the phasing were to be extended so far as to destroy the main lobe.

The "grating-lobe series" is formed by placing on every circle of the grating-lobe pattern, an impedance crater like the one peculiar to the element used in the array. The element radiation impedance is then equal to the sum of the contributions of all craters at any one point on the plane. The region of interest is usually within the central circle, with phasing limited to scanning over some part of the hemisphere. In this region, any point represents a direction of the main beam, and the sum gives the radiation impedance for this direction. In particular, it gives the entire variation of element impedance with scan angle.

The grating-lobe series of impedance craters is diagramed in Fig. 6, for the simplest case of a square array ($a = b$). It is a doubly infinite series, but the inner ring of craters contributes the greatest variation of reactance in the center circle. Also, the symmetry makes the four quadrants alike, simplifying the summation.

The formulation of the grating-lobe series is made in terms of impedance variation from broadside, in order to express the interesting phenomenon with the most rapid convergence.

The impedance crater:

$$\frac{Z_c(u, v)}{R_0}$$

The sum (element radiation impedance):

$$\frac{Z(u, v)}{R_0}$$

The variation from broadside:

$$\frac{Z(u, v)}{R_0} - \frac{Z(0, 0)}{R_0} = \sum_{-\infty \atop m}^{\infty} \sum_{-\infty \atop n}^{\infty} \left[ \frac{Z_c(-mu_1 + u, -nv_1 + v)}{R_0} - \frac{Z_c(-mu_1, -nv_1)}{R_0} \right]. \quad (9)$$

This series is essentially the double Fourier series of Stark [4], [18] and the modal series of Edelberg and Oliner [6]. Its understanding is aided by the pictorial presentation of the impedance crater and its deployment over the periodic grating-lobe pattern on the phasing coordinates of von Aulock [7].

While the nearer terms of the grating-lobe series contribute all of the resistance and the principal variation of reactance, the rate of convergence is limited by the increasing number of terms further out. If the element is an infinitesimal dipole (or quadripole), the series is not absolutely convergent in respect to reactance variation near the origin (in real space). However, it can be summed by separately evaluating a selected part that is conditionally convergent.

The convergence becomes absolute and rather rapid for an element of finite size but still small enough to be regarded as a one-port radiator to a close approximation. (Typical sizes are less than one half-wavelength in the greatest dimension.) A certain mathematical model has been found particularly helpful in this respect. It is an element having an aperture with a two-dimensional excitation tapered in accordance with the probability function. The rate of taper is specified to match the measured free-space pattern of the actual element. With this form of taper, the impedance crater has skirts which level off rapidly beyond the rim, so the series [9] converges very rapidly.

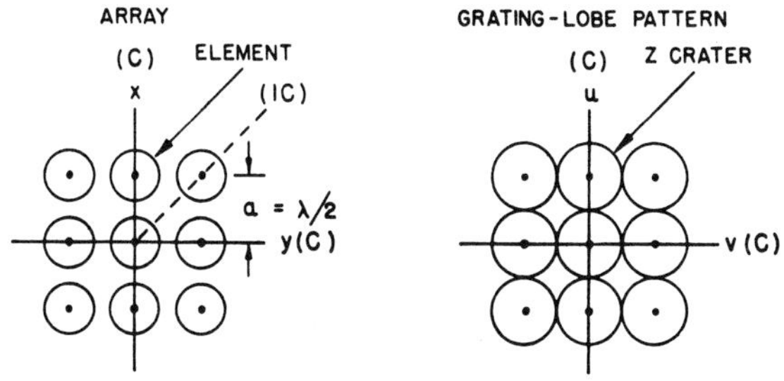


(a) Square array and its grating-lobe pattern.

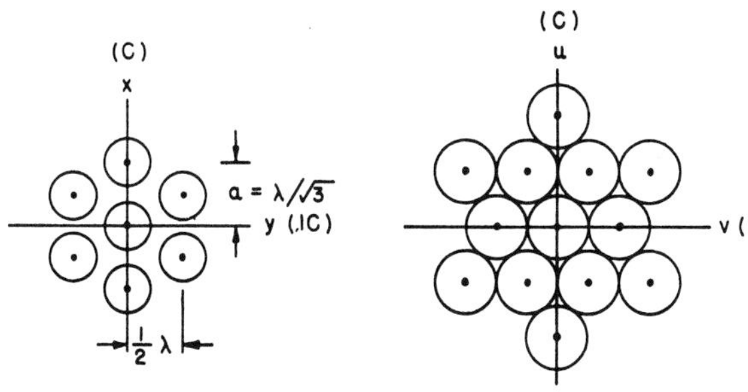


(b) Triangular array and its grating-lobe pattern.

Fig. 5. Periodic arrays and their grating-lobe patterns.

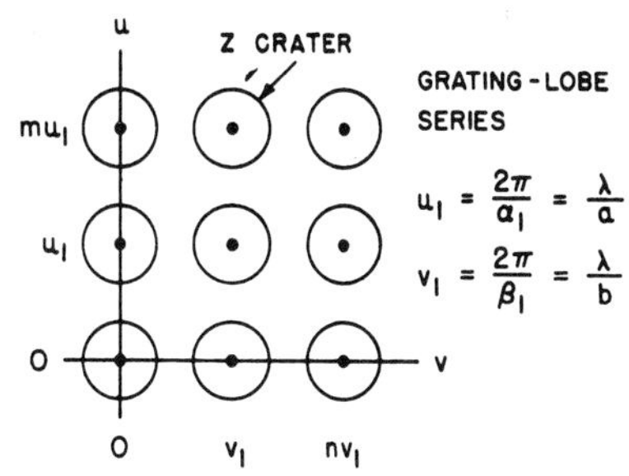


Fig. 6. The grating-lobe series presented on phasing coordinates on the sin $\theta$ plane.

If the element spacing and the phasing are within the limit for radiating only the main lobe, the center crater gives the entire resistance. The center crater may contribute also to the reactance variation, while the remaining craters contribute only reactance variation. The greatest variation of reactance is usually contributed by the nonradiating grating lobes nearest to the main lobe.

The most irregular variation of performance occurs for the condition of an "incipient grating lobe," meaning one which comes into real space (by rising over the horizon) as the main lobe is scanned beyond a certain angle from broadside. There may be an extreme change of resistance and reactance, and the incipient grating lobe may take all of the power away from the main lobe. Such effects are readily predicted from the presentation of the grating-lobe series of impedance craters. (The extreme effect is peculiar to the infinite array; a lesser effect would be experienced in a finite array.)



The theoretical basis for the simplest form of the grating-lobe series is found in two principles, both of which are implicit in the derivation by Stark [4], [18].

One principle is the relation between the radiated power and the radiation resistance, the former being readily computable on the assumption of a plane wave from an infinite array (corresponding to a narrow beam from a large broadside array). This gives the variation of resistance for all conditions of phasing, including the range of phasing that gives no main lobe but perhaps grating lobes.

The other principle is the relation between the real and imaginary parts of a complex variable which is an analytic function of a real variable. It devolves that the given variation of resistance over a limited range of phasing (main lobe in real space) with zero resistance outside of this range (main lobe in imaginary space) is accompanied by some variation of reactance, given by the same functional formula, over the entire range of phasing.

These two principles are seen in the grating-lobe series of impedance craters. If the radiating elements are contained in a plane (having zero "depth"), it is found that the impedance variation with phasing, for any one lobe in its circle of real space, is only the resistance variation. The associated reactance variation is outside of this circle. On the other hand, if the array has some depth, there may be also reactance variation within this circle. This is true of Stark's example of dipole elements spaced from a reflector. In general, it is necessary to know only the resistance variation within the circle in order to predict the reactance variation caused by the same term of the series.

## VI. Example

The square array of Fig. 5(a) is taken as a very simple example of the relations to be obtained from the grating-lobe series. It is a special case which leads to simple relations because, for any phasing along either axis ($u$ or $v$), there is always just one lobe which is radiating and therefore contributing to the resistance. The elements are taken to be small electric dipoles, as in the current sheet, but they are in a square array with spacing of one half-wavelength. In this case, the reactance variation is found to be approximately equal to the magnitude of the resistance variation. This equality appears approximately from the grating-lobe series, and has been verified to be true in the first approximation by the mutual-impedance summation. It is substantiated by some examples computed by Carter [5].

With scanning from broadside in the $H$ plane (phasing along the $u$ axis) the resistance increases. Figure 7(a) shows this profile of the center crater and of the pair of adjacent craters. Figure 7(b) shows the sum of these craters, which determine the resistance and most of the reactance variation in the scan hemisphere.

With scanning from broadside in the $E$ plane (phasing along the $v$ axis) the resistance decreases. Figure 7(c) shows the sum for this case.

These two cases individually have very different properties, but happen to give the same amount of reflection if matched with the generator resistance at broadside. This result is diagramed in Fig. 8.

Starting with the current sheet of electric dipoles, the voltage standing-wave ratio (VSWR$\equiv S$) and the reflection coefficient ($\rho$) have been derived for the principal planes [9], [14]. In this case, there is no reactance.

$H$ plane ($u$ scan, $\phi=0$):

$$\rho = (\tan \tfrac{1}{2}\theta)^2 \tag{10}$$

$E$ plane ($v$ scan, $\phi=\tfrac{1}{2}\pi$):

$$\rho = -(\tan \tfrac{1}{2}\theta)^2 \tag{11}$$

Either plane:

$$S = 1/\cos\theta. \tag{12}$$

Going to the spaced array of Fig. 5(a), the corresponding ratios are formulated on the assumption of reactance variation as noted under Figs. 7(b) and 7(c).

$H$ plane ($u$ scan, $\phi=0$):

$$\rho' = \rho\frac{(1+\rho)+j(1-\rho)}{1+\rho^2} \tag{13}$$

$E$ plane ($v$ scan, $\phi=\tfrac{1}{2}\pi$):

$$\rho' = \rho\frac{(1+\rho)-j(1-\rho)}{1+\rho^2} \tag{14}$$

Either plane:

$$|\rho'| = |\rho|\sqrt{\frac{2}{1+\rho^2}} \tag{15}$$

$$1-|\rho'| = (1-|\rho|)\frac{1+|\rho|}{1+\rho^2+|\rho|\sqrt{2(1+\rho^2)}} \tag{16}$$

$$S' = \frac{1}{2S}[\sqrt{S^2+1}+(S-1)]^2. \tag{17}$$

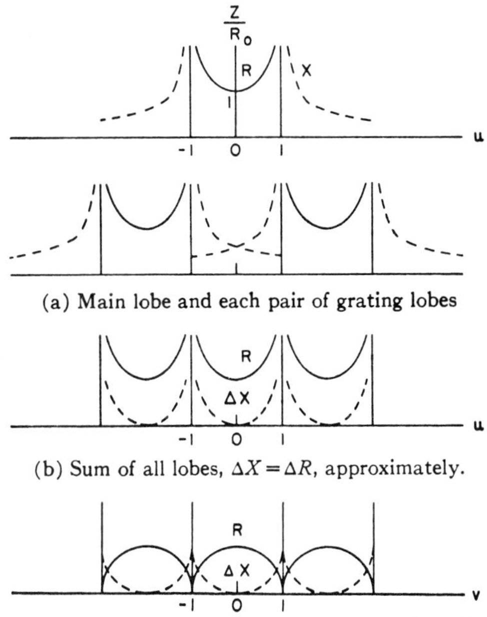


(a) Main lobe and each pair of grating lobes

(b) Sum of all lobes, $\Delta X = \Delta R$, approximately.

(c) Sum for other axis, $\Delta X = -\Delta R$, approximately.

Fig. 7. An example of the grating-lobe series, showing the sum for each of the two axes of phasing.

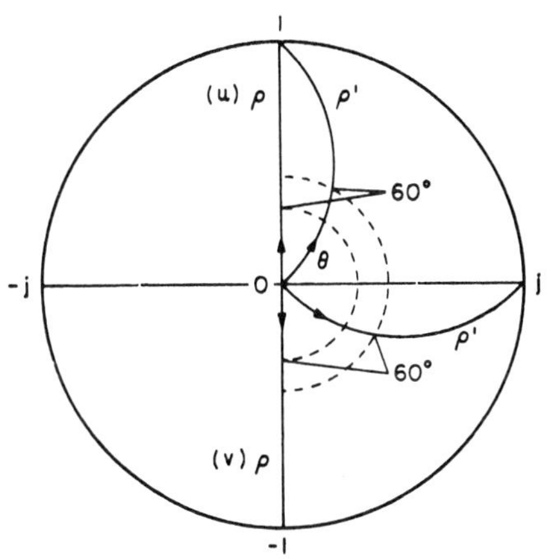


Fig. 8. The reflection-coefficient variation with phasing. ($\rho$ for current sheet, $\rho'$ for spaced array.)

The effect of the reactance variation is seen to range between the two extremes as follows.

Small reflection:

$$|\rho'| = \sqrt{2}\,|\rho| \ll 1 \tag{18}$$

$$(S' - 1) = \sqrt{2}(S - 1) \ll 1 \tag{19}$$

Large reflection:

$$(1 - |\rho'|) = \tfrac{1}{2}(1 - |\rho|) \ll 1 \tag{20}$$

$$S' = 2S \gg 1. \tag{21}$$

In Fig. 5, the points corresponding to a certain angle from broadside are marked for comparison.

$$\theta = \tfrac{1}{3}\pi = 60°: \quad |\rho| = 1/3 = 0.33 \tag{22}$$

$$|\rho'| = (1/3)\sqrt{3/2} = \sqrt{1/6} = 0.41. \tag{23}$$

It is seen in the graph that the array spacing causes the angle of reflection to change by much different amounts in the two scan planes (at 60°, for example, antitan 1/2 and antitan 2).

This exercise gives some idea of the increase of reflection caused by the usual amount of spacing in an array.

## VII. Discussion

The grating-lobe series of impedance craters gives a functional picture of the two principal factors determining the impedance variation with phasing for beam steering. The impedance crater gives the effect of element design, independent of array spacing. The grating-lobe series determines the further effect of element spacing in the array. If only the main lobe is radiated, the former entirely determines the resistance and the latter contributes only to the reactance. The validity of this approach is limited to one-port radiators, which are approximately realized by practical elements of moderate size.



This presentation is an evolutionary climax of the contributions of various workers over the past two decades. Shortly after the author's introduction [1] to the infinite array and the equivalent waveguide modes, there was a period of intensive activity sponsored by the M.I.T. Lincoln Laboratory. Some of this work carried forward the infinite-array concepts with their ultimate simplicity. While Allen's group [8] was working on all problems of phased arrays, the work of Stark [4], [18], Edelberg and Oliner [6], Carter [5], and Rabinowitz [3] provided major contributions. In the meantime, from Bell Telephone Laboratories, von Aulock [7] disclosed the phasing coordinates for presentation of grating lobes. By the time the author had arrived at the concept of the grating-lobe series of impedance craters, Rhodes [10] had independently developed the concept of an impedance crater for a theoretical application other than the array. Also, the author soon came to appreciate the fact that the grating-lobe series is essentially a graphical presentation of the double Fourier series of Stark or the waveguide-mode series of Edelberg and Oliner, obtained by deploying the impedance crater of Rhodes on the phasing coordinates of von Aulock.

This brief presentation gives only the essential concepts and a simple example of the opportunities opened up by the grating-lobe series. In the author's organization, it is already being relied on for interpolation and extrapolation from the few scan conditions that can be measured for the infinite array (by simulation in waveguide) [12], [13]. This experience adds confidence to the expectation that the series is valid as a close approximation for some practical elements in arrays. There remains to be presented much more of the studies already made, relating to impedance craters of various elements, mathematical approaches for expediting convergence, influences on practical design, and implications relative to the behavior of finite arrays of many elements.

The usefulness of the grating-lobe series resides in its rapid convergence. This results from the manner of the separation of variables and the choice of coordinates on which to deploy the terms of the series. For an appreciation of the influence of design parameters, only the center crater and the inner ring of adjacent craters are required. The radiation resistance was easily formulated, but only now has it become possible to picture the accompanying variation of reactance. The latter is found to substantially increase the impedance variation with scanning, thereby contributing further to the problems of design. In the future, this higher level of appreciation is likely to influence the design compromise between element spacing and the required number of elements.



## Acknowledgment

The development of the present concepts has benefited very much from the studies of the other workers mentioned and listed in the References. Furthermore, the author is grateful to his associates for many discussions and constructive criticism which have been helpful in arriving at a useful picture. Special mention is due to R. F. Frazita and G. G. Charlton who, working under the direction of G. H. Knittel, have made much use of the grating-lobe series. The immediate occasion for this study has been the work for Bell Telephone Laboratories under guidance of R. L. Mattingly and P. H. Smith, whose encouragement is much appreciated.[1]

## References

[1] H. A. Wheeler, "The radiation resistance of an antenna in an infinite array or waveguide," *Proc. IRE*, vol. 36, pp. 478–487, April 1948.

[2] D. Richman, "Theory of radiation from arrays and apertures," Hazeltine Research Corp., Great Neck, N. Y., Rept. 721-PW-596 (or 7666), May 24, 1949. (Two-dimensional "sin $\theta$" space coordinates.)

[3] S. J. Rabinowitz, "The conductance of a slot in an array antenna," M.I.T. Lincoln Lab., Lexington, Tech. Rept. 192, December 31, 1958. (Variation with the scan angle; dependence on the radiation pattern of an isolated element.)

[4] L. Stark, "Radiation impedance of a dipole in an infinite array," Hughes Aircraft Co., Fullerton, Calif., Rept. FL 60-230; May 1, 1960. (Double Fourier series for resistance and reactance.)

[5] P. S. Carter, Jr., "Mutual impedance effects in large beam scanning arrays," *IRE Trans. on Antennas and Propagation*, vol. AP-8, pp. 276–285, May 1960. (Variation of resistance and reactance in a planar array.)

[6] S. Edelberg and A. A. Oliner, "Mutual coupling effects in large antenna arrays: Part I—Slot arrays," *IRE Trans. on Antennas and Propagation*, vol. AP-8, pp. 286–297, May 1960. (Waveguide viewpoint, all modes.)

[7] W. H. von Aulock, "Properties of phased arrays," *Proc. IRE*, vol. 48, pp. 1715–1727, October 1960. (Phasing coordinates for two-dimensional periodic grating-lobe pattern of planar array.)

[8] J. L. Allen, "Gain and impedance variation in scanned dipole arrays," *IRE Trans. on Antennas and Propagation*, vol. AP-10, pp. 566–572, September 1962. (Many references.)

[9] H. A. Wheeler, "Simple relations derived from a phased array made of an infinite current sheet," *1964 IEEE PTGAP Internat'l. Symp., Program and Digest*, pp. 157–160, September 1964.

[10] D. R. Rhodes, "On a fundamental principle in the theory of planar antennas," *Proc. IEEE*, vol. 52, pp. 1013–1021, September 1964. (An impedance crater on a contour map.)

[11] H. A. Wheeler, "The grating-lobe series for the impedance variation in a planar phased-array antenna," Wheeler Labs., Great Neck, N. Y., Rept. 1268, December 3, 1964. (Impedance craters on the phasing coordinates.)

[12] P. W. Hannan and M. A. Balfour, "A phased array antenna simulated in waveguide," *IEEE Trans. on Antennas and Propagation*, vol. AP-13, pp. 342–353, May 1965.

[13] M. A. Balfour, "Phased array simulators in waveguide for a triangular arrangement of elements," *IEEE Trans. on Antennas and Propagation*, vol. AP-13, pp. 475–476, May 1965.

[14] H. A. Wheeler, "Simple relations derived from a phased-array antenna made of an infinite current sheet," *IEEE Trans. on Antennas and Propagation*, vol. AP-13, pp. 506–514, July 1965.

[15] ——, "The grating-lobe series for the impedance variation in a planar phased-array antenna," *1965 IEEE Internat'l. Antenna and Propagation Symp., Program and Digest*, pp. 150–156, August 1965.

[16] ——, "The grating-lobe series for the impedance variation in a planar phased-array antenna," *IEEE Trans. on Antennas and Propagation*, vol. AP-13, pp. 825–827, September 1965.

[17] P. W. Hannan, "The ultimate decay of mutual coupling in a planar array antenna," Wheeler Labs., Great Neck N. Y., Rept. 1321, September 1965. (Asymptotic exponent, based on current-sheet array.) See also, *IEEE Trans. on Antennas and Propagation*, vol. AP-14, pp. 246–248, March 1966.

[18] L. Stark, "Radiation impedance of a dipole in an infinite planar phased array," *Radio Science*, vol. 70D, pp. 361–377, March 1966. (Computed profiles of resistance and reactance by sum of series.)

[19] L. Parad, "The real and reactive power of a planar array," *IEEE Trans. on Antennas and Propagation*, vol. AP-13, pp. 990–992, November 1965.

[1] That work was supported by the Western Electric Company and the Army Ordnance Corps under prime Contract DA-30-069-ORD-1955.

# MUTUAL COUPLING ANALYSIS II

# A Generalized Approach to the Analysis of Infinite Planar Array Antennas

BLISS L. DIAMOND, MEMBER, IEEE

*Abstract*—In this paper a generalized expression for the complex power radiated by an element in an infinite planar array antenna is derived. Since this power formula applies to a large class of phased array antennas where the aperture field distribution can be completely specified (in normal mode form), it proves to be a powerful, unifying principle. The utility of this approach is illustrated by the simplicity with which previously known results can be derived; e.g., an infinite array of slots in a ground plane and an infinite array of flat dipoles with or without a ground plane. Further demonstrations of the usefulness of the power formula are provided by the systematic and straightforward solutions of the less-well-known problems of infinite arrays of crossed-dipole pairs and infinite arrays of open-ended rectangular waveguides. The waveguide array solution is particularly interesting because it reduces to a set of equations which are identical to those one would use to characterize an $N$-port network on an admittance basis ($N$ is the number of waveguide modes). Since the power formula is derived for a parallelogram element lattice, the resultant solution for a specific type of element is in its most general form.

Expressions for the scan-dependent, dominant mode radiation admittance and the element gain function for a multimode rectangular waveguide radiator are also derived. In addition, various aspects of the waveguide array solution are investigated in the light of previous studies of infinite arrays.

## I. Introduction

During the past several years, much work has been done on the characterization of the behavior of an element in an infinite planar array of identical elements. More specifically, a pioneering piece of work was done by Wheeler [1] in 1948 when he derived the radiation resistance of a dipole in an infinite array of dipoles. Edelberg and Oliner [2] later derived an expression for the radiation admittance of a slot in an infinite array of slots. Almost concurrently, Stark [3] derived an expression for the total radiation impedance of a dipole in an infinite array of dipoles. Galindo and Wu [4], [5] later presented some results for restricting scanning ($H$-plane and quasi-$E$-plane) of infinite waveguide arrays with thin walls on one or both sides of the waveguides. Most recently, Farrell and Kuhn [6], [7] have presented a solution for infinite arrays of rectangular waveguides for arbitrary scan angles and without restrictions on waveguide wall thickness.

The purpose of this paper is to present an alternate and powerful derivation which is fundamental to a large class of planar array antennas where the electric or magnetic field can be completely specified over the entire aperture plane. It serves to unify all of the preceding derivations; that is, all of the special cases mentioned above are easily obtained from the formalism to be presented. Furthermore, more complex types of array antennas can be simply treated (conceptually, but not necessarily mathematically) by direct application of this method.

The entire analysis to be presented relies very heavily on a fundamental principle of planar antennas first presented by Woodward and Lawson [8] and later extended by Rhodes [9].[1] This principle in essence states that the aperture field distribution and the antenna pattern function over all angles (real and complex) are obtainable from each other. With the aid of this principle, it is then possible to obtain a very general expression for the complex power radiated by an element in an infinite planar array antenna without placing any restrictions on the character of the radiating elements except the usual infinite array assumptions (identical elements, uniform illumination, uniform progressive phasing in two orthogonal directions across the array). This power expression automatically accounts for the periodicity of the array element lattice. Specification of a particular element amounts to simply applying appropriate constraints on the power expression. Thus, the power expression acts as a unifying principle. Furthermore, it separates the problem into two parts: 1) the half-space into which the array radiates; 2) the half-space which contains the elements and feed network. The first problem needs to be solved just once, which we do in Section II. The second problem is very simple for the types of elements previously considered, as will be demonstrated in Section III. This natural separation of the problem also facilitates the physical interpretation of the final results because parameters characteristic to the radiating elements are separable from those pertaining to the half-space above the array.

In Section II, the general expression for the complex power radiated by an element in a planar array antenna is developed.[2] Section III then examines the application of the power formula to specific cases: slot arrays, dipole arrays, and arrays of rectangular waveguides. The dipole and slot cases are only briefly treated because their solutions are well known. The waveguide array case is treated quite extensively because the solution is less well known and because the present formalism permits a different viewpoint of the solution. An expression for the element gain function for multimode elements is derived in Section IV. A discussion of the waveguide array solution and its interpretation is given in Section V. In addition, an attempt has been made to explain the manner in which higher-order modes are excited

Manuscript received April 1, 1968.

The author is with the M.I.T. Lincoln Laboratory, Lexington, Mass. (Operated with support from the U. S. Advanced Research Projects Agency.)

[1] A brief treatment of the same problem was actually presented by Borgiotti [20] prior to the appearance of Rhodes' paper.

[2] It should be noted that sinusoidal steady-state excitation is assumed throughout this paper.

with array scanning and the role of these modes on the behavior of an element in the array as the array is phased to to scan the main beam.

## II. Complex Power Radiated by an Element in an Infinite Array

It is well known that the rectangular components of the electromagnetic fields on one side of a planar aperture can be represented simply and exactly by a linear combination of plane waves [10]. The angular spectrum of these plane waves taken over all real angles is just the steady-state radiation pattern of the aperture. Furthermore, the pattern function over all angles (real and complex) and the aperture function are obtainable from each other; they are a Fourier transform pair. An additional property of this representation is that the total steady-state complex power of the aperture (radiated power plus reactive power stored about the aperture) can be determined completely from the pattern function [9]. It is this last property which we exploit in the following derivation.



For a finite aperture, the pattern function is not a particularly useful tool for the determination of the complex power of the aperture because the resulting integrals cannot be easily handled. On the other hand, the pattern function of a large array is concentrated in the vicinity of the grating lobe angles (both real and complex angles). As the number of elements approaches infinity, the pattern function approaches a weighted, period "nest" of impulse functions in direction cosine space. Consequently, the integral of the pattern function reduces to a summation which can be readily evaluated (numerically on a digital computer) to determine the real and reactive power associated with an element in the array.

The following derivation of the power expression begins with a finite array and closely parallels the derivation presented by Rhodes [9] for a planar antenna. The infinite array result is then obtained by allowing the array size and number of elements to become infinite while the element lattice is held fixed. In a sense, we are rederiving Floquet's theorem; however, this procedure sheds some light on the interpretation of the final result and indicates roughly how well the performance of a large (but finite) array can be predicted with the infinite array results.

Consider an array of holes in an infinite, perfectly conducting ground plane, as indicated in Fig. 1. Let the aperture be located in the plane $z=0$ and let the fields in the free-space region ($z>0$) be expressed as an infinite set of plane waves propagating (or attenuating) in the $+z$ direction. The electric field in free space for sinusoidal steady-state excitation at frequency $\omega$ can therefore be expressed as

$$\bar{E}(x, y, z) = \int_{-\infty}^{\infty}\int_{-\infty}^{\infty}\left[\bar{\mathscr{E}}_T(\bar{k}_T, \omega) - \bar{i}_z \frac{\bar{k}_T \cdot \bar{\mathscr{E}}_T(\bar{k}_T, \omega)}{k_z}\right] \cdot e^{-j(\bar{k}_T \cdot \bar{r}_T + k_z z)}\, dk_x dk_y, \tag{1}$$

where the wavenumber vector $\bar{k}$, the radius vector $\bar{r}$, and the transformed electric field vector $\bar{\mathscr{E}}$ have been separated

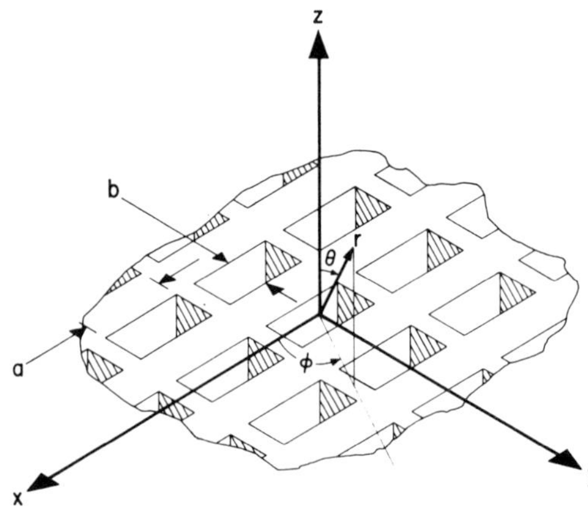


Fig. 1. Array of holes in an infinite, perfectly conducting ground plane. (Rectangular holes are shown, but this restriction is not necessary in the analysis.)

into transverse[3] and longitudinal components. We have also used the fact that $\nabla \cdot \bar{E}$ is zero for the source-free region $z>0$ to write $\mathscr{E}_z$ in terms of the transformed transverse electric field. Also, for plane waves in free space

$$|\bar{k}|^2 = k^2 = |\bar{k}_T|^2 + k_z^2 = \omega^2 \mu_o/\varepsilon_o.$$

The region $z<0$ is constrained by the radiating elements and their associated feed systems. For the sake of this discussion, let the radiating elements consist of lengths of cylindrical waveguide of arbitrary cross section. Then the aperture field will be subject to the boundary conditions imposed by the waveguide walls; that is, the aperture electric field will be a linear combination of the fields associated with the permissible waveguide modes. The strength of excitation of the various modes will depend on the scan angle of the array. Consequently, the transverse electric field at the aperture plane will be scan-dependent. For this particular part of the analysis, we do not need to know the exact form of the aperture electric field; all we need to know is that the transverse electric field at the aperture is a determinable quantity which we can represent as $\bar{E}_T(\bar{r}_T, z=0)$ for any scan angle. Then we can write

$$\bar{\mathscr{E}}_T(\bar{k}_T, \omega) = \frac{1}{4\pi^2}\int_{-\infty}^{\infty}\int_{-\infty}^{\infty} \bar{E}_T(\bar{r}_T, z=0) e^{j\bar{k}_T \cdot \bar{r}_T}\, dx dy. \tag{2}$$

The curl equation can then be used to determine the transverse magnetic field:

$$\bar{H}_T(x, y, z) = \frac{1}{kZ_o}\int_{-\infty}^{\infty}\int_{-\infty}^{\infty}\left[k_z \bar{i}_z \times \bar{\mathscr{E}}_T + (\bar{i}_z \times \bar{k}_T)\left(\frac{\bar{k}_T \cdot \bar{\mathscr{E}}_T}{k_z}\right)\right] \cdot e^{-j(\bar{k}_T \cdot \bar{r}_T + k_z z)}\, dk_x dk_y \tag{3}$$

where the free-space impedance $\sqrt{\mu_o/\varepsilon_o}$ is denoted by $Z_o$. The total complex radiated power is then given by the integral of the $z$-component of the Poynting vector over the aperture plane $z=0$:

[3] The subscript $T$ will be used to denote that part of a vector quantity which is transverse to the array normal.

$$P_r = \frac{1}{2}\iint_{\text{aperture}} \bar{E}_T \times \bar{H}_T^* \cdot i_z \, dxdy. \tag{4}$$

The integrals can be extended over the whole $z=0$ plane because the transverse electric field vanishes on the perfectly conducting plane that contains the radiating elements. If the expressions for the field components, (1) and (3), are used in (4), the complex power is given by a sextuple integral. An interchange of the order of integration results in an integral over $x$ and $y$ which can be identified as the transform of a two-dimensional impulse function of area $4\pi^2$ and located at $k'_x = k_x$ and $k'_y = k_y$. Thus, the complex power expression reduces to a double integral in transverse wavenumber space (the $k_x - k_y$ plane):

$$P_r = \frac{2\pi^2}{kZ_o}\int_{-\infty}^{\infty}\int_{-\infty}^{\infty} \cdot \frac{(k^2-k_x^2)|\mathscr{E}_y|^2 + (k^2-k_y^2)|\mathscr{E}_x|^2 + k_x k_y(\mathscr{E}_x\mathscr{E}_y^* + \mathscr{E}_y\mathscr{E}_x^*)}{k_z^*} dk_x dk_y, \tag{5}$$

where it has been assumed that $\bar{\mathscr{E}}_T$ can be functionally separated into $x$- and $y$-components. (Although this is not always possible, it is true for all cases of practical importance.) This expression relates the total complex power radiated by the array (or aperture) to the radiation pattern of the aperture over all space—both visible ($|\bar{k}_T| < k$) and invisible ($|\bar{k}_T| > k$). It is equivalent to the expression given by Rhodes [9] even though it appears quite different. (Rhodes used a coordinate system with the antenna normal in the $x$-direction and he defined the fields in space to be TE and TM relative to the $z$-coordinate of the antenna.)

Since all waves must satisfy the radiation condition at $z = +\infty$, $k_z$ is either positive real or negative imaginary. Consequently, the integral of (5) taken over visible space ($k_x^2 + k_y^2 \le k^2$) gives the radiated power while the integral over invisible space ($k_x^2 + k_y^2 \ge k^2$) gives the reactive power stored about the antenna.

The problem of interest to us is an infinite, regularly spaced, uniformly illuminated, and progressively phased array of identical elements. Unfortunately, the above formalism is not directly applicable to an infinite periodic structure because the Fourier integrals do not exist.[4] The desired result can still be achieved by the above formalism if we first consider a finite array and then compute the complex power per element as the array dimensions and number of elements approach infinity while the element lattice is held fixed. Consider an array with $(2M+1)(2N+1)$ elements arranged in a parallelogram lattice, as shown in Fig. 2(a). The element locations can be represented by the vector

$$\bar{r}_{T_{mn}} = m\bar{d}_1 + n\bar{d}_2, \qquad |m| \le M, \qquad |n| \le N, \tag{6}$$

where $\bar{d}_1$ and $\bar{d}_2$ are nonparallel vectors which define the element lattice, as indicated in Fig. 2(a). The area allotted to an element is

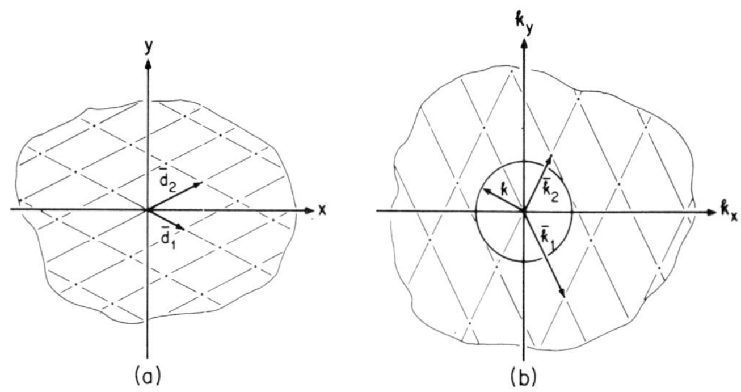


Fig. 2. Direct and reciprocal lattices: (a) element lattice; (b) grating lobe lattice ($k_x^2 + k_y^2 \le k^2$ is visible space).

$$A = i_z \cdot (\bar{d}_1 \times \bar{d}_2). \tag{7}$$

(The element need not fill the entire allotted area.) If the array is assumed to be uniformly illuminated (which it surely must be for an infinite array), the electric field in the $mn$th cell can be written as

71

$$\bar{E}_{T_{mn}}(\bar{r}_T, z=0) = \bar{f}(\bar{r}_T - \bar{r}_{T_{mn}})e^{-j\bar{k}_{Too}\cdot\bar{r}_{Tmn}}, \tag{8}$$

where $\bar{f}(\bar{r}_T)$ gives the aperture electric field distribution of the center element (which is also a typical element in the array), and $\bar{k}_{T_{oo}}$ defines the pointing direction of the main beam

$$\bar{k}_{T_{oo}} = k(ui_x + vi_y). \tag{9}$$

The direction cosines of the desired beam pointing angle $u$, $v$ are defined by:

$$u = \sin\theta\cos\phi$$
$$v = \sin\theta\sin\phi. \tag{10}$$

The total aperture electric field distribution is given by

$$\bar{E}_T(\bar{r}_T, z=0) = \sum_{m=-M}^{M}\sum_{n=-N}^{N} \bar{E}_{T_{mn}}(\bar{r}_T, z=0).$$

This expression is then substituted in (2) and the required summations performed to obtain

$$\bar{\mathscr{E}}_T(\bar{k}_T, \omega) = \bar{F}(\bar{k}_T, \omega)\left\{\frac{\sin[(M+\frac{1}{2})(\bar{k}_T - \bar{k}_{T_{oo}})\cdot\bar{d}_1]}{\sin[\frac{1}{2}(\bar{k}_T - \bar{k}_{T_{oo}})\cdot\bar{d}_1]}\right\} \cdot \left\{\frac{\sin[(N+\frac{1}{2})(\bar{k}_T - \bar{k}_{T_{oo}})\cdot\bar{d}_2]}{\sin[\frac{1}{2}(\bar{k}_T - \bar{k}_{T_{oo}})\cdot\bar{d}_2}\right\} \tag{11}$$

where

$$\bar{F}_T(\bar{k}_T, \omega) = \frac{1}{4\pi^2}\iint_{\text{cell}} \bar{f}(\bar{r}_T)e^{j\bar{k}_T\cdot\bar{r}_T} dxdy \tag{12}$$

is the Fourier transform of the aperture electric field distribution of a typical element in the array.

If the expression for $\mathscr{E}_x(\bar{k}_T, \omega)$ and $\mathscr{E}_y(\bar{k}_T, \omega)$ are used in the power formula of (5), it is found that the integrand contains the following factor:

[4] The integral of $|E_T(r_T)|$ over the entire $z=0$ plane is not finite for an infinite array, which violates the condition for the existence of a Fourier transform.

$$\left\{\frac{\sin\left[(M+\frac{1}{2})(\bar{k}_T-\bar{k}_{T_{oo}})\cdot\bar{d}_1\right]}{\sin\left[\frac{1}{2}(\bar{k}_T-\bar{k}_{T_{oo}})\cdot\bar{d}_1\right]}\right\}^2 \left\{\frac{\sin\left[(N+\frac{1}{2})(\bar{k}_T-\bar{k}_{T_{oo}})\cdot\bar{d}_2\right]}{\sin\left[\frac{1}{2}(\bar{k}_T-\bar{k}_{T_{oo}})\cdot\bar{d}_2\right]}\right\}^2. \tag{13}$$

For very large values of $M$ and $N$ (approaching infinity), this product will have negligible value except in the vicinity of points where

$$\begin{aligned}(\bar{k}_T-\bar{k}_{T_{oo}})\cdot\bar{d}_1 &= 2\pi p, \qquad p=0,\pm 1,\pm 2,\cdots,\\ (\bar{k}_T-\bar{k}_{T_{oo}})\cdot\bar{d}_2 &= 2\pi q, \qquad q=0,\pm 1,\pm 2,\cdots,\end{aligned} \tag{14}$$

If we now define

$$\bar{k}_{T_{pq}}-\bar{k}_{T_{oo}} = p\bar{k}_1+q\bar{k}_2 \tag{15}$$

with $\bar{k}_2\cdot\bar{d}_1=\bar{k}_1\cdot\bar{d}_2=0$, (14) will be satisfied when

$$\begin{aligned}\bar{k}_1 &= -\frac{2\pi}{A}(\bar{i}_z\times\bar{d}_2)\\ \bar{k}_2 &= +\frac{2\pi}{A}(\bar{i}_z\times\bar{d}_1).\end{aligned} \tag{16}$$

The vectors $\bar{k}_1$ and $\bar{k}_2$ define the reciprocal lattice for the array [11]. This reciprocal lattice has real significance in array theory because it defines the locations of all grating lobes in direction cosine space ($\bar{k}_T/k$-space). Fig. 2(b) shows a sketch of the reciprocal lattice for the array element lattice depicted in Fig. 2(a). It is evident from these figures that the reciprocal lattice is the same as the array lattice except for a scale factor and a 90° rotation. Furthermore, the area per grating lobe in $k_T$-space is just $4\pi^2/A$, which means that for a given area per element, we can control the shape of the array scan volume but not the total extent of spatial coverage in direction cosine space.

The implication of the reciprocal (or grating lobe) lattice in our problem is that the complex power radiated by a large array is concentrated in the directions of the grating lobes; that is, the integrand in the power expression of (5) has negligible value except in the vicinity of the points defined by the reciprocal lattice. Therefore, the power integral for a large array can be approximately written as a doubly infinite summation of integrals which are evaluated only in the neighborhood of the grating lobes; as the number of elements approaches infinity, the error in this approximation approaches zero.

If the remaining factors in the integrand of the power formula are slowly varying functions of $k_x$, $k_y$ as compared to the array factor of (13) in the vicinity of a grating lobe,[5] they can be evaluated at the grating lobe coordinates $k_{x_{pq}}$, $k_{y_{pq}}$ and then they may be removed from the integrand. This will be the case for very large arrays as long as a grating lobe is not too close to the edge of visible space (a few beamwidths of the array factor should be sufficient). The remaining integral to be evaluated in the vicinity of the $pq$th grating lobe is

$$\int_{-\alpha_{\max}}^{\alpha_{\max}}\int_{-\beta_{\max}}^{\beta_{\max}}\left\{\frac{\sin\left[\left(M+\frac{1}{2}\right)\alpha\frac{A}{|d_2|}\right]}{\sin\left[\frac{1}{2}\alpha\frac{A}{|d_2|}\right]}\right\}^2 \cdot\left\{\frac{\sin\left[\left(N+\frac{1}{2}\right)\beta\frac{A}{|d_1|}\right]}{\sin\left[\frac{1}{2}\beta\frac{A}{|d_1|}\right]}\right\}^2 |J|\,d\alpha\, d\beta,$$

where we have centered the integral around the $pq$th grating lobe and, in addition, we have introduced a change of variables so that the integrand is separable. The required substitution is

$$\bar{k}_T-\bar{k}_{T_{oo}} = (p\bar{k}_1+q\bar{k}_2)+\left(\alpha\frac{\bar{k}_1}{|k_1|}+\beta\frac{\bar{k}_2}{|k_2|}\right).$$

The Jacobian $|J|$ of the transformation from $k_x$, $k_y$ to $\alpha$, $\beta$ is given by

$$|J| = \frac{(\bar{i}_x\cdot\bar{k}_1)(\bar{i}_y\cdot\bar{k}_2)-(\bar{i}_x\cdot\bar{k}_2)(\bar{i}_y\cdot\bar{k}_1)}{|k_1||k_2|} = \frac{A}{|d_1||d_2|}.$$

The following approximations can then be used in the evaluation of the integrals

$$\int_{-x_{\max}}^{x_{\max}}\left\{\frac{\sin\left[\left(M+\frac{1}{2}\right)\frac{x}{x_0}\right]}{\sin\left[\frac{1}{2}\frac{x}{x_0}\right]}\right\}^2 dx \cong \int_{-\infty}^{\infty}\left\{\frac{\sin\left[\left(M+\frac{1}{2}\right)\frac{x}{x_0}\right]}{\frac{1}{2}\frac{x}{x_0}}\right\}^2 dx = 2\pi(2M+1)x_0$$

where the limits of integration are extended to $\pm\infty$ with negligible error because the integrand is small for $x>x_{\max}$. Therefore, on a per-element basis, the array factor for an infinite array can be written as a doubly infinite set of impulse functions located at the grating lobe positions (as expected, of course). Multiplying the following expression by the factor of (13)

$$\lim_{\substack{M\to\infty\\ N\to\infty}}\frac{1}{(2M+1)(2N+1)}$$

we obtain

$$\frac{4\pi^2}{A}\sum_{p=-\infty}^{\infty}\sum_{q=-\infty}^{\infty}\delta(\bar{k}_T-\bar{k}_{T_{pq}})$$

where $k_{T_{pq}}$ is defined by (15). The complex radiated power per element for an element in an infinite array (phased to point the main beam in the direction $u$, $v$), then becomes

[5] For typical array elements, $\bar{F}_T(\bar{k}_T,\omega)$ defined in (12) is a slowly varying function. The denominator factor $k_z=\sqrt{k^2-|k_T^2|}$ is a slowly varying function except when $|\bar{k}_{T_{pq}}|\to k$ (a grating lobe near end fire).

TABLE I

ELEMENT-LATTICE AND RECIPROCAL-LATTICE PARAMETERS

| Parameter | Rectangular | Isosceles-Triangle |
|---|---|---|
| $\bar{d}_1$ | $i_x d_x$ | $i_x d_x$ |
| $\bar{d}_2$ | $i_y d_y$ | $i_x \frac{d_x}{2} + i_y d_y$ |
| $A$ | $d_x d_y$ | $d_x d_y$ |
| $\bar{k}_1$ | $i_x \frac{2\pi}{d_x}$ | $i_x \frac{2\pi}{d_x} - i_y \frac{\pi}{d_y}$ |
| $\bar{k}_2$ | $i_y \frac{2\pi}{d_y}$ | $i_y \frac{2\pi}{d_y}$ |
| $\alpha_{pq} = \frac{i_x \cdot \bar{k}_{T_{pq}}}{k}$ | $\left(u + p\frac{\lambda}{d_x}\right)$ | $\left(u + p\frac{\lambda}{d_x}\right)$ |
| $\beta_{pq} = \frac{i_y \cdot \bar{k}_{T_{pq}}}{k}$ | $\left(v + q\frac{\lambda}{d_y}\right)$ | $\left(v + q\frac{\lambda}{d_y} - p\frac{\lambda}{2d_y}\right)$ |

$$P_r(u, v) = \frac{8\pi^2}{kZ_o A} \sum_p \sum_q \cdot \frac{(k^2 - k_{y_{pq}}^2)|F_x(\bar{k}_{T_{pq}})|^2 + (k^2 - k_{x_{pq}}^2)|F_y(\bar{k}_{T_{pq}})|^2}{(\sqrt{k^2 - |k_{T_{pq}}|^2})^*} + 2\frac{k_{x_{pq}} k_{y_{pq}} \operatorname{Re}\{F_x(\bar{k}_{T_{pq}}) F_y^*(\bar{k}_{T_{pq}})\}}{(\sqrt{k^2 - |k_{T_{pq}}|^2})^*}. \tag{17}$$

Again, note that the square root is taken as positive real or negative imaginary, in order to satisfy the radition condition at $z = +\infty$.

At this point, it is worthwhile to examine briefly and interpret the terms in the power expression [12]. For the main beam, $p = q = 0$, the first term in the power expression can be written as

$$\frac{(k^2 - k_{y_{oo}}^2)|F_x(\bar{k}_{T_{oo}})|^2}{(\sqrt{k^2 - |k_{T_{oo}}|^2})^*} = \frac{k(1 - \sin^2\theta \sin^2\phi)|F_x(k \sin\theta\cos\phi, k\sin\theta\sin\phi)|^2}{\cos\theta}$$

where (9) and (10) have been used. This is readily recognized as the far-field power pattern of an $x$-polarized element multiplied by the beam-broadening factor ($1/\cos\theta$). The second main beam term may be similarly interpreted in terms of a $y$-polarized element aperture distribution, while the third term represents the amount of power received by an $x$-polarized detector in the far-field due to the radiation of $y$-polarized power. All other terms in the power expression may be viewed in the same manner except now the pattern function is evaluated in the invisible region of space $|k_T| > k$, in the vicinity of a grating lobe.

Since virtually all practical arrays utilize either a rectangular element lattice or an isosceles-triangle element lattice, it is desirable to specialize the element-lattice and reciprocal-lattice parameters for these cases. The appropriate parameters are tabulated in Table I for these cases. An examination of the grating lobe locations for the isosceles-triangle lattice indicates that the same result is obtainable by defining $\alpha_{p'q'} = u + p'\lambda/d_x$ and $\beta_{p'q'} = v + q'\lambda/2d_y$ and then taking a sum of just those terms with $p' + q'$ even.

## III. APPLICATIONS OF THE POWER EXPRESSION

In the preceding section, an expression for the complex power radiated by an element in an infinite planar array was derived in considerable detail. This section will be concerned with the application of the power formula to arrays with specific types of radiating elements. In particular, it will be shown that the active admittance for narrow slots in a ground plane and the active impedance for dipoles with and without a ground plane can be simply obtained from the power formula. A third example involves dual-mode elements—crossed-dipole pairs with and without a ground plane. Finally, the power formula will be applied to the multimode problem of an infinite array of open-ended rectangular waveguides, flush mounted in a ground plane.



### *Narrow Slots in a Ground Plane*

For narrow slots of width $a$ and height $b$ polarized in the $y$-direction, the aperture electric field can be represented by

$$E_y(x, y, 0) = \sqrt{\frac{2}{ab}}\, V_{10} \cos\frac{\pi x}{a}, \qquad |x| \le \frac{a}{2}, \qquad |y| \le \frac{b}{2}$$

if we assume that only the dominant mode is excited at the aperture. The normalization factor is chosen such that the power propagated along a feed waveguide with the same cross section will be

$$P_{z10} = \frac{1}{2}\frac{\sqrt{1 - (\lambda/2a)^2}}{Z_o}|V_{10}|^2 = \frac{1}{2} Y_{10}^*|V_{10}|^2,$$

where $Y_{10}$ is the characteristic admittance of the $TE_{10}$ mode in the waveguide. If we now take the Fourier transform of the aperture electric field distribution, insert the result in the power formula of (17), and recognize that the power radiated by an element is

$$P_{r10}(u, v) = \tfrac{1}{2} Y_{D10}^*(u, v)|V_{10}|^2,$$

then the element driving admittance (radiation admittance) is readily found to be

$$Y_{D10}(u, v) = \frac{8}{\pi^2 Z_o}\frac{ab}{A}\sum_p \sum_q \left\{\frac{1 - \alpha_{pq}^2}{\sqrt{1 - \alpha_{pq}^2 - \beta_{pq}^2}}\right\} \left\{\frac{\sin\left(\pi\beta_{pq}\frac{b}{\lambda}\right)}{\pi\beta_{pq}\frac{b}{\lambda}}\right\}^2 \left\{\frac{\cos\left(\pi\alpha_{pq}\frac{a}{\lambda}\right)}{1 - \left(2\alpha_{pq}\frac{a}{\lambda}\right)^2}\right\}^2 \tag{18}$$

where we have defined $\alpha_{pq} = k_{x_{pq}}/k$ and $\beta_{pq} = k_{y_{pq}}/k$ in accordance with the definitions in Table I. For an array with the elements spaced close enough together to prevent the existence of grating lobes in real space as the beam is scanned over the design scan volume, only the $p = q = 0$

term will be real; that is, the argument of the square root will be positive only for $p=q=0$. The radiation conductance as a function of beam pointing angle under these constraints is

$$G_{D_{10}}(u, v) = \frac{8ab}{\pi^2 Z_0 A} \frac{1-u^2}{\sqrt{1-u^2-v^2}} \left\{ \frac{\cos\left(\pi u \frac{a}{\lambda}\right)}{1-\left(2u\frac{a}{\lambda}\right)^2} \right\}^2 \cdot \left\{ \frac{\sin\left(\pi v \frac{b}{\lambda}\right)}{\pi v \frac{b}{\lambda}} \right\}^2. \quad (19)$$

This result agrees exactly with the result given by Edelberg and Oliner [2]. It is interesting to note that the conductance variation with scan angle is independent of the element lattice, if the slot dimensions and area per element are fixed. The susceptance variation with scan angle is a different story because $\bar{k}_{T_{pq}}$ depends on the element lattice for $p \neq 0$ and/or $q \neq 0$.

74

The element driving admittances for a rectangular element lattice and for an isosceles-triangle element lattice are readily determined by the use of the definitions in Table I in (18).

*Flat Dipoles With and Without a Ground Plane*

Stark's results for an infinite array of flat dipoles with and without a ground plane [3] can be readily deduced from the above formalism by duality. In this case, the expansion is in terms of the magnetic field and its Fourier transform: $\bar{H}(x, y, z)$ and $\mathscr{H}(\bar{k}_T, \omega)$. The only changes in the power expression of (17) are: replacement of $\mathscr{E}$ by $\mathscr{H}$, multiplication by $2Z_o^2$ (new coefficient $16\pi^4 Z_o/k$), and complex conjugation of the integrand. (The factor of two is introduced to account for the fact that the dipole array radiates into all space whereas the above formalism only considers the half-space on one side of the array.) The boundary condition at $z=0$ is now given by

$$i_z \times \bar{H}_T(x, y, 0) = \tfrac{1}{2} \bar{K}_T(x, y, 0),$$

where $\bar{K}_T(x, y, 0)$ is the current density at the aperture plane and the factor of 1/2 accounts for the fact that we are only solving for the region $z>0$. Now,

$$\mathscr{H}_T(\bar{k}_T, \omega) = -\frac{1}{8\pi^2}\int_{-\infty}^{\infty}\int_{-\infty}^{\infty} i_z \times \bar{K}_T(x, y, 0) \cdot e^{j\bar{k}_T \cdot \bar{r}_T}\, dx dy. \quad (20)$$

Let us assume an array of $x$-directed flat dipoles (length $L$ and width $b$) with each dipole having an idealized, sinusoidal current distribution:

$$K_x(x, y, 0) = \frac{I_o}{b} \frac{\sin k\left(\frac{L}{2} - |x|\right)}{\sin k \frac{L}{2}}, \quad |x| \le \frac{L}{2}, \quad |y| \le \frac{b}{2}.$$

If we now take the Fourier transform indicated in (20), insert the result in the modified form of the power formula of (17), and recognize that the power radiated by an element is

$$P_r(u, v) = \tfrac{1}{2} Z_D(u, v) |I_o|^2,$$

then the expression for the radiation impedance is found to be

$$Z_D(u, v) = \frac{60}{\pi} \frac{\lambda^2}{A} \sum_p \sum_q \frac{1-\alpha_{pq}^2}{\sqrt{1-\alpha_{pq}^2-\beta_{pq}^2}} |F(\alpha_{pq}, \beta_{pq})|^2, \quad (21)$$

where

$$F(\xi, \eta) = \frac{\sin\left(\pi\eta \frac{b}{\lambda}\right)}{\pi\eta \frac{b}{\lambda}} \cdot \frac{\cos\left(\pi\xi \frac{L}{\lambda}\right) - \cos\left(\pi \frac{L}{\lambda}\right)}{(1-\xi_{pq}^2)\sin\left(\pi \frac{L}{\lambda}\right)}. \quad (22)$$

If the array is backed by a ground plane at $z=-s$, we can determine the appropriate impedance relation ("grating lobe series") by adding the effect of the image array at $z=-2s$. The image array is identical to the actual array except for a reversal of the current density, so the electric field of the image array is

$$\bar{E}_i(x, y, z) = -\bar{E}(x, y, z+2s).$$

Since the $z$-dependence of the field is entirely contained in an exponential term, we have

$$\bar{E}_{\text{total}}(x, y, z) = \bar{E}(x, y, z)\{1 - e^{-2j\sqrt{k^2-|\bar{k}_T|^2}\, s}\}.$$

This means that each term of our impedance expression will be modified by a multiplicative factor

$$1 - e^{-4\pi j(s/\lambda)\sqrt{1-\alpha_{pq}^2-\beta_{pq}^2}}. \quad (23)$$

When this factor is included in (21), we obtain a generalized form of the result given by Stark for a rectangular-grid array of flat, half-wavelength dipoles [3]. (Except for notation, Stark's result is obtained by setting $A=d_x d_y$, $L=\lambda/2$, $\alpha_{pq}=u+p\lambda/d_x$, and $\beta_{pq}=v+q\lambda/d_y$.) Note that for lobes in real space, the factor defined by (23) is complex; consequently, there is a contribution to the driving reactance from lobes in real space, which is not the case in the absence of a ground plane.

*Crossed Dipoles With and Without a Ground Plane*

The above procedures may also be applied to the case of an infinite array of crossed-dipole pairs. Let the dipoles be $x$- and $y$-aligned and assume sinusoidal current distributions on the dipoles with amplitudes $I_{x_o}$ and $I_{y_o}$. If both dipoles are assumed to be of length $L$ and width $b$ (not a necessary assumption), then it can be readily shown that the dipole terminal voltages are related to the dipole currents by the following relations

$$\begin{aligned} V_x &= Z_D^{xx}(u, v) I_{x_o} - Z_D^{xy}(u, v) I_{y_o} \\ V_y &= -Z_D^{xy}(u, v) I_{x_o} + Z_D^{yy}(u, v) I_{y_o} \end{aligned} \quad (24)$$

where $Z_D^{xx}(u, v)$ is given by (21), $Z_D^{yy}(u, v)$ is given by (21) with $\alpha_{pq}$ and $\beta_{pq}$ interchanged, and

$$Z_D^{xy}(u, v) = \frac{60}{\pi}\frac{\lambda^2}{A}\sum_p\sum_q \frac{\alpha_{pq}\beta_{pq}}{\sqrt{1-\alpha_{pq}^2-\beta_{pq}^2}} F(\alpha_{pq}, \beta_{pq})F(\beta_{pq}, \alpha_{pq}),$$

with $F(\xi, \eta)$ defined by (22). Equations (24) are just standard network equations for a two-port network and they apply to any dipole pair in the array.

Before an actual solution can be obtained for the performance of a dipole pair in the array environment, we must specify the type of source used to excite each dipole pair; e.g., ideal voltage sources feeding each dipole, real voltage sources feeding the individual dipoles, or a general 3- or 4-port network simultaneously feeding both dipoles. Ideal voltage sources fix the values of $V_x$ and $V_y$ independent of the dipole currents, whereas real voltage sources make $V_x$ dependent on $I_{x_o}$ and $V_y$ dependent on $I_{y_o}$. The more general case, a 3- or 4-port network feed, causes $V_x$ to depend on both $I_{x_o}$ and $I_{y_o}$ and similarly for $V_y$.

The above relations plus the source constraints appropriate to a particular feed system define the terminal behavior of a dipole pair in an infinite array without a ground plane. If the ground plane factor (23) is included in each of the impedance expressions in (24), a solution for crossed-dipole pairs above a ground plane is obtained.

*Open-Ended Rectangular Waveguide*

We turn now to the problem of an infinite array of open-ended, rectangular waveguides flush mounted in a ground plane. Since higher-order waveguide modes (propagating and evanescent) are excited at the aperture plane, we must consider their effects in the analysis. To do this, we represent each waveguide mode as an equivalent transmission line (nonpropagating for cutoff modes) feeding an aperture which has a field distribution identical to that of the particular waveguide mode. This field distribution has a "grating lobe series" of the form given by (18). It will also cross-couple to other waveguide modes which can be represented as a mutual admittance having the form of a "grating lobe series." In this context, we arrive at an equivalent $N$-port network model for an element in the array ($N$ is the number of waveguide modes considered). The $N$ input ports of the network are transmission lines corresponding to the various waveguide modes. It should be noted that the parameters (the admittance matrix) of this equivalent network vary as the array phasing is changed to steer the beam.

In the following derivation we will obtain the expressions for the admittances in the matrix defining the equivalent $N$-port network for an element in the waveguide array. In addition, we will derive the source and/or termination constraints on the equivalent network which are imposed by the characteristics of the waveguide modes involved.

For this type of problem, it is convenient to define the waveguide modes in terms of modal functions with vanishing transverse components of electric field rather than the usual TE and TM modes. Thus, we define two sets of longitudinal-section electric (LSE) modes [13]; one set with $E_y \equiv 0$ ($LSE_x$) and one set with $E_x \equiv 0$ ($LSE_y$). Since these modes can be readily combined to form all the TE and TM modes, they constitute a complete set. However, the complex power propagated by the $LSE_{xmn}$ mode is not independent of that carried by the $LSE_{ymn}$ mode; that is, the LSE modes do not form an orthogonal set. As we shall see, this nonorthogonality does not create any particular difficulties.

The expressions for the fields of the $mn$th $LSE_x$ and $LSE_y$ modes are readily obtained by summing the fields of the $TE_{mn}$ and $TM_{mn}$ modes and redefining coefficients. Appropriate expressions for the transverse fields relative to the center waveguide are

$$E_{xmn} = V_{xmn}\sqrt{\frac{2\delta_m}{ab}}\begin{bmatrix}\cos\\ j\sin\end{bmatrix}\left(m\pi\frac{x}{a}\right)\begin{bmatrix}j\sin\\ \cos\end{bmatrix}\left(n\pi\frac{y}{b}\right)$$

$$E_{ymn} = V_{ymn}\sqrt{\frac{2\delta_n}{ab}}\begin{bmatrix}j\sin\\ \cos\end{bmatrix}\left(m\pi\frac{x}{a}\right)\begin{bmatrix}\cos\\ j\sin\end{bmatrix}\left(n\pi\frac{y}{b}\right)$$

$$H_{xmn} = -\left[Y_{cmn}V_{xmn}\sqrt{\frac{2\delta_m}{ab}} + Y_{ymn}V_{ymn}\sqrt{\frac{2\delta_n}{ab}}\right] \cdot\begin{bmatrix}j\sin\\ \cos\end{bmatrix}\left(m\pi\frac{x}{a}\right)\begin{bmatrix}\cos\\ j\sin\end{bmatrix}\left(n\pi\frac{y}{a}\right) \tag{25}$$

$$H_{ymn} = +\left[Y_{xmn}V_{xmn}\sqrt{\frac{2\delta_m}{ab}} + Y_{cmn}V_{ymn}\sqrt{\frac{2\delta_n}{ab}}\right] \cdot\begin{bmatrix}\cos\\ j\sin\end{bmatrix}\left(m\pi\frac{x}{a}\right)\begin{bmatrix}j\sin\\ \cos\end{bmatrix}\left(n\pi\frac{y}{b}\right),$$

where the upper (lower) trigonometric function is used when $m$ or $n$ is even (odd). Also, $\delta_i = 1$ for $i = 0$ and $\delta_i = 2$ for $i \geq 1$. The $z$-dependence of the fields is assumed to be of the form $e^{-\gamma_{mn}z}$ where

$$\gamma_{mn} = \sqrt{\left(m\frac{\pi}{a}\right)^2 + \left(n\frac{\pi}{b}\right)^2 - k_\varepsilon^2}$$

and $k_\varepsilon$ is the wavenumber for a plane wave in a dielectric medium, $k_\varepsilon^2 = \omega^2\mu_o\varepsilon = \omega^2\mu_o\varepsilon_o\varepsilon_r = \varepsilon_r k^2$. For a lossless waveguide fully filled with a lossless, isotropic, and homogeneous dielectric material,[6] $\gamma_{mn}$ is positive real or positive imaginary. The coefficients are defined such that the *complex* power in a particular mode by itself will be given by

$$P_{z_{mn}} = \tfrac{1}{2}\,Y_{mn}^*|V_{mn}|^2,$$

with the further requirement that the characteristic admittance $Y_{mn}$ equals the transverse wave admittance for the particular mode. The transverse wave admittances are given by

[6] Dielectric loading of the waveguides is introduced because it is a simple modification of the theory and because it may sometimes be desirable in practice to reduce the waveguide dimensions by dielectric loading.

$$Y_{xmn} = \left.\frac{H_{ymn}}{E_{xmn}}\right|_{V_{ymn}=0} = -\frac{k_\varepsilon^2 - \left(n\frac{\pi}{b}\right)^2}{j\omega\mu_o\gamma_{mn}} = \frac{j}{Z_o}\frac{\varepsilon_r - \left(\frac{n\lambda}{2b}\right)^2}{\sqrt{\left(\frac{m\lambda}{2a}\right)^2 + \left(\frac{n\lambda}{2b}\right)^2 - \varepsilon_r}} \tag{26}$$

for $LSE_x$ modes and

$$Y_{ymn} = -\left.\frac{H_{xmn}}{E_{ymn}}\right|_{V_{xmn}=0} = -\frac{k_\varepsilon^2 - \left(m\frac{\pi}{a}\right)^2}{j\omega\mu_o\gamma_{mn}} = \frac{j}{Z_o}\frac{\varepsilon_r - \left(\frac{m\lambda}{2a}\right)^2}{\sqrt{\left(\frac{m\lambda}{2a}\right)^2 + \left(\frac{n\lambda}{2b}\right)^2 - \varepsilon_r}}. \tag{27}$$

76 for $LSE_y$ modes. We also define a cross-coupling admittance to account for the nonorthogonality of the $LSE_{xmn}$ and $LSE_{ymn}$ modes:

$$Y_{cmn} = \left.\frac{H_{ymn}}{E_{ymn}}\right|_{V_{xmn}=0} = -\left.\frac{H_{xmn}}{E_{xmn}}\right|_{V_{ymn}=0} = \frac{j}{Z_o}\frac{\left(\frac{m\lambda}{2a}\right)\left(\frac{n\lambda}{2b}\right)}{\sqrt{\left(\frac{m\lambda}{2a}\right)^2 + \left(\frac{n\lambda}{2b}\right)^2 - \varepsilon_r}}. \tag{28}$$

In order to simplify the remainder of this analysis, let us assume that the waveguide radiators are very long and that they are everywhere uniform; i.e., no discontinuities in cross section or dielectric loading. These assumptions assure that the fields of nonpropagating modes excited at the aperture will decay to negligible amplitude by the time they reach the end of the waveguide ($z = -l$) where the only discontinuity exists. It is also convenient to assume that nondriven, propagating modes are terminated in matched loads at the end of the waveguide. Since these nondriven modes all propagate or decay in the $-z$ direction, the appropriate transverse field components will be given by (25) with the signs of $H_{xmn}$, $H_{ymn}$, and $\gamma_{mn}$ reversed. The complex power flow at the aperture plane is given by

$$P_{zmn} = -\tfrac{1}{2}Y_{xmn}^*|V_{xmn}|^2 - \tfrac{1}{2}Y_{ymn}^*|V_{ymn}|^2 - \tfrac{1}{2}Y_{cmn}^*(V_{xmn}V_{ymn} + V_{ymn}^*V_{xmn}^*) \tag{29}$$

where the minus sign indicates power flow away from the aperture.

Equation (29) takes care of all modes except those that are excited by a source. For this last case, we make the realistic assumption that only the $LSE_{y10}$ and $LSE_{x01}$ modes ($TE_{10}$ and $TE_{01}$ modes) are excited by external sources.[7]

[7] In most practical arrays, only the $TE_{10}$ mode is driven. However, for dual linear or circular polarization, both the $TE_{10}$ and $TE_{01}$ modes will be excited.

Since the cross-coupling admittance $Y_{cmn}$ is zero when $m$ or $n$ is zero, the powers associated with these modes will be uncoupled. The power at the aperture for these driven modes can be shown to be

$$\begin{aligned} P_{z10} &= \tfrac{1}{2}\left[V_{y10}I_{y0}^* - Y_{y10}^*|V_{y10}|^2\right] \\ P_{z01} &= \tfrac{1}{2}\left[V_{x01}I_{x0}^* - Y_{x01}^*|V_{x01}|^2\right], \end{aligned} \tag{30}$$

where $I_{y0}$ and $I_{x0}$ are "short circuit currents" at the aperture plane which define the strength of excitation of the driven modes. Equations (30) are actually expressions of Norton equivalent sources in terms of power. In fact, the waveguide can be represented by equivalent sources or terminations for all modes, as indicated in Fig. 3. The admittances defined by (26), (27), and (28) appear in the equivalent circuits in Fig. 3, if there are no discontinuities in the waveguide. When there are discontinuities in the waveguides (a dielectric window, for example), it may be possible to use transmission-line theory to obtain suitable equivalent sources or terminations.

The following matrix equation can be used to summarize (29) and (30):

$$P_{z,\,\text{total}} = \tfrac{1}{2}\left[\boldsymbol{V}^t\cdot\boldsymbol{I}_s^* - \boldsymbol{V}^t\cdot\boldsymbol{Y}_s^*\cdot\boldsymbol{V}^*\right] = \tfrac{1}{2}\left[\boldsymbol{V}_x^t \,\vdots\, \boldsymbol{V}_y^t\right]\left\{\begin{bmatrix}\boldsymbol{I}_x^*\\ \boldsymbol{I}_y^*\end{bmatrix} - \begin{bmatrix}\boldsymbol{Y}_x^* & \boldsymbol{Y}_c^*\\ \boldsymbol{Y}_c^{t*} & \boldsymbol{Y}_y^*\end{bmatrix}\begin{bmatrix}\boldsymbol{Y}_x^*\\ \boldsymbol{V}_y^*\end{bmatrix}\right\} \tag{31}$$

where $t$ indicates matrix transpose and * indicates matrix conjugate. $\boldsymbol{V}_x$ and $\boldsymbol{V}_y$ are mode voltage column matrices, $\boldsymbol{I}_x$ and $\boldsymbol{I}_y$ are source current column matrices (all zeros except for the $TE_{10}$ and $TE_{01}$ entries), $\boldsymbol{Y}_x$ and $\boldsymbol{Y}_y$ are diagonal matrices containing the mode admittances defined by (26) and (27), and $\boldsymbol{Y}_c$ contains the cross-coupling admittances defined by (28). All entries in $\boldsymbol{Y}_c$ are zero except where $V_x$ and $V_y$ have identical, nonzero subscripts.

This completes the derivation of the source constraints. Now we return to the consideration of the power radiated by an element in the array. First, we take the two-dimensional Fourier transforms of the transverse electric fields:

$$\mathscr{E}_{xmn}(k_x, k_y, \omega) = \frac{\sqrt{ab}}{\pi^3}V_{xmn}F_{xmn}(k_x, k_y), \tag{32}$$

where

$$F_{xmn}(k_x, k_y) = \frac{k_x a\sqrt{2_m}}{m^2\pi^2}\left\{\frac{\begin{matrix}\cos\\ \sin\end{matrix}\left(k_x\frac{a}{2}\right)}{1 - \left(\frac{k_x a}{m\pi}\right)^2}\right\}\left\{\frac{\begin{matrix}1\\ n\end{matrix}\;\begin{matrix}\cos\\ \sin\end{matrix}\left(k_y\frac{b}{2}\right)}{1 - \left(\frac{k_y b}{n\pi}\right)^2}\right\} \tag{33}$$

and

$$\mathscr{E}_{ymn}(k_x, k_y, \omega) = \frac{\sqrt{ab}}{\pi^3}V_{ymn}F_{ymn}(k_x, k_y), \tag{34}$$

where

$$F_{ymn}(k_x, k_y) = \frac{k_y b\sqrt{2_n}}{n^2\pi^2}\;\frac{1}{m}\left\{\frac{\begin{matrix}\cos\\ \sin\end{matrix}\left(k_x\frac{a}{2}\right)}{1 - \left(\frac{k_x a}{m\pi}\right)^2}\right\}\left\{\frac{\begin{matrix}\cos\\ \sin\end{matrix}\left(k_y\frac{b}{2}\right)}{1 - \left(\frac{k_y b}{n\pi}\right)^2}\right\}. \tag{35}$$

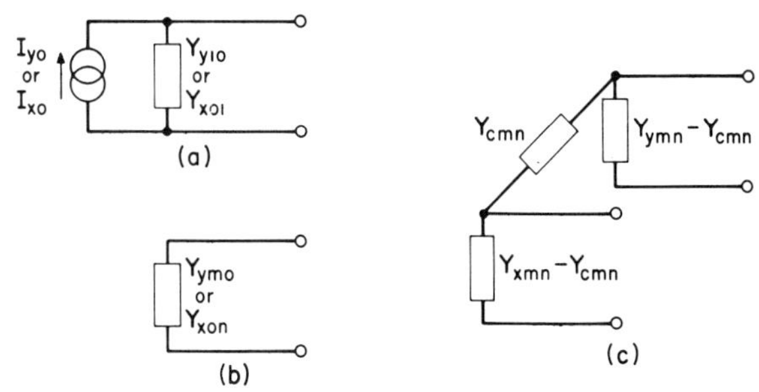


Fig. 3. Norton equivalent circuits of mode terminations: (a) driven modes; (b) uncoupled, nondriven modes; (c) coupled, nondriven modes (e.g., $LSE_{y11}$ and $LSE_{x11}$ modes).

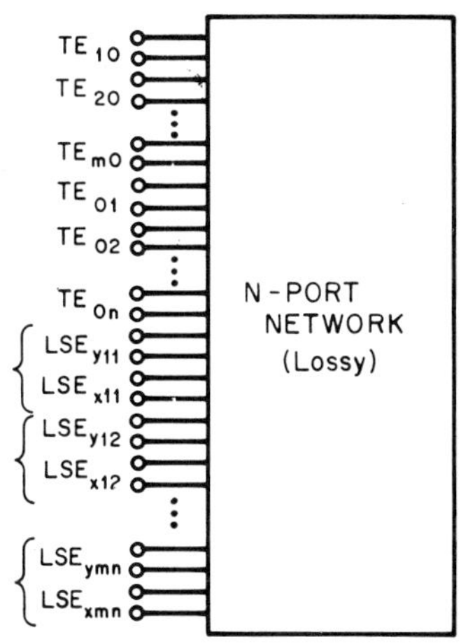


Fig. 4. $N$-port network model for a waveguide in an infinite array.

In (33) and (35), if $m(n)$ is even, we use the sine of the argument and if $m(n)$ is odd, we use the cosine of the argument. The total transverse electric field (transformed) is obtained by summing (32) and (34) over all possible waveguide modes. When the resultant expressions for the fields are inserted in (17), we find that the complex power radiated by an element in the array is given by

$$P_r(u, v) = \frac{1}{2}\sum_{m,n} V_{xmn}\left\{\sum_{r,s}[Y^{xx}_{Dmn,rs}]^* V^*_{xrs} + \sum_{r,s}[Y^{xy}_{Dmn,rs}]^* V^*_{yrs}\right\}$$
$$+ \frac{1}{2}\sum_{m,n} V_{ymn}\left\{\sum_{r,s}[Y^{xy}_{Dmn,rs}]^* V^*_{xrs} + \sum_{r,s}[Y^{yy}_{Dmn,rs}]^* V^*_{yrs}\right\} \qquad (36)$$

where we have defined the following set of scan-dependent admittance functions ("grating lobe series"):

$$Y^{xx}_{Dmn,rs}(u, v) = \frac{16}{\pi^2 Z_o}\frac{ab}{A}\sum_{p,q}\frac{1-\beta^2_{pq}}{\sqrt{1-\alpha^2_{pq}-\beta^2_{pq}}} \cdot F_{xmn}(k_{x_{pq}}, k_{y_{pq}})F_{xrs}(k_{x_{pq}}, k_{y_{pq}})$$
$$Y^{yy}_{Dmn,rs}(u, v) = \frac{16}{\pi^2 Z_o}\frac{ab}{A}\sum_{p,q}\frac{1-\alpha^2_{pq}}{\sqrt{1-\alpha^2_{pq}-\beta^2_{pq}}} \cdot F_{ymn}(k_{x_{pq}}, k_{y_{pq}})F_{yrs}(k_{x_{pq}}, k_{y_{pq}}) \qquad (37)$$
$$Y^{xy}_{Dmn,rs}(u, v) = \frac{16}{\pi^2 Z_o}\frac{ab}{A}\sum_{p,q}\frac{\alpha_{pq}\beta_{pq}}{\sqrt{1-\alpha^2_{pq}-\beta^2_{pq}}} \cdot F_{xmn}(k_{x_{pq}}, k_{y_{pq}})F_{yrs}(k_{x_{pq}}, k_{y_{pq}}).$$

If the appropriate expression for $F_{y10}(k_x, k_y)$ is inserted in (37), it is immediately found that the resulting expression for $Y^{yy}_{D10,10}(u, v)$ agrees identically with the result of (18) for slot arrays as it obviously should. The power expression of (36) can be written very compactly in the matrix form:

$$P_r(u, v) = \frac{1}{2} \boldsymbol{V}^t \cdot \boldsymbol{Y}^*_D \cdot \boldsymbol{X}^*$$
$$= \frac{1}{2}[\boldsymbol{V}^t_x \ \boldsymbol{V}^t_y] \cdot \left[\begin{array}{c|c} \boldsymbol{Y}^{xx*}_D & \boldsymbol{Y}^{xy*}_D \\ \hline \boldsymbol{Y}^{xy^t*}_D & \boldsymbol{Y}^{yy*}_D \end{array}\right]\left[\begin{array}{c} \boldsymbol{V}^*_x \\ \hline \boldsymbol{V}^*_y \end{array}\right]. \qquad (38)$$

The mode voltage matrix $\boldsymbol{V}$ is the same as the one defined in (31). The radiation admittance matrix $\boldsymbol{Y}_D$ can be partitioned as indicated such that each of the submatrices will be constructed by a systematic ordering of the appropriate admittance functions defined in (37). If we assume M-$LSE_x$ modes and N-$LSE_y$ modes, then $\boldsymbol{Y}^{xx}_D$ will be an $M \times M$ matrix, $\boldsymbol{Y}^{yy}_D$ will be an $N \times N$ matrix, and $\boldsymbol{Y}^{xy}_D$ will be an $M \times N$ matrix.

Now we have an expression for the power radiated by an element in the array. This power must equal that delivered by the sources feeding an element, as expressed by (31). Therefore, we obtain

$$\tfrac{1}{2}\boldsymbol{V}^t \cdot \boldsymbol{Y}^*_D \cdot \boldsymbol{V}^* = \tfrac{1}{2}[\boldsymbol{V}^t \cdot \boldsymbol{I}^*_s - \boldsymbol{V}^t \cdot \boldsymbol{Y}^*_s \cdot \boldsymbol{V}^*],$$

which will have a nontrivial solution, in general, only if

$$\boldsymbol{I}_s = (\boldsymbol{Y}_D + \boldsymbol{Y}_s) \cdot \boldsymbol{V}. \qquad (39)$$

It is perhaps worth noting that we are not implying a general principle of conservation of complex power. What we are actually saying is that the complex power at the aperture plane must be the same whether we compute it in terms of the fields at $z=0^-$, the waveguide region, or in terms of the fields at $z=0^+$, the free-space region, because we have satisfied the boundary conditions everywhere. If the reference planes were not coincident, the complex power expressions could not be equated due to energy storage in the enclosed volume.

Equation (39) is the expression which we have been seeking. It relates the mode excitation voltages (unknowns) to the source(s) feeding an element in the array via an admittance matrix. This admittance matrix contains contributions due to the radiation properties of the various waveguide modes and their radiative interactions plus contributions that are characteristic to the waveguide itself (the mode admittances). The analogy with network theory is apparent at this point: a waveguide in the array can be viewed as an $N$-port network where each waveguide mode corresponds to an input port, as indicated in Fig. 4; the network is described on an admittance basis where the various admittance coefficients correspond to the radiation properties of the various modes and their mutual interactions; finally, the various ports are terminated in either passive admittances (the mode admittances) or in real current sources [a current source with an internal (or generator) admittance].

A numerical solution for a specific array proceeds as follows: 1) select a set of waveguide modes which are a priori expected to be important (typically of the order of 10); 2) compute the mode admittances, the $\boldsymbol{Y}_s$ matrix; 3) specify a scan angle and compute the elements of the matrix $\boldsymbol{Y}_D$ at the particular scan angle (a few hundred terms in each of the summations in (37) are usually sufficient); 4) specify the source current matrix $\boldsymbol{I}_s$ (typically all zeros except for the entry corresponding to the $TE_{10}$ mode which is arbitrarily assigned a value of unity); 5) solve (39) for the mode voltages. This completes the solution for one scan angle. For each additional scan angle, steps 3) to 5) must be repeated.

Knowledge of the mode voltages and source currents at each scan angle is sufficient to completely define the behavior of an element in the array. However, it is usually desirable to use these known quantities to compute other parameters which are more easily measured and/or interpreted. For example, the radiation admittance seen by the dominant mode at the aperture plane is

$$Y = \frac{I_{y0}}{V_{y10}} - Y_{y10} \tag{40}$$



or, equivalently,

$$Y = \sum_{m,n} Y_{Dmn,10}^{yy} \frac{V_{ymn}}{V_{y10}} + \sum_{m,n} Y_{Dmn,10}^{xy} \frac{V_{xmn}}{V_{y10}}. \tag{41}$$

Each term in (41) corresponds to a "partial admittance" contribution to the dominant mode admittance due to the excitation of that particular higher-order mode. These "partial admittances" are sometimes quite useful in assessing the relative importance of a higher-order mode. The dominant mode reflection coefficient is also a useful parameter and is defined by

$$\Gamma = \frac{Y_{y10}^{*} - Y}{Y_{y10} + Y}, \tag{42}$$

where the possibility of a complex dominant mode admittance has been considered, which then allows the possibility of aperture matching. Other parameters may also be computed, but the only one of primary importance is the element gain function which is derived in the next section.

## IV. Gain Function of a Multimode Element in an Infinite Array

An important and easily measured characteristic of an element in an array antenna is the far-field power pattern of the element in a passive array environment; that is, all elements, except the element of interest, are terminated in their normal source impedances (admittances). Galindo and Wu [14] have shown that the pattern thus obtained is uniquely related to the transmission coefficient of the main beam in an infinite array. Since the solution given in Section III contains the information necessary to construct the aperture field distribution or transmission coefficient as a function of array phasing, we will adopt the latter approach.

Equation (1) defines the electric field in the half-space $z>0$ for any particular aperture distribution (or partial aperture distribution such as that of a single element in the array). As the array phasing is changed, the aperture field distribution changes and consequently, the fields in space change. Since we are interested only in the far-zone electric field of a single element in the direction of the main beam $\theta, \phi$, we will only evaluate the integral in the vicinity of the main beam. To do this, we express $\bar{r}$ in spherical coordinates in the direction of the main beam and we express $\bar{k}$ as follows:

$$\bar{k} = k[\sin\theta' \cos\phi' \bar{i}_x + \sin\theta' \sin\phi' \bar{i}_y + \cos\theta' \bar{i}_z].$$

Then we find

$$e^{-j\bar{k}\cdot\bar{r}} = e^{-jkr\{\sin\theta \sin\theta'[\cos(\phi-\phi')-1]+\cos(\theta-\theta')\}}.$$

Since the only significant contributions to the integral occur for $\theta' \cong \theta$ and $\phi' \cong \phi$, we can use the stationary phase argument to write

$$e^{-j\bar{k}\cdot\bar{r}} \cong e^{-jkr} e^{jkr\sin^2\theta(\phi-\phi')^2/2} e^{jkr(\theta-\theta')^2/2}.$$

The remaining part of the integrand is a slowly varying function of $\theta, \phi$; consequently, it can be removed from the integrand after it has been evaluated at $\theta, \phi(\bar{k}_T = \bar{k}_{T_{oo}})$. After a change of variables from $k_x, k_y$ to $\theta', \phi'$, the integrals reduce to two Fresnel integrals, which are readily evaluated to obtain

$$\bar{E}(r,\theta,\phi) = 2\pi jk \frac{e^{-jkr}}{r} \{k_z \bar{\mathscr{E}}_T(\bar{k}_T,\omega) - \bar{i}_z[\bar{k}_T \cdot \bar{\mathscr{E}}_T(\bar{k}_T,\omega)]\}|_{\bar{k}_T=\bar{k}_{T_{oo}}}$$

or

$$\bar{E}(r,\theta,\phi) = 2\pi jk \frac{e^{-jkr}}{r} [(\cos\theta \bar{i}_x - \sin\theta\cos\phi \bar{i}_z)\mathscr{E}_x(\theta,\phi,\omega) + (\cos\theta \bar{i}_y - \sin\theta\sin\phi \bar{i}_z)\mathscr{E}_y(\theta,\phi,\omega)] \tag{43}$$

where

$$\begin{aligned} \mathscr{E}_x(\theta,\phi,\omega) &= \frac{\sqrt{ab}}{\pi^3} \sum_m \sum_n V_{xmn} F_{xmn}(\bar{k}_T = \bar{k}_{T_{oo}}) \\ \mathscr{E}_y(\theta,\phi,\omega) &= \frac{\sqrt{ab}}{\pi^3} \sum_m \sum_n V_{ymn} F_{ymn}(\bar{k}_T = \bar{k}_{T_{oo}}). \end{aligned} \tag{44}$$

The element gain function is defined as the ratio of the power density on the peak of the beam to the power available to the element [15]. For the case at hand:

$$g(\theta,\phi) = 4\pi \frac{\left| \frac{r}{\sqrt{Z_o}} \bar{e}_p \cdot \bar{E}(r,\theta,\phi) \right|^2}{\frac{|I_{y0}|^2}{4G_{y10}} + \frac{|I_{x0}|^2}{4G_{x10}}}$$

where $G_{y10} = \mathrm{Re}\,(Y_{y10})$, $G_{x10} = \mathrm{Re}\,(Y_{x10})$, and $\bar{e}_p$ is a unit vector which defines the polarization of a receiving antenna in the far field.

If we assume that only the dominant mode is driven by a

source ($I_{x0}=0$), we can manipulate the gain expression with the aid of (40), (42), (43), and (44) to obtain

$$g(\theta,\phi) = 4\pi \frac{A}{\lambda^2}\frac{G_D}{G}(1-|\Gamma|^2)\bigg|\bar{e}_p \cdot \left\{\bar{e}_x \sum_{m,n} F_{xmn}\frac{V_{xmn}}{V_{y10}} + \bar{e}_y \sum_{m,n} F_{ymn}\frac{V_{ymn}}{V_{y10}}\right\}\bigg|^2 \tag{45}$$

where

$$G = \mathrm{Re}\,(Y), \quad G_D = \frac{16}{\pi^2 Z_o}\frac{ab}{A}, \quad \bar{e}_x = \cos\theta\bar{i}_x - \sin\theta\cos\phi\bar{i}_z,$$

and $\bar{e}_y = \cos\theta\bar{i}_y - \sin\theta\sin\phi\bar{i}_z$.

The expression for the element gain function in (45) is the one which we were initially seeking for a multimode element in an infinite-array environment. In general, (45) will have to be used in the exact form given above (with an appropriate polarization vector specified; e.g., $x$ or $y$ linear polarization, left or right circular polarization). However, when only the $TE_{10}$ mode is allowed to be nonzero (the slot array case) and there are no grating lobes in real space, $g(\theta, \phi)$ reduces to the well-known relation [16]

$$g(\theta,\phi) = 4\pi\frac{A}{\lambda^2}(1-|\Gamma|^2)\cos\theta.$$

This follows directly when we recognize that

$$G = \left[\mathrm{Re}\, Y_{D10,10}^{yy}(u,v)\right]$$

and we define $\bar{e}_p = \bar{e}_y/|\bar{e}_y|$.

## V. Some Observations on the Waveguide Array Solution

In this section, we will explore some aspects of the waveguide array solution with the intent of providing some additional insight into the solution by interpreting it in the light of previous work. An attempt is made to place quantitative bounds on the required number of waveguide modes and on the number of terms which must be summed to obtain each admittance coefficient at each scan angle.

It should be noted that the final equation (39), given above, agrees exactly with the results given by Farrell and Kuhn [8], when the definitions of parameters, coefficients, etc., are made to conform in the two solutions. This fact is not at all surprising because the same problem is solved in both cases although the methods used are quite different. The particular merit of the present formulation lies in the fact that it is derivable from a generalized power formula for infinite planar arrays and the resultant solution is obtained in a form which is directly interpretable in terms of an equivalent $N$-port network.

An interesting way to view the admittance function for a waveguide mode or a mutual admittance between two modes is afforded by Wheeler's "impedance crater" concept [17]. Wheeler introduced the concept in terms of an infinite current sheet [18] which exhibits the following impedance variation with scan angle:

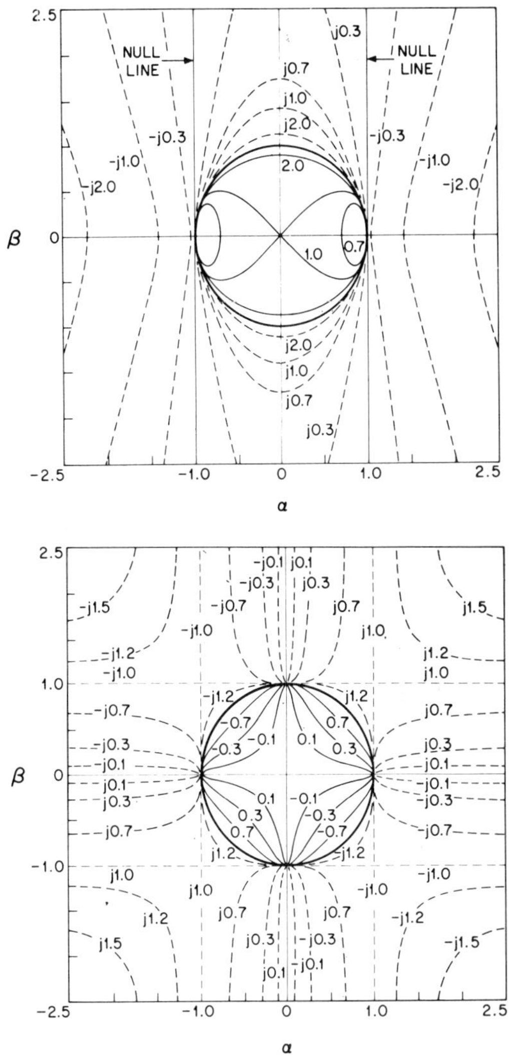


Fig. 5. Impedance crater contour plots for an infinite current sheet. *Top:* Simple current sheet [17]. *Bottom:* Cross-coupled "current sheet" multiplier.

79

$$Z(\alpha,\beta) = \frac{1-\alpha^2}{\sqrt{1-\alpha^2-\beta^2}}. \tag{46}$$

If contours of constant $Z(\alpha, \beta)$ are plotted in $\alpha, \beta$ space, then one obtains what Wheeler termed an impedance crater contour plot, as shown in Fig. 5(a). For the self-admittance of a waveguide mode, we have a function which is the product of (46) and the square of the transform of the mode electric field distribution. This function is then periodically sampled in "sin $\theta$" (or $\alpha, \beta$) space at the grating lobe locations $\alpha_{pq}$, $\beta_{pq}$. In Fig. 6 we show an admittance crater plot for the $TE_{10}$ mode in a square waveguide. (The plot is normalized to unity at broadside.) Comparison of Figs. 5(a) and 6 reveals the role played by the mode pattern: it introduces null lines for $\alpha = 3\lambda/2a, 5\lambda/2a, \cdots$, and $\beta = \lambda/b, 2\lambda/b, \cdots$, and it causes the admittance function to decay rapidly with increasing $\alpha$ and $\beta$.

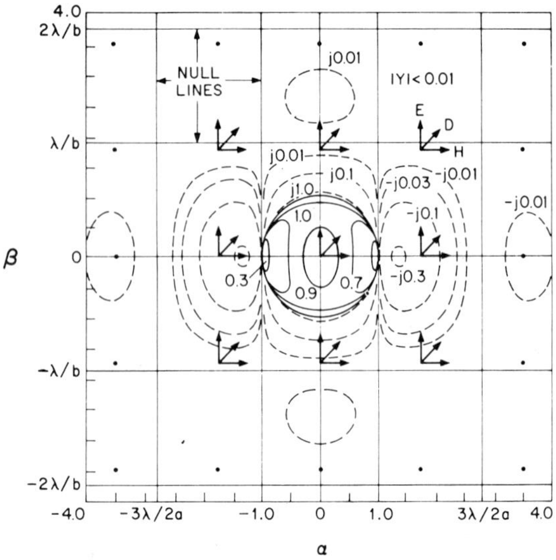


Fig. 6. Admittance crater contour plot for the $TE_{10}$ mode in a square waveguide ($a=b=0.5354\ \lambda$). The grating lobe locations for a square element lattice ($d_x=d_y=0.5714\ \lambda$) are indicated. Motions of some grating lobes for three directions of scanning are also shown.

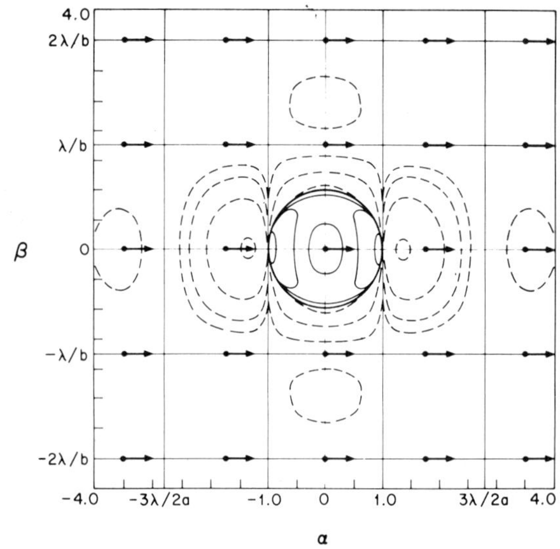


Fig. 7. Admittance crater contour plot ($TE_{10}$ mode) and grating lobe motion for $H$-plane scan of a square-grid array of waveguides with thin top and bottom walls: $a=0.5354\ \lambda$; $b=d_x=d_y=0.5714\ \lambda$.



The complete picture of a mode admittance function is found by superimposing the grating lobe locations for a particular element lattice on the admittance crater plot, as shown in Fig. 6 for a square element lattice. Each grating lobe (main beam included) samples the admittance crater at the coordinates of the grating lobe. The summation of these sample values gives precisely the mode admittance defined in (37) for a particular scan angle. As the array is phased to scan the main beam, the admittance crater remains fixed in $\alpha$, $\beta$ space, but the grating lobes all move in exactly the same direction and the same distance as the main beam, as indicated in Fig. 6. This scanning gives rise to a whole new set of admittance samples, the sum of which is, in general, different from that obtained at any other scan angle.

For the waveguide array case, we have both $x$- and $y$-polarized modes. The $y$-polarized modes have a "current sheet" multiplier given by (46) and $x$-polarized modes have the same multiplier with $\alpha$ and $\beta$ interchanged. For the cross-coupling of an $x$-polarized mode and a $y$-polarized mode, we have a different "current sheet" multiplier, which is defined by

$$Z(\alpha, \beta) = \frac{\alpha\beta}{\sqrt{1-\alpha^2-\beta^2}}.$$

This function is antisymmetric about the principal axes and symmetric about the diagonals ($\alpha=\pm\beta$), as indicated in Fig. 5(b), were we show an "impedance crater" contour plot of the function. Again, the mode patterns and the grating lobe locations can be included to give an admittance crater plot for a particular mode-pair in a typical waveguide in an array with a specific element lattice.

Let us pursue the above viewpoint a little further. In particular, consider the change in the admittance crater plot of Fig. 6 as the $E$-plane waveguide walls approach zero thickness ($b \rightarrow d_y$). The new admittance crater plot is shown in Fig. 7. It is immediately apparent that all grating lobes with $\beta \neq 0$ lie on lines where the admittance is zero. Therefore, if we scan only in the $\alpha$ direction ($H$-plane scan), the double summation for the $TE_{10}$ admittance coefficient reduces to a single summation involving only the grating lobes along the $\alpha$ axis. The coupling terms between the $TE_{10}$ ($LSE_{y10}$) mode and all other modes ($LSE_{ymn}$ and $LSE_{xmn}$) also have nulls (due to the $TE_{10}$ pattern function) occurring at all grating lobe positions with $\beta \neq 0$. Furthermore, when the subscript $n$ is nonzero (all $LSE_x$ modes and all $LSE_y$ modes except the $LSE_{ym0}$ modes), the higher-order mode pattern functions and/or the "current sheet" multiplier have nulls for $\beta=0$. Therefore, we deduce that the $TE_{10}$ mode couples only to the $TE_{m0}$ modes for $H$-plane scan of an array of waveguides where the top and bottom waveguide walls have zero thickness. In addition, each of the summations in the admittance functions of (37) reduces to a single summation involving only the grating lobes along the $\alpha$ axis. Thus, we see how the general solution reduces directly to the two-dimensional treatment given by Galindo and Wu [4], [5]. (Galindo and Wu use a residue calculus approach to obtain a solution which has a much different form than that given above; however, the final numerical results are identical in the two formulations.)

The quasi-$E$-plane scan case treated by Galindo and Wu [4], [5] can also be viewed in the above context to visualize how the general problem reduces to the two-dimensional problem when the thickness of the waveguide sidewalls becomes zero ($a=d_x$). In this case, the main beam is first scanned half-way to the first $H$-plane grating lobe and then it is scanned in the $E$-plane, as shown in Fig. 8. Now we notice that all grating lobes with $\alpha \neq \pm\lambda/2a$ lie on null lines of the $TE_{10}$ mode pattern function. Further in-

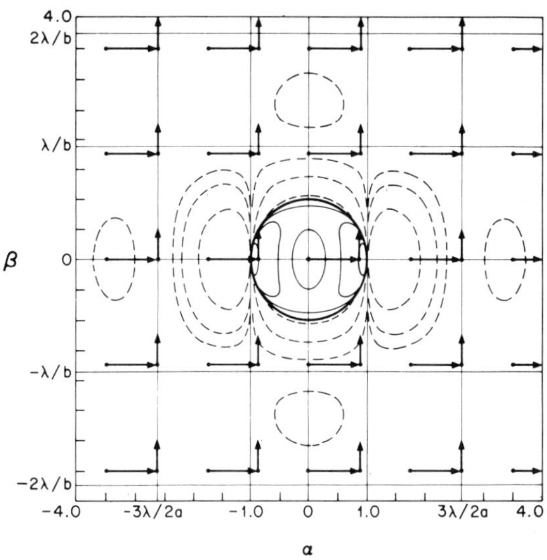


Fig. 8. Admittance crater contour plot ($TE_{10}$ mode) and grating lobe motion for quasi-$E$-plane scan of a square-grid array of waveguides with thin sidewalls: $b=0.5354\ \lambda$; $a=d_x=d_y=0.5714\ \lambda$.

vestigation of the mutual admittance terms reveals that $LSE_{x2m+1,n}$ and $LSE_{y2m+1,n}$, $m=1, 2, \cdots$, mode pattern functions introduce additional null lines at $\alpha=\pm\lambda/2a$. Furthermore, the product of the "current sheet" multiplier and the $LSE_{x2m,n}$ or $LSE_{y2m+2,n}$, $m=0, 1, 2, \cdots$, mode pattern functions are antisymmetric about $\alpha=0$. Therefore, we deduce that the $TE_{10}$ mode couples only to the $LSE_{x1n}$ and $LSE_{y1n}$ ($TE_{1n}$ and $TM_{1n}$) modes. The only grating lobes which enter into the summations are those which lie along the lines $\alpha=\pm\lambda/2a$. The summation on the grating lobes with $\alpha=-\lambda/2a$ is identical to the summation on the grating lobes with $\alpha=\lambda/2a$. Thus, we finally discover that the solution for the quasi-$E$-plane case reduces to a consideration of just the $TE_{1n}-TM_{1n}$ modes and the "grating lobe series" defined in (37) reduce to single summations involving grating lobes with $\alpha=\lambda/2a$.

For the general problem where all waveguide walls have nonzero thickness, it is still possible to reduce the number of higher-order modes involved for $E$- and $H$-plane scanning by symmetry considerations. A little study of the mutual admittance terms reveals that only $LSE_{xm,2n+2}$ and $LSE_{ym,2n}$, $n=0, 1, 2, \cdots$, modes are coupled for $H$-plane scan. Only $LSE_{x2m+1,n}$ and $LSE_{y2m+1,n}$, $m=0, 1, 2, \cdots$, modes are coupled for $E$-plane scan. For the $H$-plane scan case, the selected modes all have pattern functions which are symmetric about $\beta=0$, whereas the pattern functions are symmetric about $\alpha=0$ for the indicated modes for $E$-plane scan. If the $TE_{01}$ ($LSE_{x01}$) mode, as well as the $TE_{10}$ mode, is excited by a source, then symmetry considerations no longer apply for $E$- and $H$-plane scan. Therefore, when an array is arbitrarily scanned or when both $TE_{01}$ and $TE_{10}$ modes are excited, we must select a finite set of modes from the complete set of possible modes.

Two additional points which should be emphasized are: 1) the thin-wall solutions given by Galindo and Wu (and discussed above) only apply to arrays with rectangular lattices; and 2) $LSE_x$ and $LSE_y$ modes with identical, nonzero subscripts (e.g., $LSE_{x11}$ and $LSE_{y11}$) must either be included or excluded from the selected mode set as a pair. The second condition arises from the fact that these mode pairs are coupled in the feed waveguide by the admittance $Y_{cmn}$, defined in (28). Note that this is the only complication (and it is minor) created by the choice of a nonorthogonal mode set.

For practical array element-lattice parameters and waveguide dimensions, it is typically only necessary to consider the first few lowest-order modes to obtain quite accurate answers. The mode set will usually number less than 10 and, in general, will include the $TE_{10}$, $TE_{20}$, $TE_{30}$, $TE_{01}$, and the $TE_{11}-TM_{11}$ ($LSE_{x11}-LSE_{y11}$) modes.

There is also the question of how many terms should be used in the admittance summations ("grating lobe series") of (37). There is no definite number which can always be used; however, the data given in Table II for various $E$-plane scan angles indicate that use of relatively small numbers of terms in the summations will give quite satisfactory answers for the dominant mode radiation admittance. We can obtain a rough estimate of the minimum number of terms required

TABLE II

DEPENDENCE OF THE COMPUTED DOMINANT MODE RADIATION ADMITTANCE ON THE NUMBER OF TERMS USED IN THE "GRATING LOBE SERIES" SUMMATIONS FOR A FEW $E$-PLANE SCAN ANGLES

Square waveguide, $a=b=0.5354\ \lambda$; square element lattice, $d_x=d_y=0.5714\ \lambda$; 22 waveguide modes; $p_{max}=q_{max}=5$, 8, 10, 15, 20

| Scan Angle $\theta$ | No. of Terms $N_T$ | Computed Admittance $Y$ | Admittance Change $Y-Y_{1681}$ |
|---|---|---|---|
| 0 | 121 | $1.944211-j0.909694$ | $0.000674+j0.001151$ |
| | 289 | $1.944029-j0.909429$ | $0.000492+j0.001416$ |
| | 441 | $1.943974-j0.909393$ | $0.000437+j0.001452$ |
| | 961 | $1.943682-j0.910347$ | $0.000145+j0.000498$ |
| | 1681 | $1.943537-j0.910845$ | |
| 25 | 121 | $1.645618-j0.513046$ | $0.000033+j0.000256$ |
| | 289 | $1.645449-j0.512067$ | $-0.000136+j0.001235$ |
| | 441 | $1.645401-j0.511873$ | $-0.000184+j0.001429$ |
| | 961 | $1.645517-j0.512794$ | $-0.000068+j0.000508$ |
| | 1681 | $1.645585-j0.513302$ | |
| 45 | 121 | $1.054459+j1.291917$ | $-0.005144+j0.001868$ |
| | 289 | $1.057507+j1.291539$ | $-0.002096+j0.001490$ |
| | 441 | $1.058346+j1.291194$ | $-0.001257+j0.001145$ |
| | 961 | $1.059272+j1.290380$ | $-0.000331+j0.000331$ |
| | 1681 | $1.059603+j1.290049$ | |
| 48 | 121 | $0.437561+j4.285582$ | $-0.011162+j0.019699$ |
| | 289 | $0.445006+j4.271207$ | $-0.003717+j0.005324$ |
| | 441 | $0.446953+j4.267008$ | $-0.001770+j0.001125$ |
| | 961 | $0.448331+j4.265778$ | $-0.000392-j0.000105$ |
| | 1681 | $0.448723+j4.265883$ | |
| 50 | 121 | $5.346160+j0.979154$ | $-0.004125-j0.003871$ |
| | 289 | $5.347737+j0.984734$ | $-0.002548+j0.001709$ |
| | 441 | $5.347771+j0.986009$ | $-0.002514+j0.002984$ |
| | 961 | $5.349385+j0.984172$ | $-0.000900+j0.001147$ |
| | 1681 | $5.350285+j0.983025$ | |

by the line of reasoning which follows. A given higher-order mode ($m>1$, $n>1$) will tend to have peaks in its pattern function [(33), or (35)] in the vicinity of the points

$$(\alpha, \beta) = (\pm m\lambda/2a, \pm n\lambda/2b),$$

where the denominator factors go to zero. Beyond these lines, the pattern function decays rapidly with increasing $\alpha$ or $\beta$. If we assume that the maximum values of the mode indices are $M$ and $N$ and if we further assume that we will sum only the contributions of grating lobes which lie within a region with twice the $\alpha$ and $\beta$ extent indicated above, then we find

$$p_{\max} = \left[M \frac{d_x}{a}\right], \qquad q_{\max} = \left[N \frac{d_y}{b}\right], \tag{47}$$

where $[x]$ means the largest integer in $x$ and we have used the definitions of $\alpha_{pq}$ and $\beta_{pq}$ for a rectangular lattice. The total number of terms in a particular summation is given by

82

$$N_T = (2p_{\max} + 1)(2q_{\max} + 1). \tag{48}$$

For $M=N=4$ and relatively thin waveguide walls, $N_T=81$. This result is consistent with the data given in Table II, where we see only slight changes in the computed admittances as we use increasing numbers of terms in the summations. It should be emphasized that (47) and (48) give only rough guidelines; unsatisfactory results will usually be obtained if $p_{\max}$ or $q_{\max}$ is taken to be less than 4 or 5. Furthermore, in the vicinity of a mode resonance ($\text{Re}(Y)=G\to 0$; [6], [7], [19]), the computed value of the dominant mode radiation conductance $G$ is very sensitive to small perturbations in some of the admittance functions, which means that a much larger number of terms in the summations must be used to achieve satisfactory results.

## VI. Conclusions

The complex power formula for an element in an infinite planar array antenna, which was derived above, has been shown to be a powerful, unifying principle for infinite planar array antennas. In particular, we have shown how this generalized power formula can be simply applied to determine the well-known "grating lobe series" for the admittance (or impedance) of an element in an infinite array of slots in a ground plane or an infinite array of flat dipoles with or without a ground plane. The appropriate relations for infinite arrays of crossed-dipole pairs are also very simply derived from the power formula. The relatively complex (and less well-known) solution for the performance of an element in an infinite array of open-ended rectangular waveguides can also be derived in a very systematic and straightforward manner with the aid of the complex power formula.

The utility of the complex power formula lies in its generality; that is, it can be simply applied (conceptually, but not necessarily mathematically) to solve for the performance of an element in an infinite planar array antenna for a wide variety of radiating elements. Specification of an element type merely amounts to applying constraints on the power formula which are characteristic to the particular element. Furthermore, the resultant solution for a specific type of element will be in its most general form because the power formula was derived for an arbitrary, periodic lattice.

Another merit of this formulation is that the solution for multimode elements comes out very naturally in exactly the same form one would use in the analysis of an $N$-port network. The network in this case is defined by an admittance (or impedance) matrix which characterizes the half-space above the array. Each of the $N$-input ports corresponds to one of the modes which can exist in the radiating element. The terminations and/or sources connected to the $N$-ports are determined from the properties of the radiating element and its associated feed structure. Therefore, the problem is separated into two parts. One pertains to the half-space above the array, while the other pertains to the half-space containing the radiating elements. This separation allows us to make various changes to the radiating element and its feed system without affecting the network parameters as long as we do not perturb the mode structure of the radiating element at the aperture plane. What this means is that insertion of matching structures in the radiating elements is equivalent to changing the terminations and/or sources for the network.

We have treated just a few examples from the power formula viewpoint. However, it is not restricted to just these cases. In principal (but not necessarily in practice), it can be applied to any infinite planar array problem where the electric or magnetic field (in normal mode form) can be completely specified over the entire aperture plane; for example, flush mounted, dielectric-slab-loaded, rectangular waveguides, flush mounted circular waveguides, or flush mounted coaxial waveguides. It should also be possible to modify the power formula to accommodate the presence of a dielectric sheet placed some distance above the array aperture plane. An example of a case which cannot be treated with this formalism is an array of waveguides where the waveguides project beyond the ground plane.

## References

[1] H. A. Wheeler, "The radiation resistance of an antenna in an infinite array or waveguide," *Proc. IRE*, vol. 36, pp. 478–487, April 1948.
[2] S. Edelberg and A. A. Oliner, "Mutual coupling effects in large antenna arrays: pt. I—Slot arrays," *IRE Trans. Antennas and Propagation*, vol. AP-8, pp. 286–297, May 1960.
[3] L. Stark, "Radiation impedance of a dipole in an infinite planar phased array," *Radio Sci.*, vol. 1 (new series), pp. 361–377, March 1966.
[4] C. P. Wu and V. Galindo, "Properties of a phased array of rectangular waveguides with thin walls," *IEEE Trans. Antennas and Propagation*, vol. AP-14, pp. 163–173, March 1966.
[5] V. Galindo and C. P. Wu, "Numerical solutions for an infinite phased array of rectangular waveguides with thick walls," *IEEE Trans. Antennas and Propagation*, vol. AP-14, pp. 149–158, March 1966.
[6] G. F. Farrell, Jr., and D. H. Kuhn, "Mutual coupling effects of triangular-grid arrays by modal analysis," *IEEE Trans. Antennas and Propagation (Communications)*, vol. AP-14, pp. 652–654, September 1966.
[7] G. F. Farrell, Jr., and D. H. Kuhn, "Mutual coupling an infinite planar arrays of rectangular waveguide horns," *IEEE Trans. Antennas and Propagation*, vol. AP-16, pp. 405–414, July 1968.
[8] P. M. Woodward and J. D. Lawson, "The theoretical precision with

which an arbitrary radiation pattern may be obtained from a source of finite size," *J. IEE* (London), vol. 95, pt. III, pp. 363–370, September 1948.

[9] D. R. Rhodes, "On a fundamental principle in the theory of planar antennas," *Proc. IEEE*, vol. 52, pp. 1013–1021, September 1964.

[10] J. A. Stratton, *Electromagnetic Theory*. New York: McGraw-Hill, 1950, pp. 361–364.

[11] E. C. Dufort, "Finite scattering matrix for an infinite antenna array," *Radio Sci.*, vol. 2 (new series), pp. 19–27, January 1967.

[12] L. Parad, "The real and reactive power of a planar array," *IEEE Trans. Antennas and Propagation* (*Communications*), vol. AP-13, pp. 990–992, November 1965.

[13] R. E. Collin, *Field Theory of Guided Waves*. New York: McGraw-Hill, 1960, p. 225.

[14] V. Galindo and C. P. Wu, "The relation between the far-zone pattern of the singly excited element and the transmission coefficient of the principal lobe in an infinite array," *IEEE Trans. Antennas and Propagation* (*Communications*), vol. AP-14, pp. 397–398, May 1966.

[15] J. L. Allen, *et al.*, "Phased array radar studies, 1 January 1963 to 1 July 1964," M.I.T. Lincoln Lab., Lexington, Mass., Tech. Rept. 381, pp. 253–257, March 31, 1965.

[16] P. W. Hannan, "The element-gain paradox for a phased-array antenna," *IEEE Trans. Antennas and Propagation*, vol. AP-12, pp. 423–433, July 1964.

[17] H. A. Wheeler, "The grating-lobe series for the impedance variation in a planar phased-array antenna," *IEEE Trans. Antennas and Propagation* (*Communications*), vol. AP-13, pp. 825–827, September 1965.

[18] H. A. Wheeler, "Simple relations derived from a phased-array antenna made of an infinite current sheet," *IEEE Trans. Antennas and Propagation*, vol. AP-13, pp. 506–514, July 1965.

[19] B. L. Diamond, "Resonance phenomena in waveguide arrays, *1967 IEEE G-AP Internat'l Symp. Digest*, pp. 110–115.

[20] G. Borgiotti, "Radiation and reactive energy of aperture antennas," *IEEE Trans. Antennas and Propagation* (*Communications*), vol. AP-11, pp. 94–95, January 1963.



*Reprinted from the* PROCEEDINGS OF THE IEEE
VOL. 56, NO. 11, NOVEMBER, 1968
pp. 1837-1851

*Reprinted from* IEEE TRANSACTIONS
ON *ANTENNAS AND PROPAGATION*
Volume AP-16, Number 4, July, 1968
pp. 405-414


# Mutual Coupling in Infinite Planar Arrays of Rectangular Waveguide Horns

G. F. FARRELL, JR., MEMBER, IEEE, AND D. H. KUHN, SENIOR MEMBER, IEEE

***Abstract*—The radiation admittance and effective power transmission coefficient are derived by a field matching method for arbitrary scan of an infinite array of open rectangular waveguides. Higher order waveguide modes, as well as higher order modes in the free space cell above the array, are included in the field matching problem. The convergence of the solution was studied, and the number of external free space and internal waveguide modes necessary for an adequate answer are given. Numerical results are obtained and compared with previous theory and with experiment for rectangular grids. Comparison is also made with experiment on a triangular grid which exhibited an anomalous notch in the element pattern. The experimental results in both cases seem to substantiate the theory, at least to the extent that could be expected from the small test arrays that were used for the measurements. The comparison indicates that the method of analysis includes all of the necessary features required for the study of mutual coupling effects in infinite arrays of rectangular waveguides having thick walls.**

84

## I. INTRODUCTION

IN RECENT YEARS a very considerable effort has been made by a number of workers to determine the effects on the radiation characteristics of a radiating element when it is placed in an array of identical elements. It has been found that the radiation impedance (or admittance) of an element in an array, in addition to having a variation with the scan angle of the array, also exhibits a strong dependency upon the spacing between elements, the type of grid on which the element is placed, the size and shape of the element, and the proximity of the element to the edge of a finite array. When an array becomes sufficiently large, it has been discovered that a centrally located element does not have a measurably altered set of characteristics with further increase in the physical size of the array. For such a situation, one is free to consider the element as lying in an infinite array environment. Consequently, an infinite array analysis is useful for predicting the performance of the central portion of a finite array, and would give qualitative information for the remainder.

The power and utility of a direct analysis of electromagnetic field coupling in infinite-phased arrays has been recognized by a number of investigators. One of the first efforts to analyze an infinite-phased array in terms of the field in the aperture of an element was that of Edelberg and Oliner,[1] who considered an infinite rectangular grid array of slot radiators having an assumed $TE_{10}$ type aperture field distribution without regard to slot size or array scan angle. Amitay *et al.*,[2] as well as Wu and Galindo,[3],[4] have published results for rectangular arrays of waveguide radiators for H-plane and "quasi-E" plane scan.

It has been pointed out by Allen[5] that various workers have noted the presence of deep notches in the measured radiation pattern of a central element in an array and that these pronounced nulls occur at scan angles somewhat less than those corresponding to a visible grating lobe formation. Allen has predicted such an anomalous behavior on the basis of a phenomenological analysis of surface wave concepts. The simple grating lobe analysis of the problem[1] is inherently incapable of predicting the existence of anomalies attributable to internal multimode effects, since it is based upon an aperture field distribution for an element that remains unchanged when the array is scanned. Farrell and Kuhn[6] and Diamond[7] have presented results which include the effects of multiple modes in waveguide arrays.

It is the purpose of this paper to present a modal method of analysis that is capable of predicting the behavior of an element in an infinite-phased array of rectangular waveguides. The new features of the analysis include 1) arbitrary scan, 2) rectangular or isosceles triangle lattices, and 3) arbitrary thickness of the waveguide walls. The analysis assumes outgoing waves only, and identical generators which are sufficiently removed from the array face so that there is negligible interaction with evanescent modes generated at the array face.

## II. THEORETICAL DERIVATION OF THE RADIATION CHARACTERISTICS

The problem of determining the radiation admittance of the open end of a waveguide acting as an element of an infinite planar array, may be handled by considering the array face to be a junction between two infinite ensembles of waveguides. Those of the free-space region are assumed to be composed of virtual walls that shift the phase of the fields reflected from them by the amount of the phase shift existing between adjacent array elements.[1] This automatically accounts for the fact that the virtual walls are leaky, by allowing energy to pass through one wall into the virtual waveguide volume and then out through the opposite wall. Since the structure is completely periodic, the resulting fields are also periodic, being identical in each periodic cell except for the relative progressive phase. Consequently, the solution for the radiation characteristics of any one unit is simultaneously the solution for any member of the ensemble.

From this point, the problem may be treated by means of the usual methods for considering the junction of two dis-

Manuscript received November 15, 1967; revised March 4, 1968. This work was performed in part under subcontract with M.I.T. Lincoln Laboratory. (Operated with the support of the U. S. Advanced Research Projects Agency.)
The authors are with the Heavy Military Electronics Dept., General Electric Company, Syracuse, N. Y.

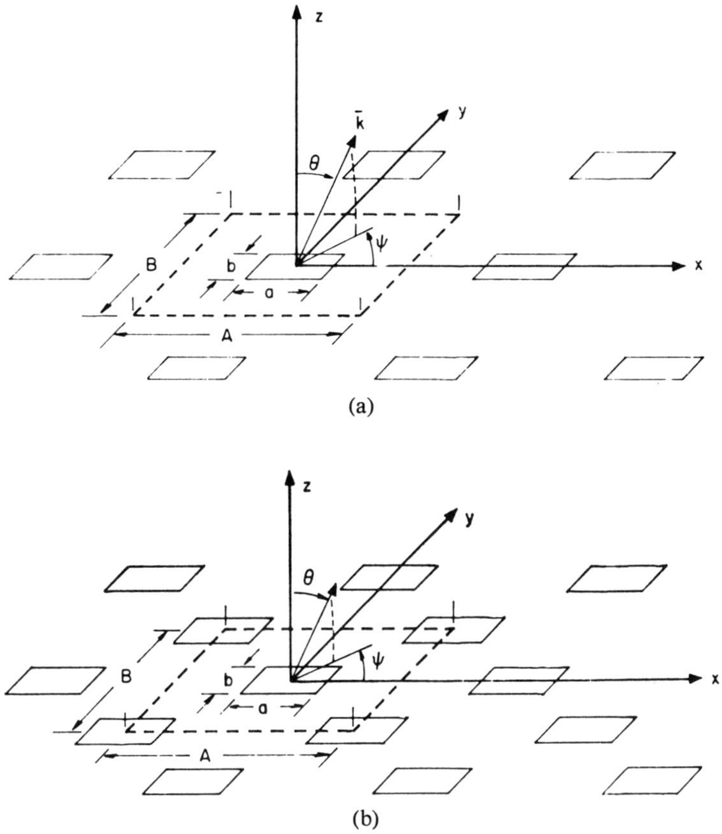


Fig. 1. Array geometries. (a) Rectangular grid. (b) Triangular grid.

similar waveguide structures. One such method is to construct the total possible field in each waveguide in terms of the characteristic eigenfunctions for that waveguide, and then to enforce continuity of the transverse fields in the plane of the junction. This method in general leads to an infinite set of simultaneous equations which may be solved, after judicious truncation, for the mode amplitude coefficients. These coefficients, when substituted into the dominant mode continuity equation for magnetic field, yield an expression for the junction admittance. This method lends itself to a wide range of problems, and is a fruitful approach for all geometries where the junction field distribution is completely unknown but can be expressed in terms of characteristic modes. This is the method that was used to derive the radiation characteristics for the waveguide array.

The radiation admittance equation as obtained by this method consists of the so-called grating lobe series for the dominant mode of the waveguide, plus correction terms. Each correction term is the product of a higher order waveguide mode amplitude coefficient and a series having a form similar to the grating lobe series. The higher order waveguide mode amplitude coefficients are obtained from the solution of a family of simultaneous equations which are derived concurrently with the admittance equation. It is obvious, therefore, that the problem may not be solved exactly, because in practice one is forced to truncate the grating lobe series. There is also a practical limit to the number of waveguide modes that one would wish to consider. Fortunately, the range of eigennumbers over which the grating lobe series needs to be summed is not excessively large, and only a few waveguide modes need to be considered in order to obtain reasonable accuracy.

The rectangular array and triangular array are shown in Fig. 1(a) and (b), respectively. In both cases the unit cell is outlined by a rectangle, the dimensions of which are the units of the lattice periodicity in the two directions. Since the lattice is periodic, each mode of the field in the free space region of the array has a form governed by Floquet's theorem. A suitable form is

$$\exp\left\{-j\left[k_x + \frac{2m\pi}{A}\right]x\right\}$$
$$\exp\left\{-j\left[k_y + \frac{2n\pi}{B}\right]y\right\} \exp\left(\mp \Gamma_{mn} z\right) \tag{1}$$

where

$$k_x = k_0 \sin\theta \cos\psi$$
$$k_y = k_0 \sin\theta \sin\psi$$
$$k_0 = 2\pi/\lambda_0$$

and where $k_x A$ and $k_y B$ are the phase shifts between adjacent lattice units in the $x$ and $y$ directions, respectively.

A phenomenological visualization of the fields above the surface of the array for the case of arbitrary scan indicates that all three components of electric and magnetic fields can be present. This condition is most easily handled by making a linear combination of TE and TM fields. Let $V_{mn}$ and $M_{mn}$ be the respective mode amplitude coefficients for the

$TE_{mn}$ and $TM_{mn}$ fields, and for ease in writing let

$$\beta_m = k_x + \frac{2m\pi}{A}$$
$$\beta_n = k_y + \frac{2n\pi}{B}. \tag{2}$$

The six components of the combined TE-$TM_{mn}$ or $EM_{mn}$ modes are

$$E_x = \sum_{-\infty}^{\infty}\sum_{-\infty}^{\infty} -(V_{mn}\beta_n - M_{mn}\beta_m) e^{-j\beta_m x} e^{-j\beta_n y} e^{\mp\Gamma_{mn} z}$$
$$E_y = \sum_{-\infty}^{\infty}\sum_{-\infty}^{\infty} (V_{mn}\beta_m + M_{mn}\beta_n) e^{-j\beta_m x} e^{-j\beta_n y} e^{\mp\Gamma_{mn} z}$$
$$E_z = \sum_{-\infty}^{\infty}\sum_{-\infty}^{\infty} \mp jM_{mn} \frac{\beta_m{}^2 + \beta_n{}^2}{\Gamma_{mn}} e^{-j\beta_m x} e^{-j\beta_n y} e^{\mp\Gamma_{mn} z}$$
$$H_x = \sum_{-\infty}^{\infty}\sum_{-\infty}^{\infty} = \left[V_{mn}\beta_m \frac{\Gamma_{mn}}{j\omega\mu} + M_{nm}\beta_n \frac{j\omega\epsilon_0}{\Gamma_{mn}}\right] e^{-j\beta_m x} e^{-j\beta_n y} e^{\mp\Gamma_{mn} z} \tag{3}$$
$$H_y = \sum_{-\infty}^{\infty}\sum_{-\infty}^{\infty} \mp \left[V_{mn}\beta_n \frac{\Gamma_{mn}}{j\omega\mu} - M_{mn}\beta_m \frac{j\omega\epsilon_0}{\Gamma_{mn}}\right] e^{-j\beta_m x} e^{-j\beta_n y} e^{\mp\Gamma_{mn} z}$$
$$H_z = \sum_{-\infty}^{\infty}\sum_{-\infty}^{\infty} jV_{mn} \frac{\beta_m{}^2 + \beta_n{}^2}{j\omega\mu} e^{-j\beta_m x} e^{-j\beta_n y} e^{\mp\Gamma_{mn} z}$$

where a time dependence of exp $(j\omega t)$ is understood and

$$\Gamma_{mn} = k_0 \sqrt{\left[\frac{\beta_m}{k_0}\right]^2 + \left[\frac{\beta_n}{k_0}\right]^2 - 1}. \tag{4}$$

$$E_y = \left[E_{pq}\frac{p\pi}{a} + H_{pq}\frac{q\pi}{b}\right] \cdot \sin\frac{p\pi x}{a} \cos\frac{q\pi y}{b} e^{\mp\gamma_{pq} z} \tag{5}$$

with the origin at the lower left-hand corner of the element waveguide and the remaining five field components are written in similar standard forms. A time dependence of exp $(j\omega t)$ is understood, and

$$\gamma_{pq} = k_0 \sqrt{\left[\frac{p\lambda}{2a}\right]^2 + \left[\frac{q\lambda}{2b}\right]^2 - \epsilon_r} \tag{6}$$

where $\epsilon_r$ is the relative dielectric constant in the waveguide.

It is assumed that energy is incident from inside the element waveguides with an amplitude of unity and in the dominant TE mode having $p=1$ and $q=0$, and so the wave has its "principal" polarization in the $y$ direction. It is further assumed that the wave is reflected from the plane of the surface of the array with a voltage reflection coefficient $R$. The total transverse field can now be written for the entire cross section represented by the array space cell. The outline of the array space cell extends in both directions from the face of the array. The waveguides are considered to be long enough so that all evanescent modes decay to negligible amplitude at the generators.

Within a unit cell, at a point $z \leq 0$, below the face of the array the transverse fields are

$$E_x = -\sum_{p=0}^{\infty}\sum_{q=1}^{\infty}\left[E_{pq}\frac{q\pi}{b} - H_{pq}\frac{p\pi}{a}\right]\sum_{i=1}^{5} U_i \cos\frac{p\pi\left[x + \frac{a}{2} - u_{xi}\right]}{a} \sin\frac{q\pi\left[y + \frac{b}{2} - u_{yi}\right]}{b} e^{-jk_x u_{xi}} e^{-jk_y u_{yi}} e^{\gamma_{pq} z}$$

$$E_y = \left[e^{-\gamma_{10} z} + Re^{\gamma_{10} z}\right]\frac{\pi}{a}\sum_{i=1}^{5} U_i \sin\frac{\pi\left[x + \frac{a}{2} - u_{xi}\right]}{a} e^{-jk_x u_{xi}} e^{-jk_y u_{yi}} + \sum_{p=1}^{\infty}\sum_{q=0}^{\infty}{}' \left[E_{pq}\frac{p\pi}{a} + H_{pq}\frac{q\pi}{b}\right]\sum_{i=1}^{5}$$
$$\cdot U_i \sin\frac{p\pi\left[x + \frac{a}{2} - u_{xi}\right]}{a} \cos\frac{q\pi\left[y + \frac{b}{2} - u_{yi}\right]}{b} e^{-jk_x u_{xi}} e^{-jk_y u_{yi}} e^{\gamma_{pq} z} \tag{7}$$

$$H_x = -\left[e^{-\gamma_{10} z} - Re^{\gamma_{10} z}\right]\frac{\pi}{a}\frac{\gamma_{10}}{j\omega\mu}\sum_{i=1}^{5} U_i \sin\frac{\pi\left[x + \frac{a}{2} - u_{xi}\right]}{a} e^{-jk_x u_{xi}} e^{-jk_y u_{yi}} + \sum_{p=1}^{\infty}\sum_{q=0}^{\infty}{}' \left[E_{pq}\frac{p\pi}{a}\frac{\gamma_{pq}}{j\omega\mu} + H_{pq}\frac{q\pi}{b}\frac{j\omega\epsilon_r}{\gamma_{pq}}\right]\sum_{i=1}^{5}$$
$$\cdot U_i \sin\frac{p\pi\left[x + \frac{a}{2} - u_{xi}\right]}{a} \cos\frac{q\pi\left[y + \frac{b}{2} - u_{yi}\right]}{b} e^{-jk_x u_{xi}} e^{-jk_y u_{yi}} e^{\gamma_{pq} z}$$

$$H_y = \sum_{p=0}^{\infty}\sum_{q=1}^{\infty}\left[E_{pq}\frac{q\pi}{b}\frac{\gamma_{pq}}{j\omega\mu} - H_{pq}\frac{p\pi}{a}\frac{j\omega\epsilon_r}{\gamma_{pq}}\right]\sum_{i=1}^{5} U_i \cos\frac{p\pi\left[x + \frac{a}{2} - u_{xi}\right]}{a} \sin\frac{q\pi\left[y + \frac{b}{2} - u_{yi}\right]}{b} e^{-jk_x u_{xi}} e^{-jk_y u_{yi}} e^{\gamma_{pq} z}$$

The upper sign of the double signs, where they occur, designates propagation in the positive $z$ direction. Only the upper sign will be retained in the remainder of the analysis because we are imposing the condition of a transmitting array.

Six components of field can also be expected inside the aperture of the element waveguide, and will be treated as a linear combination of TE and TM modes with respective mode amplitude coefficients $E_{pq}$ and $H_{pq}$. The six field components will be written in a standard form where, for example,

where

$$\sum_{p=1}^{\infty}\sum_{q=0}^{\infty}{}'$$

signifies that the mode $p=1$, $q=0$ is to be excluded from the summation. The summations with "$i$" as the index accommodate the partial waveguide apertures at the four corners of the unit cell in their proper phases relative to the waveguide at the center of the cell. The $u_{xi}$ and $u_{xi}$ give the displacement of the phase center of the $i$th partial waveguide from the center of the cell. The $U_i$ have a nonzero value only over the aperture of the $i$th partial waveguide. The value of the $U_i$ within its own aperture is $I$ except for $i=1$ which is $U_1=1$. This choice for designating the $U_i$ allows for solution of either triangular grid ($I=1$) or rectangular grid ($I=0$) with the same set of equations.

At this point we have expressions for the components of the total possible field that may exist on each side of the plane of the face of the array. In the plane of the junction ($z=0$) the field in the waveguide system must equal and be continuous with the field in the free space region over the cross section of the unit cell. Therefore,

$$\begin{aligned} E_x\,(\text{waveguide})\big|_{z=0} &= E_x\,(\text{free space})\big|_{z=0} \\ E_y\,(\text{waveguide})\big|_{z=0} &= E_y\,(\text{free space})\big|_{z=0} \end{aligned} \tag{8}$$

with a range of validity that extends over the entire cross section of the cell since, at points other than within a waveguide aperture, the electric field is parallel with a conducting surface and is therefore identically zero. A magnetic field parallel with a conducting surface is discontinuous at the surface by an amount of the surface current so the magnetic field continuity equations

$$\begin{aligned} H_x\,(\text{waveguide})\big|_{z=0} &= H_x\,(\text{free space})\big|_{z=0} \\ H_y\,(\text{waveguide})\big|_{z=0} &= H_y\,(\text{free space})\big|_{z=0} \end{aligned} \tag{9}$$

are valid only over the cross sections of the waveguide apertures (or fractional apertures) within the boundaries of the cell.

Since the electric field continuity equations are valid everywhere, let us multiply both sides of each equation by the function

$$e^{j\beta_r x} e^{j\beta_s y}$$

orthogonal to the free space set, and integrate over the cell cross section. The result of the integration on the free space side of the equations is zero except when $r=m$ and $s=n$, from which we obtain expressions for

$$V_{mn}\beta_n - M_{mn}\beta_m$$

and

$$V_{mn}\beta_m + M_{mn}\beta_n \tag{10}$$

from the $E_x$ and $E_y$ continuity equations, respectively. These are readily solved for $V_{mn}$ and $M_{mn}$.

It has been pointed out that the magnetic field continuity equations are valid only within the confines of the waveguide apertures (or partial apertures) enclosed by the cell. Therefore, let us multiply both sides of each equation by the appropriate function for each that is orthogonal to the waveguide set, and integrate over the enclosed waveguide apertures. For $H_x$, the appropriate function is

$$\sum_{i=1}^{5} U_i \sin \frac{r\pi\left(x + \frac{a}{2} - u_{xi}\right)}{a} \cdot \cos \frac{s\pi\left(y + \frac{b}{2} - u_{yi}\right)}{b} e^{jk_x u_{xi}} e^{jk_y u_{yi}}$$

and for $H_y$, it is

$$\sum_{i=1}^{5} U_i \cos \frac{r\pi\left(x + \frac{a}{2} - u_{xi}\right)}{a} \cdot \sin \frac{s\pi\left(y + \frac{b}{2} - u_{yi}\right)}{b} e^{jk_x u_{xi}} e^{jk_y u_{yi}}.$$

Before performing the integration, let us replace $p$ and $q$ by $P$ and $Q$, respectively, in order to distinguish the $P$, $Q$th equation in an infinite family of simultaneous equations from the $p$, $q$ modes in any equation, therefore avoiding confusion later on in the development. The results of the integration on the waveguide side of the equations are zero except when $r=P$ and $s=Q$. From the $H_x$ continuity equation we obtain

$$1 - R = \frac{j\omega\mu}{\gamma_{10}} a \sum_{-\infty}^{\infty}\sum_{-\infty}^{\infty} \frac{\delta_{mn}}{1+I}\left[V_{mn}\beta_m \frac{\Gamma_{mn}}{j\omega\mu} + M_{mn}\beta_n \frac{j\omega\epsilon_0}{\Gamma_{mn}}\right]\frac{\beta_n b}{2} S_m(1)S_n(0) \tag{11}$$

and

$$E_{PQ}\frac{P\pi}{a}\frac{\gamma_{PQ}}{j\omega\mu} + H_{PQ}\frac{Q\pi}{b}\frac{j\omega\epsilon_r}{\gamma_{PQ}} = j2\epsilon_{0Q}\frac{P\pi}{2}\sum_{-\infty}^{\infty}\sum_{-\infty}^{\infty}\frac{\delta_{mn}}{1+I}\left[V_{mn}\beta_m\frac{\Gamma_{mn}}{j\omega\mu} + M_{mn}\beta_n\frac{j\omega\epsilon_0}{\Gamma_{mn}}\right]\frac{\beta_n b}{2} \cdot S_m(P)S_n(Q)e^{j(P\pi/2)}e^{j(Q\pi/2)} \tag{12}$$

where

$$\begin{aligned} \epsilon_{0Q} &= 1 \quad \text{if } Q = 0 \\ &= 2 \quad \text{otherwise} \end{aligned}$$

$$S_m(p) = \frac{\sin\left[\frac{\beta_m a}{2} - \frac{p\pi}{2}\right]}{\left[\frac{\beta_m a}{2}\right]^2 - \left[\frac{p\pi}{2}\right]^2}$$

$$S_n(q) = \frac{\sin\left[\frac{\beta_n b}{2} - \frac{q\pi}{2}\right]}{\left[\frac{\beta_n b}{2}\right]^2 - \left[\frac{q\pi}{2}\right]^2}. \tag{13}$$

From the $H_y$ continuity equation we get

$$E_{PQ}\frac{Q\pi}{b}\frac{\gamma_{PQ}}{j\omega\mu} - H_{PQ}\frac{P\pi}{a}\frac{j\omega\epsilon_r}{\gamma_{PQ}} = j2\epsilon_{0P}\frac{Q\pi}{2}\sum_{-\infty}^{\infty}\sum_{-\infty}^{\infty}\frac{\delta_{mn}}{1+I}\left[V_{mn}\beta_n\frac{\Gamma_{mn}}{j\omega\mu} - M_{mn}\beta_m\frac{j\omega\epsilon_0}{\Gamma_{mn}}\right]\frac{\beta_m a}{2} \cdot S_m(P)S_n(Q)e^{j(P\pi/2)}e^{j(Q\pi/2)} \tag{14}$$

where

$$\epsilon_{OP} = 1 \quad \text{if } P = 0$$
$$= 2 \quad \text{otherwise.}$$

Let us now substitute the expressions for $V_{mn}$ and $M_{mn}$ into (11), (12), and (14). The expressions for $V_{mn}$ and $M_{mn}$, although not written out here, are easily obtained by simultaneous solution of the equations represented by (10). Before performing the substitution it is advantageous to define a new pair of waveguide amplitude coefficients.

$$C_{pq} = \frac{a}{\pi} \frac{1}{1+R} \left[ E_{pq} \frac{q\pi}{b} - H_{pq} \frac{p\pi}{a} \right] \cdot e^{-j(p\pi/2)} e^{-j(q\pi/2)} \tag{15}$$

$$D_{pq} = \frac{a}{\pi} \frac{1}{1+R} \left[ E_{pq} \frac{p\pi}{a} + H_{pq} \frac{q\pi}{b} \right] \cdot e^{-j(p\pi/2)} e^{-j(q\pi/2)}. \tag{16}$$

An investigation of (7) would show that $C_{pq}$ and $D_{pq}$ are the amplitude coefficients for the so called "longitudinal-section-electric" LSE modes having their transverse electric field polarization in the $x$ and $y$ directions, respectively. As a consequence of the usual waveguide dimension nomenclature as used here, $C_{pq}$ and $D_{pq}$ represent the amplitudes of "cross" polarized and "dominant" polarized modes, respectively. (The incident field ($D_{10}$) from the generators is oriented parallel with the $y$ axis.) Further, it is expeditious to define the following summations.

$$J[(p,P)(q,Q)] = \sum_{-\infty}^{\infty} \sum_{-\infty}^{\infty} \frac{\delta_{mn}{}^2}{1+I} \frac{\left[\frac{\beta_m}{k_0}\right] - 1}{\left[\frac{\Gamma_{mn}}{k_0}\right]} \left[\frac{\beta_n b}{2}\right]^2 S_m(p) S_m(P) S_n(q) S_n(Q)$$

$$K[(p,P)(q,Q)] = \sum_{-\infty}^{\infty} \sum_{-\infty}^{\infty} \frac{\delta_{mn}{}^2}{1+I} \frac{\left[\frac{\beta_m}{k_0}\right]\left[\frac{\beta_n}{k_0}\right]}{\left[\frac{\Gamma_{mn}}{k_0}\right]} \frac{\beta_m a}{2} \frac{\beta_n b}{2} S_m(p) S_m(P) S_n(q) S_n(Q) \tag{17}$$

$$L[(p,P)(q,Q)] = \sum_{-\infty}^{\infty} \sum_{-\infty}^{\infty} \frac{\delta_{mn}{}^2}{1+I} \frac{\left[\frac{\beta_n}{k_0}\right]^2 - 1}{\left[\frac{\Gamma_{mn}}{k_0}\right]} \left[\frac{\beta_m a}{2}\right]^2 S_m(p) S_m(P) S_n(q) S_n(Q).$$

In these definitions

$$\delta_{mn} = 1 + I \cos m\pi \cos n\pi$$

where as before "$I$" is unity for a triangular grid and zero for a rectangular grid.

If use is made of the definition for the admittance of a function in a waveguide in terms of the reflection coefficient,

$$Y = \frac{1-R}{1+R}$$

and the indicated substitutions are performed, (11) becomes

$$Y = 2\left[\frac{\pi}{2}\right]^2 \frac{\frac{ab}{AB}}{\frac{\gamma_{10}}{k_0}} \Big[ J[(1,1)(0,0)] + j \sum_{p=1}^{\infty} \sum_{q=0}^{\infty}{}' \, p D_{pq} J[(p,1)(q,0)] + j \sum_{p=0}^{\infty} \sum_{q=1}^{\infty} q C_{pq} K[(p,1)(q,0)] \Big]. \tag{18}$$

This equation is the desired expression for the radiation admittance. This admittance is normalized to the $TE_{10}$ mode admittance. The term $J[(1, 1)(0, 0)]$ is the result one would have if only the dominant mode had been assumed to exist in the waveguide aperture. The other two terms, containing $D_{pq}$ and $C_{pq}$, respectively, are correction terms and for some cases they can be major contributors to the radiation admittance. The coefficients $C_{pq}$ and $D_{pq}$ are determined by solving the double family of infinite simultaneous equations that are obtained from (12) and (14) by substituting the values for $V_{mn}$ and $M_{mp}$. The resulting pair is

$$D_{PQ}\left[\frac{\gamma_{PO}}{k_0}\right]^2 + 2\epsilon_{OQ}\left[\frac{\pi}{2}\right]^2 \frac{ab}{AB} P \left[\frac{\gamma_{PQ}}{k_0}\right] \sum_{p=1}^{\infty} \sum_{q=0}^{\infty}{}' \, p D_{pq} J[(p,P)(q,Q)] + \frac{\lambda^2}{4ab} PQC_{PQ}$$
$$+ 2\epsilon_{OQ}\left[\frac{\pi}{2}\right]^2 \frac{ab}{AB} P \left[\frac{\gamma_{PQ}}{k_0}\right] \sum_{p=0}^{\infty} \sum_{q=1}^{\infty} q C_{pq} K[(p,P)(q,Q)] = j2\epsilon_{OQ}\left[\frac{\pi}{2}\right]^2 \frac{ab}{AB} P \left[\frac{\gamma_{PQ}}{k_0}\right] J[(1,P)(0,Q)] \tag{19}$$

and

$$\frac{\lambda^2}{4ab} PQD_{PQ} + 2\epsilon_{OP}\left[\frac{\pi}{2}\right]^2 \frac{ab}{AB} Q \left[\frac{\gamma_{PQ}}{k_0}\right] \sum_{p=1}^{\infty} \sum_{q=0}^{\infty}{}' \, p D_{pq} K[(p,P)(q,Q)] + C_{PQ}\left[\frac{\gamma_{OQ}}{k_0}\right]^2$$
$$+ 2\epsilon_{OP}\left[\frac{\pi}{2}\right]^2 \frac{ab}{AB} Q \left[\frac{\gamma_{PQ}}{k_0}\right] \sum_{p=0}^{\infty} \sum_{q=1}^{\infty} q C_{pq} L[(p,P)(q,Q)] = j2\epsilon_{OP}\left[\frac{\pi}{2}\right]^2 \frac{ab}{AB} Q \left[\frac{\gamma_{PQ}}{k_0}\right] K[(1,P)(0,Q)]. \tag{20}$$

Note that each of these two general equations provides a specific equation for each eigennumber pair $P$ and $Q$.

## III. Convergence Considerations

In the preceding sections, an expression has been obtained for the radiation admittance. The admittance equation (18) is in the form of a double sum over the waveguide mode numbers of terms which are composed of the waveguide mode amplitude coefficient multiplied by expressions which are double sums over the external eigennumbers. The amplitude coefficients are obtained as solutions of a double set of infinite simultaneous equations, (19) and (20), having terms which are double sums over the external eigennumbers. To effect a solution it is obviously necessary to limit both the internal and the external eigennumbers to finite sets. The problem then is to determine how extensive the range of eigennumbers must be in order to obtain a reasonably accurate answer.

For the purpose of preliminary programming of the equations for computer solution, and considering the unknown but possibly large size of the matrix for the simultaneous equations, it was decided to limit the maximum range of the external mode numbers ($m$ and $n$) to $\pm 30$ in order not to exceed the memory of the available machine. It was subsequently discovered that the range of $\pm 30$ for $m$ and $n$ was generally in excess of those necessary to obtain an adequate answer. A simple formula for the upper and lower limit ($M$) for $m$ was developed by a trial and error investigation on the function $K[(p, P)(q, Q)]$. The effort resulted in the formula

$$M \begin{cases} = IFIX\left\{15 * \left[\frac{PA}{2a}\right]\right\} \\ \quad \text{if } p \neq P \text{ (use the larger of } p \text{ and } P) \\ = 30 \quad \text{if } p = P = 0 \\ = IFIX\left\{15 * \left[\frac{\left[\frac{PA}{2a}\right]}{\sqrt{P}}\right]\right\} \\ \quad \text{if } p = P \neq 0 \\ \leq 30 \quad \text{all } p \text{ and } P \end{cases} \tag{21}$$

where $IFIX(Z)$ indicates the operation of taking the nearest lower integer to the quantity in the brackets. This same basic formula is used to determine the range of $N$ by merely replacing $M, P, A, a$ by $N, Q, B, b$, respectively.

For the solution for the waveguide mode amplitude coefficients, a finite set of simultaneous equations must be selected from the infinite set available. The problem is to include enough of the higher order modes nearest cutoff for the solution of the truncated set to give values that converge to very nearly the right answer, but within a reasonable computation time. Bearing in mind the computer memory limitations, it was decided to select the first 17 "dominant" polarization ($D_{pq}$) and the first 17 "cross" polarization ($C_{pq}$) modes having the lowest propagation constant ($\gamma_{pq}$) since the power content in a higher order mode is roughly in inverse proportion to the propagation constant.

All 34 of these "lowest order" modes were used to determine the admittance at various points in sine theta space for various grid sizes and types. In each of the cases, the mode admittance contribution and the mode power content were tabulated. Then only those modes were chosen which had a power content (absolute value) equal to or greater than an arbitrarily selected fixed percentage of the power in the incident $TE_{10}$ mode. A number of trail computations indicated that power content is a satisfactory criterion for mode selection providing the threshold percentage is established at a sufficiently low value. A level of one-half of one percent was found to be a satisfactory threshold for all grid sizes considered for both grid types, and for all scan directions. Since the cases considered included grids having rather extreme dimensions, this criterion is believed to be safe for all practical cases. The number of modes and their order of importance resulting from this criterion are given in Table I as a function of scan direction and grid type. The criteria given in Table I were tested by varying the number of modes from the recommended number. It was found that the modes, when used in the order given and in the number recommended, will provide results sufficiently accurate for engineering purposes. It was noted that a reduction in the number of modes will, in many cases, cause a marked deterioration in accuracy.

89

## IV. Computed Results

Equations (19) and (20) have been solved using formula (21) to establish the range of the external modes and Table I to set the number and order of internal waveguide modes according to the grid type and scan direction. The resulting internal waveguide mode amplitude coefficients were then substituted into (18) to obtain the radiation admittance. The reflection coefficient was calculated from the admittance and this in turn was used to calculate the element realized pattern. The results are presented in Figs. 2 through 10.

Solutions obtained by the use of the modal theory, in comparison with the H-plane solutions obtained by Galindo and Wu,[4] are given in Figs. 2 and 3 for the variations in the absolute value of the reflection coefficient with scan angle in the H plane of a rectangular grid. Fig. 3 is for a square grid (0.5714×0.5714) composed of square waveguides having zero thickness walls. It can be seen that the results of the modal theory, and those obtained by Galindo and Wu by a function theoretic technique, very nearly coincide. The grating lobe series is included for reference, and it predicts the correct shape for the curve but not the correct magnitude. Fig. 3 is for the same square grid as Fig. 2, except that the side walls of the waveguide are thick, but the top and bottom walls are still of zero thickness. Again the modal theory solution and the solution given by Galindo and Wu nearly coincide, and again the grating lobe series only predicts the proper shape. It appears, therefore, that the theory given here is in agreement with that of Galindo and Wu for the H-plane scan of a rectangular grid array.

A comparison of the computed power pattern of a square grid made up of square waveguides with the measured power pattern[1] is shown in Fig. 4. The grating lobe series is in-

[1] These experimental data were furnished by the Research and Development Center, General Electric Co., Schenectady, N. Y.

## TABLE I

MODE ORDER AND NUMBER OF MODES AS A FUNCTION OF SCAN DIRECTION AND GRID TYPE BASED ON PERCENTAGE MODE POWER CONTENT

| Scan Direction | H-Plane | | E-Plane | | Intercardinal | |
|---|---|---|---|---|---|---|
| Mode | Rectangular | Triangular | Rectangular | Triangular | Rectangular | Triangular |
| Order | PQ | PQ | PQ | PQ | PQ | PQ |
| | | | Dominant Incident Mode | | | |
| 0 | 10 | 10 | 10 | 10 | 10 | 10 |
| | | | Dominant Polarization Higher Order Modes ($D_{pq}$) | | | |
| 1 | 20 | 20 | 11 | 11 | 11 | 20 |
| 2 | 30 | 12 | 12 | 30 | 12 | 11 |
| 3 | | 22 | 13 | 12 | 20 | 12 |
| 4 | | 14 | 14 | 31 | 13 | 22 |
| 5 | | 42 | 30 | | 21 | 30 |
| 6 | | 40 | 31 | | 14 | 21 |
| 7 | | 30 | 32 | | 30 | 14 |
| 8 | | 24 | | | | |
| | | | Cross Polarization Higher Order Modes ($C_{pq}$) | | | |
| 1 | | 22 | 11 | 11 | 11 | 01 |
| 2 | | 02 | 12 | 31 | 01 | 21 |
| 3 | | 42 | 31 | | 21 | 11 |
| 4 | | 24 | 32 | | 31 | 22 |
| 5 | | | 14 | | 12 | 12 |
| 6 | | | 13 | | 02 | |
| 7 | | | 33 | | 41 | |

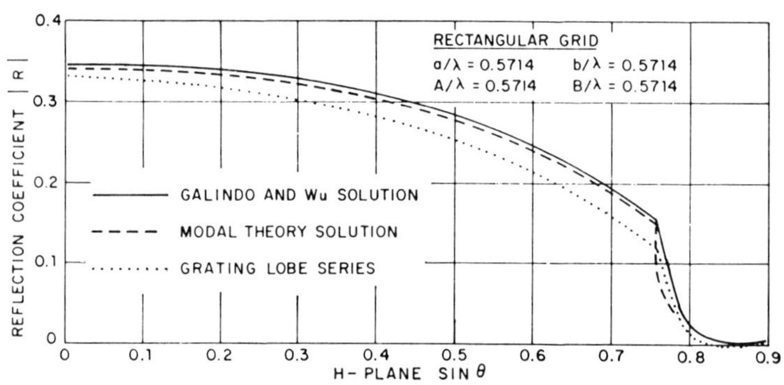


Fig. 2. Comparison of modal theory with solution given by Galindo and Wu, thin wall case.

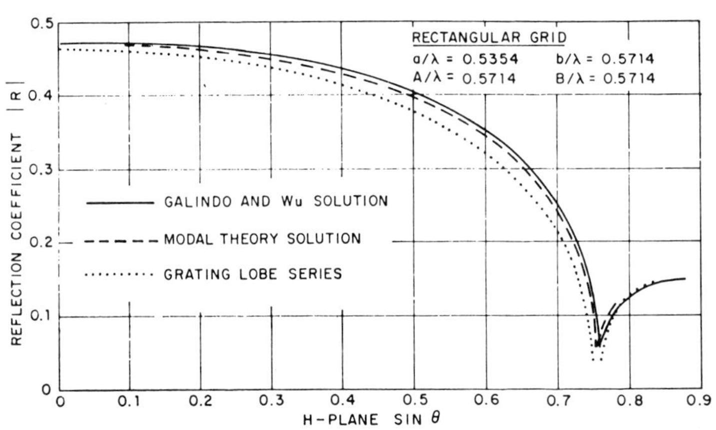


Fig. 3. Comparison of modal theory with solution given by Galindo and Wu, thick side wall case.

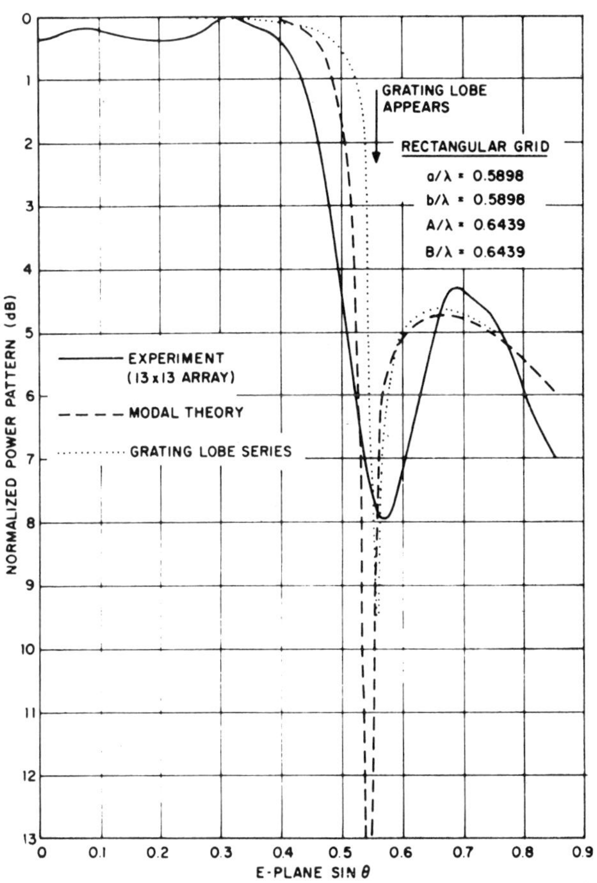


Fig. 4. Comparison of theory and experiment for a square grid [$A/\lambda=0.6439$].

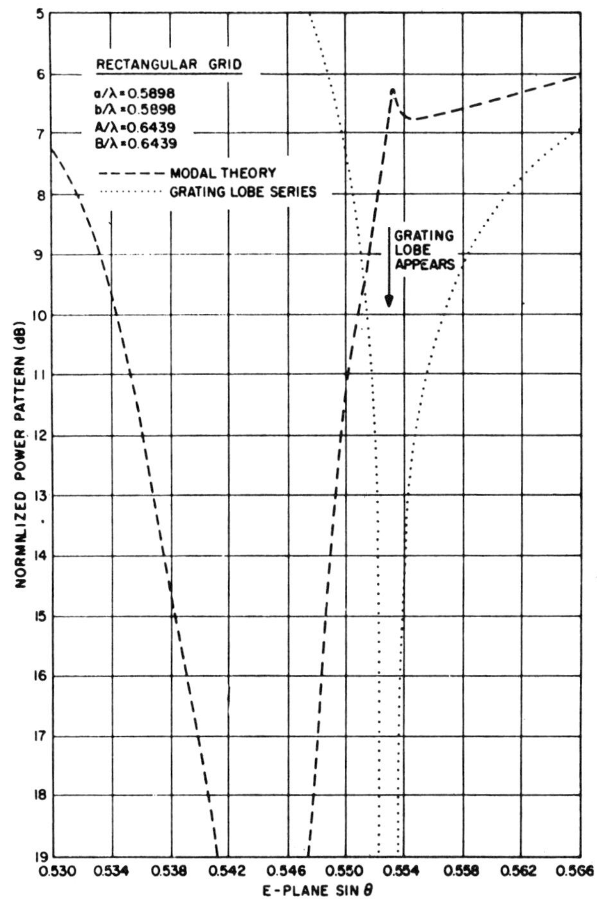


Fig. 5. Magnification of the region containing the null for a square grid [$A/\lambda=0.6439$].

cluded for reference. The modal theory prediction and the grating lobe series prediction appear to be very much alike. The agreement between theory and experiment is reasonably good. The modal theory prediction is slightly better than that of the grating lobe series in the main lobe region. The disagreement with the experiment in the grating lobe onset region is attributed to the small size of the experimental array (13×13). A "blow-up" of the region $0.530<\sin\theta<0.566$ from Fig. 4 is shown in Fig. 5. Here it is easily seen that the modal theory and the grating lobe series predictions are quite different. The grating lobe series predicts a null at the grating lobe onset position. The modal theory predicts a null inside the grating lobe onset position. The latter prediction is a very peculiar one because at the null position the radiation conductance is zero, whereas the null according to the grating lobe series is due to an infinite susceptance. This unexpected situation is shown in Table II at $\sin\theta=0.545$ where the conductance has a nearly zero value (0.018989) according to the modal theory but is 1.075194 according to the grating lobe series. The grating lobe onset position is very nearly $\sin\theta=0.553$ at which the grating lobe series would predict an infinite susceptance, whereas the modal theory gives a finite value of −6.685541. A comparison of the absolute value of the reflection coefficient according to the two theories through the same region as Fig. 5 is given in Fig. 6.

The theoretical and experimental response of a very interesting triangular grid array made up from large rectangular waveguides are shown in Figs. 7 through 9. The experimental power patterns were measured on a 95-element planar array.[2] The H-plane measured pattern (Fig. 7) exhibits a 32-dB null at $\sin\theta\doteq0.45$. For this array the H-plane onset position of the grating lobe is at $\sin\theta\doteq0.866$. Such a null is remarkable since it is so strongly pronounced, and also because it occurs so far inside the grating lobe onset position. Comparison of this experimental curve with the grating lobe series prediction shows that the grating lobe series is completely inadequate for describing this physical situation. Also shown in Fig. 7 is the modal theory prediction which does predict a null well inside the grating lobe onset position and which is in reasonable agreement with the measurement. The difference is probably attributable to the small size of the experimental array. This null, as was the case for the rectangular grid, is due to a zero radiation conductance. Comparisons of the predictions of the modal theory with experiment for a 45° cut in intercardinal space are shown in Fig. 8, and comparisons for a cut along the E-plane are shown in Fig. 9. Both figures show good agreement of the modal theory with experiment except that the E-plane case shows a null which the experiment does not

[2] These experimental data were furnished by courtesy of M.I.T. Lincoln Laboratory, Lexington, Mass.

TABLE II

COMPARISON OF ADMITTANCE PREDICTION ACCORDING TO MODAL THEORY AND GRATING LOBE SERIES ALONG E PLANE

| E Plane sin θ | Normalized Admittance | | | |
|---|---|---|---|---|
| | Modal Theory | | Grating Lobe Series | |
| | G | B | G | B |
| 0.525 | 0.346229 | 1.554860 | | |
| 0.530 | 0.269700 | 1.885190 | 1.084332 | 1.038385 |
| 0.535 | 0.174163 | 2.378268 | | |
| 0.540 | 0.059932 | 3.221669 | 1.078238 | 1.598947 |
| 0.545 | 0.018989 | 5.109655 | 1.075194 | 2.204214 |
| 0.550 | 4.415969 | 13.844779 | 1.072152 | 3.942191 |
| 0.551 | 14.627278 | 16.449801 | 1.071544 | 4.934046 |
| 0.552 | 22.769887 | −1.884326 | 1.070936 | 7.137518 |
| 0.553 | 7.071423 | −6.685541 | 1.070328 | 47.164832 |
| 0.554 | 6.084207 | −2.119235 | 8.901583 | −0.440792 |
| 0.555 | 5.462520 | −1.324210 | 6.588497 | −0.440560 |
| 0.560 | 4.058891 | −0.422704 | 4.051292 | −0.439413 |
| 0.570 | | | 3.036932 | −0.437177 |



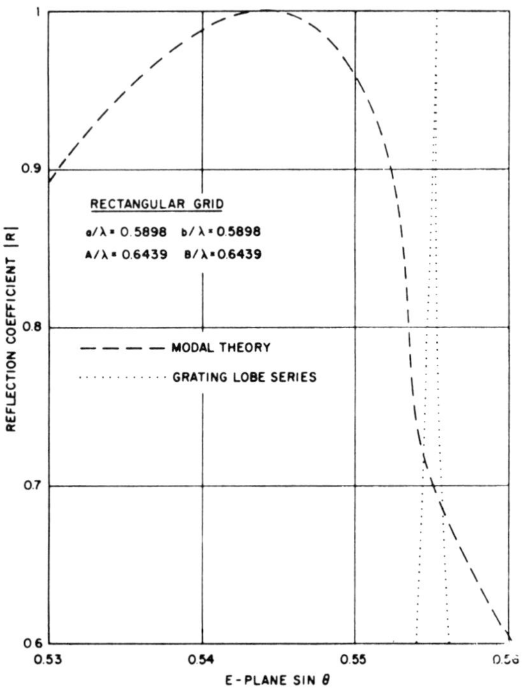


Fig. 6. The reflection coefficient through the null of a square grid [$A/\lambda = 0.6439$].

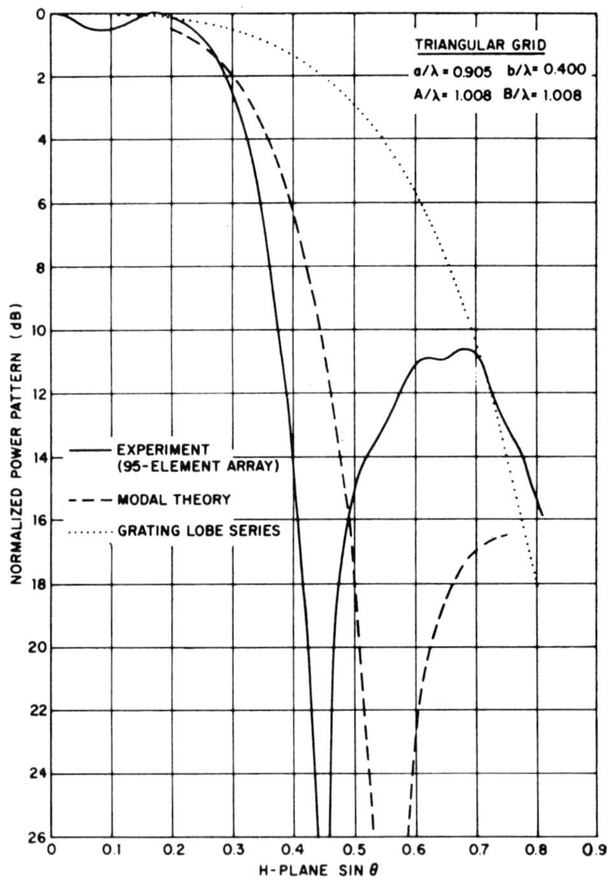


Fig. 7. Comparison of theory and experiment for a triangular grid, H-plane scan.

indicate. In both cases the theory predicts the null to occur inside the grating lobe onset position.

A contour map of the normalized radiation pattern over one quadrant of sin $\theta$ space for the same grid as Figs. 7, 8, and 9 is shown in Fig. 10. It shows a large triangular region near the H-plane axis over which the response is down by more than 25 dB. It also shows a line notch lying along, but inside, the grating lobe circle which is indicated by the dotted line. The dashed lines represent contours that have been estimated rather than actually computed. This map was obtained by using the E-plane modes listed in Table I to calculate the response for points in region A. The H-plane modes were used for points in region B and the intercardinal modes for region C. The implied assumption in doing this is that the E-plane modes, for example, although strictly valid only for points actually on the E-plane axis would still be adequate for points not too far off the axis.

## V. CONCLUSIONS

A general theory has been developed which describes the radiation characteristics of infinite arrays of rectangular waveguides arranged as either a rectangular or a triangular grid. The theory is capable of handling an arbitrary scan angle. The solution was obtained by matching a finite number of modes selected from the infinite set of characteristic

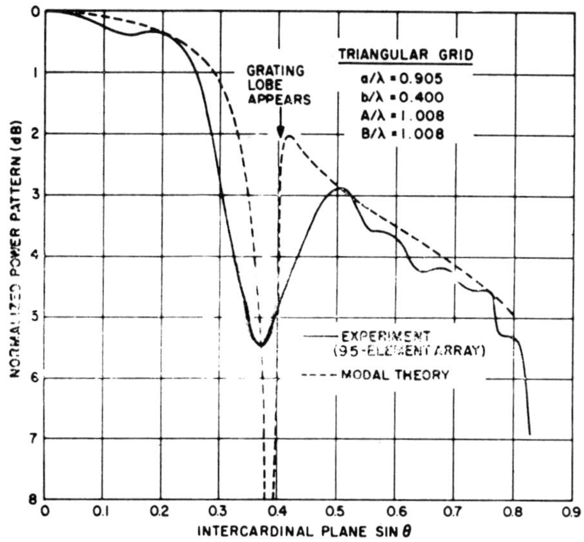


Fig. 8. Comparison of theory and experiment for a triangular grid, intercardinal plane scan.

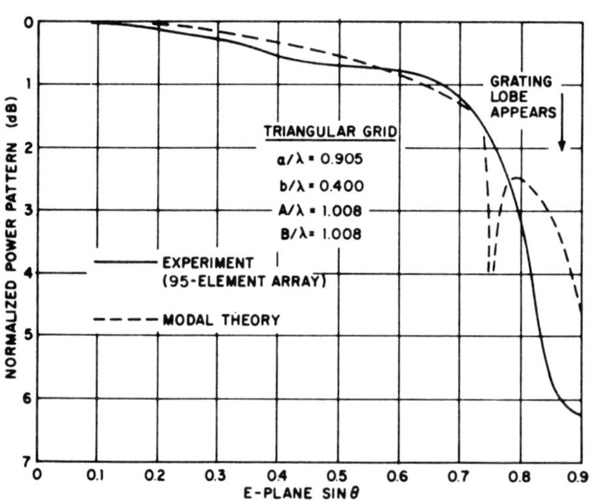


Fig. 9. Comparison of theory and experiment for a triangular grid, E-plane scan.

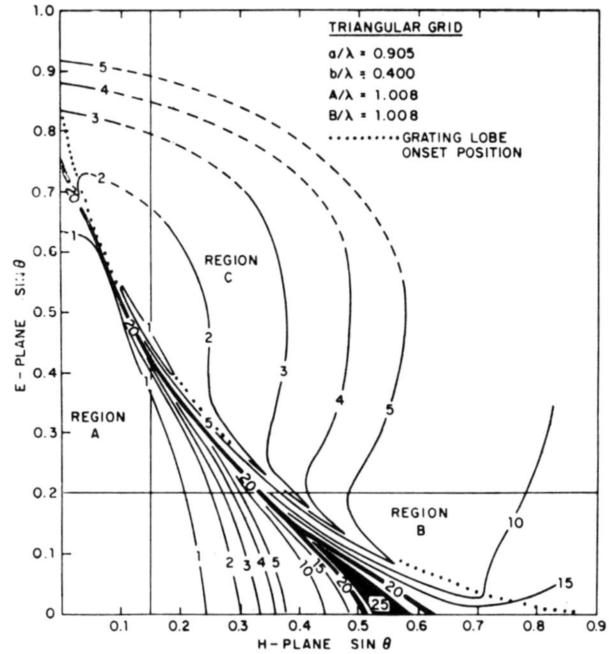


Fig. 10. Contour map (in decibels) of the radiation pattern of a triangular grid.

waveguide modes, with a likewise restricted number of modes that are characteristic to a free space cell of the infinite array. Obviously, the problem cannot be solved exactly, because in practice one is forced to limit the total number of modes that are used. It has been found, however, that the range of eigennumbers over which the external modes must be taken is not excessively large, and only a few waveguide modes need to be considered in order to obtain reasonable accuracy.

Numerical results were obtained and comparisons were made with previous theory and with experiment for a square grid array of square waveguides. Comparisons of theory and experiment were also made for a triangular grid array of rectangular waveguides for H-plane, E-plane, and intercardinal plane cuts through the radiation pattern. The measured H-plane pattern exhibited a 32-dB anomalous notch located well inside the grating lobe onset position. Good agreement with experiment was obtained using the modal theory. Some features predicted by the theory, such as the E-plane notch for the triangular grid, were not apparent in the data measured on small test arrays.

The above results have considerable significance. It has long been known that deep notches can occur in array-element radiation patterns inside the grating lobe onset position and may not be apparent except in the very large array environment. The analysis technique described here predicts such conditions. These results indicate that the method of analysis presented here includes all of the necessary features required for the study of mutual coupling effects in infinite arrays of rectangular waveguides having thick walls.

## Acknowledgment

We wish to gratefully acknowledge the substantial efforts, suggestions, and encouragement of B. Diamond of M.I.T. Lincoln Lab., Lexington, Mass.

## References

[1] S. Edelberg and A. A. Oliner, "Mutual coupling effects in large antenna arrays: Part I—slot arrays," *IRE Trans. Antennas and Propagation*, vol. AP-8, pp. 286–297, May 1960.

[2] N. Amitay, J. S. Cook, R. G. Pecina, and C. P. Wu, "On mutual coupling and matching conditions in large planar phased arrays," *IEEE G-AP Internat'l Symp. Digest*, 1964.

[3] C. P. Wu and V. Galindo, "Properties of a phased array of rectangular waveguides with thin walls," *IEEE Trans. Antennas and Propagation*, vol. AP-14, pp. 163–173, March 1966.

[4] V. Galindo and C. P. Wu, "Numerical solutions for an infinite phased array of rectangular waveguides with thick walls," *IEEE Trans. Antennas and Propagation*, vol. AP-14, pp. 149–158, March 1966.

[5] J. L. Allen, "On surface-wave coupling between elements of large arrays," *IEEE Trans. Antennas and Propagation (Communications)*, vol. AP-13, pp. 638–639, July 1965.

[6] G. F. Farrell, Jr., and D. H. Kuhn, "Mutual coupling effects in infinite planar arrays of rectangular waveguide horns," *IEEE G-AP Internat'l Symp. Digest*, 1966.

[7] B. L. Diamond, "Resonance phenomena in waveguide arrays," *IEEE G-AP Internat'l Symp. Digest*, 1967.

RADIO SCIENCE, Vol. 1 (New Series), No. 3, March 1966

# Radiation Impedance of a Dipole in an Infinite Planar Phased Array

L. Stark

TP 72–14–22

Hughes Ground Systems, Hughes Aircraft Company, Fullerton, Calif. 92634, U.S.A.

(Received June 11, 1965; revised November 1, 1965)

The radiation impedance of a dipole in an infinite array phased for any angle of radiation is calculated. The calculation is made by means of plane wave expansions and gives a very simple result. The radiation resistance is the product of the broadside resistance times a scanning factor which is unity on broadside and changes the resistance as the beam is scanned. The scanning factor is identified as the product of the element pattern and a factor equal to the ratio of the broadside aperture to the projected aperture. The radiation reactance is also calculated and consists of a leading term due to reflection of the wave front from the ground plane plus an infinite series of higher order-terms contributed by the cutoff waves. Results are plotted for the variation of radiation resistance and reactance with changes in radiation angle. A comparison is also made between results calculated by the method of plane wave expansions and those obtained from the method employing mutual impedance coefficients.



## 1. Introduction

The radiation resistance of a dipole in an infinite array phased for broadside radiation has been calculated previously by Wheeler [1948]. Certain symmetry conditions which are obeyed by the fields in the array permit the transformation of this problem to that of a typical isolated wave channel or waveguide, as shown in figure 1. The waveguide consists of opposing electric and opposing magnetic walls, and in it is located a single radiating element. The results of this calculation may be applied, approximately, to obtain the resistance of an element in a finite antenna array, provided that attention is centered on elements sufficiently far in from the edge of the array. The extent of the edge effect was studied by Blasi [1953], who used P. S. Carter's mutual impedance formulation and results [1932] to show that the impedance of a dipole in a half-wave spaced array is not affected by elements more than approximately three spaces removed from the element in question. Therefore in a large array the edge effect should be confined to an edge ring approximately three elements wide.

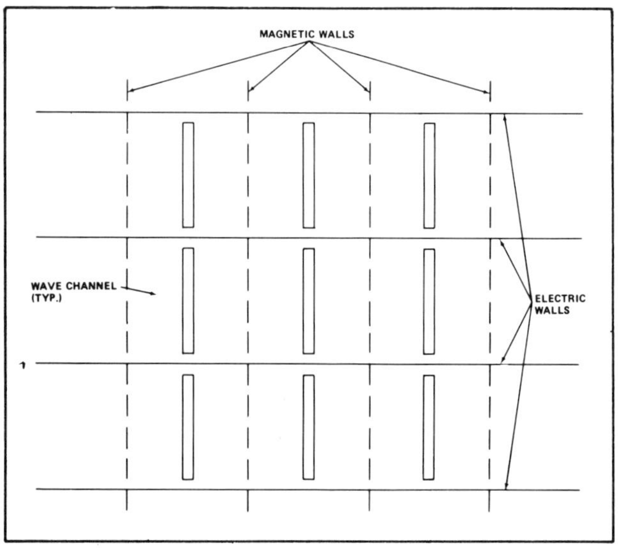


FIGURE 1. *Symmetry of fields in array of dipoles allows placement of electric and magnetic walls to form wave channels.*

P. S. Carter, Jr. [1960] considered the problem of a finite array phased for an arbitrary firing angle and it was shown that the edge effect engrosses a larger number of elements when the beam is scanned and depends upon the scan angle. This phenomenon can be related to the approach of a grating lobe of the array to real space. When the grating lobe just enters real space, the element phasing produces in-phase addition of the mutual coupled signals along the surface of the array, giving rise to slow convergence of the coupling series and producing a large edge effect. A larger edge effect will also be manifested in arrays of wider element spacing even when phased for broadside radiation, also due to an impending grating lobe.

To the extent that the edge effect can be ignored in a practical array, the infinite-array model can be used to provide many results very simply. The present paper will be concerned with radiation impedance in an infinite array and will extend the work of Wheeler to the case of arbitrary radiation angle. As in the case of Wheeler's results, the expression for radiation resistance proves to be extremely simple to evaluate and one therefore has a means of estimating radiation resistance for a physical problem without a large matrix calculation. The result proves to have simple physical significance in that it embodies the notions of element pattern and projected aperture in the direction of radiation. Radiation reactance obtained from the infinite-array model is not formally as simple a result since it is determined from an infinite number of evanescent modes of the structure and must be calculated by summing an infinite series. The result has been formally derived, however, and numerical results have been calculated for some special cases of interest.

## 2. Symbols

$\omega$ = angular frequency

$k = \dfrac{2\pi}{\lambda}$ where $\lambda$ = free-space wavelength

$Z$ = impedance

$\bar{E}$ = complex vector electric-field intensity

$\bar{K}$ = complex vector surface current density

$\bar{H}$ = complex vector magnetic-field intensity

$dS$ = element of area, scalar

$I_0$ = current amplitude at center of element, peak value

$a$ = element spacing in $y$-dimension

$b$ = element spacing in $z$-dimension

$L$ = length of current element

$\delta$ = width of current element

$\beta$ = fundamental propagation constant of current wave along $y$

$h$ = fundamental propagation constant of current wave along $z$

$\bar{i}_x, \bar{i}_y, \bar{i}_z$ = unit vectors along $x$, $y$, and $z$, respectively

$K_z$ = $z$-component of surface current density

$K_{m,n}$ = Fourier coefficient in expansion of $K_z$

$E_x, E_y, E_z$ = complex electric-field components along $x$, $y$, and $z$, respectively

$H_x, H_y, H_z$ = complex magnetic-field components along $x$, $y$, and $z$, respectively

$E_z^{m,n}$ = Fourier coefficient in expansion of $E_z$

$\gamma_{m,n}$ = propagation constant of a plane wave along $x$-coordinate

$\mu$ = magnetic permeability of free space = $4\pi \times 10^{-7}$ MKS units

$\epsilon$ = permittivity of free space = $\dfrac{1}{36\pi} \times 10^{-9}$ MKS units

$s$ = array to reflector spacing

$P_c$ = complex power flowing out of array element

$K_{0,0}$ = Fourier coefficient in expansion of $K_z$ for $m=0$, $n=0$.

## 3. Method of Solution

The problem is solved by the technique of a Fourier series expansion of the field in plane waves. The infinite extent of the array is the factor that allows for this simplicity. It is also required that all elements be excited with equal amplitude and that the phase shift per element of the array be constant along each coordinate of the array. These restrictions meet a number of physical problems quite well. The effect of tapering the amplitude distribution cannot be included by this method, but this should exert a very small effect on impedance when the amplitude difference per element is small, as in the case of a large array [Blasi, 1953].

The current distribution is first expanded in a set of current sheets which are waves traveling across the surface of the array. The radiated and evanescent fields excited by this distribution are then calculated. The electric field is then traced back to the current elements and the impedance of an element is calculated by a method equivalent to the induced emf method,

$$Z_{\text{element}} = \frac{\displaystyle\iint_{\text{element}} -\frac{\bar{E}\cdot\bar{K}^*}{2}\,dS}{\frac{1}{2}\,|I_0|^2}.$$

By Poynting's theorem the integral in the numerator is the complex power flowing out of the radiating element [Stratton, 1941]. Therefore the impedance is given by dividing the integral by the square of the rms input current. It can be shown by taking the imaginary part of the complex Poynting vector that positive reactive power corresponds to excess magnetic energy stored or inductive reactance.

It is not possible to solve this type of problem exactly. Various current distributions on the element can be assumed, each producing a certain variation of electric field surrounding the element. The correct distribution is one that produces zero tangential electric field on the surface of the element, remaining finite at the generator gap. It is not in general possible to derive the exact form of the current wave but it has been shown that a sinusoidal distribution is very nearly correct for a thin element near half-wave resonance [King and Sandler, 1964]. The displaced cosine component of the current waveform, which is normally the first correction term, provides no correction to the current distribution at half-wave resonance and only a small correction in this vicinity. The results to be obtained for element impedance are, moreover, insensitive to the current distribution.

## 4. Formal Solution for Input Impedance

Figure 2 shows the array geometry. The array consists of current strips separated a distance $a$ in the $y$-dimension and $b$ in the $z$-dimension. The strips are of length $L$ and width $\delta$ and are infinitely thin. To begin with, the fields will be calculated without a backup reflector present. The effect of a reflector will then be included by adding the fields of the image distribution. Let the phase shift along the $y$-co-

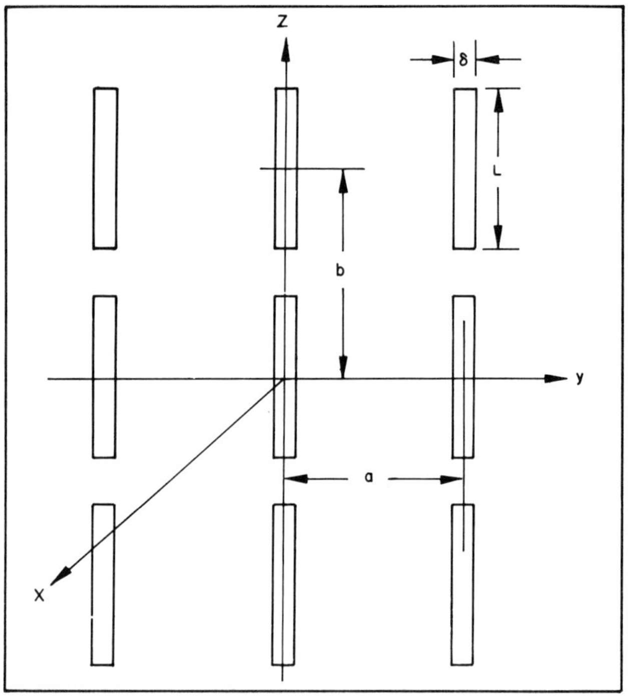


FIGURE 2. *Geometry of the dipole array.*

ordinate be $\beta a$ radians per dipole, and along the $z$-coordinate, $hb$ radians per dipole. The phase will lag increasingly along the positive coordinate directions. The surface current density in the plane of the array ($x=0$) can then be represented by the complex Fourier series,

$$\bar{K}=\bar{i}_z K_z=\bar{i}_z \sum_{n=-\infty}^{\infty} \sum_{m=-\infty}^{\infty} K_{m,n} e^{-j\frac{2\pi m}{a}y} e^{-j\frac{2\pi n}{b}z} e^{-j\beta y} e^{-jhz}. \quad (1)$$

It is seen that the phase shifts increase by $\beta a$ in a displacement $a$ along the $y$-axis and by $hb$ in a displacement $b$ along the $z$-axis. Furthermore, a complete set of Fourier harmonic terms is available to represent an arbitrary current distribution on the strips. The Fourier coefficients $K_{m,n}$ will be determined by assuming a sinusoidal loop of current on the strips. By combining the coefficients of $y$ and $z$ in (1), eq (1) can be written

$$K_z=\sum_{m=-\infty}^{\infty} \sum_{n=-\infty}^{\infty} K_{m,n} e^{-j\left(\beta+\frac{2\pi m}{a}\right)y} e^{-j\left(h+\frac{2\pi n}{b}\right)z}. \quad (2)$$

A time variation $e^{j\omega t}$ is understood. In this form each term is shown to be a traveling current wave along the surface having a propagation constant

$$\beta_m=\left(\beta+\frac{2\pi m}{a}\right) \text{ along } y, \text{ and } h_n=\left(h+\frac{2\pi n}{b}\right) \text{ along } z.$$

To derive the fields which arise from the current distribution it is noted that no $z$-component of $\bar{H}$ is required since there is no component of current along $y$ or $x$ to excite it. Therefore $H_z \equiv 0$ and the fields are TM with respect to the $z$-axis. All fields can be derived from $E_z$. Corresponding to the current-excitation eq (2), the field $E_z$ is

$$E_z=\sum_{m=-\infty}^{\infty} \sum_{n=-\infty}^{\infty} E_z^{m,n} e^{-j\beta_m y} e^{-jh_n z} e^{-j\gamma_{m,n}x}. \quad (3)$$

The $x$-variation of the field $e^{-j\gamma_{m,n}x}$ is not arbitrary but must be determined such that $E_z$ satisfies the wave equation, $(\nabla^2+k^2)E_z=0$. This yields $-\beta_m^2-h_n^2-\gamma_{m,n}^2+k^2=0$, which determines the value of $\gamma_{m,n}$. Equation (3) is an expansion in plane waves of $E_z$, excited by the prescribed current distribution. The remaining field components are determined in a well-known way [Collin, 1960]:

$$\bar{i}_x H_x+\bar{i}_y H_y=j\sqrt{\frac{\epsilon}{\mu}}\frac{k}{p^2}\left\{\bar{i}_z \times \nabla_T E_z\right\}$$

$$=j\sqrt{\frac{\epsilon}{\mu}}\frac{k}{p^2}\left\{\bar{i}_y \frac{\partial E_z}{\partial x}-\bar{i}_x \frac{\partial E_z}{\partial y}\right\} \quad (4a)$$

$$\bar{i}_x E_x+\bar{i}_y E_y=j\frac{\Gamma}{p^2}\nabla_T E_z=j\frac{\Gamma}{p^2}\left\{\bar{i}_x \frac{\partial E_z}{\partial x}+\bar{i}_y \frac{\partial E_z}{\partial y}\right\} \quad (4b)$$

where $\Gamma$ is the propagation constant in the $z$-direction and $p^2=\Gamma^2-k^2$.

Equation (4a) in particular allows the calculation of $H_y$,

$$H_y=j\sqrt{\frac{\epsilon}{\mu}}\frac{k}{p^2}\frac{\partial E_z}{\partial x}. \quad (5)$$

Now by relating $E_z$ to $H_y$ through (5) and then relating $H_y$ to the current distribution on the array the fields are completely specified. By using (3) and (5) one gets

$$H_y=j\sqrt{\frac{\epsilon}{\mu}} \sum_{m=-\infty}^{\infty} \sum_{n=-\infty}^{\infty} \frac{k}{h_n^2-k^2}(-j\gamma_{m,n}) E_z^{m,n}$$

$$e^{-j\beta_m y} e^{-jh_n z} e^{-j\gamma_{m,n}x}. \quad (6)$$

In the plane of the array ($x=0$) there is a boundary condition to apply on $H_y$ which states that the discontinuity in magnetic field is equal to the surface current density. By symmetry it is assumed that the magnetic field is equal and opposite on the two sides of the array. One therefore has

$$H_y\big|_{x=0+}=\tfrac{1}{2}K_z. \quad (7)$$

Applying the above condition to (6) and (2) one has

$$E_z^{m,n} = \sqrt{\frac{\mu}{\epsilon}} \frac{h_n^2 - k^2}{k\gamma_{m,n}} \frac{K_{m,n}}{2} \tag{8}$$

which relates the plane wave amplitudes of $E_z$ to the exciting current coefficients.

In the more useful physical problem, the array is not allowed to radiate in both directions. A ground plane or reflector is placed behind at some distance, determined by impedance considerations. If the reflector is placed a distance $s$ behind the array in the plane $x=-s$, an image array is formed at $x=-2s$ as shown in figure 3. The field produced by the image array $E_{zi}$ is given in terms of the primary field by

$$E_{zi} = -E_z(x+2s). \tag{9}$$

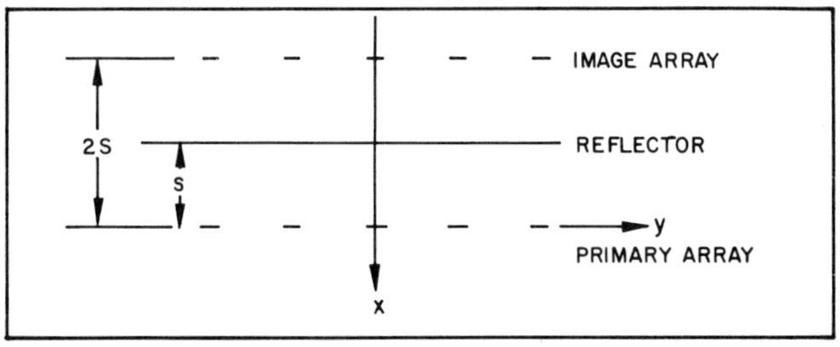


FIGURE 3. *Geometry of image array and reflector.*

The minus sign occurs because of the inversion of the current elements into the image. Thus the total field is the sum of the primary field plus the image field and is equal to

$$E_z = \sum_{m=-\infty}^{\infty} \sum_{n=-\infty}^{\infty} \sqrt{\frac{\mu}{\epsilon}} \frac{K_{m,n}}{2} \frac{h_n^2 - k^2}{k\gamma_{m,n}} (1 - e^{-j\gamma_{m,n}2s})$$
$$e^{-j\beta_m y} e^{-jh_n z} e^{-j\gamma_{m,n} x}, \tag{10}$$

which one obtains by using (8) and (9) in (3).

Equation (10) gives the correct value of $E_z$ everywhere in front of the reflector. By taking the field at the array and forming the integral of $\frac{-\overline{E}\cdot\overline{K}^*}{2}$ over the area of a current strip one can compute the complex power flowing out of the strip. The expression for $\overline{K}^*$ will contain exponentials similar to that of (10). When the product of the two series is formed and the integration is performed over a cell $(a, b)$ it is found that most of the terms vanish due to the orthogonality of the exponentials over the area of integration. Only products of terms having the same periods contribute, with the result that the complex power

$$P_c = -\frac{1}{2}\sqrt{\frac{\mu}{\epsilon}}\, ab \sum_{m=-\infty}^{\infty} \sum_{n=-\infty}^{\infty} \frac{K_{m,n}K_{m,n}^*}{2} \frac{h_n^2 - k^2}{k\gamma_{m,n}}$$
$$(1 - e^{-2j\gamma_{m,n}s}). \tag{11}$$

Radiation resistance is contributed by the real part of the complex power. All factors in (11) are always real with the exception of $\gamma_{m,n}$ and $e^{-2j\gamma_{m,n}s}$. When $\gamma_{m,n}$ is imaginary the $(m, n)$th term of the sum is imaginary. In the case of practical importance the real part of (11) comes from the first term only, the term for which $m=0$, $n=0$, for which $\gamma_{m,n}$ is real. This corresponds to one radiating wave and all higher waves evanescent. To insure that only one wave radiates one must have

$\gamma_{m,n} = \sqrt{k^2 - (\beta_m^2 + h_n^2)}$ is imaginary for all $m$, $n$ except $m=0$, $n=0$,

or

$$\beta_m^2 + h_n^2 > k^2 \text{ for all } m, n \text{ except } m=0, n=0.$$

This can be insured for all angles of radiation in the forward hemisphere by spacing the elements a half-wavelength in the two principle directions. A graphical presentation of maximum allowable element spacing as dictated by the contour of maximum radiation angle and single-wave radiation has been given by Blass and Rabinowitz [1957] and Sharp [1961].

The evaluation of the Fourier coefficients $K_{m,n}$ for use in (11) proceeds from (1) after assuming a current distribution for the strips. The distribution for a half-wave dipole at the origin having a current amplitude $I_0$ at the center is

$$K_z = \frac{I_0}{\delta}\cos\frac{\pi z}{L} \left\{ \begin{array}{l} -\frac{\delta}{2} < y < \frac{\delta}{2} \\ \\ -\frac{L}{2} < z < \frac{L}{2} \end{array} \right. \tag{12}$$

$$= 0, \text{ otherwise in area } a, b.$$

To evaluate the coefficients, both sides of (1) are first multiplied by $e^{+j\beta y}e^{+jhz}$. The calculation then proceeds by making use of the orthogonality of the exponentials over the area $a$, $b$. The result of carrying this out gives

$$K_{m,n} = \frac{2}{\pi}\frac{I_0}{a}\frac{L}{b}\frac{\sin\beta_m\delta/2}{\beta_m\delta/2}\frac{\cos h_nL/2}{1-\left(\frac{h_nL}{\pi}\right)^2}. \tag{13}$$

These Fourier coefficients apply to the case of half-wave dipoles and one may therefore substitute $\lambda/2$ for $L$. The computation of radiation resistance comes from (11) by taking the real part of the $m=0$, $n=0$ term and dividing by $\frac{1}{2}|I_0|^2$. Using (13) for $K_{0,0}$ one gets radiation resistance

$$R = \sqrt{\frac{\mu}{\epsilon}} \frac{L^2}{ab}\left(\frac{2}{\pi}\right)^2 \frac{k^2-h^2}{k\gamma_{0,0}} \left[\frac{\cos hL/2}{1-\left(\frac{hL}{\pi}\right)^2}\right]^2 \sin^2\gamma_{0,0}s. \tag{14}$$

In obtaining (14) the $\frac{\sin x}{x}$ factor in (13) was approximated by unity. This approximation is good for much higher $m$ also, since the dipole width is small in terms of cell width. The radiation reactance is contributed by the imaginary part of the 0, 0 term plus all higher terms. For quarterwave spacing of the reflector and broadside radiation angle the contribution of the 0, 0 term to reactance vanishes.

The physical significance of the propagation factors $\beta$, $h$, and $\gamma_{0,0}$ which appear throughout is readily established by reference to the radiating wave and its direction of propagation. The radiating wave is

$$E_z^{0,0}=A_{0,0}\,e^{-j\beta y}\,e^{-jhz}\,e^{-j\gamma_{0,0}x};$$

whereas the general plane-wave form which shows the direction cosines of a vector in the direction of propagation is

$$E_z=A\,e^{-jky\cos\theta_y}\,e^{-jkz\cos\theta_z}\,e^{-jkx\cos\theta_x}.$$

By direct comparison, one has

$$\cos\theta_y=\frac{\beta}{k}$$

$$\cos\theta_z=\frac{h}{k}\qquad\text{Also }\cos^2\theta_x+\cos^2\theta_y+\cos^2\theta_z=1.\tag{16}$$

$$\cos\theta_x=\frac{\gamma_{0,0}}{k}.$$

By making these substitutions in (14) and, further, substituting $L=\frac{\lambda}{2}$, one gets

$$R=\sqrt{\frac{\mu}{\epsilon}}\left(\frac{2}{\pi}\right)^2\frac{\lambda^2}{4ab}\sin^2 ks\left\{\left[\frac{\cos\left(\frac{\pi}{2}\cos\theta_z\right)}{\sin\theta_z}\right]^2\frac{\sin^2(ks\cos\theta_x)}{\sin^2 ks}\right\}\frac{1}{\cos\theta_x}.\tag{17a}$$

The expression for radiation reactance may also be written

$$X=\sqrt{\frac{\mu}{\epsilon}}\left(\frac{2}{\pi}\right)^2\frac{\lambda^2}{4ab}\left\{\left[\frac{\cos\left(\frac{\pi}{2}\cos\theta_z\right)}{\sin\theta_z}\right]^2\frac{1}{2}\frac{\sin(2ks\cos\theta_x)}{\cos\theta_x}+\sum_{m=-\infty}^{\infty}{}'\sum_{n=-\infty}^{\infty}{}'F_m^2\,G_n^2\,H_{m,n}\right\}$$

where $\sum'$ means zero term excluded and

$$F_m=\frac{\sin\left[\frac{\pi\delta}{\lambda}\left(\cos\theta_y+\frac{m\lambda}{a}\right)\right]}{\frac{\pi\delta}{\lambda}\left(\cos\theta_y+\frac{m\lambda}{a}\right)}\tag{17b}$$

$$G_n=\frac{\cos\left[\frac{\pi}{2}\left(\cos\theta_z+n\frac{\lambda}{b}\right)\right]}{1-\left[\cos\theta_z+\frac{n\lambda}{b}\right]^2}$$

$$H_{m,n}=\frac{\left[\cos\theta_z+\frac{n\lambda}{b}\right]^2-1}{\sqrt{\left(\cos\theta_z+\frac{n\lambda}{b}\right)^2+\left(\cos\theta_y+\frac{m\lambda}{a}\right)^2-1}}$$

Equation (17) is a detailed form for resistance and reactance, which was obtained by dividing (11) by the square of the rms current at the dipole center. It has been assumed in obtaining (17) that over the range of radiation angles and dipole spacings of interest, only the $m=0$, $n=0$ wave radiates and all others are evanescent. The factor $e^{-2j\gamma_{m,n}s}$ in (11) has been neglected compared with unity for all but the 0, 0 wave, implying that all higher waves are well below cutoff and are essentially damped out in a distance $2s$. This equation therefore may not be employed when a grating lobe is imminent or existent. The analysis may include the case of grating lobes out to any order by retaining the accuracy of (11). If the array parameters are such that one grating lobe can arise in scanning along either the $y$- or $z$-coordinates, then the $m=0$, $n=-1$, and the $m=-1$, $n=0$ wave will become radiative in addition to the $m=0$, $n=0$ wave. Starting with (11), a revised form of (17) may be derived accordingly. As the radiation condition is approached for a grating lobe the damping factor $\exp -j2\gamma_{m,n}s$ may not be ignored compared with unity, since $\gamma_{m,n}$ becomes small and then zero. The correct limiting form for $P_c$ when the grating lobe just occurs ($\gamma_{m,n}=0$) is obtained by grouping the terms $\frac{1-e^{-2j\gamma_{m,n}s}}{\gamma_{m,n}}$ in $P_c$, all other terms remaining well-behaved and finite. Letting $\gamma_{m,n}=-j\alpha_{m,n}$, this factor takes the form

$$\frac{1-e^{-2j\gamma_{m,n}s}}{\gamma_{m,n}}=j\frac{1}{\alpha_{m,n}}\left[2\alpha_{m,n}s+\frac{(2\alpha_{m,n}s)^2}{2}+\ldots\right]$$

$$\sim 2js\text{ as }\alpha_{m,n}\rightarrow 0.\tag{18}$$

Thus the complex power, and hence the impedance, remains finite at the grating-lobe phasing condition. It can also be seen from (10) that if the image field (reflector) is absent, the factor $e^{-2j\gamma_{m,n}s}$ is absent and the complex power [eq (11)] becomes infinite at the grating lobe phasing condition with no reflector present. In one special case, corresponding to $E$-plane scanning in the array ($\beta=0$), the numerator factor $h_n^2-k^2$ in $P_c$ cancels $\gamma_{m,n}$ in the denominator for the $m=0$, $n=-1$ grating lobe and the contribution of the grating-lobe term to $P_c$ goes to zero rather than infinity.

For many practical cases the results of (17) are useful as they stand. From (17) the radiation resistance can be written in the form

$$R=R_{bs}\,(SF) \tag{19}$$

where $R_{bs}$ is the resistance on broadside and $(SF)$ is the factor by which this is multiplied as the direction of radiation is scanned from broadside. The factor $SF$ is normalized to unity on broadside. Thus,

$$R_{bs}=\frac{120}{\pi}\,\frac{\lambda^2}{ab}\,\sin^2 ks \tag{20a}$$

$$SF=\left\{\left[\frac{\cos\left(\frac{\pi}{2}\cos\theta_z\right)}{\sin\theta_z}\right]^2\frac{\sin^2(ks\cos\theta_x)}{\sin^2 ks}\right\}\frac{1}{\cos\theta_x}. \tag{20b}$$

Equation (20a) agrees with the result previously obtained by Wheeler [1948] for radiation resistance. The extension of Wheeler's result which includes the effect of beam scanning is given by (20b). It is interesting to note that the scanning factor can be identified as the product of the isolated element factor and a projected aperture factor, the element factor in braces { } being the pattern of a half-wave dipole spaced a distance $s$ in front of a ground plane. The projected aperture factor is $\frac{1}{\cos\theta_x}$. This type of product relation for the scanning factor was also exhibited in a result obtained by Rabinowitz in the analysis of a slot array [Rabinowitz, 1958]. In that case the element factor was that of a slot in a conducting surface.

This type of result can be simply deduced from antenna-array theory by a consideration of the total power in the radiation pattern when fixed current sources are used to drive the current elements (fixed voltage sources in the case of slots). Let it be assumed that the array of figure 2 is scanned in the $E$-plane by changing the phase of otherwise constant current sources. The peak radiation-pattern amplitude decreases according to the isolated antenna element pattern and the antenna beam becomes wider according to the projected aperture rule, as shown in figure 4. The total power in the pattern, proportional to the area under the pattern, therefore goes as

$$P_T=\text{const.}\ x\,\frac{F_{\text{isol.}}}{\cos\theta},$$

where $F_{\text{isol.}}$ = isolated element power pattern, and

$\theta$ = scan angle,

and the radiation resistance therefore follows the same variation since constant current sources have been assumed. To show that the array pattern amplitude follows the isolated element pattern it is only necessary to apply the principle of superposition. At the scan angle $\theta$ the total field may be considered to be the sum of separate contributions of rows of radiators along the $y$-axis. When each row is energized the current sources in all other rows are reduced to zero. The field intensity at angle $\theta$ is therefore determined as though the other rows were missing from the array. This is exactly the case for an array of hypothetical current strips and very nearly the case for an array of halfwave conducting strips when the strips are open-circuited at the driving gaps. The previous calculation culminating in (20b) gives precisely the isolated element pattern as the factor in resistance change. In truth, the current distribution on the conducting strips will change slightly with scan angle to produce a small modification to this result. It should be pointed out that in the infinite array the element driving-point impedance is independent of the type of source used and therefore the result obtained above is not restricted to the use of current sources.

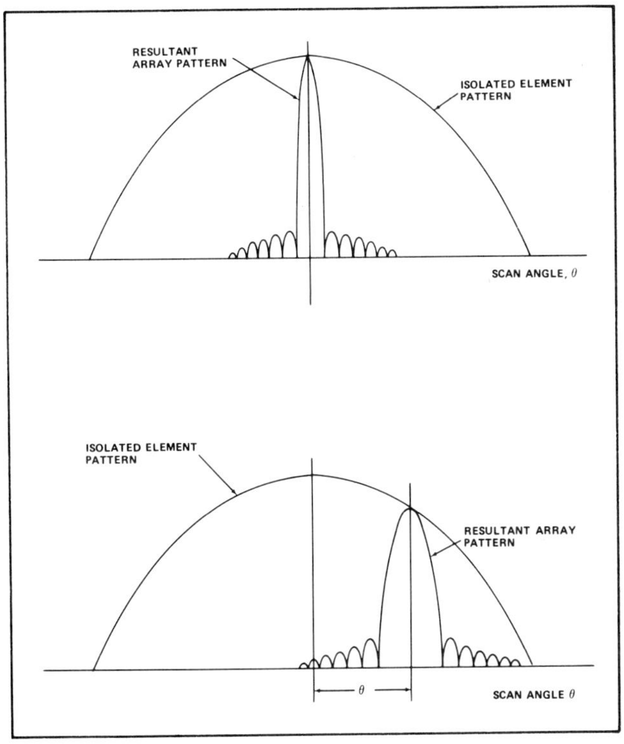


FIGURE 4. *Properties of antenna array pattern when scanned by changing phase of constant current sources.*

The expression obtained for dipole reactance in (17) is a doubly infinite series. The first term of the series is due to the reflected wave from the ground plane interacting with the dipole. This term is the dominant term of the series. The series converges slowly, however, and additional terms must be summed to obtain good accuracy.

The factor ($SF$) by which the resistance is changed due to scan, (20b), was calculated for a number of array-to-reflector spacings. The results have been presented by plotting contours of constant ($SF$) in the cos $\theta_y$, cos $\theta_z$ plane, where cos $\theta_y$ and cos $\theta_z$ are the direction cosines of a vector pointing in the main beam direction. The dipoles are oriented as shown in figure 2, and thus $H$-plane scanning is along the horizontal axis and $E$-plane scanning along the vertical axis. The results are shown in figures 5, 6, and 7. It is noted that when the array-to-reflector spacing is 0.125λ the resistance drops monotonically as the beam is scanned from broadside. For larger spacing, resistance peaks occur, first along the $H$-plane axis, and, as the spacing is further increased, a peak develops along the $E$-plane axis. The second peak is just beginning to develop for the spacing of figure 7.



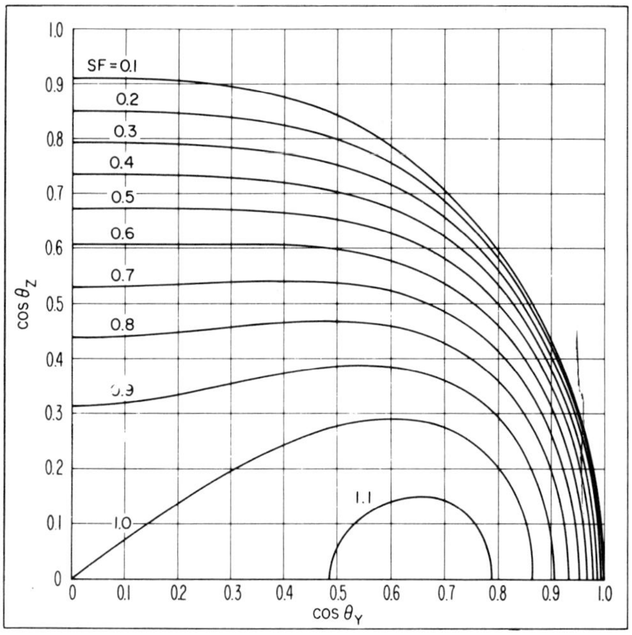


FIGURE 6. *Contours of constant* SF *in the cos $\theta_y$, cos $\theta_z$ plane.* Array-reflector spacing is 0.25λ.

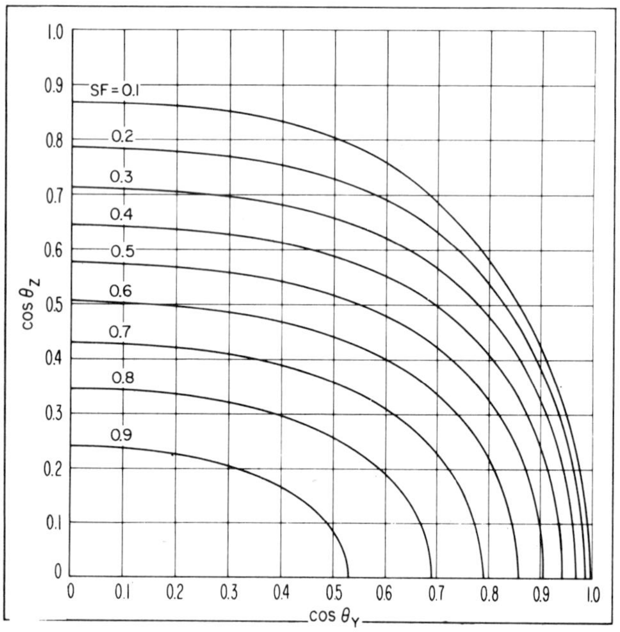


FIGURE 5. *Contours of constant* SF *in the cos $\theta_y$, cos $\theta_z$ plane.* Array-reflector spacing is 0.125λ.

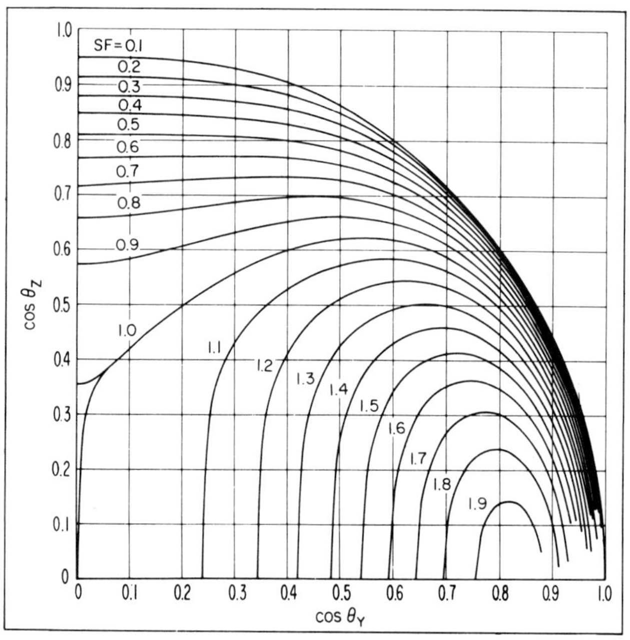


FIGURE 7. *Contours of constant* SF *in the cos $\theta_y$, cos $\theta_z$ plane.* Array-reflector spacing is 0.33λ.

These contour plots assume that the array elements are sufficiently closely spaced that no grating lobes occur throughout the scan range. Radiation resistance has also been calculated for element spacings and scan angles permitting the formation of grating lobes. Figures 8, 9, and 10 show resistance versus scan angle for scanning in the *E*-plane, *H*-plane, and *D*-(diagonal) plane for element spacings of 0.5λ, 0.6λ, and 0.7λ, respectively. The array-reflector spacing is 0.25λ for all of these cases. For the latter two cases, grating lobes occur when the beam is scanned along the principle *E*- and *H*-planes. The results show that the grating-lobe discontinuity is much less pronounced when scanning in the *E*-plane and this may be attributed to the wide null in the element pattern centered at endfire along the *Z*-axis. The crossed data points on the curves show results calculated from a mutual impedance formulation of the element impedance problem, employing P. S. Carter's mutual impedance coefficients, and the agreement is seen to be excellent. A very large array was used for the mutual impedance calculation, one consisting of 65 collinear by 149 parallel elements.

The radiation reactance has also been calculated numerically from (17), modified to account for grating lobes when necessary. Reactance as a function of scan angle is shown in figures 11, 12, and 13 for element spacings of 0.5λ, 0.6λ, and 0.7λ, respectively. The array-to-reflector spacing is 0.25λ for these cases. The grating-lobe effect is again least noticeable in the case of *E*-plane scan. As shown for the general case by (11) and (18) the impedance remains finite when the grating lobe phasing condition occurs. The change in reactance is always greater when scanning in the *H*-plane than when scanning in the *E*-plane. The change in resistance obeys the inverse rule, manifesting greater change for *E*-plane scanning. Crossed data points show element reactance calculated from a mutual impedance formulation, and again the agreement is seen to be good.

The rate of convergence of the series expression for reactance, (17), was studied by summing the first $N$ terms, where $N=1$, 9, 121, and 441. These results are plotted in figure 14. A very small increase in reactance values is evident between the last two cases. It is also evident that the curves for all but the smallest $N$ values have nearly the same shape so that reactance changes from the broadside value may be accurately computed with a relatively small number of terms.

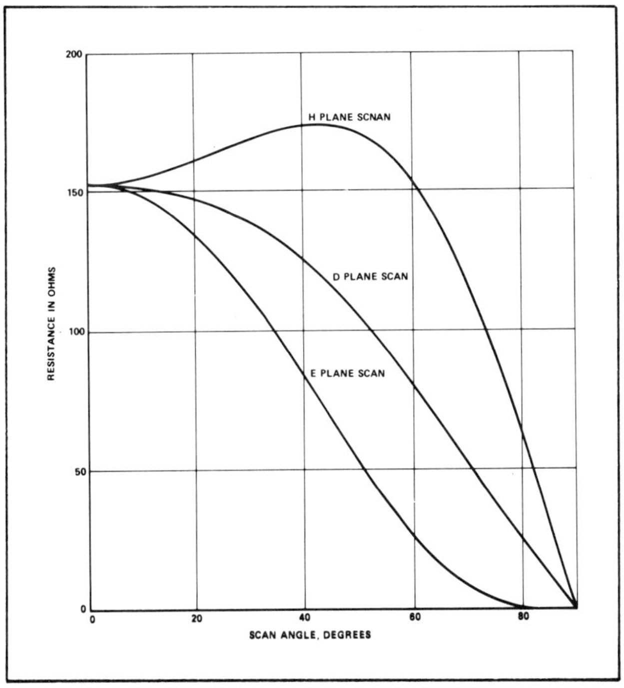


FIGURE 8. *Element driving resistance versus scan angle as obtained from (17).*

Half-wave dipoles, $a=b=0.5$ λ, $s=0.25$ λ.

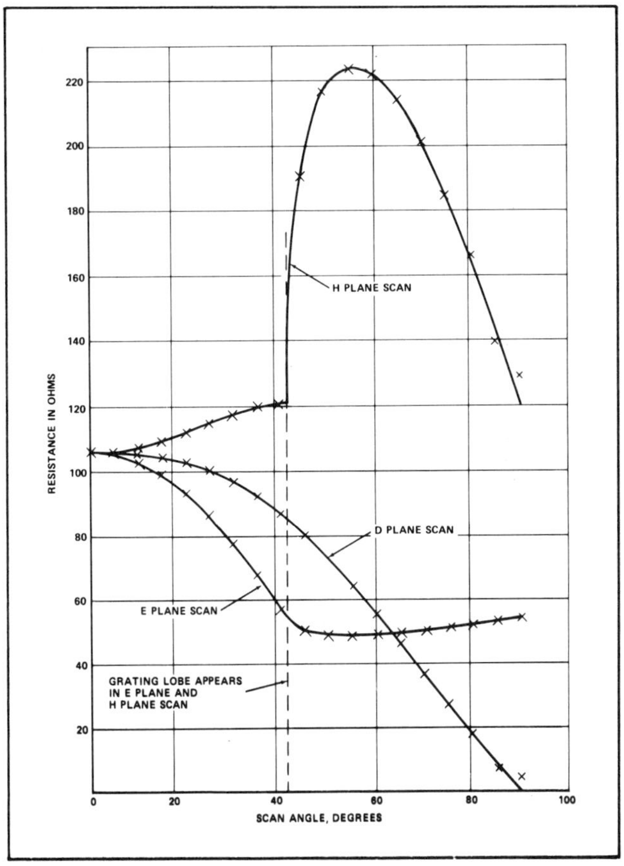


FIGURE 9. *Element driving resistance versus scan angle as obtained from modified equation (17).*

Half-wave dipoles, $a=b=0.6$ λ, $s=0.25$ λ. The $x$'s are data points obtained from the mutual impedance solution for a 65 (collinear) × 149 (parallel) array.

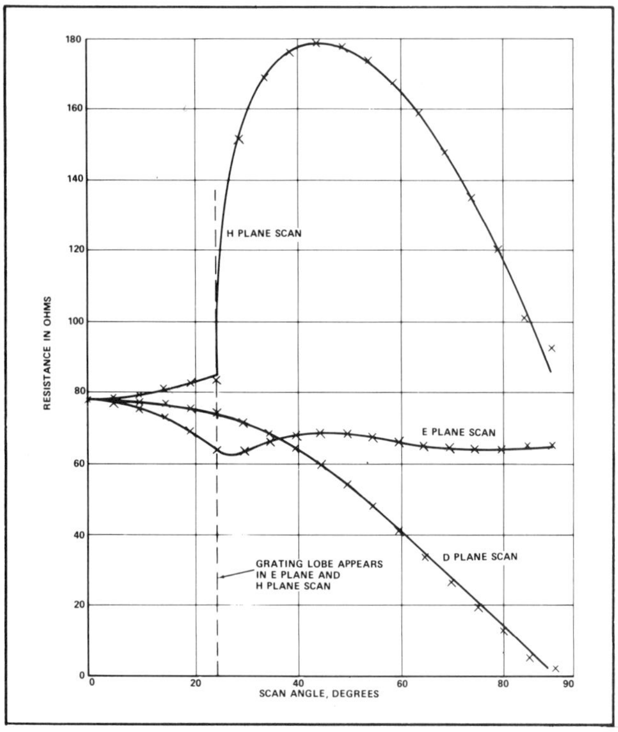


FIGURE 10. *Element driving resistance versus scan angle as obtained from modified eq (17).*

Half-wave dipoles, $a=b=0.7\ \lambda$, $s=0.25\ \lambda$. The $x$'s are data points obtained from the mutual impedance solution for a 65 (collinear) × 149 (parallel) array.

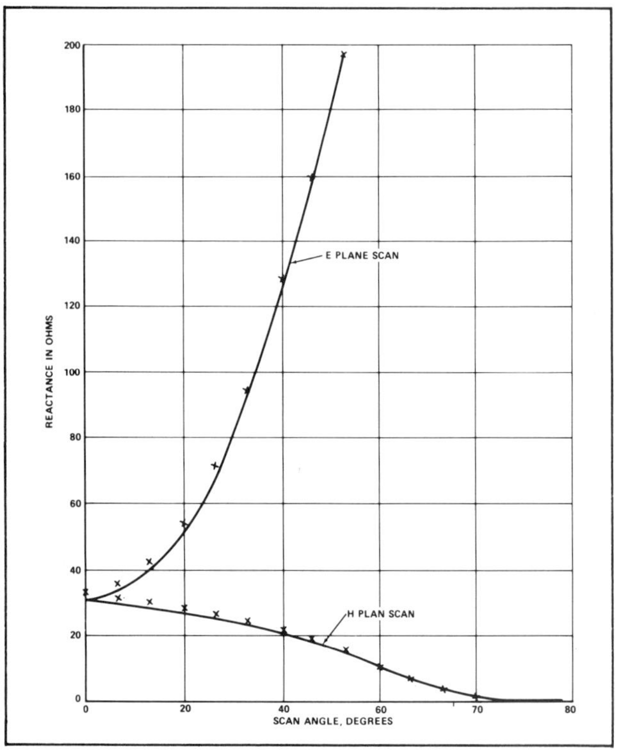


FIGURE 11. *Element driving reactance versus scan angle as obtained from (17).*

The series is truncated at $m=\pm 10$, $n=\pm 10$. Half-wave dipoles, $a=b=0.5\ \lambda$, $s=0.25\ \lambda$. The $x$'s are data points obtained from the mutual impedance solution for a 65 (collinear) × 149 (parallel) array.

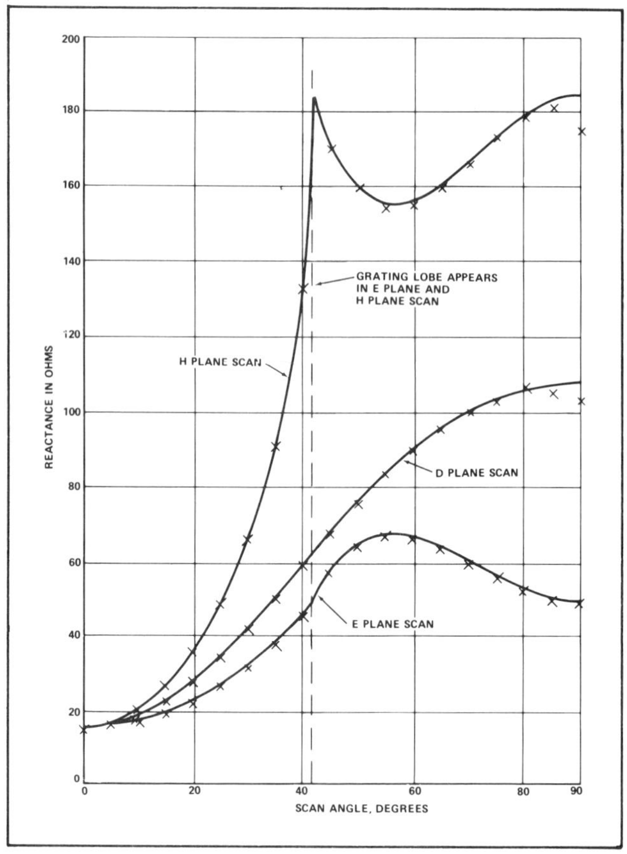


FIGURE 12. *Element driving reactance versus scan angle as obtained from modified eq (17).*

The series is truncated at $m=\pm 75$, $n=\pm 25$. Half-wave dipoles, $a=b=0.6\ \lambda$, $s=0.25\ \lambda$. The $x$'s are data points obtained from the mutual impedance solution for a 65 (collinear) × 149 (parallel) array.

## 5. Conclusions

Results of previous work dealing with element impedance in arrays has been cited. This has shown the need and the applicability of the present calculation of radiation impedance in an infinite array under the conditions of variable direction of radiation. The calculation has been carried out in sufficient detail to present the method of calculation, as well as the results. Both resistance and reactance were obtained. The expression for resistance is of simple form and it was pointed out that the change in resistance from the value on broadside is the product of the element factor and the projected aperture factor. The reactance has also been calculated by summing a series, and plots of resistance and reactance versus scan angle have been given. The effect of the grating lobe on impedance has been shown. A comparison has been made with element impedance calculated from a mutual impedance formulation and the agreement shown to be excellent.

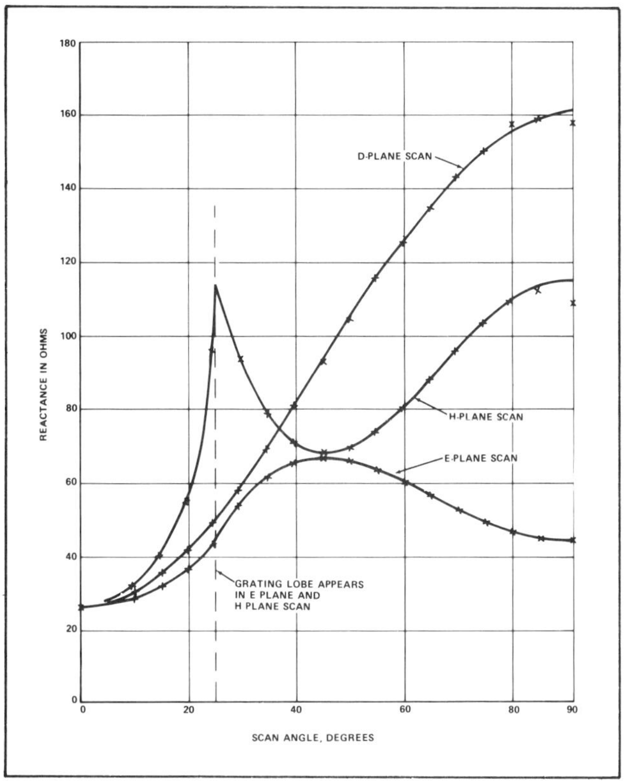


FIGURE 13. *Element driving reactance versus scan angle as obtained from modified eq (17).*

The series is truncated at $m=\pm 40$, $n=\pm 20$. Half-wave dipoles, $a=b=0.7\ \lambda$, $s=0.25\ \lambda$. The $x$'s are data points obtained from the mutual impedance solution for a 65 (collinear) $\times$ 149 (parallel) array.

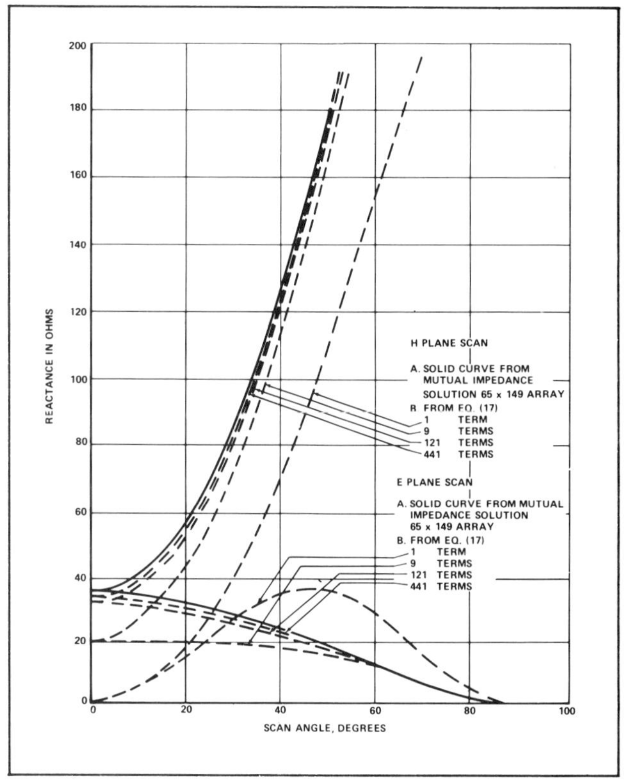


FIGURE 14. *Plot of element reactance versus scan angle for scanning in the* E*-plane and* H*-plane.*

$a=b=0.5\ \lambda$, $s=0.25\ \lambda$. Plot shows rate of convergence when an increasing number of terms from (17) are used. A comparison with reactance calculated from P. S. Carter's mutual impedance data is also shown. For this case a $65 \times 149$ element array was assumed.

---

The author acknowledges the encouragement of John Allen to publish these results. He also thanks Harold Wheeler for a helpful review of the manuscript and for encouragement to put this work in publishable form. The numerical computation of element impedance using eq (17) and also using the mutual impedance formulation was carried out independently by B. L. Diamond for array studies at MIT Lincoln Laboratory and kindly furnished to the author for inclusion in this paper.

A large part of this work was carried out on a program between Hughes Aircraft Company and MIT Lincoln Laboratory in 1955 on Contract AF19 (122)–458.

## 6. References

Blasi, E. A. (Dec. 1953), Effects of mutual interactions on the design of various dipole arrays, TM–336, Hughes Aircraft Co.

Blass, J., and S. J. Rabinowitz (1957), Mutual coupling in two dimensional arrays, IRE Wescon, pt. 1, 134–150.

Carter, P. S. (June 1932), Circuit relations in radiating systems and applications to antenna problems, Proc. IRE **20,** 1004–1041.

Carter, P. S., Jr. (May 1960), Mutual impedance effects in large beam scanning arrays, IRE Trans. Ant. Prop. **AP–8,** No. 3, 276–285.

Collin, R. E. (1960), Field Theory of Guided Waves, p. 174 (McGraw-Hill Book Co., Inc., New York, N.Y.).

King, R. W. P., and S. S. Sandler (May 1964), A theory of broadside arrays, IEEE Trans. Ant. Prop. **AP–12,** No. 3, 269–275.

Rabinowitz, S. J. (31 Dec. 1958), The conductance of a slot in an antenna array, TR–192, MIT Lincoln Lab., p. 6.

Sharp, E. D. (Mar. 1961), A triangular arrangement of planar-array elements that reduces the number needed, IRE Trans. Ant. Prop. **AP–9,** No. 2, 126–129.

Stratton, J. A. (1941), Electromagnetic Theory, p. 132 (McGraw-Hill Book Co., Inc., New York, N.Y.).

Wheeler, H. A. (Apr. 1948), The radiation resistance of an antenna in an infinite array or waveguide, Proc. IRE **36,** 478–487.

(Paper 1–3–51)

RADIO SCIENCE, Vol. 3 (New Series), No. 5, May 1968

# A Scattering Matrix Method for Solving Waveguide Array Impedance Problems

**E. C. DuFort**

Hughes Aircraft Company, Ground Systems Group, Fullerton, Calif. 92632, U.S.A.

(Received August 12, 1967; revised October 11, 1967)

A new method for solving rectangular waveguide array impedance problems is described in which the waveguide-space junction is regarded as two junctions separated by a parallel plate region. The scattering matrices for each junction must first be determined by solving the scalar problems for the $E$ and $H$ modes. An equivalent circuit for the array consists of the interconnection of the ports of these two junctions using the parallel plate transmission line modes. The reflection coefficient for all scan angles can be found explicitly for arbitrary lengths of the plates. The method is particularly suitable for arrays where the parallel plate region is substantial in length, and where no other method of solution is available, but in any case valuable physical insight into the array impedance problem is gained by this formulation. Several types of arrays that can be analyzed using this method are described. An array of maximally spaced thin walled waveguides in a triangular lattice and with $H$ plane baffles is analyzed numerically and shown to possess a fundamental surface wave effect mismatch. The same type of array with the elements more closely spaced is matched by a suitable choice of the baffle length to a VSWR of 1.58 or better within the scan range of 45° from broadside.



## 1. Introduction

The advent of large planar phased arrays has created a need for simple radiating elements operating in an essentially infinite array that is well matched for wide scan angles. The performance of some simple, inexpensive elements, such as slots and dipoles, has been analyzed by several authors using the induced emf method (Hansen, 1966). The emf method is particularly applicable for certain elements of small cross sectional areas. The method can be used for waveguide elements of substantial cross sectional area, but the accuracy is questionable, and there is no systematic approach to improving the result. In the particular case of a waveguide array in a ground plane, an alternative approach is an integral equation method for the aperture field, where the array reflection coefficient is an eigenvalue. The solution usually is obtained approximately, but the accuracy can be improved with additional labor, and the method is convenient for the flush mounted waveguide array. Some numerical results have been reported (Farrell and Kuhn, 1966a, b); however, a complete study of this type of array has not yet been published. The integral equation approach is extremely difficult for waveguide arrays with such common matching aids as corrugations and baffles. The method proposed below is capable of solving a broad class of waveguide array problems including the array in a simple ground plane. A hint of how this can be done emerges from the work of Wu and Galindo (1966) who noted that a thin walled waveguide array arranged in a rectangular lattice and scanned in the $H$ plane can be replaced by a semi-infinite parallel plate array excited by a $TE_{10}$ mode. This problem has been solved exactly by Heins and Carlson (1947), and Whitehead (1951), and the reflection coefficient can be calculated exactly. However, there is no known way to adapt this approach to other scan planes. The purpose of the present work is to describe a new method of solving such problems from a generalized scattering matrix viewpoint and to give a simple approximate application to illustrate the method.

The present method is useful for arrays of rectangular waveguides each having a substantial cross sectional area where there is a parallel plate region (which may be arbitrarily short) into which the waveguides radiate. The radiation subsequently passes from the parallel plates into space. Examples are arrays of thin walled guides arranged in a rectangular lattice with either the $E$ or $H$ plane walls extended to form baffles as in figure 1. Or the waveguides may be arranged in a triangular lattice form, such as shown in figure 2, with the $H$ plane walls extended to form baffles. Other more complex structures, such as that shown in figure 3, can be solved using the method described below, and the results of these investigations will be reported in a subsequent paper.

## 2. Generalized Scattering Matrix for Parallel Plates

The essential ingredient of the present method is a matrix describing the scattering properties of an interface between free space and a semi-infinite array of parallel plates as shown in figure 4. The fields can be described in terms of $E$ modes and $H$ modes derived only from the electric field component parallel to the edge in the former case, or the magnetic field component parallel to the edge in the latter case. If the plates are thin and the incident field is a propagating plane wave in space of either type with the propagation vector in a plane perpendicular to the edges but otherwise at any angle, then the scattering coefficients into all other propagating modes have been found by Heins and Carlson (1947) and Whitehead (1951). The extension of their results to arbitrary angles of incidence and for sources inside the plates is elementary. If the waves are properly normalized this is the usual scattering matrix that is symmetrical and unitary. With some additional effort the coefficients for incident propagating waves scattering into evanescent modes also can be extracted from the results of the aforementioned authors. Furthermore, by reformulating the problem for the various possible incident evanescent modes the scattering into all other modes again can be calculated. Although these coefficients have not been published, the procedure for obtaining them is clearly a straightforward one proceeding along the lines used by Whitehead (1951). Note that this problem is amenable to an exact solution by the Wiener Hopf or other equivalent techniques. Generalized scattering matrices involving $E$ or $H$ type evanescent or propagating modes can be constructed for a large class of parallel plate structures by Wiener Hopf (Heins and Carlson, 1947; Whitehead, 1951) scalar integral equation (Galindo and Wu, 1966) or other methods (DuFort, 1968), and hereafter it will be assumed that these matrices are known. Similar generalized matrices have been constructed and used in the study of waveguide discontinuity problems (Pace and Mittra, 1966) and periodic structures (Collin, 1960).

## 3. A Simple Example

The general method is easier to describe in terms of an extension of the following simple problem.



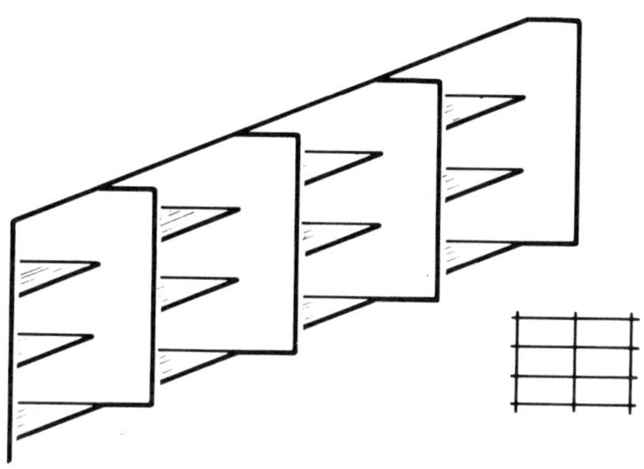

FIGURE 1. *Rectangular array of waveguides with* E *plane baffles.*

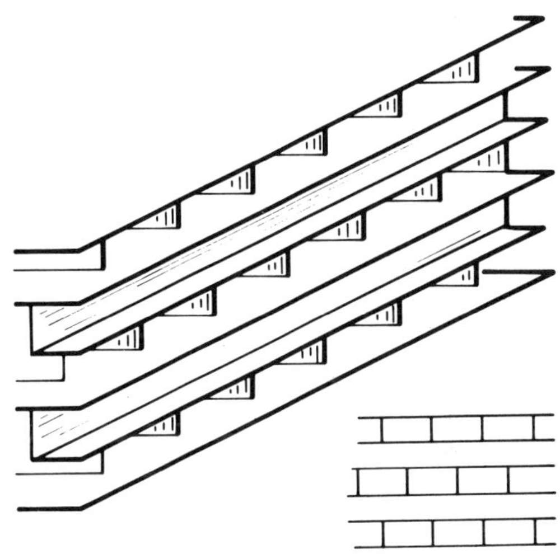

FIGURE 3. *Triangular array of waveguides with corrugations.*

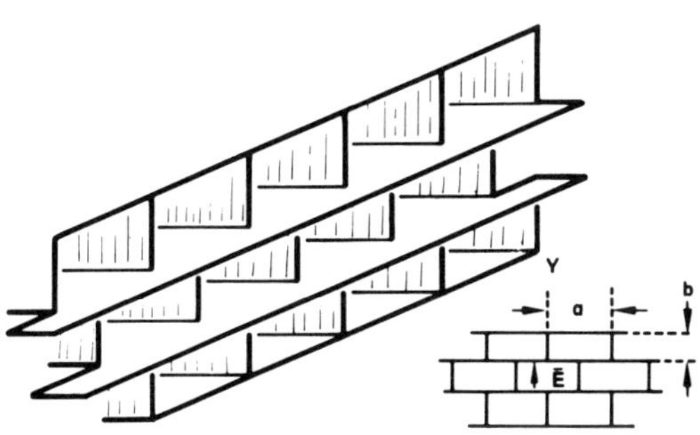


FIGURE 2. *Triangular array of waveguides with* H *plane baffles.*

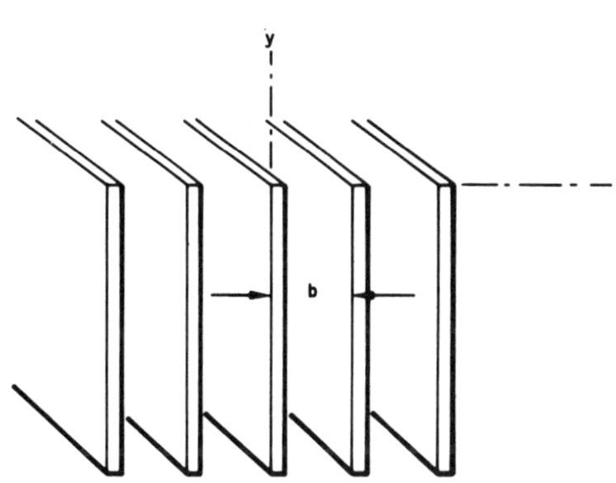


FIGURE 4. *Basic parallel plate array.*

Thin walled rectangular waveguides with dimensions $a$ and $b$ are stacked in a triangular lattice, as depicted in figure 2. The waveguides are excited to launch a wave in space with propagation vector components $k_x$, $k_y$, $k_z$. The broad walls of the waveguides are extended to form $H$ plane baffles, hence a parallel plate region, and the junction between a typical waveguide and the plate region is described by a scattering matrix $S^{(y)}$. For simplicity, we neglect all but the propagating modes in this example; therefore, the incident $TE_{10}$ mode in the waveguide launches a mode in the plates derived solely from $E_y$ (hence an $E_y$ mode) as follows:

$$E_x=0 \quad (1a) \qquad E_y=e^{-j(k_x x+K_z z)} \quad (1b)$$

$$E_z=0 \quad (1c) \qquad H_x=-K_z E_y/\omega\mu \quad (1d)$$

$$H_y=0 \quad (1e) \qquad H_z=\frac{k_x}{\omega\mu}E_y \quad (1f)$$

$$k_x^2+K_z^2=k^2 \quad (1g) \qquad k_x^2+k_y^2+k_z^2=k^2. \quad (1h)$$

If the interguide phase shift $k_x a$ is sufficiently large, a quasi-grating lobe also will be launched in the plate region with wave number components $k'_x$, $K'_z$,

$$k'_x=k_x-2\pi/a \quad (2a) \qquad K'_z=\sqrt{k^2-k_x^{2\prime}}. \quad (2b)$$

The principle wave given by (1) is labeled the 0 wave, and the quasi-grating lobe in (2) is termed the $(-1)$ wave in accordance with the $x$ component wave numbers. There are three distinct ports here, and a $3\times 3$ scattering matrix $S^{(y)}$ is needed to describe the junction. If the waves are properly normalized, the matrix is unitary as well as symmetrical. The matrix elements can be found in the work of Whitehead (1951) and are summarized in the appendix. If $|k'_x|$ exceeds $k$, the quasi-grating lobe is evanescent, and $S^{(y)}$ for the propagating modes reduces to a $2\times 2$ matrix.

Both the (0) and $(-1)$ waves, which are purely $E_y$ modes with respect to the waveguide-parallel plate junction, are simultaneously purely $H_x$ modes with respect to parallel plate-space junctions. Due to the phasing and displacement $a/2$ between adjacent waveguide rows, the $H_x$ modes in space (including the evanescent modes) are of the form (DuFort, 1967),

$$e^{-j\left\{\left(k_x+\frac{2n\pi}{a}\right)x+\left[k_y+\frac{2\pi}{b}\left(m+\frac{n}{2}\right)\right]y+\Gamma_{nm}z\right\}} \quad (3a)$$

$$\left[k_x+\frac{2n\pi}{a}\right]^2+\left[k_y+\frac{2\pi}{b}\left(m+\frac{n}{2}\right)\right]^2=k^2-\Gamma_{nm}^2. \quad (3b)$$

The (0) waves in the parallel plate regions launch an $H_x$ plane wave in space with propagation vector components $k_x$, $k_y$, $k_z$, i.e., $n=m=0$ in (3). The transmission and reflection coefficients for this wave are given by a $2\times 2$ matrix $S^{(x)}$ whose elements can be found in the appendix. The $(-1)$ waves in the parallel plate regions tend to launch an $H_x$ wave with $x$ and $y$ propagation vector components $\left(k_x-\frac{2\pi}{a}\right)$, $\left(k_y-\frac{\pi}{b}\right)$, i.e., $n=-1$, $m=0$ in (3). Since only one propagating wave is desired in space, $\Gamma_{-1,0}$ in (3) must be imaginary or $\Gamma_{-1,0}^2<0$, which is the usual grating lobe condition. With this condition satisfied, the $(-1)$ wave in each parallel plate region is totally reflected from the aperture, and the phase of the reflection coefficient, which can be extracted from Whitehead's results, is listed in the appendix. A simple equivalent network for this array is the $T$ junction shown in figure 5. The parallel plate or baffle length is denoted by $L$ and the (0) wave travels along a transmission line of length $L$ with a wave number $K_z$, which feeds a 2-port network $S_0^{(x)}$ terminated in a matched load. The $(-1)$ wave with the wave number $K'_z$ travels along a transmission line also of length $L$, but terminated in a length of transmission line shorted a distance $\Delta L$ from the aperture. In the event that $|k'_x|>k$, the $(-1)$ wave is evanescent in the parallel plates as well as in space, and it is assumed that $L$ is large such that port 3 can be ignored, and $S^{(y)}$ can be replaced by a simple $2\times 2$ matrix. This is a good approximation near $k_x=0$, but it would appear to fail just prior to the onset of propagation of the $(-1)$ mode; however, the amplitude of the scattering coefficient into port 3 vanishes at $K'_z=0$ (see the appendix). Therefore, the error is nullified to a large extent, and the additional complexity accompanying the exact treatment of the $(-1)$ wave for all $k_x$ is not justified in this simple example. The result is exact for all $k_x$ in the limit of large $L$, and approximates the exact solution provided the $e$-folding distance of the dominant evanescent cross polarized $E_x$ mode is smaller than $L$. This simple treatment of a complicated problem contains the essential features of the general, exact method described in the following section.

## 4. Exact Solution

The previous problem can be solved exactly for all values of baffle length $L$ using the generalized scattering matrix described in section 2. Other problems, such as those sketched in figures 1 and 3, can be treated in a similar manner. The most general waves that can exist at the waveguide parallel plate junction are $E_y$ and $H_y$ modes with wave numbers in $x$ and $k_x+2n\pi/a$ and trigonometric $y$ dependence as sin $m\pi y/b$ or cos $m\pi y/b$. The modes are characterized by two indices $n$ and $m$ as well as the $E_y$ or $H_y$ type, and the modes are orthogonal over the unit cell of the array. The waveguide-parallel plate junction is depicted in figure 6 by the scattering matrix $S^{(y)}$. For each index $n$ there are modes of either the $E_y$ or $H_y$ type characterized by a second index $m$ representing the $y$ variation in the modes; however, for simplicity the $m$ indices are omitted in the figure. Also, the $E_y$ and $H_y$ modes do not couple through $S^{(y)}$, but this refinement

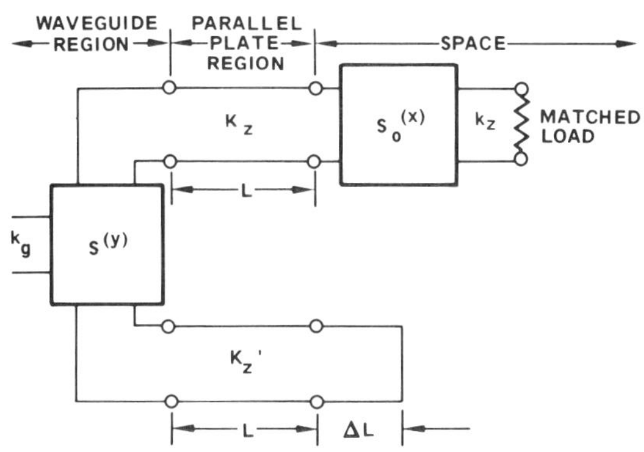


FIGURE 5. *Approximate equivalent circuit for the array.*

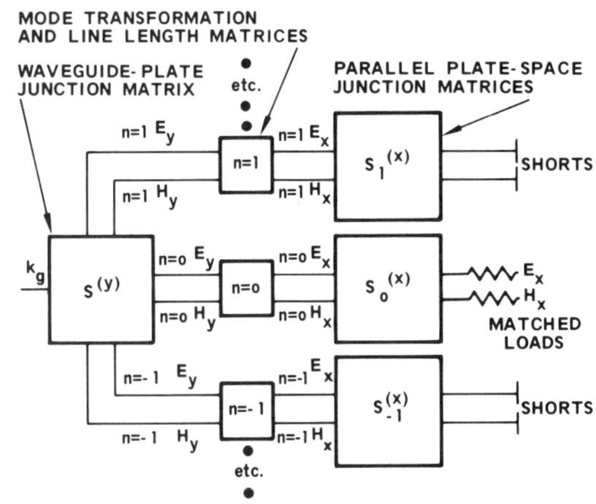


FIGURE 6. *Portion of exact equivalent circuit for the array.*

also is omitted. The excitation of the infinite waveguide array initially will launch $E_y$ modes in the parallel plates. All but a few of these modes are evanescent. The $E_y$ and $H_y$ modal description is convenient for the waveguide-parallel plate junction, but is not appropriate to the parallel plate-space junction. Any $E_y$ mode can be transformed into at most one $E_x$ and one $H_x$ mode with the same indices ($n$, $m$), and similarly for the $E_x$ and $H_x$ modal transformations back to $E_y$ and $H_y$ modes. The transformation networks for the various values of $n$ as well as the phase shift and amplitude corrections due to the length of the plate region are combined into the center blocks in figure 6. Now $E_x$ and $H_x$ modes are incident on the aperture, which has a scattering matrix $S^{(x)}$. Waves travel away from the aperture in the various $x$ modes, and waves are transmitted to space in only one direction corresponding to the waveguide phasing, provided the element spacing suppresses grating lobes. The grating lobes are evanescent in space and are not shown in the figure. Generally, two orthogonal waves, an $E_x$ type and an $H_x$ type propagate in space in the same direction.

The manner in which cross polarized radiation arises in waveguide phased arrays is easy to see with the aid of figure 6. The waveguide array initially excites only the dominant $n=m=0$ $E_y$ mode as in (1). This is equivalent to one dominant $H_x$ mode, $n=m=0$, which is incident on the junction described by $S^{(x)}$, and part of this mode is transmitted to space in the desired $H_x$ mode. If the phasing between parallel plate regions and the plate length are such that there are variations of $H_x$ in the $y$ direction of the aperture, the incident $H_x$ mode with $n=m=0$ is partially backscattered into $H_x$, $n=0$, $m=0$, 1, 2, . . . modes. These transform back into $E_y$ and $H_y$ modes now incident on the junction described by $S^{(y)}$, and are again scattered into the parallel plate region with a new relative amplitude distribution since the reflection coefficients for $E_y$ and $H_y$ modes are different. These modes now transform into several $E_x$ and $H_x$ modes incident on $S^{(x)}$ and these secondary waves will be partly transmitted to space in both $E_x$ and $H_x$ modes. Thus, cross polarization arises essentially from mode conversion as a result of coupling between the two junctions via the evanescent modes. For a sufficient long parallel plate region, the evanescent modes decay so rapidly between the junctions described by $S^{(y)}$ and $S^{(x)}$ that only propagating modes need be considered, and no significant cross polarization is transmitted to space.

The reflection coefficient, $R$, looking into the waveguide array can be calculated exactly using all modes simultaneously. A simplified diagram of the various waves is shown in figure 7. The symbol $b_n$ (for $b_1$, $b_2$, . . .) represents all the $E_y$ and $H_y$ modes traveling away from the junction $S^{(y)}$, and similarly $a_n$ (for $a_1$, $a_2$, . . .) represents all those modes incident on that junction from the plate region. The $b_n$'s are transformed by a transformation matrix (without reflection) to $E_x$ and $H_x$ modes represented by $a_n'$ (for $a_1'$, $a_2'$, . . .), traveling toward the $S^{(x)}$ junction:

$$a' = S_1 b. \tag{4}$$

Similarly, $b_n'$ (for $b_1'$, $b_2'$, . . .) are $E_x$ and $H_x$ modes which are converted to $E_y$ and $H_y$ modes $a_n$ via the matched matrix $S_3$:

$$a = S_3 b'. \tag{5}$$

Since there are no waves incident from space, $a_n'$ is the sole source of $b_n'$ via a matrix relation,

$$b' = S_2 a', \tag{6}$$

where $S_2$ is derived from the matrix $S^{(x)}$ by deleting the first two rows and first two columns, corresponding to terminating the $E_x$ and $H_x$ ports in space with matched loads. In addition to these relations, the $S^{(y)}$ matrix yields the following relations:

$$\begin{pmatrix} b_0 \\ b_1 \\ b_2 \\ \cdot \\ \cdot \\ \cdot \end{pmatrix} = \begin{pmatrix} S^{(y)} \end{pmatrix} \begin{pmatrix} a_0 \\ a_1 \\ a_2 \\ \cdot \\ \cdot \\ \cdot \end{pmatrix}. \tag{7}$$

The procedure for finding $R$ is to eliminate the unknown waves $b'$, $a'$, and $b$ to obtain a homogeneous set of equations in the $a$'s with $R$ as a parameter. To do this, substitute $b'$ from (6) into (5) to obtain $a = S_3S_2a'$, and eliminate $a'$ using (4) yielding $a = S_3S_2S_1b$

or

$$\begin{pmatrix} a_1 \\ a_2 \\ \cdot \\ \cdot \\ \cdot \end{pmatrix} = (S_3S_2S_1) \begin{pmatrix} b_1 \\ b_2 \\ \cdot \\ \cdot \\ \cdot \end{pmatrix}. \tag{8}$$

To eliminate $b$, multiply both members of (7) by a matrix composed of $S_3S_2S_1$ augmented by a new row and column with unity in the first diagonal position and with zeros in the remaining positions of the first row and column. Then, with the aid of (8), (7) becomes

$$\begin{pmatrix} 1 & 0 \quad 0 \quad \cdot \quad \cdot \\ 0 & \\ \cdot & \\ \cdot & S_3S_2S_1 \\ \cdot & \end{pmatrix} \begin{pmatrix} b_0 \\ b_1 \\ \cdot \\ \cdot \\ \cdot \end{pmatrix} = \begin{pmatrix} b_0 \\ a_1 \\ a_2 \\ \cdot \\ \cdot \end{pmatrix}$$

$$= \begin{pmatrix} 1 & 0 \\ 0 & \\ \cdot & S_3S_2S_1 \end{pmatrix} \begin{pmatrix} S^{(y)} \end{pmatrix} \begin{pmatrix} a_0 \\ a_1 \\ \cdot \\ \cdot \\ \cdot \end{pmatrix}. \tag{9}$$

Define two new matrices, $S$ and a diagonal matrix $D$, as follows:

$$S = \begin{pmatrix} 1 & 0 \\ 0 & \\ 0 & S_3S_2S_1 \end{pmatrix} \begin{pmatrix} S^{(y)} \end{pmatrix} \tag{10}$$

$$D = \begin{pmatrix} R & & & \bigcirc \\ & 1 & & \\ & & 1 & \\ \bigcirc & & & \ddots \end{pmatrix}. \tag{11}$$

Using these definitions in (9), and remembering that $b_0 = Ra_0$, results in (9) becoming the homogeneous set of equations in the unknowns $a$ with $R$ as a parameter:

$$Da = Sa. \tag{12}$$

A solution for $a$ exists only if $R$ takes on a value such that the determinant of $S - D$ vanishes:

$$\|S - D\| = 0. \tag{13}$$

In practice it is necessary to truncate the set of equations in (12) at some finite level such that (13) is a finite determinant. Ideally, the variables $a_n$ are numbered in order of decreasing influence on the value of $R$. One reasonable way of doing this is to number the set $a_n$ according to the $e$-folding distance of the corresponding evanescent parallel plate mode, accounting for degeneracies in an orderly fashion. Using only the first $N$ components of $a$, one may write the corresponding reflection coefficient $R_N$ from (13):

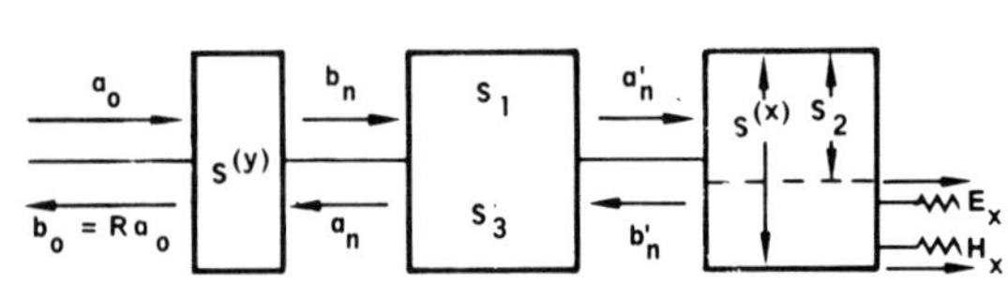


FIGURE 7. *Simplified general equivalent circuit.*

$$R_N = \frac{\begin{vmatrix} S_{00} & S_{01} & \cdots & S_{0N} \\ S_{10} & S_{11}-1 & \cdots & S_{1N} \\ S_{20} & \cdot & S_{22}-1 & \cdot \\ S_{N0} & S_{N1} & \cdots & S_{NN}-1 \end{vmatrix}}{\begin{vmatrix} S_{11}-1 & & S_{1N} \\ & S_{22}-1 & \\ \cdot & & \\ \cdot & & \\ \cdot & & \\ S_{N1} & & S_{NN}-1 \end{vmatrix}}. \tag{14}$$

The exact solution is obtained by letting $N$ tend to infinity. When the parallel region is of nonzero length, this approach will surely yield a rapidly converging value of $R_N$ as $N$ increases, because ultimately the $e$-folding distance corresponding to $a_n$ becomes so small that the junctions corresponding to $S^{(x)}$ and $S^{(y)}$ are essentially decoupled in these modes, and the corresponding transmission lines in figure 6 may be cut without affecting the results. Cutting the lines corresponds to forcing $a_{N+1}=a_{N+2} \ldots =0$ as seen in figures 6 or 7, and is not the same approximation as neglecting higher order evanescent modes since an infinite set $b_n$ is retained. As the parallel plate region becomes large, the rows of $S_3S_2S_1$ go to zero except for the first one or two rows corresponding to propagating modes. For large parallel plate regions, a small value of $N$ such as two or three may suffice. So long as $N$ is large enough that the $e$-folding distance of the mode corresponding to $a_N$ is substantially smaller than the plate length, then $R_N$ will be a good approximation to the exact answer. This physical reasoning fails when the plate length goes to zero. In this case the rate of convergence is determined by the form of the elements in the determinants in (14) that define $R_N$ (Kantorovich and Krylov, 1958). These elements depend primarily on $S^{(x)}$ and $S^{(y)}$ when the plate length vanishes. In particular, if the rows of $S^{(x)}$ and $S^{(y)}$ rapidly go to zero, (14) will converge rapidly for all plate lengths.

## 5. A Numerical Example

Consider the array of waveguides with baffles shown in figure 2. For convenience the lattice may be chosen to be a square rotated by 45° so that the broad dimension $a$ and the narrow dimension $b$ of the waveguide are related by the common two-to-one aspect ratio, $a=2b$. For scanning within a 45° angle from broadside, (3) shows that

$$b=\lambda/(1+\sqrt{2}) \tag{15}$$

barely suppresses the grating lobes. This gives an $e$-folding distance of $\lambda/5$ for the leading evanescent mode between parallel plates; consequently, if the baffles exceed $\lambda/5$ in length, the simple procedure described in section 3 and the three-port equivalent circuit in figure 5 are applicable with good accuracy. The scattering matrices required, $S^{(x)}$ and $S^{(y)}$, are listed in the appendix as $S^{(E)}$ and $S^{(H)}$. The reflection coefficient as a function of $\theta$, the angle from broadside, for several $\varphi$ cuts ($\varphi=0$ is the $H$ plane, $\varphi=\pi/2$ is the $E$ plane) is shown in figure 8 for a baffle length of $\lambda/5$. The array is well matched in the $E$ plane, but the match deteriorates rapidly as $\varphi$ decreases, and in the $\varphi=45°$ plane total reflection occurs at $\sin\theta \simeq .4$, whereas total reflection occurs for $\sin\theta \simeq .5$ in the $\varphi=0$ plane. The reflection coefficient varies with baffle length $L$, but there is always a beam position not necessarily in the $H$ plane for which total reflection occurs; consequently, the result in figure 8 is typical. Although the accuracy of the solution is suspect for baffle lengths less than $\lambda/5$, the calculation can be performed and is meaningful as a first approximation even for $L=0$. When $L=0$, the total mismatch can be found in the $\varphi=45°$ plane. Indeed, this severe mismatch, often termed the surface wave effect in arrays, appears to be fundamental for this structure, and impossible to circumvent for any choices of baffle length within the framework of the first approximation; however, there is a simple explanation for this difficulty. The surface wave effect in arrays has been studied in a general way for arrays with a three-port equivalent junction even when the scattering matrices are unsymmetrical (DuFort, 1968). In the present case the scattering matrices are symmetrical, and the three port is equivalent to a waveguide $T$ junction with one arm short circuited. As the beam is scanned, the scattering matrix $S^{(y)}$ changes and the effective short circuit position also changes because of the change in the quantities $K_z'$ and $\Delta L$ shown in figure 5. If the reflection coefficient looking into the shorted arm in figure 5 equals a critical value $G$ related to $S^{(y)}$ as follows:

$$G=\frac{S_{33}^{*(y)}+g}{1+S_{33}^{(y)}g}, \qquad g=\left[\frac{S_{22}^{(y)}}{S_{23}^{(y)}}-\frac{S_{12}^{(y)}}{S_{13}^{(y)}}\right] S_{23}^{*(y)}, \tag{16}$$

then total reflection will occur. Note $|G|=1$ since the unitary conditions on $S^{(y)}$ yield $|g|=1$. Condition (16) can always be satisfied for some beam position by a deliberate choice of $L$. What is not so obvious is that there is no fixed value of $L$ for which the surface wave condition described by (16) can be avoided. This is due to the fundamental form of the scattering matrix $S^{(y)}$ which enters into (16) and the length $\Delta L$ which depends on the scattering matrix $S^{(x)}$. Other arrays have been studied where a surface wave effect occurs as a result of an unfortunate choice of the parameters, but can be avoided by a judicious choice. In the present case there is no choice of the baffle length that circumvents the difficulty; however, the present calculations are

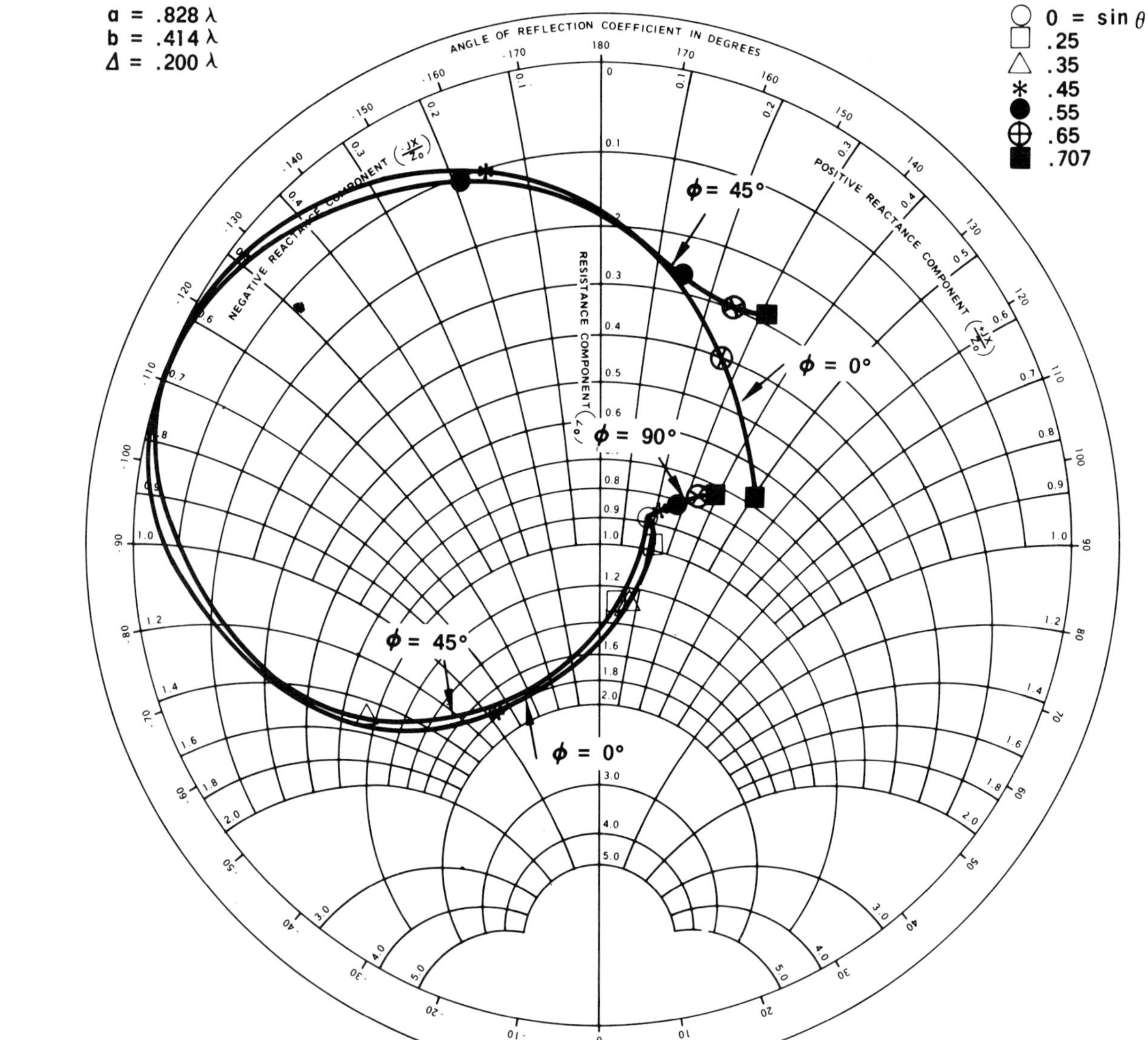


FIGURE 8. *Reflection coefficient for various θ and φ = 0, 45°, 90°.*

based on a first approximation that is accurate only if $L \gtrsim \lambda/5$. In the limit of $L=0$ an accurate treatment of the higher order modes may alter this conclusion. Since our purpose is to develop a method and display simple examples, the more elaborate calculations using interacting evanescent modes as described in section 4 will not be attempted here. Instead, a second example, which results in a good match over wide scan angles and for which the first approximation is accurate, will be presented.

To achieve a good match according to the first approximation, it is necessary that the matrices $S^{(x)}$ and $S^{(y)}$ be such that the conditions described by (16) et seq., are avoided. Thus, a basic change in the structure is necessary. One possible change is to decrease the element spacing because this surely affects the matrices $S^{(x)}$ and $S^{(y)}$, which determine $G$ and $\Delta L$. To ensure a good match, the broad dimension of the waveguide may be chosen so that the quasi-grating lobe or $(-1)$ wave is always evanescent; then a surface wave effect is not likely to arise when using long baffles because $S^{(y)}$ for the propagating modes in figure 5 reduces to a two port junction, i.e., $K_z'$ is imaginary. For scanning within 45° of the array normal, the $(-1)$ wave is cut off if $a/\lambda=\sqrt{2}/(\sqrt{2}+1)$, and cross polarization is adequately suppressed if $b/\lambda=2/5$ and

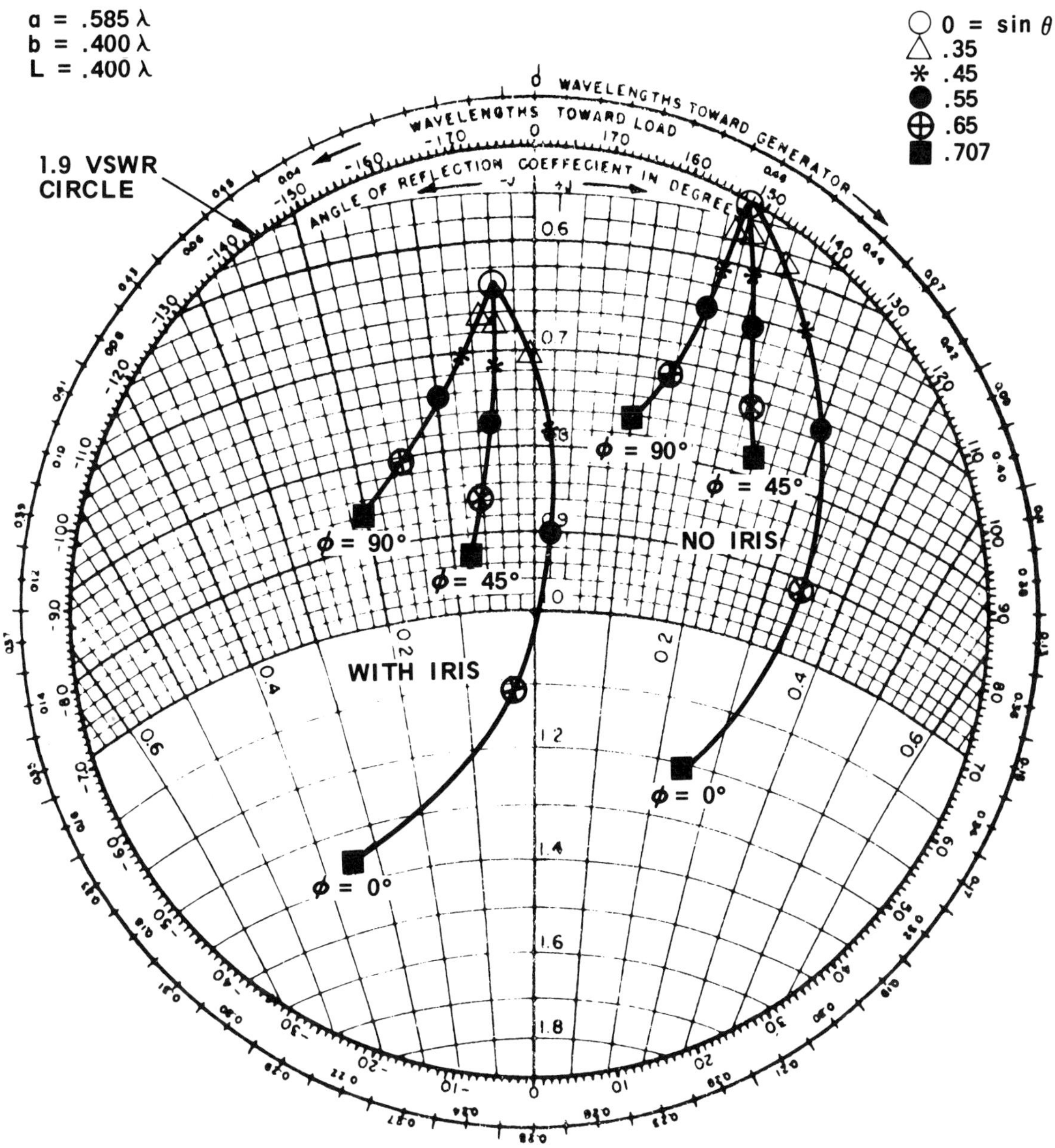


FIGURE 9. *Reflection coefficient for various θ, φ for a closely spaced array (expanded Smith Chart).*

$L > \lambda/5$. Since the baffles are matched to space in the $H$ plane, this structure is similar to that studied by Wu and Galindo (1966) for $H$ plane scan; consequently, the present method permits extension of results similar to theirs to other planes of scan for arrays with sufficiently long baffles. Finally, a suitable value of $L$ is one that yields the poorest match in the $H$ plane where the match is virtually independent of $L$. The length $L = 2\lambda/5$ is adequate for this purpose, and the approximate equivalent circuit in figure 5 also becomes quite accurate. The reflection coefficient as a function of scan angle measured from broadside is shown in an expanded Smith chart in figure 9 for several cuts of constant $\varphi$. The worst match is in the $H$ plane where the VSWR reaches 1.9. When an iris is placed in each waveguide, the worst match is reduced to a 1.58

VSWR for a beam anywhere within 45° of broadside.

This excellent performance has been obtained at the expense of having more than the minimum number of elements; however, there is no fundamental reason foı adding elements to obtain a match. Hannan (1967) has shown that by interconnecting the elements properly, a match can be obtained. These interconnections are not always convenient or economical; therefore, further analysis of simple structures would be valuable. The analytical methods described above can be applied to some simple structures that are maximally spaced and that show promise of being well matched also. In particular, the corrugated structure in figure 3 has a matrix $S^{(x)}$ that can be adjusted to bear little resemblance to the $S^{(x)}$ for simple baffles. This candidate structure is being investigated.

## 6. Summary and Conclusion

A new method for solving a class of waveguide array problems has been developed. Two generalized scattering matrices $S^{(x)}$ and $S^{(y)}$ must be known completely before the method can be employed to give exact solutions. These matrices involve the solution of four types of scalar boundary value problems for the $E$ and $H$ modes for the two structures corresponding to $S^{(x)}$ and $S^{(y)}$. In the event that $S^{(x)}$ and $S^{(y)}$ correspond to a thin parallel plate-space junction, these matrices are essentially the same and can be found exactly using the Wiener Hopf method. When the parallel plate region connecting these junctions vanishes, the method involves the manipulation of relatively large matrices to obtain good accuracy and is simply an alternative to the integral equation approach. If the interconnecting plate region is substantial in length, only a few terms of the matrices need to be retained to obtain good accuracy. Furthermore, the present method is particularly suited to analyzing thick structures whose scattering matrices are known (see fig. 5), but which may not otherwise be amenable to solution.

Many structures of the type under consideration exhibit a surface wave effect whose presence can be predicted from the analysis of simple multiport junctions. The present method furnishes valuable insight into the cause of this problem. One particular waveguide structure chosen as an example was maximally spaced in a triangular lattice so that grating lobes were suppressed anywhere within 45° of broadside. This structure exhibited a fundamental surface wave effect that could be analyzed using an equivalent three-port junction. There was no choice of baffle length, which avoided the surface wave mismatch; however, by spacing the elements closer together the array was matched to a VSWR of 1.58 or better.

## 7. Appendix

The $H$ modes are interrelated via a scattering matrix $S^{(H)}$ where the fields are all derived from the magnetic field component $H_y$ parallel to the edges of the plates shown in figure 4. For the present purpose it is necessary to know only the relations for the dominant $H$ mode. These are listed in appendix 7.1. Similarly the $E$ modes are interrelated via a matrix $S^{(E)}$ where the fields are derived from the electric field component $E_y$ parallel to the plate edges. Only the dominant $E$ mode solutions are required at present and these are listed in appendix 7.2. All these results can be extracted from the literature (Heins and Carlson, 1947; Whitehead, 1951), but not without considerable effort; therefore, for convenience the essential results are listed here.

### 7.1. Scattering Coefficients for the Dominant $H$ Mode

Consider the parallel plate array of figure 4 for $H$ modes where the fields are derived from $H_y$ only. For plate spacings $b$ less than $\lambda/2$, only waves with no $x$ variation propagate in the plates, and only an $H_y$ plane wave propagates in space. The scattering matrix relating the propagating modes is a $2\times 2$ symmetrical matrix where port 1 is the plate region and port 2 is space. Let $k_x$, $k_y$, $k_z$ denote components of the propagation vector in space, and $k_y$, $K_z$ the components of the propagation vector in the plates, where

$$k_y^2+K_z^2=k^2 \text{ and } k_x^2+k_y^2+k_z^2=k^2. \tag{17}$$

The scattering matrix in this case is identified with $S^{(x)}$, described in section 3, with a 90° rotation of the plates.

Define quantities $P_n$, $p_n$, and $\psi(K)$ as follows:

$$P_n=\sqrt{\left(k_x+\frac{2n\pi}{b}\right)^2-K_z^2} \tag{18a}$$

$$p_n=\sqrt{\left(\frac{n\pi}{b}\right)^2-K_z^2} \tag{18b}$$

$$\psi(K)=\sum_1^\infty \tan^{-1}\frac{K}{P_n}+\tan^{-1}\frac{K}{P_{-n}}-\tan^{-1}\frac{K}{p_n}. \tag{18c}$$

If $P_0^2\leq 0$, a plane wave characterized by $k_x$, $k_y$, $k_z$ propagates in space, and the amplitude of the matrix elements are

$$|S_{11}|=\frac{K_z-k_z}{K_z+k_z}=|S_{22}|$$

$$|S_{12}|=2\sqrt{k_zK_z}/(K_z+k_z). \tag{19}$$

The phase angles of the matrix elements are

$$\angle S_{11}=-\pi+2\frac{K_zb}{\pi}\ln 2-2\psi(K_z) \tag{20}$$

$$\angle S_{12}=(K_z-k_z)\frac{b}{\pi}\ln 2+\psi(k_z)-\psi(K_z) \tag{21}$$

$$\angle S_{22}=\frac{-2k_zb}{\pi}\ln 2+2\psi(k_z). \tag{22}$$

In the event that the interelement phase shift $k_x b$ is so large that $P_0^2 > 0$ (as for the $(-1)$ wave described in sec. 3), there are no propagating modes in space, and the only quantity of interest then is the reflection coefficient as seen from the plates.

$$|S_{11}| = 1 \tag{23a}$$

$$\angle S_{11} = 2\tan^{-1}\frac{P_0}{K_z} - \pi + 2\frac{K_z b}{\pi}\ln 2 - 2\psi(K_z). \tag{23b}$$

Thus, the waveguide-space junction as seen from the plates is equivalent to a short circuit at a position $\Delta L$ from the aperture where

$$S_{11} = -e^{-2jK_z\Delta L}. \tag{24}$$

### 7.2. Scattering Coefficients for the Dominant $E$ Modes

When $b$ exceeds $\lambda/2$, an $E$ mode propagates in the plates, and either one or two waves may radiate into space. The scattering matrix for this case corresponds to the matrix $S^{(y)}$ of section 3. When $0 \leq k_x < (2\pi/b) - k$, only one wave propagates in space, and the scattering matrix is a symmetrical $2\times 2$ matrix with port 1 representing the plate region and port 2 representing space. There is need only for the case where the $E$ mode in the plates travels normal to the plate edges ($k_y = 0$) with the usual guide wave number $k_g$,

$$k_g = \sqrt{k^2 - \frac{\pi^2}{b^2}}. \tag{25}$$

The wave number along $z$ for the wave in space is termed $K_z$,

$$K_z = \sqrt{k^2 - k_x^2}. \tag{26}$$

Define quantities $\Gamma_n$, $\gamma_n$, and $\psi(K)$ as follows:

$$\Gamma_n = \sqrt{\left(k_x + \frac{2n\pi}{b}\right)^2 - k^2} \tag{27a}$$

$$\gamma_n = \sqrt{\left(\frac{n\pi}{b}\right)^2 - k^2} \tag{27b}$$

$$\psi(K) = \sum_{2}^{\infty} \tan^{-1}\frac{K}{\Gamma_{-n}} + \tan^{-1}\frac{K}{\Gamma_n} - \tan^{-1}\frac{K}{\gamma_n}. \tag{27c}$$

The matrix elements may be written in terms of these quantities as follows:

$$|S_{11}| = (K_z - k_g)/(K_z + k_g) \tag{28}$$

$$|S_{12}| = 2\sqrt{K_z k_g}/(K_z + k_g) \tag{29}$$

$$\angle S_{11} = \pi + 2k_g\frac{b}{\pi}\ln 2 - 2\left[\tan^{-1}\frac{k_g}{\Gamma_1} + \tan^{-1}\frac{k_g}{\Gamma_{-1}}\right] - 2\psi(k_g)$$

$$\angle S_{12} = (k_g - K_z)\frac{b}{\pi}\ln 2 + \left[\tan^{-1}\frac{K_z}{\Gamma_1} + \tan^{-1}\frac{K_z}{\Gamma_{-1}}\right] \tag{30}$$

$$-\left[\tan^{-1}\frac{k_g}{\Gamma_1} + \tan^{-1}\frac{k_g}{\Gamma_{-1}}\right] + \psi(K_z) - \psi(k_g). \tag{31}$$

The coefficient $S_{22}$ can be written in terms of $S_{11}$ and $S_{12}$ using the symmetry and unitary property of the matrix:

$$S_{22} = -\frac{S_{11}^* S_{12}}{S_{12}^*}. \tag{32}$$

In the event that $k_x$ exceeds $(2\pi/b) - k$, there will be a second plane wave, a grating lobe, in space (termed the $(-1)$ wave or quasi-grating lobe in sec. 3) with propagation vector components $k_x'$ and $K_z'$,

$$k_x' = k_x - \frac{2\pi}{b} \tag{33}$$

$$K_z' = \sqrt{k^2 - k_x'^2}. \tag{34}$$

The scattering matrix becomes a symmetrical $3\times 3$ unitary matrix, and the third port is identified with the grating lobe where the wave number along $z$ is $K_z'$. The amplitudes of four of the matrix elements for this case are:

$$|S_{11}| = \frac{(K_z - k_g)(k_g - K_z')}{(K_z + k_g)(k_g + K_z')} \tag{35}$$

$$|S_{12}| = 2\left[\left(\frac{K_z + K_z'}{K_z - K_z'}\right)\left(\frac{k_g - K_z'}{k_g + K_z'}\right)\right]^{1/2}\frac{[K_z k_g]^{1/2}}{K_z + k_g} \tag{36}$$

$$|S_{13}| = 2\left[\left(\frac{K_z + K_z'}{K_z - K_z'}\frac{K_z - k_g}{K_z + k_g}\right)\right]^{1/2}\frac{[K_z' k_g]^{1/2}}{K_z' + k_g} \tag{37}$$

$$|S_{22}| = \left(\frac{K_z + K_z'}{K_z - K_z'}\right)\left(\frac{K_z - k_g}{K_z + k_g}\right). \tag{38}$$

The corresponding phase angles are:

$$\angle S_{11} = 2k_g\frac{b}{\pi}\ln 2 - 2\tan^{-1}\frac{k_g}{\Gamma_1} - 2\psi(k_g) \tag{39}$$

$$\angle S_{12} = (k_g - K_z)\frac{b}{\pi}\ln 2 + \left[\tan^{-1}\frac{K_z}{\Gamma_1} - \tan^{-1}\frac{k_g}{\Gamma_1}\right]$$

$$+ [\psi(K_z) - \psi(k_g)] \tag{40}$$

$$\angle S_{13} = (k_g - K_z')\frac{b}{\pi}\ln 2 + \left[\tan^{-1}\frac{K_z'}{\Gamma_1} - \tan^{-1}\frac{k_g}{\Gamma_1}\right] + [\psi(K_z') - \psi(k_g)] \tag{41}$$

$$\angle S_{22} = \pi - 2K_z\frac{b}{\pi}\ln 2 + 2\tan^{-1}\frac{K_z}{\Gamma_1} + 2\psi(K_z). \tag{42}$$

The remaining elements can be related to those above by the symmetry and unitary conditions:

$$S_{23} = -(S_{11}^* S_{12} + S_{12}^* S_{22})/S_{13}^* \tag{43}$$

$$S_{33} = -(S_{12}^* S_{13} + S_{22}^* S_{23})/S_{23}^*. \tag{44}$$

## 8. References


Collin, R. E. (1960), Field Theory of Guided Waves, 390–401 (McGraw-Hill Book Co., Inc., New York, N.Y.).

DuFort, E. C. (1967), Finite scattering matrix for an infinite antenna array, Radio Sci. **2** (New Series), No. 1, 19–27.

DuFort, E. C. (1968), Design of corrugated plates for phased array matching, IEEE Trans. Ant. Prop. **AP-16,** No. 1, 37–46.

Farrell, G. F., and D. H. Kuhn (1966a), Mutual coupling effects of triangular arrays by modal analysis, IEEE Trans. Ant. Prop. **AP-14,** No. 5, 652–654.

Farrell, G. F., and D. H. Kuhn (1966b), Mutual coupling effects in infinite planar arrays of rectangular waveguide horns, IEEE International Antennas and Propagation Symposium Digest, 392–393 (IEEE, New York, N.Y.).

Galindo, V., and C. P. Wu (1966), Numerical solutions for an infinite phased array of rectangular waveguides with thick walls, IEEE Trans. Ant. Prop. **AP-14,** No. 2, 149–158.

Hannan, P. W. (1967), Proof that a phased array can be impedance matched for all scan angles, Radio Sci. **2** (New Series), No. 3, 361–369.

Hansen, R. C. (1966), Microwave Scanning Antennas, 195–335 (Academic Press, New York, N.Y.).

Heins, A. E., and J. F. Carlson (1947), Reflection of an electromagnetic plane wave by an infinite set of plates, Quart. Appl. Math. **5,** No. 1, 82–88.

Kantorovich, L. V., and V. I. Kylov (1958), Approximate Methods of Higher Analysis, 20–141 (Interscience Publ., New York, N.Y.).

Pace, J. R., and R. Mittra (1966), The trifurcated waveguide, Radio Sci. **1** (New Series), No. 1, 117–122.

Whitehead, E. A. N. (1951), Theory of parallel plate media for microwave lenses, Proc. IEE **98,** Part III, 133–140.

Wu, C. P., and V. Galindo (1966), Properties of phased arrays of rectangular waveguides with thin walls, IEEE Trans. Ant. Prop. **AP-14,** No. 2, 163–173.


(Paper 3-5-386)

IEEE TRANSACTIONS ON ANTENNAS AND PROPAGATION, VOL. AP-18, NO. 2, MARCH 1970

# Theory of Mutual Coupling Among Minimum-Scattering Antennas

WASYL WASYLKIWSKYJ, MEMBER, IEEE, AND WALTER K. KAHN, FELLOW, IEEE

*Abstract*—A novel and rigorous formulation for mutual impedance among a class of idealized but realizable antennas is presented. Unlike past treatments of mutual coupling, which have proceeded from structurally specified antennas, the present treatment deals with a class of antennas, all the electromagnetic properties of which are rigorously expressed explicitly in terms of their radiation patterns alone. Employing a network formulation based on a scattering representation of electromagnetic fields in terms of spherical (or cylindrical) modes, the mutual impedances between such antennas are computed exactly. Several alternate representations for the mutual impedance are derived, one of which is an integral representation involving the power pattern function for both real and complex angles. In special cases, these exact results agree with published results obtained by applying well-known approximate techniques to structurally specified antennas. An important and illuminating example is provided by an infinite planar array of antennas radiating into a half-space for which the present formulation yields the well-known grating-lobe series representation of the active impedance without an explicit reference to structural properties of the radiating elements.

## I. Introduction

A NOVEL and rigorous formulation for the mutual impedance among a class of idealized but realizable antennas is presented. The subject of mutual coupling among antennas has, of course, been studied extensively. Past treatments have proceeded from particular antenna structures, specified configurations of dielectrics, and conductors in which the electromagnetic properties of the antennas are implicit. Since the associated boundary-value problems can generally not be solved exactly, in order to extract these implicit electromagnetic properties, including mutual impedance, resort must be had to approximation techniques. In place of such approximate analyses of mutual coupling among structurally specified antennas, the present treatment deals with a class of idealized antennas, the electromagnetic properties of which are rigorously expressed explicitly in terms of their radiation patterns alone. Such idealized antennas are termed minimum-scattering (MS) antennas [1]–[5]. They do not scatter, i.e., they become electromagnetically "invisible" when their ports are terminated in a particular set of reactive loads. Should these reactive loads be pure open circuits, the antennas are termed canonical minimum-scattering (CMS) antennas. For such antennas, the invisibility property may be employed directly to compute the elements of the open-circuit impedance matrix of an array. An assumption essentially equivalent to the above invisibility property has been employed in the past to simplify antenna calculations and, indeed, appears to be inherent in many elementary approximate techniques. These include the calculation of mutual impedance between elementary dipoles, as well as the use of the so-called "induced EMF" method for mutual impedance calculations among linear filamentary antennas. The validity of such techniques is usually inferred qualitatively from the physical structures and quantitatively by comparison with experimental results. Thus, for example, it is recognized that a (tuned) short dipole becomes invisible when its terminals are open circuited. Generally, approximations of this sort are considered reasonable for antennas that are small in terms of wavelength. As mentioned above, this paper departs from the conventional viewpoint of considering the invisibility property to hold approximately for structurally specified antennas, but instead employs this property to define a class of idealized antennas. This idealization is consistent with the Maxwell equations and major conservation laws. The result is a self-consistent theory of mutual coupling, which is independent of the structural properties of the antennas. The theory is exact in that the mathematical construct of an MS antenna correctly takes account of and employs both its radiative and scattering properties. This formulation, when applied to predict the interactions among structurally specified antennas, may be interpreted an an approximating technique, which is based on a single unifying assumption and is applicable to antennas that may be structurally quite dissimilar. Although the degree to which an MS antenna represents a "physical" antenna having the same pattern function is not examined quantitatively in this paper, the fact that many results derived herein are substantially identical to those obtained for structurally specified antennas establishes the utility of the MS antenna as a model. A noteworthy example is provided by the mutual impedance between two coplanar apertures. For this structure, Borgiotti [6] has derived a formula that yields the mutual admittance in terms of far-field pattern functions based on the assumption of a single-mode field in the waveguides below the aperture plane. In this paper, equivalent formulas are obtained for more general arrangements of MS antennas without reference to the physical structures, i.e., aperture configurations. Thus, once a constraint like invisibility of the radiating elements is assumed (either implicitly as in the case of

Manuscript received May 22, 1969; revised October 22, 1969. This paper is based on the dissertation submitted by W. Wasylkiwskyj in partial fulfillment of the requirements for the Ph.D. degree in electrical engineering, Polytechnic Institute of Brooklyn, Brooklyn, N. Y.

W. Wasylkiwskyj is with the Institute for Defense Analyses, Arlington, Va. 22202.
W. K. Kahn is with the School of Engineering and Applied Science, George Washington University, Washington, D. C. 20006.

Borgiotti's derivation or explicitly as in the definition of an MS antenna), the physical shape of the radiators is in a sense irrelevant as far as its effect on mutual coupling is concerned. Such formulas for mutual admittance (impedance) in terms of far-field patterns are of particular importance since they lead directly to the well-known grating-lobe series representation of the active admittance (impedance) for an infinite planar phased array [6]. It develops that the active admittance (impedance) for an infinite planar phased array is given exactly by a grating-lobe series representation if and only if the elements in question are MS antennas [4]. Thus, while on the one hand reference to the structures of the elementary radiators is eliminated in deriving the grating-lobe series representation, on the other hand, a limitation on the class of admissible element pattern functions is established. For, it may be shown that given any polar radiation pattern displaying point symmetry with respect to the origin, one can, in principle, realize a reciprocal MS antenna with that pattern [2]–[4]. Conversely, for a pattern not possessing the above symmetry, it is impossible to realize a reciprocal MS antenna. Thus, the grating-lobe series representation cannot be exact for an array whose elements do not possess the correct pattern symmetry. Since the grating-lobe series has been shown to represent the active impedance quite accurately for a variety of radiating elements, the description of elementary radiators in an array as MS antennas appears to be quite useful as well as general. Although this paper deals exclusively with MS antennas, the formulation can be extended to general, lossless antennas [4].

The properties of an antenna system with respect to an arbitrary number of local ports and outside some sphere or cylinder enclosing the antenna system are described by a scattering matrix. The basis for this scattering representation at the local ports is entirely conventional, while at the radiation ports it is a (denumerably) infinite set of spherical or cylindrical vector mode functions. Consequently, the scattering matrix comprises an infinite number of rows and columns. The mutual impedances of two MS antennas are computed exactly. The scattering representation employed herein has been described in [3], [4], and a concise summary of the main results required in this paper is given in Section II and the Appendix.

MS antennas radiating into two-dimensional (parallel-plate) regions and into three-dimensional spherical regions are treated separately. The simpler two-dimensional case forms the best introduction to the more complex three-dimensional vector case. Employing the scattering representation together with suitable addition theorems for cylindrical or spherical harmonics, the mutual impedance of two MS antennas is expressed in the form of an infinite series involving the radiation patterns of the antennas. The series is then transformed into an integral representation, which contains explicitly the far-field pattern function of each antenna. The integral representation can be cast into several alternate forms, one of which is applicable to aperture antennas.

The three-dimensional vector case is considered next. Here, instead of employing addition theorems for the spherical vector mode functions directly, a somewhat simpler technique is used, which involves a plane-wave expansion of the electromagnetic fields. This alternate approach yields integral representations for mutual impedance similar to those obtained in the two-dimensional scalar case. These representations are compared with similar expressions derived for structurally specified antennas by means of well-known approximation techniques.

## II. Scattering Representation of Antennas

### *General Lossless Antennas*

For reference we present a concise summary of the scattering representation for antennas in terms of spherical or cylindrical modes as given in [3], [4]. The radiation and scattering properties (as well as mutual coupling among antenna ports of a finite, lossless, but otherwise arbitrary antenna) are related by the matrix equation

$$\mathbf{b} = S\mathbf{a} = \begin{bmatrix} \mathbf{b}_\alpha \\ \hline \mathbf{b}_\beta \end{bmatrix} = \left[\begin{array}{c|c} S_{\alpha\alpha} & S_{\alpha\beta} \\ \hline S_{\beta\alpha} & S_{\beta\beta} \end{array}\right] \begin{bmatrix} \mathbf{a}_\alpha \\ \hline \mathbf{a}_\beta \end{bmatrix}, \tag{1}$$

where $S^+S = 1$ and the superscript plus sign denotes the complex conjugate transpose. For convenience, the scattering matrix $S$ is normalized to a set of 1-ohm resistances. The $\mathbf{a}_\alpha$ and $\mathbf{b}_\alpha$ are column matrices each containing, respectively, incident and reflected wave amplitudes at the $N$ local accessible antenna ports; $\mathbf{a}_\beta$ and $\mathbf{b}_\beta$ are infinite-dimensional column matrices denoting incident and reflected wave amplitudes of the spherical (cylindrical) modes in free space outside a sphere (cylinder) completely enclosing the antenna. An input $\mathbf{a}_\alpha$ at the antenna ports produces a radiated wave $\mathbf{b}_\beta = S_{\beta\alpha}\mathbf{a}_\alpha$; the wave reflected back into the antenna ports is $\mathbf{b}_\alpha = S_{\alpha\alpha}\mathbf{a}_\alpha$. Alternatively, when the $N$ antenna ports are connected to matched receivers, $\mathbf{a}_\alpha = \mathbf{0}$. An incident wave (from free space) $\mathbf{a}_\beta$ gives rise to a wave $\mathbf{b}_\alpha = S_{\alpha\beta}\mathbf{a}_\beta$ at the antenna ports and causes a wave $\mathbf{b}_\beta = S_{\beta\beta}\mathbf{a}_\beta$ to be reflected into surrounding space. Thus the $N$-dimensional submatrix $S_{\alpha\alpha}$ describes the mutual and self-coupling among the accessible antenna ports, while $S_{\alpha\beta}, S_{\beta\alpha}, S_{\beta\beta}$ describe, respectively, the receiving, transmitting, and scattering properties of the antenna. The quantities $S_{\alpha\beta}, S_{\beta\alpha}$ will be termed modal receiving and transmitting patterns, respectively, to distinguish them from conventional far-field pattern quantities, which depend explicitly on the angular coordinates. For a reciprocal antenna, $S_{\alpha\beta} = \tilde{S}_{\beta\alpha}$ (the tilde denotes the transpose). An $N$-element array of antennas may be considered as a single $N$-port antenna in which case $S_{\alpha\alpha}$ describes the mutual and self-coupling among the array elements. The mutual interaction among the elements of the array can also be expressed by the associated impedance matrix $\bar{Z}$, $\bar{Z} = (1 + S_{\alpha\alpha})(1 - S_{\alpha\alpha})^{-1}$.

The correspondence between the column matrices $\mathbf{a}_\beta, \mathbf{b}_\beta$ and the electromagnetic fields in free space is established by properly normalizing the cylindrical (spherical) mode amplitudes and arranging them in an infinite linear sequence in some agreed order. For a cylindrical two-dimensional region, the $z$ component of the electric field (a dual

116

representation exists for $H_z$) is

$$E_z = \frac{(\omega\mu_0)^{1/2}}{2}\left[a_0 H_0^{(1)}(kr) + b_0 H_0^{(2)}(kr)\right] + \left(\frac{\omega\mu_0}{2}\right)^{1/2}\sum_{n=1}^{\infty}\{[a_n^{(e)}\cos n\theta + a_n^{(o)}\sin n\theta]H_n^{(1)}(kr) + [b_n^{(e)}\cos n\theta + b_n^{(o)}\sin n\theta]H_n^{(2)}(kr)\}, \quad (2)$$

where $r$, $\theta$ are circular cylinder coordinates and $H_n^{(1)}$, $H_n^{(2)}$ are Hankel functions. The time dependence $\exp[+j\omega t]$ is assumed so that the $H_n^{(2)}$ are associated with outgoing waves from the origin and the $H_n^{(1)}$ with incoming waves. The elements of $\mathbf{a}_\beta$ and $\mathbf{b}_\beta$ are then defined by

$$\tilde{\mathbf{a}}_\beta = [a_0 \quad a_1^{(e)} \quad a_1^{(o)} \quad a_2^{(e)} \quad a_2^{(o)} \quad \cdots], \quad (3a)$$

$$\tilde{\mathbf{b}}_\beta = [b_0 \quad b_1^{(e)} \quad b_1^{(o)} \quad b_2^{(e)} \quad b_2^{(o)} \quad \cdots]. \quad (3b)$$

For a spherical region, we employ the representation of the transverse fields $\mathbf{E}_t, \mathbf{H}_t$, as described in the Appendix:

$$r\mathbf{E}_t = \zeta^{1/2}\sum_{n=1}^{\infty}\sum_{m=0}^{n}\sum_{(e,o)}\{[a_{nm}'^{(e,o)}\dot{\hat{H}}_n^{(1)}(kr) + b_{nm}'^{(e,o)}\dot{\hat{H}}_n^{(2)}(kr)]\mathbf{e}_{nm}'^{(e,o)}(\theta,\varphi) + [a_{nm}''^{(e,o)}\hat{H}_n^{(1)}(kr) + b_{nm}''^{(e,o)}\hat{H}_n^{(2)}(kr)]\mathbf{e}_{nm}''^{(e,o)}(\theta,\varphi)\}, \quad (4a)$$

$$r\mathbf{H}_t = j(\eta)^{1/2}\sum_{n=1}^{\infty}\sum_{m=0}^{n}\sum_{(e,o)}\{[a_{nm}'^{(e,o)}\hat{H}_n^{(1)}(kr) + b_{nm}'^{(e,o)}\hat{H}_n^{(2)}(kr)]\mathbf{e}_{nm}''^{(e,o)}(\theta,\varphi) + [a_{nm}''^{(e,o)}\dot{\hat{H}}_n^{(1)}(kr) + b_{nm}''^{(e,o)}\dot{\hat{H}}_n^{(2)}(kr)]\mathbf{e}_{nm}'^{(e,o)}(\theta,\varphi)\}, \quad (4b)$$

where $\mathbf{e}_{nm}^{(e,o)}(\theta,\varphi)$ are vector mode functions and $\hat{H}_n^{(1)(2)}$ are spherical Hankel functions. For the elements of $\mathbf{a}_\beta$ and $\mathbf{b}_\beta$ we have

$$\tilde{\mathbf{a}}_\beta = [\cdots \quad a_{nm}'^{(e)} \quad a_{nm}'^{(o)} \quad a_{nm}''^{(e)} \quad a_{nm}''^{(o)} \quad \cdots], \quad (5a)$$

$$\tilde{\mathbf{b}}_\beta = [\cdots \quad b_{nm}'^{(e)} \quad b_{nm}'^{(o)} \quad b_{nm}''^{(e)} \quad b_{nm}''^{(o)} \quad \cdots]. \quad (5b)$$

The ordering of the mode amplitudes in (3) and (5) is logically arbitrary. Once a particular ordering has been adopted, each mode amplitude may be distinguished by a single index. In spherical coordinates this single index will then specify 1) mode type ($E$ or $H$), 2) the index $n$, 3) the index $m$, and 4) even or odd dependence on $\varphi$.

The normalization of the mode amplitudes is such that in the absence of an antenna $\mathbf{a}_\beta = \mathbf{b}_\beta$, i.e., the scattering matrix of free space in the unit matrix [2]–[4]. Also, for a wave $\mathbf{a}_\beta$ incident from free space, the power absorbed by the antenna is

$$P = \mathbf{a}_\beta^+\mathbf{a}_\beta - \mathbf{b}_\beta^+\mathbf{b}_\beta. \quad (6)$$

For an antenna in transmission, the outgoing wave $\mathbf{b}_\beta$ can be related directly to the far-field radiation pattern. In a two-dimensional cylindrical region we define a far-field quantity $f(\theta)$,

$$f(\theta) = \lim_{r\to\infty} E_z(r,\theta)(kr/\omega\mu_0)^{1/2}\exp(-j\pi/4)\exp(jkr). \quad (7)$$

From (2) with $a_n = 0$ and the asymptotic forms of the Hankel functions for large argument, we obtain

$$f(\theta) = (1/\sqrt{\pi})\sum_{n=-\infty}^{\infty}\sigma_n(j)^n\exp(jn\theta), \quad (8)$$

where

$$\sigma_n = \tfrac{1}{2}(b_n^{(e)} - jb_n^{(o)})$$

$$\sigma_{-n} = \tfrac{1}{2}(b_n^{(e)} + jb_n^{(o)})(-1)^n, \quad n > 0$$

$$\sigma_0 = (1/\sqrt{2})b_0. \quad (9)$$

The power radiated by the antenna may now be written

$$P_r = \mathbf{b}_\beta^+\mathbf{b}_\beta = \int_{-\pi}^{\pi}|f(\theta)|^2\,d\theta. \quad (10)$$

Similarly, for a spherical region, we define a far-field (vector) pattern quantity $\mathbf{F}(\theta,\varphi)$, proportional to the electric field and such that the power radiated by the antenna is given by

$$P_r = \mathbf{b}_\beta^+\mathbf{b}_\beta = \int_0^{2\pi}d\varphi\int_0^{\pi}|\mathbf{F}(\theta,\varphi)|^2\sin\theta\,d\theta. \quad (11)$$

Employing (4) together with asymptotic forms of the spherical Hankel functions for large argument (see Appendix), we obtain

$$\mathbf{F}(\theta,\varphi) \equiv F_\theta\boldsymbol{\theta}_0 + F_\varphi\boldsymbol{\varphi}_0 = \lim_{r\to\infty}\eta^{1/2}[r\exp jkr]\mathbf{E}_t(r,\theta,\varphi) = \sum_{n=1}^{\infty}\sum_{m=0}^{n}\sum_{(e,o)}\{(j)^n b_{nm}'^{(e,o)}\mathbf{e}_{nm}'^{(e,o)}(\theta,\varphi) + (j)^{n+1}b_{nm}''^{(e,o)}\mathbf{e}_{nm}''^{(e,o)}(\theta,\varphi)\}. \quad (12)$$

*Canonical Minimum-Scattering Antennas*

The scattering matrix of a CMS antenna has the form [2]:

$$S = \left[\begin{array}{c|c} 0 & S_{\beta\alpha}^+ \\ \hline S_{\beta\alpha} & I_{\beta\beta} - S_{\beta\alpha}S_{\beta\alpha}^+ \end{array}\right]. \quad (13)$$

This antenna is the unique MS antenna (with the given modal radiation pattern $S_{\beta\alpha}$), which does not scatter when the local ports are open circuited. If the antenna is reciprocal, its modal radiation pattern must be pure real. It may be shown that a pure-real modal radiation pattern constrains the power pattern function to have point symmetry about the origin [1], [2]. In addition, when the excitations at the local ports of a reciprocal CMS antenna are pure real, the radiation fields defined by (8) and (12) must possess the following symmetry properties [4]:

$$f(\theta) = f^*(\theta + \pi), \tag{14}$$

$$F_\theta^*(\pi - \theta, \varphi + \pi) = -F_\theta(\theta,\varphi),$$
$$F_\varphi^*(\pi - \theta, \varphi + \pi) = F_\varphi(\theta,\varphi). \tag{15}$$

The property of invisibility of a CMS antenna on open circuit has already been employed to compute the mutual coupling between line antennas and elementary dipoles [3], [7]. In the following, this technique will be extended to arbitrary CMS antennas.

## III. Mutual Impedance of CMS Antennas Radiating into Two-Dimensional Regions

Consider two one-port CMS antennas with the coordinate system as shown in Fig. 1. Denote by column matrices $\mathbf{s}_1$ and $\mathbf{s}_2^*$ the modal transmitting patterns, respectively, of antennas 1 and 2. By (13), the scattering representations are



$$S_1 = \left[\begin{array}{c|c} 0 & \mathbf{s}_1^+ \\ \hline \mathbf{s}_1 & I_{\beta\beta} - \mathbf{s}_1\mathbf{s}_1^+ \end{array}\right], \tag{16a}$$

$$S_2 = \left[\begin{array}{c|c} 0 & \tilde{\mathbf{s}}_2 \\ \hline \mathbf{s}_2^* & I_{\beta\beta} - \mathbf{s}_2^*\tilde{\mathbf{s}}_2 \end{array}\right]. \tag{16b}$$

The scattering matrices $S_1$ and $S_2$ are referred, respectively, to origins $O_1$ and $O_2$ in Fig. 1, i.e., the cylindrical modes entering into the scattering representation of each antenna are defined relative to its local coordinate system. Both scattering matrices are unitary since the antennas are lossless. Hence

$$\mathbf{s}_1^+\mathbf{s}_1 = \mathbf{s}_2^+\mathbf{s}_2 = 1. \tag{17}$$

Let antenna 1 be excited at its local port with a unit amplitude wave when antenna 2 is open circuited (and hence indistinguishable from free space). Then the fields radiated at every point in space are not affected by the presence of the second antenna. Referring to (1), (2), and (16a), the electric field at $P$ of Fig. 1 is

$$E_z(r_1,\theta_1) = [(\omega\mu_0)^{1/2}/2]s_0^{(1)}H_0^{(2)}(kr_1) + (\omega\mu_0/2)^{1/2}\sum_{n=1}^{\infty}[s_n^{(1)(e)}\cos n\theta_1 + s_n^{(1)(o)}\sin n\theta_1]H_n^{(2)}(kr_1), \tag{18}$$

where the $s_n^{(1)(e,o)}$ are elements of $\mathbf{s}_1$.

Each cylindrical harmonic in (18) can be expanded by means of well-known addition theorems [8] in terms of harmonics involving the coordinates $r_2,\theta_2$. Since the field must be finite at $O_2$, the required expansion valid for $r_2 < D$ must contain Bessel functions with arguments $kr_2$, and hence must be of the form

$$E_z(r_2,\theta_2) = [(\omega\mu_0)^{1/2}/2]a_0^{(1)}2J_0(kr_2) + (\omega\mu_0/2)^{1/2}\sum_{n=1}^{\infty}[a_n^{(1)(e)}\cos n\theta_2 + a_n^{(1)(o)}\sin n\theta_2]2J_n(kr_2), \tag{19}$$

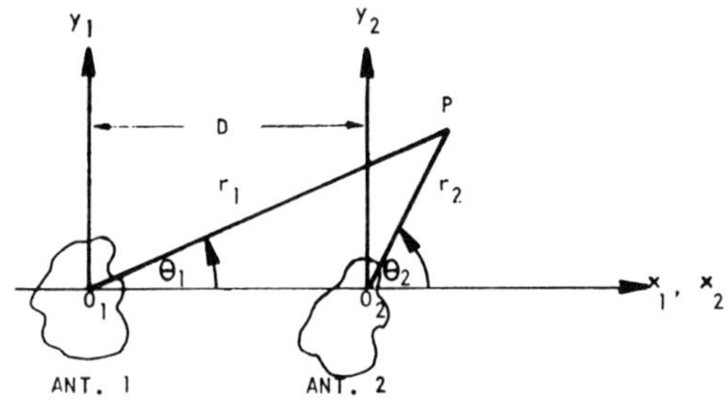


Fig. 1. Coordinates for two CMS antennas radiating into a two-dimensional region.

where the coefficients $a_n^{(1)(e,o)}$ (which are functions of $\mathbf{s}_1$ and $D$) are yet to be determined. The identity $2J_n = H_n^{(1)} + H_n^{(2)}$ and (2) identify the coefficients $a_n^{(1)(e,o)}$ in (19) as amplitudes of incoming and outgoing cylindrical modes relative to $O_2$. Arranging these in a column matrix $\mathbf{a}^{(1)}$,

$$\tilde{\mathbf{a}}^{(1)} = [a_0^{(1)} \quad a_1^{(1)(e)} \quad a_1^{(1)(o)} \quad a_2^{(1)(e)} \quad a_2^{(1)(o)} \quad \cdots], \tag{20}$$

and denoting the incident wave (equals reflected wave on open circuit) at the port of antenna 2 by $b_2$, we obtain for antenna 2

$$\left[\begin{array}{c} b_2 \\ \hline \mathbf{a}^{(1)} \end{array}\right] = \left[\begin{array}{c|c} 0 & \tilde{\mathbf{s}}_2 \\ \hline \mathbf{s}_2^* & I_{\beta\beta} - \mathbf{s}_2^*\tilde{\mathbf{s}}_2 \end{array}\right]\left[\begin{array}{c} b_2 \\ \hline \mathbf{a}^{(1)} \end{array}\right]. \tag{21}$$

Since the antennas are tuned to free space, the reflected wave $b_1$ at the port of antenna 1 is zero. The incident wave $a_1$ is unity. Hence the mutual impedance is

$$\bar{Z}_{21} = \left.\frac{V_2}{I_1}\right]_{I_2=0} = \frac{2b_2}{a_1 + b_1} = 2b_2. \tag{22}$$

For the open-circuit input impedances we have

$$\bar{Z}_{11} = \bar{Z}_{22} = 1. \tag{23}$$

Employing (21), (22) may be written

$$\bar{Z}_{21} = 2\tilde{\mathbf{s}}_2\mathbf{a}^{(1)} = 2[a_0^{(1)}s_0^{(2)} + \sum_{n=1}^{\infty} a_n^{(1)(e)}s_n^{(2)(e)} + a_n^{(1)(o)}s_n^{(2)(o)}], \tag{24}$$

where the $s_n^{(2)(e,o)}$ are elements of $\mathbf{s}_2$. We will express the mode amplitudes $a_n^{(1)(e,o)}$ in terms of the modal radiation pattern of antenna 1. This is more readily accomplished by using equivalent complex forms of (18) and (19). These are

$$E_z(r_1,\theta_1) = (\omega\mu_0/2)^{1/2}\sum_{n=-\infty}^{\infty}\sigma_n^{(1)}H_n^{(2)}(kr_1)\exp(jn\theta_1), \tag{25a}$$

$$E_z(r_2,\theta_2) = (\omega\mu_0/2)^{1/2}\sum_{n=-\infty}^{\infty}\alpha_n^{(1)}2J_n(kr_2)\exp(jn\theta_2). \tag{25b}$$

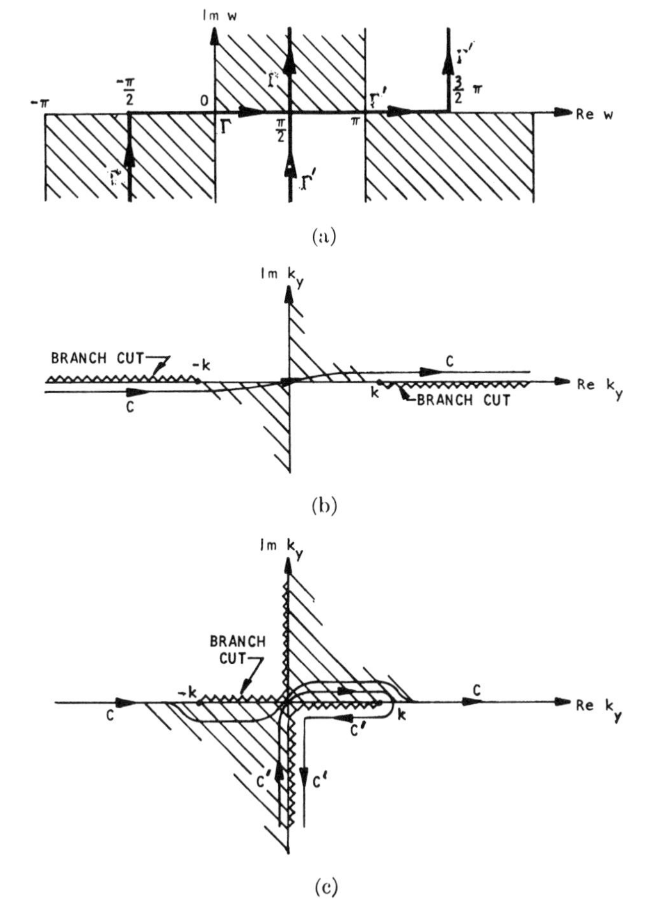


Fig. 2. Contours of integration for evaluation of mutual impedance. (a) Complex $w$ plane; Im cos $w < 0$ in shaded regions. (b) Complex $k_y$ plane; Im $(k^2 - k_y^2)^{1/2} < 0$ in shaded quadrants. (c) Alternate contours in the $k_y$ plane; Im $(k^2 - k_y)^{1/2} < 0$ on entire top Riemann sheet; Re $(k^2 - k_y^2)^{1/2} > 0$ in shaded quadrants; Re $(k^2 - k_y^2)^{1/2} < 0$ in unshaded quadrants.

The mode amplitudes $\sigma_n^{(1,2)}$ and $\alpha_n^{(1)}$ are defined by

$$\begin{aligned} \sigma_n^{(1,2)} &= \tfrac{1}{2}(s_n^{(1,2)(e)} - js_n^{(1,2)(o)}) \\ \sigma_{-n}^{(1,2)} &= \tfrac{1}{2}(s_n^{(1,2)(e)} + js_n^{(1,2)(o)})(-1)^n, \quad n > 0 \\ \sigma_0^{(1,2)} &= (1/\sqrt{2})s_0^{(1,2)} \end{aligned} \tag{26a}$$

$$\begin{aligned} \alpha_n^{(1)} &= \tfrac{1}{2}(a_n^{(1)(e)} - ja_n^{(1)(o)}) \\ \alpha_{-n}^{(1)} &= \tfrac{1}{2}(a_n^{(1)(e)} + ja_n^{(1)(o)})(-1)^n, \quad n > 0 \\ \sigma_0^{(1)} &= (1/\sqrt{2})a_0^{(1)}. \end{aligned} \tag{26b}$$

With the aid of (26), the mutual impedance takes on the symmetric form

$$\bar{Z}_{21} = 4 \sum_{n=-\infty}^{\infty} (-1)^n \alpha_n^{(1)} \sigma_{-n}^{(2)}. \tag{27}$$

To solve for $\alpha_n^{(1)}$ in terms of the $\sigma_n^{(1)}$, we employ the addition theorem [8]

$$H_n^{(2)}(kr_1) \exp(jn\theta_1) = \sum_{m=-\infty}^{\infty} (-1)^m H_m^{(2)}(kD) J_{n+m}(kr_2) \cdot \exp[j(n+m)\theta_2], \quad r_2 < D. \tag{28}$$

Substituting the above in (25a) yields

$$\begin{aligned} E_z(r_1,\theta_1) &\equiv E_z(r_2,\theta_2) \\ &= (\omega\mu_0/2)^{1/2} \sum_{m=-\infty}^{\infty} \sum_{n=-\infty}^{\infty} (-1)^m \sigma_{n-m}^{(1)} \\ &\quad \cdot H_m^{(2)}(kD) \exp(jn\theta_2) J_n(kr_2). \end{aligned} \tag{29}$$

Upon comparing (29) with (25b) we have

$$\alpha_n^{(1)} = \tfrac{1}{2} \sum_{m=-\infty}^{\infty} (-1)^m \sigma_{n-m}^{(1)} H_m^{(2)}(kD). \tag{30}$$

Combining (30) with (27) yields

$$\bar{Z}_{21} = 2 \sum_{n=-\infty}^{\infty} \sum_{m=-\infty}^{\infty} (-1)^{m+n} \sigma_{n-m}^{(1)} \sigma_{-n}^{(2)} H_m^{(2)}(kD). \tag{31}$$

By an entirely parallel procedure, one can find $\bar{Z}_{12}$ as well. The only difference is that the addition theorem for expanding the field of antenna 2 about $O_1$ in Fig. 1 differs from (28) by the absence of the factor $(-1)^m$. Omitting this factor in (31) and interchanging superscripts 1 and 2, we obtain

$$\bar{Z}_{12} = 2 \sum_{n=-\infty}^{\infty} \sum_{m=-\infty}^{\infty} (-1)^n \sigma_{n-m}^{(2)} \sigma_{-n}^{(1)} H_m^{(2)}(kD). \tag{32}$$

Equations (31) and (32), together with (23), give the complete coupling impedance matrix for two arbitrary CMS antennas in terms of modal receiving and transmitting patterns. Note that $\bar{Z}_{21}$ is expressed in terms of the transmitting pattern of antenna 1 and the receiving pattern of antenna 2 [see (16) and (26a)]. If both antennas are reciprocal, the modal radiation patterns are real, or, from (26a), $\sigma_{-n} = \sigma_n^*(-1)^n$. Employing this relation in (31) and (32), one readily verifies that $\bar{Z}_{12} = \bar{Z}_{21}$, as expected. In the special case of two reciprocal line antennas, each capable of radiating only the lowest order cylindrical mode [i.e., the first term in (18)] we have, in virtue of (17), $\sigma_0^{(1,2)} = 1/\sqrt{2}(\sigma_n^{(1,2)} = 0, n \neq 0)$. Thus (31) and (32) yield

$$\bar{Z}_{12} = \bar{Z}_{21} = H_0^{(2)}(kD), \tag{33}$$

in agreement with the result in [3].

Usually the receiving and transmitting patterns of an antenna would be given directly as quantities proportional to the electric field in free space, e.g., $f(\theta)$ in (8). The modal quantities required for computation of mutual impedance may be obtained from $f(\theta)$ by solving for the $\sigma_n^{(1,2)}$ with the aid of orthogonality properties of trigonometric functions. This intermediate step may be eliminated by expressing the mutual impedance explicitly in terms of far-field pattern quantities. For this purpose, we employ the Sommerfeld integral for the Hankel function, viz.,

$$H_m^{(2)}(kD) = \frac{1}{\pi} \int_\Gamma \exp(-jkD\cos w) \cdot \exp[jm(w + \pi/2)]\, dw, \tag{34}$$

where the path of integration $\Gamma$ is shown in Fig. 2(a).

Inserting (34) into (31) and interchanging the order of summation and integration, yields

$$\bar{Z}_{21} = \frac{2}{\pi}\int_{\Gamma} \exp(-jkD\cos w) \cdot \Big\{ \sum_{n=-\infty}^{\infty} \sum_{m=-\infty}^{\infty} (-1)^n \sigma_{n-m}^{(1t)} \sigma_{-n}^{(2r)} \cdot \exp(-jm\pi/2)\exp(jmw) \Big\}\, dw, \quad (35)$$

where we have added superscripts $t$ and $r$ to differentiate, respectively, between the modal receiving and transmitting pattern. The far-field quantities related to $\sigma_n^{(t)}$ and $\sigma_n^{(r)}$ by (8) will be denoted, respectively, by $f_t(\theta)$ and $f_r(\theta)$. We now form the function $f_{1t}(-\theta)f_{2r}(-\theta+\pi)$ and expand it in a Fourier series as follows:

$$f_{1t}(-\theta)f_{2r}(-\theta+\pi) = \sum_{m=-\infty}^{\infty} p_m \exp(jm\theta), \quad |\theta| \leq \pi. \quad (36)$$

Taking account of (8), we have

$$p_m = \frac{1}{2\pi}\int_{-\pi}^{\pi} \exp(-jm\theta) f_{1t}(-\theta) f_{2r}(-\theta+\pi)\, d\theta = (1/\pi) \sum_{n=-\infty}^{\infty} (-1)^n \exp(-jm\pi/2)\sigma_{n-m}^{(1t)}\sigma_{-n}^{(2r)}. \quad (37)$$

By (36) and (37), (35) becomes

$$\bar{Z}_{21} = 2\int_{\Gamma} \exp(-jkD\cos w) f_{1t}(-w) f_{2r}(-w+\pi)\, dw = 2\int_{\Gamma} \exp(-jkD\cos w) f_{1t}(w) f_{2r}(w+\pi)\, dw, \quad (38)$$

where $f_{1t}(w)$ and $f_{2r}(w+\pi)$ are analytic continuations of $f_{1t}(\theta)$ and $f_{2r}(\theta+\pi)$ into the complex $w$ plane. To obtain $\bar{Z}_{12}$ we employ an alternate integral representation for the Hankel function, viz.,

$$H_m^{(2)}(kD) = \frac{1}{\pi}\int_{\Gamma'} \exp(jkD\cos w)\exp[jm(w-\pi/2)]\, dw, \quad (39)$$

taken over the path $\Gamma'$ in Fig. 2(a). Substituting (39) into (32) and using a Fourier series expansion for $f_{2t}(-\theta)\cdot f_{1r}(-\theta+\pi)$ we proceed as above and obtain

$$\bar{Z}_{12} = 2\int_{\Gamma'} \exp(jkD\cos w) f_{2t}(w) f_{1r}(w+\pi)\, dw. \quad (40)$$

We can now change the variables of integration to shift the path $\Gamma'$ to $\Gamma$ and obtain

$$\bar{Z}_{12} = 2\int_{\Gamma} \exp(-jkD\cos w) f_{1r}(w) f_{2t}(w+\pi)\, dw. \quad (41)$$

For reciprocal antennas $f_t = f_r$, and (41) is identical with (38) as expected. Since the elements of the modal receiving pattern are complex conjugates of the elements of the transmitting pattern [see (16)] we have, in virtue of (26a), $\sigma_n^{*(r)} = (-1)^n \sigma_{-n}^{(t)}$. Coupled with (8) this yields, for real $\theta$,

$$f_r(\theta) = f_t^*(\theta+\pi). \quad (42)$$

Therefore (38) and (41) may be written as follows:

$$\bar{Z}_{21} = 2\int_{\Gamma} \exp(-jkD\cos w) f_{1t}(w) f_{2t}^*(w^*)\, dw, \quad (43a)$$

$$\bar{Z}_{12} = 2\int_{\Gamma} \exp(-jkD\cos w) f_{1r}(w) f_{2r}^*(w^*)\, dw. \quad (43b)$$

For two identical and reciprocal antennas, the mutual impedance becomes

$$\bar{Z}_{12} = \bar{Z}_{21} = 2\int_{\Gamma} \exp(-jkD\cos w) P(w)\, dw, \quad (44)$$

where $P(w)$ is a power pattern function defined by

$$P(w) = f(w) f(w+\pi). \quad (45)$$

Reference to (14) or (42) shows that, for real $w$, $P(w)$ reduces to the conventional power pattern function of an isolated antenna. Also, in virtue of (10) and (17), $P(w)$ integrated over all real angles is unity. Equation (44) states that the mutual impedance is determined by integrating the power pattern over real and complex angles. For a line antenna, the power pattern function is isotropic and $P(w) = 1/2\pi$. Substituting this into (44) yields $\bar{Z}_{12} = H_0^{(2)}(kD)$, in agreement with our previous result (33).

An alternative representation for $\bar{Z}_{12}$ may be obtained by changing variables in (44) via $k_y = k \sin w$. The path of integration in the $w$ plane maps into the path $C$ in the $k_y$ plane as shown in Fig. 2(b). The branch cuts in the $k_y$ plane have been chosen to ensure $\mathrm{Im}\,(k^2 - k_y^2)^{1/2} < 0$ in the two shaded quadrants. Equation (44) may now be written

$$\bar{Z}_{12} = 2\int_{-\infty}^{\infty} \frac{\exp[-j(k^2-k_y^2)^{1/2}D]}{(k^2-k_y^2)^{1/2}} P(k_y)\, dk_y. \quad (46)$$

The integration over $|k_y| < k$ (visible region) contributes only to the real part of $\bar{Z}_{12}$ while integration over $|k_y| > k$ (invisible region) contributes only to the imaginary part. Equation (46) can also be expressed as a Fourier transform. Choosing the branch cuts in the $k_y$ plane as in Fig. 2(c), we can deform the path $C$ into $C'$, provided $P(k_y)$ has no singularities for $\mathrm{Im}\,k_y < 0$. Taking account of the sign of $\mathrm{Re}\,(k^2 - k_y^2)^{1/2}$ in Fig. 2(c), integration around the branch cut $C'$ yields

$$\bar{Z}_{12} = 2\int_{-\infty}^{\infty} \frac{\exp(-jk_x D)}{(k^2-k_x^2)^{1/2}} P(k_x)\, dk_x, \quad (47)$$

where $k_x = k\cos w$. Equation (47) bears a strong resemblance to the mutual impedance between two one-dimensional "apertures" in a perfectly conducting magnetic screen, paralleling the results of Borgiotti [6].

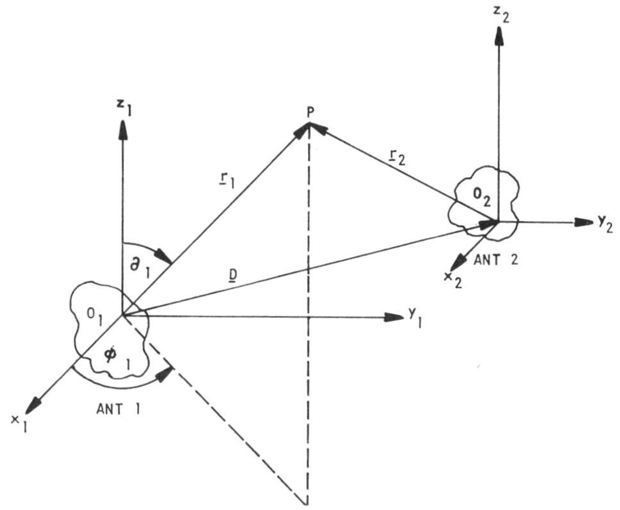


Fig. 3. Coordinates for two CMS antennas radiating into a three-dimensional region.

## IV. Mutual Impedance of CMS Antennas Radiating into Three-Dimensional Regions

The technique for computing the mutual impedance between antennas in three-dimensional regions in terms of spherical modes is similar to the one employed above for cylindrical modes. We locate an antenna at $O_1$ in Fig. 3 and excite it with a unit amplitude wave, leaving the antenna at $O_2$ open circuited. The scattering matrices of the antennas are as in (16) with

$$\tilde{\mathbf{s}}_1 = [\cdots \; s_{1nm}{}'^{(e)} \; s_{1nm}{}'^{(o)} \; s_{1nm}{}''^{(e)} \; s_{1nm}{}''^{(o)} \; \cdots],$$

$$\tilde{\mathbf{s}}_2 = [\cdots \; s_{2nm}{}'^{(e)} \; s_{2nm}{}'^{(o)} \; s_{2nm}{}''^{(e)} \; s_{2nm}{}''^{(o)} \; \cdots]. \quad (48)$$

The transverse electric field at $P$ expressed in terms of coordinates $r_1,\theta_1,\varphi_1$ is

$$r_1\mathbf{E}_t(\mathbf{r}_1) = \zeta^{1/2} \sum_{n=1}^{\infty} \sum_{m=0}^{n} \sum_{e,o} \{s_{1nm}{}'^{(e,o)} \hat{H}_n{}^{(2)}(kr_1)\, \mathbf{e}_{nm}{}'^{(e,o)}(\theta_1,\varphi_1) + s_{1nm}{}''^{(e,o)} \hat{H}_n{}^{(2)}(kr_1)\, \mathbf{e}_{nm}{}''^{(e,o)}(\theta_1,\varphi_1)\}. \quad (49)$$

By means of addition theorems developed by Stein [9] and Cruzan [10] for spherical vector mode functions, the above can be expanded about $O_2$ in terms of coordinates $r_2,\theta_2,\varphi_2$. Since the fields must be finite at $O_2$, the resulting expression for the transverse electric field will be of the form

$$r_2\mathbf{E}_t(\mathbf{r}_2) = \zeta^{1/2} \sum_{n=1}^{\infty} \sum_{m=0}^{n} \sum_{e,o} \{a_{1nm}{}'^{(e,o)} 2\hat{J}_n(kr_2)\, \mathbf{e}_{nm}{}'^{(e,o)}(\theta_2,\varphi_2) + a_{1nm}{}''^{(e,o)} 2\hat{J}_n(kr_2)\, \mathbf{e}_{nm}{}''^{(e,o)}(\theta_2,\varphi_2)\}, \quad (50)$$

where the $a_{1nm}{}^{(e,o)}$ are amplitudes of incoming and outgoing spherical modes with respect to $O_2$ and the $\hat{J}_n$ are spherical Bessel functions. Thus, by analogy with the two-dimensional case, the mutual impedance is

$$\bar{Z}_{21} = 2 \sum_{n=1}^{\infty} \sum_{m=0}^{n} \sum_{e,o} \{a_{1nm}{}'^{(e,o)} s_{2nm}{}'^{(e,o)} + a_{1nm}{}''^{(e,o)} s_{2nm}{}''^{(e,o)}\}. \quad (51)$$

One could now use the addition theorems in [9] to express the $a_{1nm}{}^{(e,o)}$ explicitly in terms of the modal radiation pattern of antenna 1 and arrive at an expression similar to (31). This procedure has been previously employed for elementary dipoles [3]. Its generalization to arbitrary CMS antennas is cumbersome and will not be employed here. The approach we shall take will be to express first the fields of antenna 1 as a superposition of plane waves. We shall then compare the resulting expressions with the corresponding expansions in spherical modes. This will enable us to identify the $a_{1nm}{}^{(e,o)}$ in (51) as integrals involving the far-field pattern of antenna 1.

Referring to Fig. 3, the fields at $P$ due to the antenna at $O_1$ (when antenna 2 is open circuited) may be written as two-dimensional Fourier transforms:

$$\mathbf{E}(\mathbf{r}_1) = \frac{1}{2\pi} \int_{-\infty}^{\infty} \int_{-\infty}^{\infty} dk_x\, dk_y\, \boldsymbol{\mathcal{E}}(k_x,k_y) \exp(-j\mathbf{k}\cdot\mathbf{r}_1), \quad (52a)$$

$$\mathbf{H}(\mathbf{r}_1) = \frac{1}{2\pi} \int_{-\infty}^{\infty} \int_{-\infty}^{\infty} dk_x\, dk_y\, \boldsymbol{\mathcal{H}}(k_x,k_y) \exp(-j\mathbf{k}\cdot\mathbf{r}_1), \qquad z_1 \geq 0, \quad (52b)$$

$$\boldsymbol{\mathcal{H}} = \frac{1}{\omega\mu_0} \mathbf{k} \times \boldsymbol{\mathcal{E}},$$

$$\mathbf{k} = k_x\mathbf{x}_{01} + k_y\mathbf{y}_{01} + k_z\mathbf{z}_{01}. \quad (52c)$$

The fields will be decomposed into $E$ and $H$ modes along $\mathbf{z}_{01}$. Let $\Pi'$ and $\Pi''$ be, respectively, the scalar $E$- and $H$-mode Hertz potentials. Then

$$\mathbf{E}(\mathbf{r}_1) = \nabla \times \nabla \times \mathbf{z}_{01}\Pi' - j\omega\mu_0 \nabla \times \mathbf{z}_{01}\Pi'', \quad (53a)$$

$$\mathbf{H}(\mathbf{r}_1) = j\omega\epsilon_0 \nabla \times \mathbf{z}_{01}\Pi' + \nabla \times \nabla \times \mathbf{z}_{01}\Pi''. \quad (53b)$$

Denote by $f',f''$ the Fourier transforms of the Hertz potentials, defined by

$$f'(k_x,k_y) = \frac{1}{2\pi} \int_{-\infty}^{\infty} \int_{-\infty}^{\infty} \Pi'(x_1,y_1,0) \cdot \exp[j(k_x x_1 + k_y y_1)]\, dx_1\, dy_1, \quad (54a)$$

$$f''(k_x,k_y) = \frac{1}{2\pi} \int_{-\infty}^{\infty} \int_{-\infty}^{\infty} \Pi''(x_1,y_1,0) \cdot \exp[j(k_x x_1 + k_y y_1)]\, dx_1\, dy_1. \quad (54b)$$

The transformed fields $\boldsymbol{\mathcal{E}}, \boldsymbol{\mathcal{H}}$ in (52) are related to $f'$ and $f''$ by

$$\boldsymbol{\mathcal{E}} = -\mathbf{k} \times (\mathbf{k} \times \mathbf{z}_{01})f' - \omega\mu_0\mathbf{k} \times \mathbf{z}_{01} f'', \quad (55a)$$

$$\boldsymbol{\mathcal{H}} = \omega\epsilon_0\mathbf{k} \times \mathbf{z}_{01} f' - \mathbf{k} \times (\mathbf{k} \times \mathbf{z}_{01})f''. \quad (55b)$$

The connection between the plane-wave representation and the spherical mode representation can be made through the radial components of the fields. As is well known, the electromagnetic fields can also be derived from potential functions $\mathbf{r}_1\cdot\mathbf{E}$ and $\mathbf{r}_1\cdot\mathbf{H}$ (sometimes referred to as Debye potentials), the former yielding $E$ modes in spherical coordinates, the latter $H$ modes. Employing the identity $\mathbf{r}_1\cdot\mathbf{E} = x_1E_x + y_1E_y + z_1E_z$ together with (52a) and (55a), we obtain

$$\mathbf{r}_1\cdot\mathbf{E}(\mathbf{r}_1) = \frac{1}{2\pi}\int_{-\infty}^{\infty}\int_{-\infty}^{\infty} dk_x\, dk_y[-\psi_1 x_1(k_x k_z f' + \omega\mu_0 k_y f'') - \psi_1 y_1(k_y k_z f' - \omega\mu_0 k_x f'') + \psi_1 z_1(k^2 - k_z^2)f'], \quad (56)$$

where

$$\psi_1 = \exp(-j\mathbf{k}\cdot\mathbf{r}_1). \quad (57)$$

Upon expressing the rectangular components of $\mathbf{k}$ in spherical coordinates $(k,\alpha,\beta)$, (56) may be put into the form

$$\mathbf{r}_1\cdot\mathbf{E}(\mathbf{r}_1) = \frac{1}{2\pi}\int_{-\infty}^{\infty}\int_{-\infty}^{\infty} dk_x\, dk_y \cdot[\omega\mu_0 f''(\partial\psi_1/\partial\beta) - k\sin\alpha f''(\partial\psi_1/\partial\alpha)]. \quad (58)$$

After evaluating (52) by the method of stationary phase and employing (55), $f'$ and $f''$ may be related to the far fields radiated by antenna 1. Using the far-field pattern quantity defined in (12) the result is

$$f'(\theta_1,\varphi_1) = \zeta^{1/2}\frac{F_{1\theta_1}(\theta_1,\varphi_1)}{-jk^3\sin\theta_1\cos\theta_1}, \quad (59a)$$

$$f''(\theta_1,\varphi_1) = \eta^{1/2}\frac{F_{1\varphi_1}(\theta_1,\varphi_1)}{jk^3\sin\theta_1\cos\theta_1}, \quad (59b)$$

where $\mathbf{F}_1 = F_{1\theta_1}\boldsymbol{\theta}_{01} + F_{1\varphi_1}\boldsymbol{\varphi}_{01}$ and $\boldsymbol{\theta}_{01}, \boldsymbol{\varphi}_{01}$ are unit vectors in spherical coordinates. $\mathbf{F}_1$ is normalized such that $P_r$ in (11) is unity. Substituting (59) into (58), and changing the variables of integration to the angular coordinates $\alpha,\beta$, we obtain

$$\mathbf{r}_1\cdot\mathbf{E}(\mathbf{r}_1) = \frac{\zeta^{1/2}}{2\pi}\int_0^{2\pi} d\beta\int_\Gamma d\alpha[F_{1\beta}(\alpha,\beta)(\partial\psi_1/\partial\beta) + \sin\alpha F_{1\alpha}(\alpha,\beta)(\partial\psi_1/\partial\alpha)]. \quad (60)$$

The path of integration $\Gamma$ in the complex $\alpha$ plane is shown in Fig. 4. For the magnetic field we make the usual duality replacements in (60), and obtain

$$\mathbf{r}_1\cdot\mathbf{H}(\mathbf{r}_1) = \frac{\eta^{1/2}}{2\pi}\int_0^{2\pi} d\beta\int_\Gamma d\alpha[F_{1\alpha}(\alpha,\beta)(\partial\psi_1/\partial\beta) - \sin\alpha F_{1\beta}(\alpha,\beta)(\partial\psi_1/\partial\alpha)]. \quad (61)$$

The integral representations in (60) and (61) express the Debye potentials in terms of the coordinate system $(r_1,\theta_1,\varphi_1)$. These potentials can also be expressed in terms of the translated coordinate system $(r_2,\theta_2,\varphi_2)$ in Fig. 3. By noting that translation of coordinates corresponds to a multiplication of all far-field quantities by $\exp[-j\mathbf{k}\cdot\mathbf{D}]$, we have, upon changing $\mathbf{r}_1$ into $\mathbf{r}_2$ in (60) and (61),

$$\mathbf{r}_2\cdot\mathbf{E}(\mathbf{r}_2) = \frac{\zeta^{1/2}}{2\pi}\int_0^{2\pi} d\beta\int_\Gamma d\alpha\exp(-j\mathbf{k}\cdot\mathbf{D}) \cdot[F_{1\beta}(\alpha,\beta)(\partial\psi_2/\partial\beta) + \sin\alpha F_{1\alpha}(\alpha,\beta)(\partial\psi_2/\partial\alpha)], \quad (62a)$$

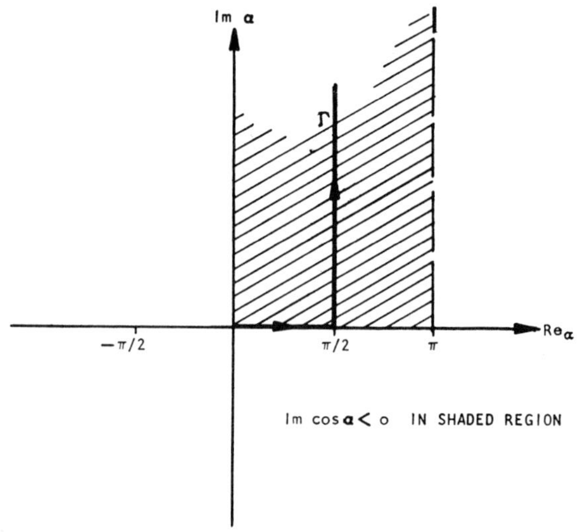


Fig. 4. Contour of integration in the complex $\alpha$ plane.

$$\mathbf{r}_2\cdot\mathbf{H}(\mathbf{r}_2) = \frac{\eta^{1/2}}{2\pi}\int_0^{2\pi} d\beta\int_\Gamma d\alpha\exp(-j\mathbf{k}\cdot\mathbf{D}) \cdot[F_{1\alpha}(\alpha,\beta)(\partial\psi_2/\partial\beta) - \sin\alpha F_{1\beta}(\alpha,\beta)(\partial\psi_2/\partial\alpha)], \quad (62b)$$

where

$$\psi_2 = \exp(-j\mathbf{k}\cdot\mathbf{r}_2). \quad (63)$$

Choosing $\mathbf{D} = D_x\mathbf{x}_{02} + D_y\mathbf{y}_{02} + D_z\mathbf{z}_{02}$ with $D_z > 0$, we note that each integrand in (62) contains a decaying exponential for $r_2 = 0$, since $\operatorname{Im}\cos\alpha < 0$ on $\Gamma$. Consequently, the integral representations in (62) yield a set of potential functions that are finite at $r_2 = 0$. It follows that all field components derived from these potential functions will also be finite at $r_2 = 0$ and the transverse fields must possess series expansions in terms of spherical vector mode functions as in (50). The corresponding series expansion for $\mathbf{r}_2\cdot\mathbf{E}(\mathbf{r}_2)$ and $\mathbf{r}_2\cdot\mathbf{H}(\mathbf{r}_2)$ in terms of scalar mode functions $\Phi_{nm}$ are given in (100). After comparing the mode voltages $V_{nm}$ and mode currents $I_{nm}$ in (97) and (100) with the expansion coefficients in (50), we obtain

$$\mathbf{r}_2\cdot\mathbf{E}(\mathbf{r}_2) = -2(\zeta)^{1/2}\sum_{n=1}^{\infty}\sum_{m=0}^{n}\sum_{e,o} n(n+1)a_{1nm}'^{(e,o)}(k\mathbf{D}) \cdot\Phi_{nm}^{(e,o)}(\theta_2,\varphi_2)[\hat{J}_n(kr_2)/kr_2], \quad (64a)$$

$$\mathbf{r}_2\cdot\mathbf{H}(\mathbf{r}_2) = -j2(\eta)^{1/2}\sum_{n=1}^{\infty}\sum_{m=0}^{n}\sum_{e,o} n(n+1)a_{1nm}''^{(e,o)}(k\mathbf{D}) \cdot\Phi_{nm}^{(e,o)}(\theta_2,\varphi_2)[\hat{J}_n(kr_2)/kr_2]. \quad (64b)$$

Equations (64) and (62) are alternate representations for the Debye potentials. Employing the expansion of $\exp(-j\mathbf{k}\cdot\mathbf{r}_2)$ in terms of scalar mode functions [11] in (62), and equating the coefficients of the resulting series to the corresponding coefficients of $\Phi_{nm}$ in (64), one can solve for the spherical mode amplitudes $a_{1nm}$. The result is

$$a_{1nm}'^{(e,o)} = \int_0^{2\pi} d\beta \int_\Gamma d\alpha (-j)^n \exp(-j\mathbf{k}\cdot\mathbf{D}) \cdot \sin\alpha \mathbf{F}_{1t}(\alpha,\beta)\cdot \mathbf{e}_{nm}'^{(e,o)}(\alpha,\beta), \quad (65a)$$

$$a_{1nm}''^{(e,o)} = \int_0^{2\pi} d\beta \int_\Gamma d\alpha (-j)^{n+1} \exp(-j\mathbf{k}\cdot\mathbf{D}) \cdot \sin\alpha \mathbf{F}_{1t}(\alpha,\beta)\cdot \mathbf{e}_{nm}''^{(e,o)}(\alpha,\beta), \quad (65b)$$

where we have added the subscript $t$ to indicate that $\mathbf{F}_{1t}$ is a transmitting pattern quantity. The mutual impedance is now obtained by combining (65) with (51). After interchanging summation and integration, we have

$$\bar{Z}_{21} = 2\int_0^{2\pi} d\beta \int_\Gamma d\alpha \exp(-j\mathbf{k}\cdot\mathbf{D}) \sin\alpha \mathbf{F}_{1t}(\alpha,\beta) \cdot \{\sum_{nm}\sum_{e,o} (-j)^n s_{2nm}'^{(e,o)} \mathbf{e}_{nm}'^{(e,o)}(\alpha,\beta) + (-j)^{n+1} s_{2nm}''^{(e,o)} \mathbf{e}_{nm}''^{(e,o)}(\alpha,\beta)\}. \quad (66)$$

The received and transmitted normalized far-field pattern quantities for antenna 2 will be denoted, respectively, by $\mathbf{F}_{2r}$ and $\mathbf{F}_{2t}$. From (12) and (16b) they are

$$\mathbf{F}_{2r}(\theta,\varphi) = \sum_{nm}\sum_{e,o} \{(j)^n s_{2nm}'^{(e,o)} \mathbf{e}_{nm}'^{(e,o)}(\theta,\varphi) + (j)^{n+1} s_{2nm}''^{(e,o)} \mathbf{e}_{nm}''^{(e,o)}(\theta,\varphi)\}, \quad (67a)$$

$$\mathbf{F}_{2t}(\theta,\varphi) = \sum_{nm}\sum_{e,o} \{(j)^n s_{2nm}^{*\prime(e,o)} \mathbf{e}_{nm}'^{(e,o)}(\theta,\varphi) + (j)^{n+1} s_{2nm}^{*\prime\prime(e,o)} \mathbf{e}_{nm}''^{(e,o)}(\theta,\varphi)\}. \quad (67b)$$

In addition, applying the transformation $\theta \to \pi - \theta$, $\varphi \to \pi + \varphi$ to the spherical mode functions we obtain the following [4]:

$$\hat{\mathbf{F}}_{2r}(\theta,\varphi) \equiv -F_{2r\theta}(\pi-\theta,\varphi+\pi)\boldsymbol{\theta}_0 + F_{2r\varphi}(\pi-\theta,\varphi+\pi)\boldsymbol{\varphi}_0$$
$$= \sum_{nm}\sum_{e,o} \{(-j)^n s_{2nm}'^{(e,o)} \mathbf{e}_{nm}'^{(e,o)}(\theta,\varphi) + (-j)^{n+1} s_{2nm}''^{(e,o)} \mathbf{e}_{nm}''^{(e,o)}(\theta,\varphi)\}. \quad (68)$$

Thus the sum in the integrand of (66) is identified as $\hat{\mathbf{F}}_{2r}(\alpha,\beta)$ and the mutual impedance may be written[1]

$$\bar{Z}_{21} = 2\int_0^{2\pi} d\varphi \int_\Gamma \sin\alpha\, d\alpha \exp(-j\mathbf{k}\cdot\mathbf{D}) \mathbf{F}_{1t}(\alpha,\varphi)\cdot\hat{\mathbf{F}}_{2r}(\alpha,\varphi), \quad (69a)$$

$$\bar{Z}_{12} = 2\int_0^{2\pi} d\varphi \int_\Gamma \sin\alpha\, d\alpha \exp(-j\mathbf{k}\cdot\mathbf{D}) \mathbf{F}_{1r}(\alpha,\varphi)\cdot\hat{\mathbf{F}}_{2t}(\alpha,\varphi). \quad (69b)$$

[1] Note that $\hat{\mathbf{F}}$ is formed from $\mathbf{F}$ by an inversion of the coordinate system.

In the above we have set $\beta = \varphi$. If the antennas are reciprocal, $\mathbf{F}_r = \mathbf{F}_t$ and $\bar{Z}_{12} = \bar{Z}_{21}$. Since the spherical mode functions employed in (67) are real, we have

$$\hat{\mathbf{F}}_{2r}(\theta,\varphi) = \mathbf{F}_{2t}^*(\theta,\varphi),$$
$$\hat{\mathbf{F}}_{2t}(\theta,\varphi) = \mathbf{F}_{2r}^*(\theta,\varphi), \quad (70)$$

for all real $\theta$. Hence (69) may be written in an alternate form

$$\bar{Z}_{21} = 2\int_0^{2\pi} d\varphi \int_\Gamma \sin\alpha\, d\alpha \exp(-j\mathbf{k}\cdot\mathbf{D}) \cdot \mathbf{F}_{1t}(\alpha,\varphi)\cdot\mathbf{F}_{2t}^*(\alpha^*,\varphi), \quad (71a)$$

$$\bar{Z}_{12} = 2\int_0^{2\pi} d\varphi \int_\Gamma \sin\alpha\, d\alpha \exp(-j\mathbf{k}\cdot\mathbf{D}) \cdot \mathbf{F}_{1r}(\alpha,\varphi)\cdot\mathbf{F}_{2r}^*(\alpha^*,\varphi). \quad (71b)$$

The mutual impedance between identical, reciprocal CMS antennas may be written in terms of a power pattern function, in parallel with (44). In the present case, the pattern function is defined by

$$P(\alpha,\varphi) = \mathbf{F}(\alpha,\varphi)\cdot\hat{\mathbf{F}}(\alpha,\varphi). \quad (72)$$

From (70) we see again that for real $\alpha$, $P(\alpha,\varphi)$ is just the conventional power pattern of an isolated antenna.

We now consider some special cases of (71) for identical reciprocal antennas. For antennas lying on the $z$ axis we have $\mathbf{D} = \mathbf{z}_0 D$, and letting $t = \cos\alpha$,

$$\bar{Z}_{21} = 2\int_0^{2\pi} d\varphi \int_{-j\infty}^{1} \exp(-jkDt) P(t,\varphi)\, dt. \quad (73)$$

Alternatively, we can let both antennas lie in the $xy$ plane. In this case, $\mathbf{D} = \mathbf{x}_0 D_x + \mathbf{y}_0 D_y$. Upon changing the variables of integration in (71) from the angular coordinates to $k_x, k_y$, we obtain

$$\bar{Z}_{21} = \frac{2}{k}\int_{-\infty}^{\infty}\int_{-\infty}^{\infty} \frac{\exp[-j(k_x D_x + k_y D_y)]}{(k^2 - k_x^2 - k_y^2)^{1/2}} P(k_x,k_y)\, dk_x\, dk_y, \quad (74)$$

where the negative square root is chosen for $k^2 < k_x^2 + k_y^2$. The above is the three-dimensional counterpart of (47). Equation (74) will diverge if $P(k_x,k_y)/k_z$ fails to have a Fourier transform as, for example, in the case of elementary dipoles or antennas radiating a single higher order spherical mode. In this case, one can use (73) or the more general form, (71), with $D_z > 0$. Thus with antenna 1 at $z = 0$ and antenna 2 at $z = D$,

$$\bar{Z}_{21} = 2\int_0^{2\pi} d\varphi \int_{-j\infty}^{1} \exp(-jkDt) \mathbf{F}_1(t,\varphi)\cdot\hat{\mathbf{F}}_2(t,\varphi)\, dt, \quad (75)$$

where

$$\mathbf{F}(\theta,\varphi) = \begin{cases} (j)^n \mathbf{e}_{nm}'^{(e,o)}(\theta,\varphi), \\ (j)^{n+1} \mathbf{e}_{nm}''^{(e,o)}(\theta,\varphi), \end{cases}$$

$$\hat{\mathbf{F}}(\theta,\varphi) \equiv \mathbf{F}^*(\theta,\phi). \quad (76)$$

For $z$-directed elementary electric dipoles

$$\mathbf{F} = -j(3/8\pi)^{1/2} \sin\theta . \theta_o$$

Thus with $\mathbf{F}_1 = \mathbf{F}_2$, (75) together with a well-known integral representation for spherical Hankel functions in terms of Legendre polynomials [11], yields

$$\bar{Z}_{21} = 3[\hat{H}_1^{(2)}(kD)/(kD)^2]. \tag{77}$$

The above is in agreement with the result obtained by a direct application of addition theorems [3] and (apart from a normalization constant) with the well-known formula of Carter [12].

As another example, consider two identical CMS antennas (again on the $z$ axis) with power patterns identical to those produced by collinear half-wave dipoles with sinusoidal current distributions. The power pattern is

$$P(\theta,\varphi) = 7.65806 \frac{\cos^2[(\pi/2)\cos\theta]}{\sin^2\theta} \tag{78}$$

and the mutual impedance becomes

$$\bar{Z}_{21} = (4\pi)(7.65806) \int_{-j\infty}^{1} \frac{\cos^2(\pi/2)t}{1-t^2} \exp(-jkDt)\, dt. \tag{79}$$

The results of numerical integration of (79) are plotted in Fig. 5. These results may be shown to agree with the mutual impedance between collinear half-wave dipoles (with assumed sinusoidal current distributions) as computed by the induced EMF method [23], provided account is taken of the fact that in Fig. 5 each antenna is tuned to free space.

## V. Extension of Mutual Impedance Formulas to Antennas Radiating into a Half-Space

In the preceding formulation for mutual impedance we assumed that each antenna could be enclosed by a sphere (cylinder) of finite radius outside of which the fields were expanded in terms of spherical (cylindrical) mode functions characteristic of free space. For antennas radiating into a half-space bounded by a perfectly conducting electric or magnetic screen of infinite extent (aperture antennas), one can also develop a scattering representation in terms of mode functions characteristic of the half-space. This scattering representation differs from the corresponding representation in free space in no essential way [4], and relates the fields outside any hemisphere (or half-cylinder), which completely encloses the antenna, to the incident and reflected waves at the antenna ports. A CMS antenna radiating into a half-space may be defined as an antenna that scatters like the perfectly conducting (electric or magnetic) plane that bounds the half-space, whenever the antenna ports are open circuited. It may be shown that with this alternate definition of "invisibility" the mutual impedance formulas derived in the preceding sections carry over to antennas in a half-space with only minor modifications. In particular, if the perfectly conducting plane is chosen as the $xy$ plane, we have

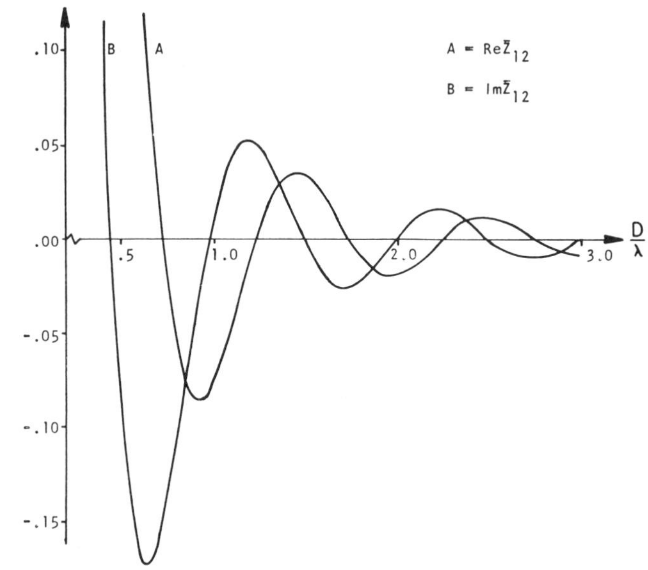


Fig. 5. Real and imaginary parts of mutual impedance of CMS antennas with the same radiation patterns as collinear dipoles.

$$\bar{Z}_{12} = \pm \int_{-\infty}^{\infty}\int_{-\infty}^{\infty} kk_z(k_x^2+k_y^2)[\eta f_1'(k_x,k_y) \cdot f_2'(-k_x,-k_y) + \zeta f_1''(k_x,k_y) f_2''(-k_x,-k_y)] \cdot \exp[-j(k_xD_x+k_yD_y)]\, dk_x\, dk_y, \tag{80}$$

where the plus or minus sign applies, respectively, to antennas in a half-space bounded by a perfectly conducting electric or magnetic screen. The partial pattern functions $f', f''$ are related to the components of the electric far field as in (59) with $0 < \theta < \pi/2$. The constraint of invisibility on open circuit imposes the following constraints on $f'$ and $f''$ [4]:

$$\begin{aligned} k_z f'(k_x,k_y) - k_z^* f^{*\prime}(-k_x,-k_y) &= 0, \\ f''(k_x,k_y) - f^{*\prime\prime}(-k_x,-k_y) &= 0, \end{aligned} \tag{81}$$

for an electric screen and

$$\begin{aligned} k_z f''(k_x,k_y) + k_z^* f^{*\prime\prime}(-k_x,-k_y) &= 0, \\ f'(k_x,k_y) + f^{*\prime}(-k_x,-k_y) &= 0, \end{aligned} \tag{82}$$

for a magnetic screen. In the above $k_z = -j|k_z|$ for $k_x^2 + k_y^2 > k^2$. For identical antennas, (80), (81), and (82) lead to the following:

$$\bar{Z}_{21} = \frac{1}{k}\int_{-\infty}^{\infty}\int_{-\infty}^{\infty} \frac{P(k_x,k_y)}{(k^2-k_x^2-k_y^2)^{1/2}} \cdot \exp[-j(k_xD_x+k_yD_y)]\, dk_x\, dk_y, \tag{83}$$

where

$$\frac{P(k_x,k_y)}{k^2k_z(k_x^2+k_y^2)} = \begin{cases} \eta k_z^* |f'|^2 + \zeta k_z |f''|^2, & \text{(84a)} \\ \eta k_z |f'|^2 + \zeta k_z^* |f''|^2. & \text{(84b)} \end{cases}$$

Equation (84a) applies to an electric screen while (84b) applies to a magnetic screen. The pattern function $P(k_x,k_y)$ in the visible region $(k_x^2 + k_y^2 < k^2)$ corresponds to the real power radiated by each isolated antenna. Since we have been assuming a unit excitation at the antenna input port, the net radiated power is unity. Hence

$$\int_{k_x^2+k_y^2<k^2}\int \frac{P(k_x,k_y)}{kk_z}\,dk_x\,dk_y = 1. \qquad (85)$$

The integral of $P(k_x,k_y)$ over the invisible region, i.e.,

$$\int_{k_x^2+k_y^2>k^2}\int \frac{P(k_x,k_y)}{kk_z}\,dk_x\,dk_y$$

may be shown to be related to the energy stored in the half-space $z \geq 0$ [4], [13], [14], [20].

In the derivation of (83) as well as of the corresponding formula for antennas in free space, (74), no reference was made to the physical structures that produce the given radiation patterns. Indeed, our method of analysis provides no information about the physical configuration of each radiator. By applying a well-known approximating technique, Borgiotti [6] has derived a formula for mutual admittance between planar apertures in an infinite perfectly conducting plane. Apart from a constant multiplier, Borgiotti's formula is identical with our mutual impedance formula in (80) for CMS antennas in a perfectly conducting magnetic screen. The fact that our impedance formula corresponds to Borgiotti's admittance formula arises from our constraint of invisibility on open circuit. (It may be shown that for MS antennas invisible on short circuit, (80) becomes the formula for mutual admittance, provided one also interchanges the electric and magnetic screens [4].) Thus, Borgiotti's approximate result for mutual admittance of aperture antennas is an exact relation for MS antennas.

## VI. Application to an Infinite Planar Array

Consider an infinite planar array of identical, uniformly spaced CMS antennas in the $xy$ plane. Denote the element separation in the $x$ and $y$ directions, respectively, by $D_x$ and $D_y$. By symmetry, the mutual impedance between element $(n,m)$ and $(n+\nu,m+\mu)$ is

$$Z_{\nu\mu} = \frac{1}{k}\int_{-\infty}^{\infty}\int_{-\infty}^{\infty} \frac{P(k_x,k_y)}{(k^2-k_x^2-k_y^2)^{1/2}} \cdot \exp\left[-j(\nu k_x D_x + \mu k_y D_y)\right] dk_x\,dk_y, \quad \nu,\mu \neq 0 \qquad (86)$$

$$\bar{Z}_{00} = 1. \qquad (87)$$

When the array is excited by equal-amplitude waves with progressive phasing, the input impedance at the port of each array element (i.e., the active impedance) is [7], [15],

$$Q(\alpha,\beta) = \sum_{\nu=-\infty}^{\infty}\sum_{\mu=-\infty}^{\infty} \bar{Z}_{\nu\mu}\exp\left[j(\nu\alpha+\mu\beta)\right] \qquad (88)$$

where $\alpha$, $\beta$ are phasing constants, respectively, in the $x$ and $y$ directions. Substituting (86) and (88) and employing the Dirac delta function representation

$$\sum_{\nu=-\infty}^{\infty}\sum_{\mu=-\infty}^{\infty} \exp\left\{-j[(k_xD_x-\alpha)\nu + (k_yD_y-\beta)\mu]\right\} = (2\pi)^2 \sum_{n=-\infty}^{\infty}\sum_{m=-\infty}^{\infty} \delta(k_xD_x-\alpha-2n\pi) \cdot \delta(k_yD_y-\beta-2m\pi),$$

we obtain

$$Q(\alpha,\beta) = \frac{4\pi^2}{k^2D_xD_y}\sum_{(n,m)\in v} \frac{P[(\alpha+2n\pi)/D_x,(\beta+2m\pi)/D_y]}{\{1-[(\alpha+2n\pi)/kD_x]^2-[(\beta+2m\pi)/kD_y]^2\}^{1/2}}$$
$$+ j\left\{\int_{k_x^2+k_y^2>k^2}\int \frac{-P(k_x,k_y)}{(k_x^2+k_y^2-k^2)^{1/2}}\,dk_x dk_y + \frac{4\pi^2}{k^2D_xD_y}\sum_{(n,m)\in Cv} \frac{P[(\alpha+2n\pi)/kD_x,(\beta+2m\pi)/kD_y]}{\{[(\alpha+2n\pi)/kD_x]^2+[(\beta+2m\pi)/kD_y]^2-1\}^{1/2}}\right\}, \qquad (89)$$

where the first sum is taken over all $n,m$ in the visible region $v$

$$\left(\frac{\alpha+2n\pi}{kD_x}\right)^2 + \left(\frac{\beta+2m\pi}{kD_y}\right)^2 < 1$$

and the second sum over the complementary region $Cv$

$$\left(\frac{\alpha+2n\pi}{kD_x}\right)^2 + \left(\frac{\beta+2m\pi}{kD_y}\right)^2 > 1.$$

Apart from a constant (i.e., independent of $\alpha,\beta$) term in the imaginary part, (89) is the well-known grating-lobe series representation for the active impedance of a planar infinite array [16]–[20]. The additional constant term is simply a consequence of the assumption made herein that each element is tuned to free space.

Past treatments of infinite planar arrays have dealt with structurally specified radiating elements (e.g., half-wave dipoles [16] and planar apertures [18], [20]) for which grating-lobe series representations have been derived under various implicit or explicit approximating assumptions. The present network formulation of mutual coupling yields a grating-lobe series representation for the active impedance of an infinite array, which is exact for CMS antennas. It depends exclusively on the isolated far-field pattern function $P(k_x,k_y)$ and makes no reference to structural properties of the radiating elements. Conversely, it may be shown that the active impedance of a planar infinite array is given precisely by the grating-lobe series (89) only if the elementary radiators are CMS antennas [4]; for more general antennas, correction terms involving the reflection coefficients and the scattering properties on open circuit of the individual elements appear in the formulas for mutual impedance.

## Appendix
## Spherical Modes

Using a notation analogous to that used for uniform waveguides [21], [22], the electromagnetic fields in spherical coordinates $r,\theta,\varphi$ are decomposed into transverse components $\mathbf{E}_t,\mathbf{H}_t$ and radial components $\mathbf{r}_0E_r,\mathbf{r}_0H_r$. The total fields are then written

$$\mathbf{E} = \mathbf{E}_t + \mathbf{r}_0E_r$$
$$\mathbf{H} = \mathbf{H}_t + \mathbf{r}_0H_r \qquad (90)$$

where $\mathbf{r}_0$ is the unit vector in the radial direction.

The transverse fields can be expanded in terms of a complete set of orthonormal vector mode functions of two types: the $E$-mode functions $\mathbf{e}_i'(\theta,\varphi)$, $\mathbf{h}_i'(\theta,\varphi)$ and the $H$-mode functions $\mathbf{e}_i''(\theta,\varphi)$, $\mathbf{h}_i''(\theta,\varphi)$. The representations of the fields in terms of these functions are

$$r\mathbf{E}_t = \sum_i \{V_i'(r)\mathbf{e}_i'(\theta,\varphi) + V_i''(r)\mathbf{e}_i''(\theta,\varphi)\},$$
$$r\mathbf{H}_t = \sum_i \{I_i'(r)\mathbf{h}_i'(\theta,\varphi) + I_i''(r)\mathbf{h}_i''(\theta,\varphi)\}, \qquad (91)$$

where $V_i'(r),V_i''(r),I_i'(r),I_i''(r)$ are radially dependent expansion coefficients.

The mode functions $\mathbf{e}_i$ and $\mathbf{h}_i$ are related by

$$\mathbf{r}_0 \times \mathbf{e}_i{}^{','''} = \mathbf{h}_i{}^{','''} \qquad (92)$$

and their orthonormality expressed by

$$\iint_\Omega \mathbf{e}_i'\cdot\mathbf{e}_j''\, d\Omega = 0$$

$$\iint_\Omega \mathbf{h}_i'\cdot\mathbf{h}_j''\, d\Omega = 0$$

$$\iint_\Omega \mathbf{e}_i{}^{','''}\cdot\mathbf{e}_j{}^{','''}\, d\Omega = \delta_{ij}$$

$$\iint_\Omega \mathbf{h}_i{}^{','''}\cdot\mathbf{h}_j{}^{','''}\, d\Omega = \delta_{ij} \qquad (93)$$

where $d\Omega$ is an element of solid angle.

For regions defined over a complete sphere

$$\mathbf{e}_i' = \mathbf{h}_i''$$
$$\mathbf{h}_i' = -\mathbf{e}_i'' \qquad (94)$$

and in view of (91) and (92) the fields can be written in terms of one set of mode functions $\mathbf{e}_i \equiv \mathbf{e}_i'$ alone. The summation index $i$ in (91) denotes the double index $n,m$ and parity associated with the spherical harmonics. It is convenient to work with purely real mode functions defined by

$$\mathbf{e}_i(\theta,\varphi) \equiv \mathbf{e}_{nm}{}^{(e,o)} = -\frac{1}{N_{nm}}\left\{\boldsymbol{\theta}_0\frac{\partial}{\partial\theta}P_n{}^m(\cos\theta)\begin{bmatrix}\cos m\varphi\\ \sin m\varphi\end{bmatrix} + \boldsymbol{\varphi}_0\frac{1}{\sin\theta}\cdot\frac{\partial}{\partial\varphi}P_n{}^m(\cos\theta)\begin{bmatrix}\cos m\varphi\\ \sin m\varphi\end{bmatrix}\right\}, \qquad (95)$$

where $P_n{}^m(\cos\theta)$ are associated Legendre polynomials and

$$N_{nm}^2 = \frac{4\pi(n+m)!\,n(n+1)}{(n-m)!\,(2n+1)\epsilon_{nm}}$$
$$\epsilon_{nm} = 1, \quad m = 0$$
$$= 2, \quad m \neq 0. \qquad (96)$$

Taking account of (92), (94), and (95), (91), may be written

$$r\mathbf{E}_t = \sum_{n=1}^{\infty}\sum_{m=0}^{n}\{V_{nm}{}'^{(e)}(r)\,\mathbf{e}_{nm}{}^{(e)}(\theta,\varphi) + V_{nm}{}'^{(o)}(r)\,\mathbf{e}_{nm}{}^{(o)}(\theta,\varphi) + V_{nm}{}''^{(e)}(r)[\mathbf{e}_{nm}{}^{(e)}(\theta,\varphi)\times\mathbf{r}_0] + V_{nm}{}''^{(o)}(r)[\mathbf{e}_{nm}{}^{(o)}(\theta,\varphi)\times\mathbf{r}_0]\}, \qquad (97a)$$

$$r\mathbf{H}_t = \sum_{n=1}^{\infty}\sum_{m=0}^{n}\{I_{nm}{}'^{(e)}(r)[\mathbf{r}_0\times\mathbf{e}_{nm}{}^{(e)}(\theta,\varphi)] + I_{nm}{}'^{(o)}(r)[\mathbf{r}_0\times\mathbf{e}_{nm}{}^{(o)}(\theta,\varphi)] + I_{nm}{}''^{(e)}(r)\,\mathbf{e}_{nm}{}^{(e)}(\theta,\varphi) + I_{nm}{}''^{(o)}(r)\,\mathbf{e}_{nm}{}^{(o)}(\theta,\varphi)\}. \qquad (97b)$$

Explicitly, the dependence of the coefficients on the radial coordinate is

$$V_{nm}{}'^{(e,o)} = j\zeta[\alpha_{nm}{}^{(e,o)}\dot{\hat{H}}_n{}^{(1)}(kr) + \beta_{nm}{}^{(e,o)}\dot{\hat{H}}_n{}^{(2)}(kr)],$$
$$V_{nm}{}''^{(e,o)} = \gamma_{nm}\hat{H}_n{}^{(1)}(kr) + \delta_{nm}\hat{H}_n{}^{(2)}(kr),$$
$$I_{nm}{}'^{(e,o)} = \alpha_{nm}{}^{(e,o)}\hat{H}_n{}^{(1)}(kr) + \beta_{nm}{}^{(e,o)}\hat{H}_n{}^{(2)}(kr),$$
$$I_{nm}{}''^{(e,o)} = j\eta[\gamma_{nm}{}^{(e,o)}\dot{\hat{H}}_n{}^{(1)}(kr) + \delta_{nm}{}^{(e,o)}\dot{\hat{H}}_n{}^{(2)}(kr)], \qquad (98)$$

where $\hat{H}_n{}^{(1)(2)}(kr)$ and $\dot{\hat{H}}_n{}^{(1)(2)}(kr)$ denote, respectively, spherical Hankel functions and their derivative with respect to the argument and the $\alpha_{nm}$, $\beta_{nm}$, $\gamma_{nm}$, and $\delta_{nm}$ are constants. The asymptotic forms of the spherical Hankel functions[2] for large arguments are

$$\hat{H}_n{}^{(1)(2)}(kr) \sim (\mp j)^{n+1}\exp(\pm jkr)$$
$$\dot{\hat{H}}_n{}^{(1)(2)}(kr) \sim (\mp j)^{n}\exp(\pm jkr). \qquad (99)$$

[2] This definition of spherical Hankel functions agrees with that given in [21], [22]. These functions are related to the more commonly accepted spherical Hankel functions $h_n(kr)$ by $\hat{H}_n(kr) = (kr)h_n(kr)$.

The radial components may be expressed as follows:

$$j\omega\epsilon(\mathbf{r}\cdot\mathbf{E}) = \sum_{n=1}^{\infty}\sum_{m=0}^{n}\sum_{(e,o)}[n(n+1)/r]I_{nm}{}'^{(e,o)} \cdot \Phi_{nm}{}^{(e,o)}(\theta,\varphi) \quad (100a)$$

$$j\omega\mu(\mathbf{r}\cdot\mathbf{H}) = \sum_{n=1}^{\infty}\sum_{m=0}^{n}\sum_{(e,o)}[n(n+1)/r]V_{nm}{}''^{(e,o)} \cdot \Phi_{nm}{}^{(e,o)}(\theta,\varphi) \quad (100b)$$

where the $\Phi_{nm}$ are scalar mode functions defined by

$$\Phi_{nm}{}^{(e,o)}(\theta,\varphi) = (1/N_{nm})P_n{}^m(\cos\theta)\begin{bmatrix}\cos m\varphi \\ \sin m\varphi\end{bmatrix}. \quad (101)$$

## References


[1] C. G. Montgomery, R. H. Dicke, and E. M. Purcell, *Principles of Microwave Circuits*, M.I.T. Rad. Lab. Ser., vol. 8. New York: McGraw-Hill, 1948, pp. 317–333.

[2] W. K. Kahn and H. Kurss, "Minimum-scattering antennas," *IEEE Trans. Antennas and Propagation*, vol. AP-13, pp. 671–675, September 1965.

[3] W. K. Kahn and W. Wasylkiwskyj, "Coupling, radiation, and scattering by antennas," *Proc. Symp. on Generalized Networks*, vol. 16. Brooklyn, N. Y.: Polytechnic Press, 1966, pp. 83–114.

[4] W. Wasylkiwskyj, "A network theory of coupling, radiation and scattering of antennas," Ph.D. dissertation, Polytechnic Institute of Brooklyn, Brooklyn, N. Y., June 1968.

[5] A. G. Gately, Jr., D. J. R. Stock, and B. Ru-Shao Cheo, "A network description for antenna problems," *Proc. IEEE*, vol. 56, pp. 1181–1193, July 1968.

[6] G. V. Borgiotti, "A novel expression for the mutual admittance for planar radiating elements," *IEEE Trans. Antennas and Propagation*, vol. AP-16, pp. 329–333, May 1968.

[7] W. Wasylkiwskyj and W. K. Kahn, "Mutual coupling and element efficiency for infinite linear arrays," *Proc. IEEE*, vol. 56, pp. 1901–1907, November 1968.

[8] J. A. Stratton, *Electromagnetic Theory*. New York: McGraw-Hill, 1941, pp. 372–374.

[9] S. Stein, "Addition theorems for spherical wave functions," *Quart. Appl. Math.*, vol. 19, pp. 15–24, 1961.

[10] O. R. Cruzan, "Translational addition theorems for spherical vector wave functions," *Quart. Appl. Math.*, vol. 20, pp. 33–40, 1961.

[11] J. A. Stratton [8], pp. 409–410.

[12] P. S. Carter, "Circuit relations in radiating systems and application to antenna problems," *Proc. IRE*, vol. 20, pp. 1004–1041, June 1932.

[13] D. R. Rhodes, "On a fundamental principle in the theory of planar apertures," *Proc. IEEE*, vol. 52, pp. 1013–1021, September 1964.

[14] G. V. Borgiotti, "On the reactive energy of an aperture," *IEEE Trans. Antennas and Propagation* (Communications), vol. AP-15, pp. 565–566, July 1967.

[15] J. L. Allen and B. L. Diamond, "Mutual coupling in array antennas," M.I.T. Lincoln Lab., Lexington, Mass., Tech. Rept. 424, October 4, 1966.

[16] L. Stark, "Radiation impedance of a dipole in an infinite planar phased array," *Rad. Sci.*, vol. 1, pp. 361–378, March 1966.

[17] H. A. Wheeler, "The grating-lobe series for the impedance variation in a planar phased-array antenna," *IEEE Trans. Antennas and Propagation*, vol. AP-14, pp. 707–714, November 1966.

[18] L. Parad, "The real and reactive power of a planar array," *IEEE Trans. Antennas and Propagation* (Communications), vol. AP-13, pp. 990–992, 1965.

[19] R. C. Hansen, Ed., *Microwave Scanning Antennas*, vol. 2. New York: Academic Press, 1966.

[20] B. L. Diamond, "A generalized approach to the analysis of infinite planar array antennas," *Proc. IEEE*, vol. 56, pp. 1837–1851, November 1968.

[21] L. B. Felsen, "Spherical transmission line theory," Microwave Research Inst., Polytechnic Institute of Brooklyn, Brooklyn, N. Y., Rept. R-253-51, P.I.B.-194, January 1952.

[22] N. Marcuvitz, *Waveguide Handbook*, M.I.T. Rad. Lab. Ser., vol. 10. New York: McGraw-Hill, 1951, sec. 2.B.

[23] J. D. Kraus, *Antennas*. New York: McGraw-Hill, 1950, p. 269.




*Reprinted by permission from* IEEE TRANSACTIONS ON ANTENNAS AND PROPAGATION
Vol. AP-18, No. 2, March 1970
pp. 204-216

# Characteristics of Dielectric Loaded and Covered Circular Waveguide Phased Arrays

NOACH AMITAY, MEMBER, IEEE, AND VICTOR GALINDO, MEMBER, IEEE

128

*Abstract*—The use of dielectric materials for the hardening and matching of phased-array antennas in recent years has shown that a more complete understanding of the effects of these materials upon the array performance is necessary. The characteristics of fully loaded, plugged, and sheath covered circular waveguide phase arrays are analyzed and discussed. Numerical solutions of the boundary-value problem are verified by experimental and convergence tests.

Particular emphasis is placed on the study of (forced) surface wave resonance effects. Three different cases for surface wave resonances were obtained. These include the case in which surface wave resonances are present in the absence of dielectrics, the case in which they are trapped by the presence of dielectric plugs, as well as the case in which waves are trapped by the presence of a dielectric sheath. The surface wave resonance due to the plug is shown to vanish for certain "bandpass" ranges of plug thickness which repeat periodically for a single trapped waveguide mode. On the other hand, the surface wave trapped in the sheath exhibits no "bandpass" characteristics. Instead, multiple surface wave resonances occur with increasing sheath thickness.

Finally, the surface wave resonances observed here appear at isolated points in the scan plane.

## I. Introduction

THE USE OF dielectric materials for the hardening and matching of phased-array antennas in recent years has shown that a more complete understanding of the effects of these materials upon the array performance is necessary. Many significant characteristics of dielectric-free circular waveguide phased arrays have been discussed for both hexagonal [1] and rectangular [2] grids excited by linear and circular incident polarizations. These results were determined by the use [1] of the Ritz–Galerkin method, which is readily applicable to the same arrays when they are covered, loaded, or plugged with dielectric materials. In this paper, the effects of the dielectrics on the impedance and radiation characteristics of a circular waveguide array have been studied. The numerically obtained results show close agreement with experiments.

Fully dielectric loaded waveguides are studied as a function of the waveguide radius, spacing, and dielectric constant. For various hexagonal grids, the effects of dielectric sheaths (covers) and plugs as a function of their thickness, and dielectric constant are studied.

Three cases of forced surface wave resonances were observed. These phenomena are manifested by total internal reflection and consequently no radiation of power by the antenna. In one case surface waves exist in the absence of dielectrics, and the scan angle at which total reflection occurs shifts towards broadside with increasing dielectric constant. In a second case, surface waves are "trapped" in the dielectric cover and are shown to require a minimum sheath thickness for propagation. In the third case, surface waves are trapped in the dielectric plugs and, in contrast to the second case, propagate only for certain ranges (passbands) of plug thickness. Finally, it is found that the forced surface wave resonances observed for this array occur at isolated beam pointing directions. This is a different result from that found by other investigators [3] for different kinds of elements.

Manuscript received January 6, 1969; revised May 12, 1969. This work was supported by the U. S. Army Materiel Command under Contract DAHC-60-69-C-0008.

The authors are with Bell Telephone Laboratories, Inc., Whippany, N. J. 07981.

## II. Analysis

Fig. 1 shows an infinite planar array of open-ended circular waveguide elements in a conducting ground plane ($z = 0$). The elements are arranged in a doubly periodic grid along the skewed (nonorthogonal) coordinates $s_1$ and $s_2$. The $s_1$ axis coincides with the $x$ axis, while the $s_2$ axis makes an angle $\alpha$ with the $x$ axis. The element location is defined by two indices $(p,q)$ corresponding to a physical location

$$\boldsymbol{\varrho}_{pq} = pb\hat{s}_1 + qd\hat{s}_2 \tag{1}$$

where $\hat{s}_1$ and $\hat{s}_2$ are unit vectors along the $s_1$ and $s_2$ axes, respectively, while $b$ and $d$ represent the basic periods of the two-dimensional grid.

The tangential electric field $\boldsymbol{E}_t$ at the aperture $A$ of an element is determined from a solution of a vector two-dimensional integral equation. A detailed derivation of this integral equation for the dielectric-free array of Fig. 1 is given in [1] and found as

$$2\sum_{j=1}^{J} A_jY_j\boldsymbol{\Phi}_j = \sum_{j=1}^{\infty} Y_j\boldsymbol{\Phi}_j \iint_A \boldsymbol{\Phi}_j\cdot\boldsymbol{E}_t\,da$$
$$+\sum_{p=1}^{2}\sum_{m=-\infty}^{+\infty}\sum_{n=-\infty}^{+\infty} Y_{mnp}'\boldsymbol{\Psi}_{mn}{}^{p}\iint_A (\boldsymbol{\Psi}_{mn}{}^{p})^*\cdot\boldsymbol{E}_t\,da. \tag{2}$$

The $\{\boldsymbol{\Phi}_j\}$ and $\{\boldsymbol{\Psi}_{mn}{}^{p}\}$ are, respectively, the circular waveguide modes [1], [2], [4] and the periodic free-space modes [1] required by Floquet's theorem ($p = 1$ denotes a TE mode and $p = 2$ denotes a TM mode). The $Y_j$ and $Y_{mnp}'$ are the respective modal admittances. The mode of excitation of the array is specified by the given coefficients $\{A_j\}$ which are the amplitudes of the incident waveguide modes. The time convention is chosen as $\exp(-i\omega t)$.

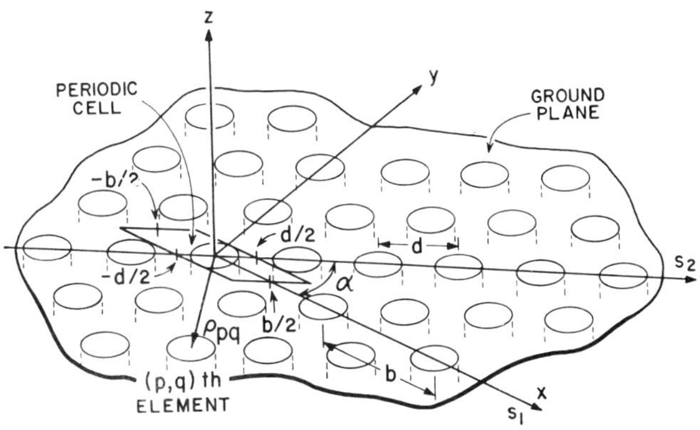


Fig. 1. Circular waveguide array geometry.

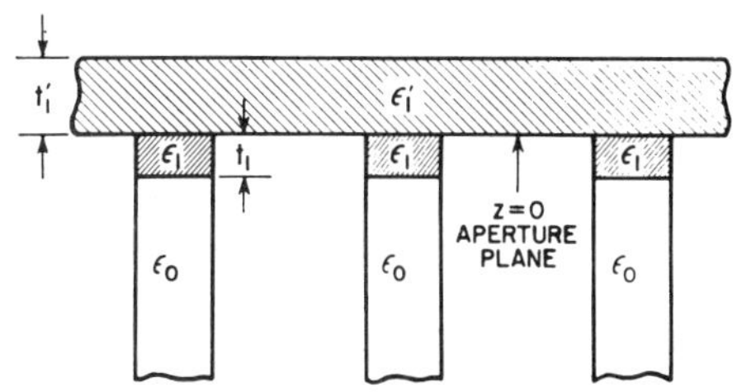


Fig. 2. Dielectric sheath and plug geometry.

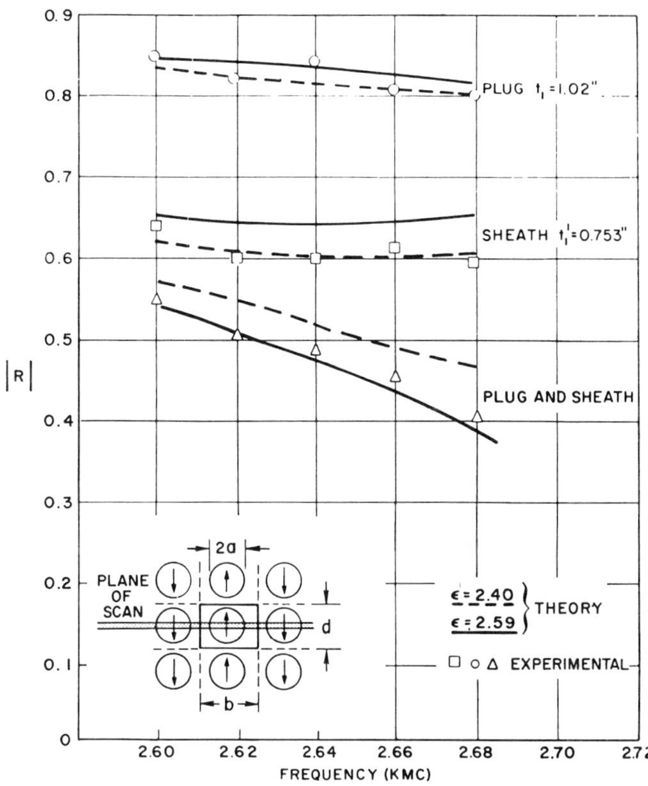


Fig. 3. Experimental check with dielectrics; $H$-plane simulator. $a$ = 1.4 inches, $b$ = 6.5 inches, $d$ = 3.25 inches, $\epsilon_1 = \epsilon_1'$ = 2.4 and 2.59.

The reflection coefficient $R_j$ of the $j$th incident mode is readily found from the aperture field solution by

$$A_j + R_j = \iint_A \mathbf{\Phi}_j \cdot \mathbf{E}_t \, da, \quad j = 1, \cdots, J. \tag{3}$$

The far-field radiation pattern (including axial ratio and tilt angle) are readily obtainable also [2], once the solution for $\mathbf{E}_t$ is found.

As was pointed out in [1], the addition of either a dielectric sheath, plug, or both to the array (Fig. 2) does not alter the functional form of the integral equation (2). As was shown earlier [1], [5], [6], the modal admittances in (2) have to be modified to account for the dielectrics. For the case when dielectric plugs are used, each $Y_j$ on the right-hand side of (2) is *replaced* by

$$Y_j \rightarrow y_j \frac{Y_j - iy_j \tan \gamma_j t_1}{y_j - iY_j \tan \gamma_j t_1} \tag{4}$$

where $y_j$ and $\gamma_j$ are the modal admittance and propagation constant, respectively, of the $\mathbf{\Phi}_j$ mode in the dielectric plug (with thickness $t_1$). The left-hand side of (2), which is proportional to the incident magnetic field, must be replaced accordingly

$$2 \sum_{j=1}^{J} A_j Y_j \mathbf{\Phi}_j \rightarrow 2 \sum_{j=1}^{J} A_j \frac{y_j Y_j \exp(-i\Gamma_j t_1)}{y_j \cos \gamma_j t_1 - iY_j \sin \gamma_j t_1} \mathbf{\Phi}_j \tag{5}$$

where $A_j$ now represents the amplitude of the $j$th incident electric field waveguide mode in the dielectric-free region preceding the plug ($\Gamma_j$ being the corresponding propagation constant). The reflection coefficients of the propagating modes in the $\epsilon_0$ region are related to the aperture field by

$$R_j = \frac{y_j \iint_A \mathbf{E}_t \cdot \mathbf{\Phi}_j \, da - A_j \exp(-i\Gamma_j t_1) \cdot [y_j \cos \gamma_j t_1 + iY_j \sin \gamma_j t_1]}{\exp(i\Gamma_j t_1)[y_j \cos \gamma_j t_1 - iY_j \sin \gamma_j t_1]} \tag{6}$$

the phase of $R_j$ being referred to the aperture ($z = 0$). Similarly, for the exterior dielectric sheath covering the array, each exterior modal admittance, $Y_{mnp}'$, is replaced by

$$Y_{mnp}' \rightarrow y_{mnp}' \frac{Y_{mnp}' - iy_{mnp}' \tan \beta_{mn} t_1'}{y_{mnp}' - iY_{mnp}' \tan \beta_{mn} t_1'} \tag{7}$$

where $y_{mnp}'$ and $\beta_{mn}$ are the modal admittance and propagation constant, respectively, of the $\mathbf{\Psi}_{mn}{}^p$ mode in the dielectric sheath (with the thickness $t_1'$). The coefficient of the $\mathbf{\Psi}_{mn}{}^p$ mode in the dielectric-free region above the sheath $F_{mnp}$ (with propagation constant $B_{mn}$) is related to the aperture field by

$$F_{mnp} = \frac{y_{mnp}' \exp(-iB_{mn} t_1')}{y_{mnp}' \cos \beta_{mn} t_1' - iY_{mnp}' \sin \beta_{mn} t_1'} \iint_A \mathbf{E}_t \cdot (\mathbf{\Psi}_{mn}{}^p)^* \, da \tag{8}$$

the phase of $F_{mnp}$ being referred to the aperture ($z = 0$).

The results to be discussed have been obtained from a numerical solution of (2) by the Ritz–Galerkin method

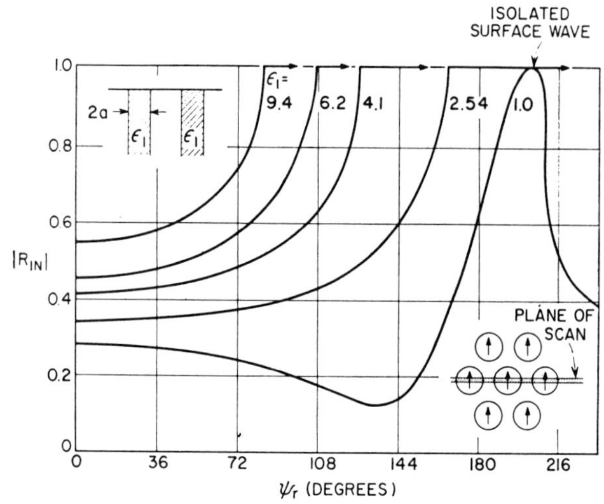


Fig. 4. Fully loaded array; $H$-plane scan. $\epsilon_1 a^2$ = constant = 0.118 $\lambda_0^2$, $b/a = d/a$ = constant = 2.08.

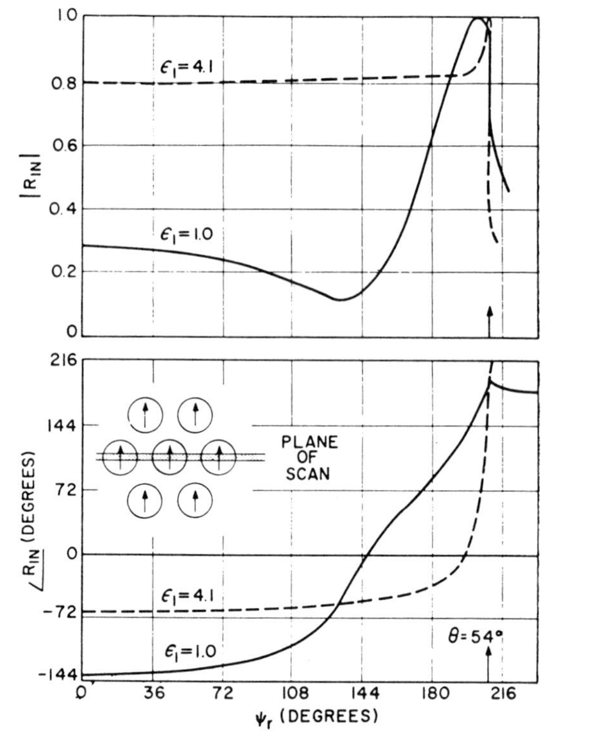


Fig. 5. Fully loaded array; $H$-plane scan. $\epsilon_1 a^2$ = constant = 0.118 $\lambda_0^2$, $b = d = 0.714\ \lambda_0$, $a = 0.169\ \lambda_0$, $\epsilon = 4.1$, and $\epsilon = 1.0$.



[1], [6], [7]. In this method, we approximate $\boldsymbol{E}_t$ by a finite set $\{\boldsymbol{\Phi}_j: j = 1,\cdots,N\}$ with unknown coefficients. This approximation reduces the integral equation to a matrix equation from which the unknown coefficients are determined. From this approximate solution, the array reflection and far-field characteristics were obtained. The array was excited by either the horizontal ($x$ direction) or vertical polarized $TE_{11}$ modes. $R_{1N}$ will designate the reflection coefficient of the particular exciting $TE_{11}$ mode, while $R_{2N}$ is the reflection coefficient of the other (orthogonal) $TE_{11}$ mode.

For convenience, the numerical results will be presented as a function of a radial scan parameter, $\psi_r$. This quantity, $\psi_r$, is related to the differential phase shifts between elements $\psi_x$ and $\psi_y$ (in the $x$ and $y$ directions, respectively) and the beam pointing directional cosines $T_x$ and $T_y$ by

$$\psi_x = \frac{2\pi b}{\lambda_0} T_x$$

$$\psi_y = \frac{2\pi d \sin \alpha}{\lambda_0} T_y$$

$$\psi_r = \sqrt{\psi_x^2 + \psi_y^2} \tag{9}$$

where $\lambda_0$ is the free-space wavelength.

Certain numerical and experimental checks [1] have been carried out to ascertain the validity of the numerical results. Convergence tests as a function of the number of waveguide and free-space modes used in the numerical analysis have been carried out, with the results indicating that several percent (usually less than 2 percent) accuracy in the magnitude of the reflection coefficients $R_j$ was obtained. An exception to this accuracy occurs near very sharp changes in $|R_j|$, where we find that the accuracy in determining the angular position of the sharp change is several degrees in $\psi_r$. As many as 40 waveguide and 882 free-space modes are necessary to obtain this accuracy at certain critical scan angles.

Experimental checks were carried out with the aid of an $H$-plane simulator [1], both for a dielectric plug and a sheath. Theoretical results for a range of dielectric constant as well as experimental results are shown in Fig. 3. The agreement is satisfactory and within the bounds of experimental error in determining the dielectric constant which was to fall between 2.40 and 2.59.

## III. Results

The results which will be discussed in this section have been obtained for an equilateral triangular grid. In earlier work [1], it was shown that forced surface wave resonances can occur in these and other types of arrays [2], [5], [8], [9], [13] without the presence of dielectrics. The effects of dielectrics upon these surface waves as well as the appearance of surface waves that are bound or "trapped" in the dielectrics [5], [6], [8], in either the sheath or plugs, are discussed here. The principal difference between the nontrapped and trapped surface waves is that the nontrapped waves merely change angular position $\psi_r$ as $\epsilon \rightarrow \epsilon_0$ (for the plug or sheath), whereas the trapped waves disappear as $\epsilon \rightarrow \epsilon_0$.

### A. Fully Loaded Arrays

The waveguide radius $a$ was varied with the dielectric constant $\epsilon$ of the loading material so that $\epsilon a^2$ = constant. This relationship insures a fixed ratio between the waveguide operating and cutoff frequencies. Fig. 4 shows typical reflection characteristics of a fully dielectric loaded array in the $H$-plane scan. The grid dimensions were varied, with wavelength kept fixed, so that $d/a = b/a =$

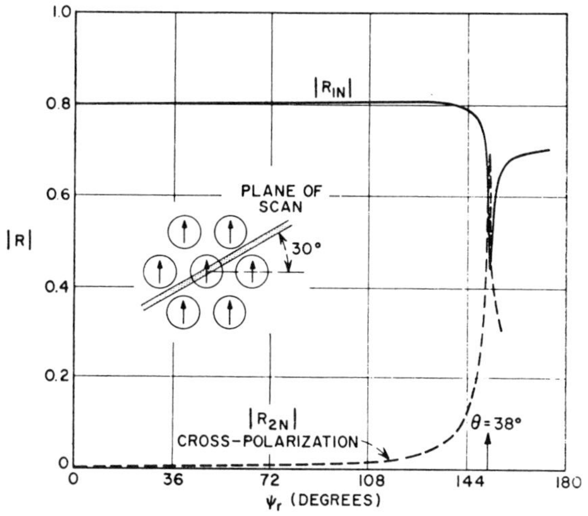


Fig. 6. Fully loaded array; 30° plane of scan. $\epsilon a^2$ = constant = 0.118 $\lambda_0^2$, $b = d = 0.714\ \lambda_0$, $a = 0.169\ \lambda_0$ for $\epsilon = 4.1$.

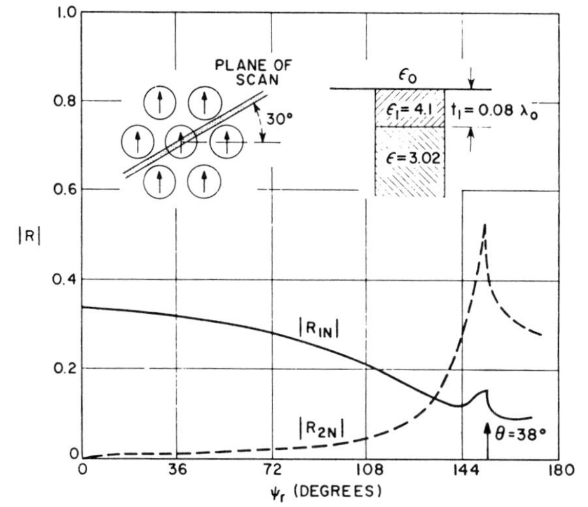


Fig. 7. Fully loaded array; 30° plane of scan. Matching by dielectric plug. $b = d = 0.714\ \lambda_0$, $a = 0.169\ \lambda_0$, $\epsilon = 3.02$, $\epsilon_1 = 4.1$, $t_1 = 0.08\ \lambda_0$.

constant, in addition to $\epsilon a^2$ = constant. Under these conditions, the array is more closely packed with increasing $\epsilon$, while the aperture filling factor ($\pi a^2/bd \sin \alpha$), the ratio between the waveguide apertures to the unit cell, is constant.

As can be expected due to closer element packing, continuous regions of total reflection appear. The isolated surface wave resonance which exists in the absence of dielectrics ($\epsilon = 1.0$) merges with the continuous regions of total reflections and does not reappear with increasing values of $\epsilon$.

Fig. 5 shows the reflection characteristics in the $H$-plane scan of a fully loaded array (with $\epsilon a^2$ = constant) when the grid dimensions relative to the wavelength $\lambda_0$ were maintained fixed ($b = d$ = constant and $\lambda_0$ = constant). In this case the aperture filling factor decreases with increasing $\epsilon$. As can be seen, the magnitude of $R_{IN}$ increases greatly in real space with increasing dielectric constant, compared to the relatively small changes shown in Fig. 4. Note, also, that $R_{IN}$ becomes very flat as a function of $\psi_r$. The flat reflection characteristic (in *phase* and amplitude) was observed in other planes of scan. The only sharp variations noticed were localized around grating lobe incipience (vertical arrows) in Figs. 5 and 6. Note that the crossed polarized mode for the 30° scan plane (Fig. 6) is negligible except near grating lobe incipience where a sharp localized peak is present. We may also observe that due to the vector symmetries of an equilateral triangular grid array of circularly symmetric elements, the reflection coefficient at broadside is independent of the polarization of the exciting $TE_{11}$ modes. This, together with the flat reflection characteristics, allows us, in principle, to realize a good match over a wide scan region. Of course, similarly flat transmission coefficients (both in phase and amplitude) were also observed. This characteristic yields a flat polarization, e.g., circular, pattern over the same region of scan. Fig. 7 demonstrates how an improved match of the array can be achieved when a plug of $\epsilon_1 = 4.1$ with thickness $t_1 = 0.08\ \lambda_0$ is followed by a dielectric loading material of $\epsilon = 3.02$. Similar performance was observed in all other planes of scan. The frequency band performance of such a relatively small aperture with a simple matching device is narrow, but with a judicious synthesis [10], [11] of the matching network, one can possibly obtain a good match characteristic as a function of scan over a 10 to 15 percent bandwidth.

### *B. Dielectric Plugged Phased Arrays*

The introduction of dielectric plugs into a previously unloaded array for matching, hardening, or weather protection can seriously affect the array performance [6], [8], [12]. Higher order modes which are below cutoff in the unloaded waveguides can possibly propagate in the dielectric plug. If the plugs are placed flush with the aperture, these modes cannot propagate beyond the plug and are "trapped" in it. They can strongly influence the *existence* and *position* (with scan) of the surface waves. Different higher order modes will be excited in different planes of scan due to the symmetry dictated by the array element and grid. Therefore, an improvement in performance that is achieved in one plane of scan by use of a dielectric plug may, in fact, degrade the array performance in another plane of scan.

Fig. 8 shows $|R_{IN}|$ as a function of scan for various plug thicknesses in the $E$ plane of scan. With the dimensions as indicated and $\epsilon_1 = 1.8$, the circular $TM_{01}$ mode is "trapped" in the plug (in the $H$ plane of scan this mode will not be excited). By increasing the plug thickness $t_1$ from zero (no plug), we observe that the location of the surface wave shifts towards broadside. For a critical range of plug thickness, we observe that there are no surface waves, while they reappear for still higher values of thickness. In other words, the dielectric plug exhibits an equivalent "bandpass" characteristic as a function of thickness. If we assume that the evanescent modes in the

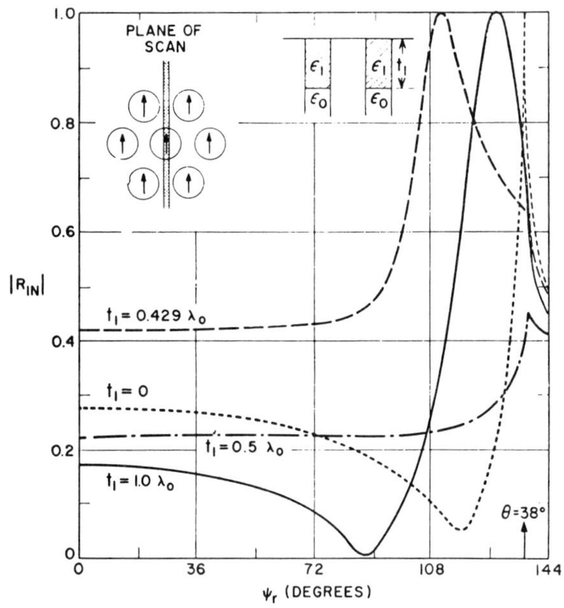


Fig. 8. Dielectric plugged array; $E$-plane scan. One trapped mode in plug. $b = d = 0.714\ \lambda_0$, $a = 0.340\ \lambda_0$, $\epsilon_1 = 1.8$, $\epsilon = \epsilon_0$.

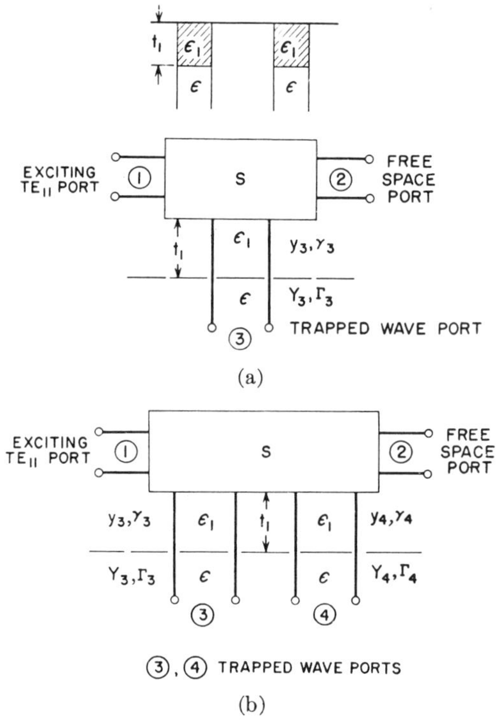


Fig. 9. Scattering representation of the array with trapped modes. (a) One trapped mode. (b) Two trapped modes.

plug (generated at the aperture, $z = 0$) decay sufficiently at $z = -t_1$, we may use a transmission-line approach [16] to evaluate the surface-wave angular location $\psi_{sw}$ as a function of the thickness.

Fig. 9(a) shows the equivalent lossless (and consequently unitary, with appropriate normalization) 3-port scattering matrix $S$. Port 3 corresponds to the trapped $TM_{01}$ mode. The elements $\{s_{ij}\}$ of this matrix are obtained from the solution of the integral equation (2) when the dielectric material totally loads the waveguide, and the various ports are excited alternatively. The exciting incident field is used as the left-hand side of (2). The matrix inversion needs to be performed only once for all incident fields. Since a surface wave is characterized by a zero transmission between ports 1 and 2, this condition is imposed so that we may obtain an equation that relates the surface wave scan location $\psi_{sw}$ to the dielectric thickness $t_1$

$$\frac{y_3 - Y_3}{y_3 + Y_3} \exp(2i\gamma_3 t_1) = \frac{s_{12}}{s_{12}s_{33} - s_{23}s_{13}} \tag{10}$$

where $\{s_{ij}\}$ are the known elements of the scattering matrix $S$ (which is a function of the scan angle), $y_3$ and $\gamma_3$ are the modal admittance and propagation constant of the $TM_{01}$ mode in the plug, while $Y_3$ is the modal admittance in the air. It can be easily shown that the absolute value of either side of (10) is unity since $y_3$ is real and $Y_3$ is imaginary. Once the boundary-value problem for the totally loaded waveguide is solved at a given scan angle $\psi_{sw}$, we can determine from (10) the thickness $t_1$ which will produce a surface wave at this angle. Results for the 3-port case are shown in Fig. 10. Note that there are "bandpass" regions of thickness for which no surface wave is excited. This occurs because it is possible that as $\psi_r$ varies from $0 \to \pi$, the phase of the right-hand side of (10), $\arg[s_{12}/(s_{12}s_{33} - s_{23}s_{13})]$, lies within an interval which is smaller than $2\pi$. Since $y_3$, $Y_3$, and $\gamma_3$ are independent of $\psi_r$, there are certain ranges of $t_1'$ for which (10) will not be satisfied. Except for very thin plugs, wherein the evanescent modes excited at the aperture cannot be neglected, $\psi_{sw}$ is a periodic function of $t_1$ with a period of $\pi/\gamma_3$ (half of a $TM_{01}$ wavelength). As can be seen in Fig. 10, the transmission-line approximation holds well even for relatively thin plugs.

The same transmission-line approximation can be used when $\epsilon_1$ is increased to a value such that two modes are trapped in the plug (the $TM_{01}$ and $TE_{21}$ modes). Fig. 11 shows $|R_{IN}|$ versus scan for various plug thicknesses in the $E$ plane of scan for a value of $\epsilon_1 = 2.54$. The element and grid dimensions are identical to the ones discussed previously. Again the dielectric-free forced surface wave shifts towards broadside with increasing values of thickness. In contrast to the previous case, multiple surface waves may exist as shown for $t_1 = 0.286\ \lambda_0$. The $TM_{01}$ and $TE_{21}$ modes in the plug are of the same order of magnitude for both resonances. As $t_1$ increases in value, the surface wave disappears and then reappears again for higher values of $t_1$. We observe, again, that certain "bandpass" regions of $t_1$ exist for which no surface wave is excited. The equivalent network for a transmission-line approximation of this case is shown in Fig. 9(b), where an additional fourth port is added to account for the trapped $TE_{21}$ mode. The equation that relates $\psi_{sw}$ and $t_1$ is a more complicated transcendental form given by

$$1 = G_{34} \exp[2i(\gamma_3 + \gamma_4)t_1] + G_3 \exp(2i\gamma_3 t_1) + G_4 \exp(2i\gamma_4 t_1) \tag{11}$$

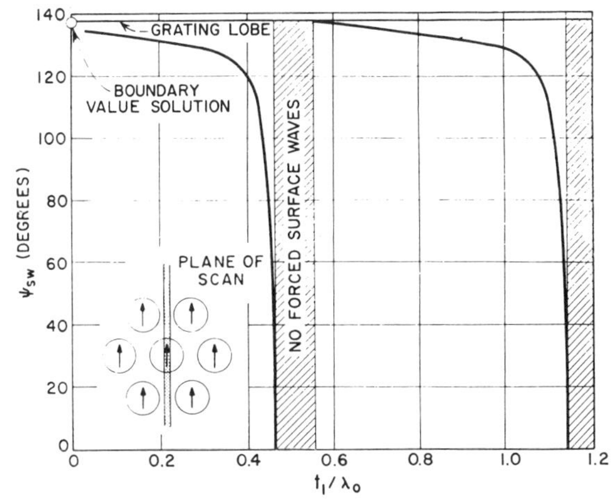


Fig. 10. Dielectric plugged array; $E$-plane scan. Surface wave scan location ($\psi_{sw}$) versus plug thickness. One trapped mode in plug. $b = d = 0.714\ \lambda_0$, $a = 0.340\ \lambda_0$, $\epsilon_1 = 1.8$, $\epsilon = \epsilon_0$.

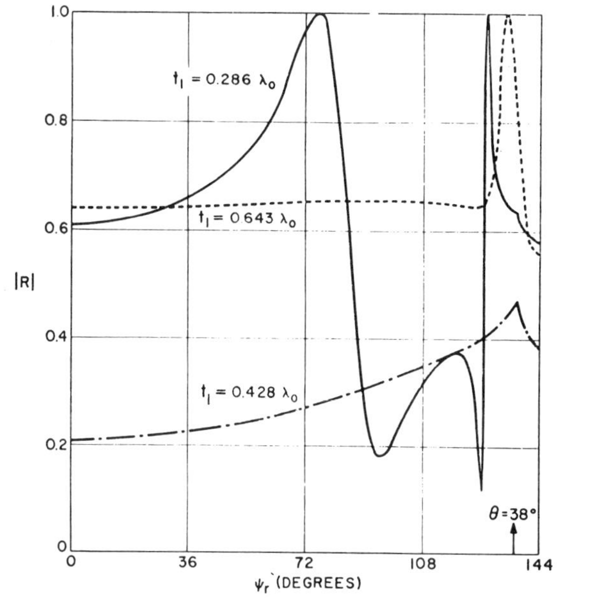


Fig. 11. Dielectric plugged array; $E$-plane scan. Two trapped modes in plug. $b = d = 0.714\ \lambda_0$, $a = 0.340\ \lambda_0$, $\epsilon_1 = 2.54$, $\epsilon = \epsilon_0$.

where

$$G_{34} = +\frac{s_{12}^*}{s_{12}} \det \| S \| \frac{y_3 - Y_3}{y_3 + Y_3} \frac{y_4 - Y_4}{y_4 + Y_4}$$

$$G_3 = \frac{s_{12}s_{33} - s_{13}s_{23}}{s_{12}} \frac{y_3 - Y_3}{y_3 + Y_3}$$

$$G_4 = \frac{s_{12}s_{44} - s_{14}s_{24}}{s_{12}} \frac{y_4 - Y_4}{y_4 + Y_4}.$$

Note that $y_3$, $Y_3$, $y_4$, $Y_4$, $\gamma_3$, and $\gamma_4$ are the appropriate modal admittances and propagation constants as indicated in Fig. 9(b). As before, the elements of the scattering matrix are obtained from a solution of (2), with appropriate incident fields chosen. Due to the fact that the ratio of $\gamma_3$ to $\gamma_4$ is, in general, not a rational number, $\psi_{sw}$ is not necessarily periodic with $t_1$ [see (11)]. In Fig. 12 we observe the aperiodicity of $\psi_{sw}$ as a function of $t_1$. Note, again, the "bandpass" characteristics as well as the accuracy of the transmission-line approximation for very thin plugs.

### C. Dielectric Covered Arrays

The presence of a dielectric sheath or cover over the array, flush with the aperture (see Fig. 2), can also strongly influence the array performance [5], [6]. Trapped modes (the periodic free-space type: $\psi_{mn}{}^p$) can exist in the dielectric sheath, but their influence on the array performance differs markedly from the effects of the trapped waveguide modes in the plugs. The trapped surface waves[1] in the sheath do *not* exhibit the "bandpass" characteristics that occur in dielectric plugged arrays. With increasing values of the sheath thickness $t_1'$ or dielectric constant $\epsilon_1'$, the angle $\psi_r$, at which a dielectric-free surface wave resonance occurs, shifts towards broadside but does not vanish. In fact, additional or multiple surface waves appear at larger $\psi_r$ values as either $t_1'$ or $\epsilon_1'$ is increased. As a result, the usable scan region becomes seriously limited for higher $t_1'$ or $\epsilon_1'$. In Fig. 13, we observe the reflection coefficient magnitude as a function of scan $\psi_r$ for an $E$-plane scan of a dielectric covered array with dielectric constant $\epsilon_1' = 2.00$ and grid and element dimensions identical to the array considered for the plugs. We observe the effect of increasing the sheath thickness $t_1'$. As can be seen, for a thin sheath, $t_1' = 0.1\ \lambda_0$, the dielectric-free ($t_1' = 0$) surface wave resonance shifts towards broadside while the magnitude of the $R_{IN}$ curve around the surface wave maxima broadens considerably. As the sheath thickness increases to $t_1' = 0.5\ \lambda_0$, a further shift towards broadside takes place together with an appearance of a second surface wave that occurs due to a trapped mode within the sheath.

A transmission-line analysis can also be carried out for this case. If only one trapped mode exists in the dielectric, and, assuming that the evanescent modes in the sheath decay sufficiently at $z = t_1'$, port 3 of Fig. 9(a) can represent this trapped mode. One important difference between this case and that for the plugs is that for the sheath the Floquet modal admittances and propagation constants $y_3$, $Y_3$, and $\gamma_3$ are functions of $\psi_r$. The scattering matrix elements $\{s_{ij}\}$ are obtained from the appropriate solutions of (2) when the whole space, $z \geq 0$, is loaded with dielectric $\epsilon_1'$. In (10), the dominant contribution to the phase variation of the left-hand side comes from the term $\exp(2i\gamma_3 t_1')$, where for the $E$-plane scan we have

$$\gamma_3 = \sqrt{\epsilon_1' \left(\frac{2\pi}{\lambda_0}\right)^2 - \left(\frac{2\pi - \psi_r}{d \sin \alpha}\right)^2}, \quad 0 \leq \psi_r \leq \pi. \quad (12)$$

Note that $\gamma_3$ increases with increasing values of $\psi_r$. If we assume, also, that the right-hand side of (10) varies slowly with $\psi_r$ by comparison, then a slight increase

[1] The "trapped" surface waves are due to either the trapped waveguide or the trapped periodic free-space modes.

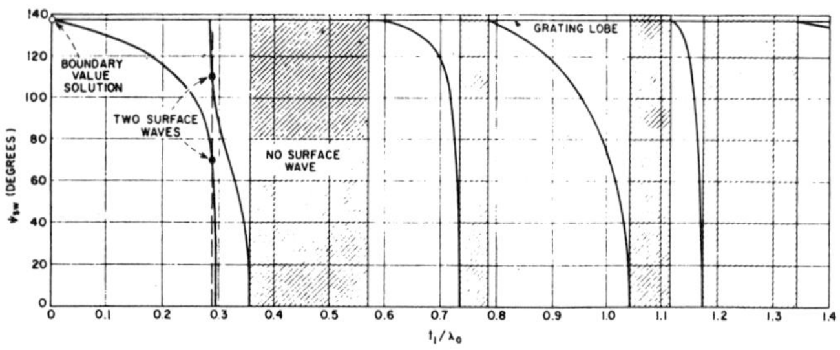


Fig. 12. Dielectric plugged array; $E$-plane scan. Surface wave scan location ($\psi_{sw}$) versus plug thickness. Two trapped modes in plug. $b = d = 0.714\ \lambda_0$, $a = 0.340\ \lambda_0$, $\epsilon_1 = 2.54$, $\epsilon = \epsilon_0$.

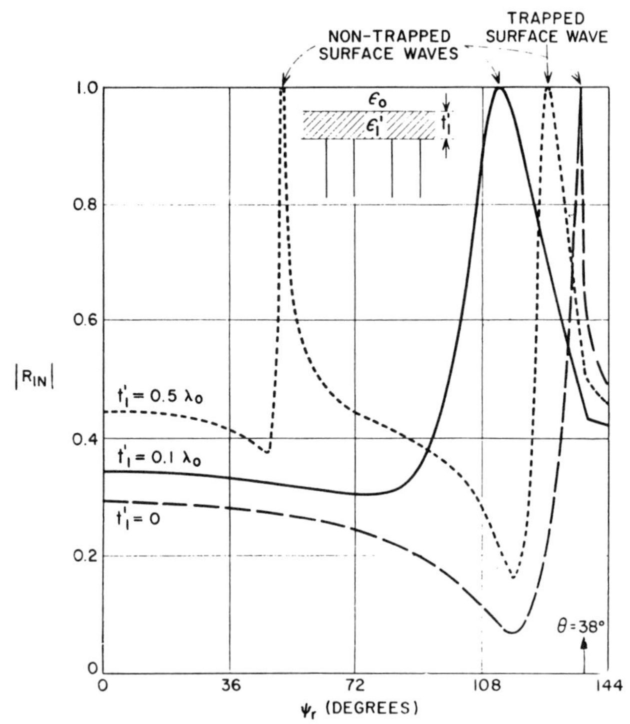


Fig. 13. Dielectric covered array; $E$-plane scan. One trapped mode in sheath. $b = d = 0.714\ \lambda_0$, $a = 0.340\ \lambda_0$, $\epsilon_1' = 2.0$, $\epsilon' = \epsilon_0$.

in the thickness $t_1'$ will result in a decrease in $\psi_{sw}$, as shown in Fig. 14. We note also that for a sufficiently large value of $t_1'$, the phase $2\gamma_3 t_1'$ may go through many multiples of $2\pi$ for the indicated range of $\psi_r$. Consequently, multiple surface waves exist for large thicknesses. For $t_1/\lambda_0 > 0.9$, as many as three forced surface waves are seen to exist. Note that with increased values of $t_1'$, $\psi_{sw}$ does not reach broadside but asymptotically approaches a fixed value. The discrepancy (as the sheath becomes thin) between the boundary-value solutions and the transmission-line approximation exists over a bigger range of thickness values than exists in the corresponding plug case. This illustrates the usefulness of the boundary-value solution for thin sheaths. In the region where multiple surface waves exist, the approximations improve.

### *D. Forced Surface Wave Isolation*

In previous works [1], [2] it was found that the surface wave resonances do occur at isolated points in the

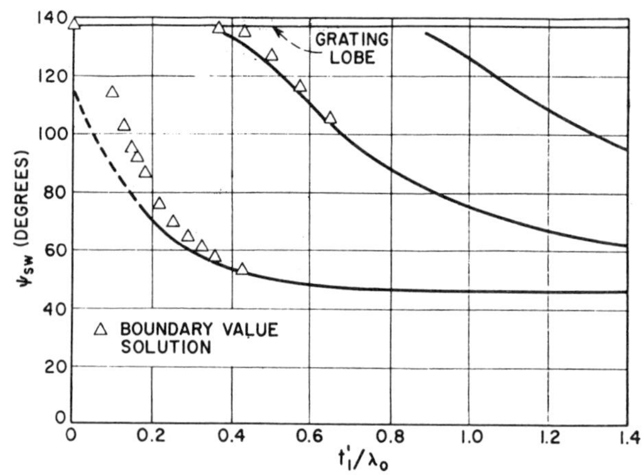


Fig. 14. Dielectric covered array; $E$-plane scan. Surface wave scan location ($\psi_{sw}$) versus sheath thickness. One trapped mode in sheath. $b = d = 0.714\ \lambda_0$, $a = 0.340\ \lambda_0$, $\epsilon_1' = 2.0$, $\epsilon' = \epsilon_0$.

scan plane ($\psi_x, \psi_y$). This isolation phenomenon was also observed in both dielectric covered and loaded arrays by observing the numerical results in a neighborhood of the surface wave. The surface wave resonances observed for this type of array, with dielectrics present, were also found to be isolated points in the scan plane.

## IV. Conclusions

A study of the effects of dielectric loading, plugging, or covering a phased array of circular waveguides was made. Results obtained from a numerical solution of the integral equation for a hexagonal grid array were presented. Experimental and other verification of the numerical results were also discussed.

The fully dielectric loaded waveguide array was studied as a function of waveguide radius, spacing, and dielectric constant. A situation in which the reflection coefficient was flat over a wide scan region for all planes of scan and all polarizations was discussed.

A discussion of three situations in which forced surface wave resonances (total reflection of the incident mode) occur was presented. In the first, surface wave resonances exist in the absence of dielectric materials and are shifted in position towards broadside with increasing dielectric constant, plug, or sheath thickness.

In the second situation, surface waves are observed to be trapped in the dielectric plug. An approximate transmission-line analysis (corroborated by accurate solutions of the integral equation) shows that when a single mode is trapped in the plug, then periodic "bandpass" regions of plug thickness exist for which no surface wave is excited. When more than one waveguide mode is trapped in the plug, the "bandpass" regions occur aperiodically with increasing plug thickness.

The third situation occurs when surface waves are observed to be trapped in the dielectric sheath. An analysis of this situation shows that as the sheath thickness is increased, the surface waves shift towards broadside and increase in number. No bandpass regions were observed in this case.

Finally, it was observed that all surface waves discussed occurred or peaked at isolated or singular points in the scan $(\psi_x,\psi_y)$ plane.

An understanding of the effects of dielectrics on these arrays with particular regard to the occurrence of surface waves, taken together with other studies on the subject [10], [11], [14], [15], will enable the designer to avoid or, if desired, enhance the occurrence of these phenomena.

## References


[1] N. Amitay and V. Galindo, "The analysis of circular waveguide phased arrays," *Bell Sys. Tech. J.*, vol. 47, pp. 1903–1932, November 1968.

[2] ——, "A note on the radiation characteristics and forced surface wave phenomena in triangular-grid circular waveguide phased arrays," *IEEE Trans. Antennas and Propagation* (Communications), vol. AP-16, pp. 760–762, November 1968.

[3] R. F. Frazita, "Surface-wave behavior of a phased array analyzed by the grating-lobe series," *IEEE Trans. Antennas and Propagation* (Communications), vol. AP-15, pp. 823–824, November 1967.

[4] N. Marcuvitz, Ed., *Waveguide Handbook*, M.I.T. Rad. Lab. Ser., vol. 10. New York: McGraw-Hill, 1951.

[5] C. P. Wu and V. Galindo, "Surface wave effects on dielectric sheathed phased arrays of rectangular waveguides," *Bell Sys. Tech. J.*, vol. 47, pp. 117–142, January 1968.

[6] V. Galindo and C. P. Wu, "Dielectric loaded and covered rectangular waveguide phased arrays," *Bell Sys. Tech. J.*, vol. 47, pp. 93–116, January 1968.

[7] R. F. Harrington, *Field Computation by Moment Methods.* New York: Macmillan, 1968.

[8] C. P. Wu and V. Galindo, "Surface wave effects on phased arrays of rectangular waveguides loaded with dielectric plugs," *IEEE Trans. Antennas and Propagation* (Communications), vol. AP-16, pp. 358–360, May 1968.

[9] G. F. Farrell, Jr., and D. H. Kuhn, "Mutual coupling effects of triangular-grid arrays by modal analysis," *IEEE Trans. Antennas and Propagation* (Communications), vol. AP-14, pp. 652–654, September 1966.

[10] N. Amitay, P. E. Butzien, and R. C. Heidt, "Match optimization of a two-port phased array antenna element," *IEEE Trans. Antennas and Propagation*, vol. AP-16, pp. 47–57, January 1968.

[11] N. Amitay, "Improvement of planar array match by compensation through contiguous element coupling," *IEEE Trans. Antennas and Propagation*, vol. AP-14, pp. 580–586, September 1966.

[12] B. L. Diamond, "A generalized approach to the analysis of infinite planar array antennas," *Proc. IEEE* (Special Issue on Electronic Scanning), vol. 56, pp. 1837–1851, November 1968.

[13] N. Amitay, J. S. Cook, R. G. Pecina, and C. P. Wu, "On mutual coupling and matching conditions in large planar phased arrays," *Proc. 1964 IEEE Internatl. Symp. on Antennas and Propagation* (Long Island, N. Y.), pp. 150–156.

[14] V. Galindo, "A generalized approach to a solution of aperiodic arrays and modulated surfaces," *IEEE Trans. Antennas and Propagation*, vol. AP-16, pp. 424–429, July 1968.

[15] E. C. DuFort, "Design of corrugated plates for phased array matching," *IEEE Trans. Antennas and Propagation*, vol. AP-16, pp. 37–46, January 1968.

[16] C. P. Wu, "Determination of resonance conditions in dielectric-sheathed or plug-loaded phase arrays," *IEEE Trans. Antennas and Propagation* (Communications), vol. AP-16, pp. 753–755, November 1968.




*Reprinted by permission from* IEEE TRANSACTIONS ON ANTENNAS AND PROPAGATION
Vol. AP-17, No. 6, November 1969
pp. 722-729

# BLIND SPOTS II

# Element Pattern Nulls in Phased Arrays and Their Relation to Guided Waves

GEORGE H. KNITTEL, SENIOR MEMBER, IEEE, ALEXANDER HESSEL, SENIOR MEMBER, IEEE, AND ARTHUR A. OLINER, FELLOW, IEEE

*Abstract*—**In recent years, unusual dips or nulls have been observed in the element patterns of certain phased-array antennas at angles closer to broadside than that for which an endfire grating lobe exists. These nulls are known to bear a direct and simple relation to a total reflection condition which occurs when all elements of the array are excited. The purposes of this paper are to investigate the basic cause of these element pattern nulls, and to clarify the relationship between these nulls and possible guided waves which may be supported by the array face.**

**It is shown herein that, for a certain class of arrays, these nulls (actually the total reflection condition) are caused by the forced resonance of a higher external mode in the vicinity of the array aperture. The nature of this resonance is demonstrated on a simple infinite array representative of a class of structures which permit external resonances. A detailed analysis of this array (of waveguide-fed slots covered by a dielectric layer) yields a simple equivalent network from which the existence, approximate angular location, and qualitative behavior of the element pattern null are determined by inspection.**

**There are two types of guided waves which are of concern here, surface waves and leaky waves. It is shown that the actual array cannot support a true surface wave, and that the surface wave customarily associated with element pattern nulls is really a driven surface-wave-like field which appears on the array face only at precisely the total reflection condition, and which is indistinguishable in its properties from a true surface wave on a modified (corrugated) structure, not on the actual array. The leaky wave, whose role is explained here for the first time, is a true mode of the actual array and radiates at an angle close to that for the element pattern null. Its influence on the element pattern null is akin to the relation between free and forced resonances, and its attenuation constant yields the width of this null.**

## I. Introduction

DURING the past ten years, phased-array antennas have seen increasing use in radar systems, and have been the subject of intensive study. The usual kind of array for radar, and the kind with which this paper is concerned, has been that in which the radiating elements are arranged in a periodic lattice in a plane or in a line and are each excited by an individual generator. The phases of the generators are varied linearly from one element to the next to form a beam which can scan about the broadside direction to the plane or line. As early as 1954 [1], but more notably in the past five years, there has been a good deal of interest in an unusual phenomenon which occurs in certain of these phased arrays. It has been observed that the element pattern (the radiation pattern of one element in the array with all others passively terminated in their generator admittances) exhibits a dip or null at an angle closer to broadside than that at which an endfire grating lobe can exist. Although the possibility of such a dip was expected at the latter angle, it was not expected at angles closer to broadside and was somewhat of a surprise. This effect is important since the gain of the phased array at a particular angle is proportional to the element pattern gain at that angle, and the array is "blind" at those angles for which the element pattern has a null.

Although this effect is by now well known, comparatively few reliable and accurate experimental results are available. The element pattern of an array of coaxial horns covered by individual dielectric radomes was obtained by Lechtreck [2], [3]. Bates [4] used a linear array of open-ended waveguides in a conducting plane covered with a plexiglass dielectric slab. Farrell and Kuhn [5] showed a pattern, measured at the Bendix Corporation, of an array of open-ended waveguides arranged in a triangular lattice ("brick" array) in a conducting plane; measurements on a similar array were reported by Diamond [6]. Each of these element patterns has a dip at an angle substantially closer to broadside than that at which the first endfire grating lobe appears. The shallowness of the dips obtained in these measurements is due to the fact that the arrays were small; if the arrays were infinite in extent, the dips would become nulls.

There are also other kinds of evidence for the existence of this unusual null in element patterns, from theoretical calculations and from waveguide simulator experiments. Farrell and Kuhn [5] have performed theoretical calculations which verified the experimental behavior of "brick" arrays. Diamond [6] has computed an element pattern for an array of open-ended square waveguides in a conducting plane which shows the null. Frazita [7] has found the null for an array of circular waveguides in a conducting plane covered by a dielectric slab from computations of the admittance with the grating-lobe series [8]. Hannan [9] detected this type of dip by measurement in a phased-array waveguide simulator.

Ever since these element pattern dips or nulls were first discovered, they have been associated with the presence of a "surface wave" on the array face. However, the nature of these surface waves and the role they played was not clearly understood. The early history of speculations in this context has been summarized by Oliner and Malech [10]. These

Manuscript received April 1, 1968; revised July 1, 1968. This work was supported in part by the Air Force Office of Scientific Research, Office of Aerospace Research, USAF, under Contract AF 49(638)-1402. It is taken from a dissertation submitted by G. H. Knittel to the Faculty of the Polytechnic Institute of Brooklyn in partial fulfillment of the requirements for the degree of Doctor of Philosophy (Electrophysics), June 1969.

The authors are with the Department of Electrophysics, Polytechnic Institute of Brooklyn, Long Island Graduate Center, Farmingdale, N. Y. 11735

authors also presented a "ghost-mode" hypothesis to explain the cause of such nulls on dielectric-coated slot arrays, and showed the relation between these nulls and a type of surface wave. The presentation was incomplete, but they were the first to demonstrate two important features: 1) they showed that the surface wave was a mode of a modified, short-circuited structure and not a mode of the actual, complete array, and 2) their ghost-mode viewpoint[1] indicated that the element pattern null corresponded to a resonance of a higher mode in the dielectric region, one that could propagate in dielectric but not in air.

These early efforts are greatly expanded in the present paper, and certain new concepts are presented herein. We attempt here to:

1) present a simple network picture for an array typical of a certain class of arrays, so that from this network one can ascertain in a simple fashion all of the basic features of the radiation behavior
2) demonstrate the relationships between the element pattern null, the surface wave everyone speaks of, and a leaky wave which is supportable by the actual array; the connection between this leaky wave and element pattern nulls was recently introduced by the present authors [11], and the current paper contains its first elaboration in print.



It will be shown that the aforementioned surface wave is present on the array face only when the array is scanned at exactly the angle corresponding to the element pattern null, and that it has no meaning elsewhere. Furthermore, it is a mode of a modified structure and not one of the actual array, and the actual array cannot support a true surface-wave mode. Information regarding this modified-structure surface wave may be useful, however, in indicating the location in angle of the null, but it offers no aid in ascertaining the width of the null.

It is also shown that the actual array can support a complex-wave mode of the leaky-wave type, which is guided along the array face and radiates power at an angle from the array face. This angle is very near to the angle corresponding to the element pattern null and the attenuation constant of this leaky wave provides a measure of the width of this null. All of these aspects are explored below in Section V.

It has become increasingly clear from various theoretical analyses that in all of these examples of element pattern nulls a key role is played by one (or more) of the higher modes. Depending on the situation, this higher mode will resonate or will partake in some cancellation effect which changes rapidly with change in scan angle, and which thus resembles a resonance. It is therefore appropriate to distinguish between two basic types of such resonances,

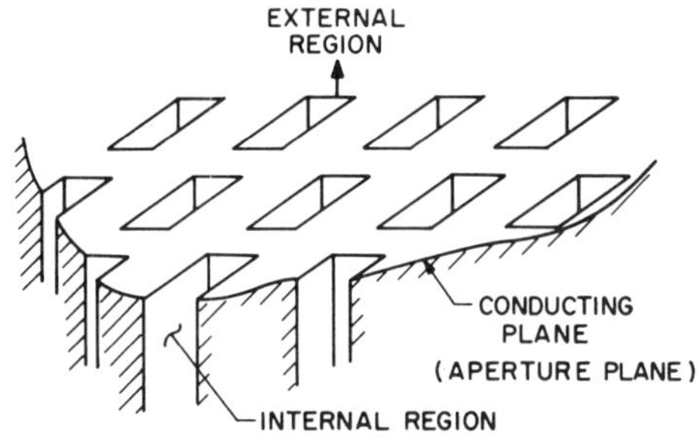


Fig. 1. Array of aperture-type elements in conducting plane.

*external resonances* and *internal resonances*, according as the relevant higher mode is in the region outside of the array aperture or within the feed guide region, respectively.[2] This division into types is useful for phased arrays, such as that shown in Fig. 1, where one can define an aperture plane. Configurations which exhibit external resonances contain a region which can support the propagation of a higher external mode, even though that mode is below cutoff in the air space beyond. For internal resonances, a higher internal mode plays a key role.

For some structures, the aperture field is strongly scan-dependent, while for others the aperture field remains essentially constant except in the immediate vicinity of the null. The latter is the case for certain external resonances for which the properties of the element pattern null are dependent essentially on the behavior of a higher mode in the external region, which usually contains some dielectric material or other kind of slow-wave loading. If, in these external resonances, the aperture field also varies considerably with scan, modifications will occur in the radiation behavior but the key features of the external resonance will be similar. In order to study the basic characteristics of external resonances with as simple a prototype structure as possible, therefore, the array chosen for detailed analysis in Section II is a two-dimensional array of narrow slots fed by parallel-plate waveguides. The simplicity of the structure avoids unnecessary complications in description and yet retains most of the basic behavior features.

Except for various pertinent comments, this paper will be concerned primarily with the study and explanation of external resonances. In order to simplify the analysis, a simple lossless infinite structure[3] is chosen for examination; the qualitative features of the resonance thus determined are valid for all structures for which the aperture field is essentially scan-invariant. A simple equivalent network is derived for the chosen structure from which one may determine many of the characteristics of the resonance by inspection. It is shown that the qualitative features of the radiation behavior are obtained from only three modes, the dominant

[1] This viewpoint refers to the situation when all the elements are excited and a beam is being radiated into space at a particular (scan) angle. The resonance causes a complete reflection of power incident in each element which, as discussed in Section III, is equivalent to a null in the element pattern at that angle.

[2] Actually in both types of resonance the higher mode produces a selective enhanced coupling between internal and external modes; this coupling changes the aperture field distribution, producing a cancellation effect and resulting in a null. This feature is general to all nulls and will be discussed more fully in a subsequent paper.

[3] The choice of an infinite structure is usual for this kind of study, the results being applicable for all elements in a large finite array except those very near the edges.

internal and external modes, and the first higher external mode. The inclusion of only one additional higher external mode is sufficient to determine this behavior rather accurately over the complete range of scan angles.

The array structure which is analyzed in detail herein is described in Section II, together with a simple equivalent network for it. The assumptions underlying the form of this network are discussed there, but its actual derivation is presented in the Appendix. Many qualitative features of the radiation behavior can be obtained by inspection from this simple equivalent network, including the location in angle of the element pattern null; these aspects are treated in Section III and in Section IV-A. The remainder of Section IV presents additional information on the radiation behavior, including the demonstration that very accurate results valid over the complete range of scan angles may be obtained from calculations employing only three external modes. Section V is devoted to the relations between the properties of the element pattern nulls and two types of guided waves—surface waves and leaky waves—and rather general conclusions are drawn. Section VI gives the conclusions and discussion, summarizes certain results, and presents some additional observations.

## II. Array of Slots and Equivalent Network

The physical structure chosen for a detailed examination of the external resonance phenomenon is shown in Fig. 2. It consists of an array of slots or apertures in a conducting plane, fed from below with parallel-plate waveguides and covered above by a dielectric slab. The structure is infinite in the $x$- and $y$-directions. It is assumed that a TEM mode is incident from below in each parallel-plate or feed waveguide. These modes are excited by generators which lie many wavelengths below the slots and which are matched to the respective waveguides. The generators are identical except for their phases so that the incident mode in each guide is of the same amplitude but of phase $\psi$ relative to the preceding guide. A plane wave (the main beam in a finite array) is then radiated into the space above the array at an angle $\theta$. This scan angle is in the $xz$-plane, as is the $E$-field of the radiated wave ($E$-plane scan).

There are several reasons for the choice of the array in Fig. 2 for a detailed investigation of the external resonance phenomenon. The array is infinite, which permits use of the powerful unit-cell technique for infinite periodic structures (discussed in detail by Oliner and Malech [12]). It is two-dimensional, thus confining scanning to one plane and simplifying the analysis. It is the simplest kind of array utilizing "flush-type" or aperture radiating elements, thereby permitting a clear distinction between the "external" and "internal" portions of the array. If the slot width is kept small enough, say $d/B<0.1$, an equivalent network is easily derived which clearly shows the existence of the resonance and its important characteristics. The structure may also be approximated physically so that experimental verification of the theoretical results is possible by construction of a finite array. It is capable of being matched at an arbitrary scan angle both theoretically and physically.

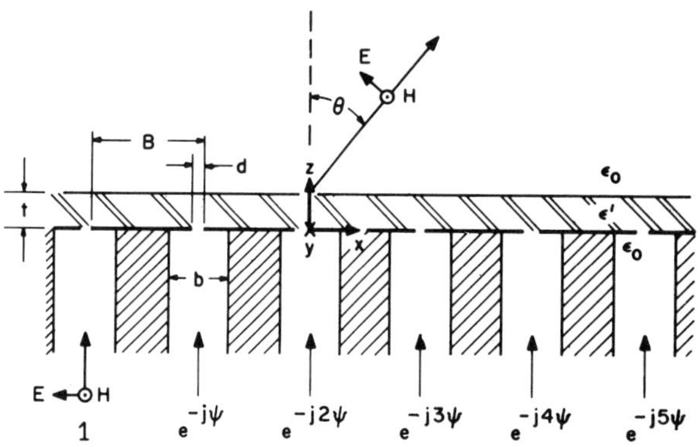


Fig. 2. Array of narrow slots fed by parallel-plate waveguides and covered by dielectric slab; $E$-plane scan.

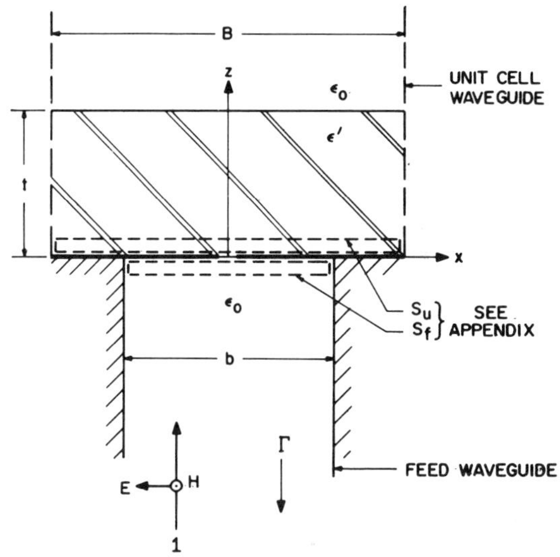


Fig. 3. Unit cell for array in Fig. 2.

Finally, its performance is representative of that of many finite three-dimensional planar arrays when scanned in the $E$-plane. The $E$-plane is the most useful scan plane to choose for a study of external resonances since one finds an $E$-plane scan resonance for thin dielectric slabs or only light "slow-wave" loading. Such loading is often found in practical arrays. In the $H$-plane, however, heavier slow-wave loading or a thicker dielectric slab is required for a resonance.

To derive an equivalent network for the structure in Fig. 2 one first forms a unit cell by drawing the phase-shift walls of the unit-cell waveguide on the $z>0$ side of the aperture, as in Fig. 3. The problem is then reduced to that of finding an equivalent network for the junction of two waveguides with one of them loaded with a dielectric layer of finite thickness. If one knows the *shape or form* of the electric field in the aperture at all scan angles he may then derive the equivalent network for the structure. Details of this derivation are given in the Appendix, and only the results and the assumptions are discussed here.

If the slots in the array in Fig. 2 are chosen to be sufficiently narrow, it is known that the electric field in the aperture may be approximated by

$$E_{ap} = x_0 C \tag{1}$$

where $x_0$ is a unit vector in the $x$-direction and $C$ is an amplitude which is independent of $x$ but which varies with scan. Expression (1) is assumed to be valid regardless of whether the aperture is excited from above or below and is equivalent to approximating the aperture electric field with that of the dominant mode of the aperture. The actual slot width, $d/B=0.05$, used for all results obtained in succeeding sections, is well within the range for which this approximation is valid. The slots of the array are thus particularly simple examples of elements for which the form of the aperture field is essentially invariant with scan, and the results presented here should be representative of all such elements, as mentioned in Section I.

From the Appendix, the input admittance of the array at the aperture plane when all elements are excited is

$$Y=\sum_{m=-\infty}^{\infty}\frac{b}{B}\left[\frac{\sin\left(k_{xm}\frac{d}{2}\right)}{k_{xm}\frac{d}{2}}\right]^2 Y_{m\varepsilon}\frac{jY_{m\varepsilon}+Y_m\cot(\kappa_{m\varepsilon}t)}{Y_{m\varepsilon}\cot(\kappa_{m\varepsilon}t)+jY_m} \qquad (2)$$

$$+jB_{in}$$

where $Y$ is the admittance seen by the TEM mode in the feed guide in the positive $z$-direction. The quantities $Y_{m\varepsilon}$ and $Y_m$ are the modal admittances for the $m$th mode in the unit-cell waveguide in dielectric and air, respectively, and are given by



$$Y_{m\varepsilon}=\frac{\omega\varepsilon'\varepsilon_0}{\kappa_{m\varepsilon}},\qquad Y_m=\frac{\omega\varepsilon_0}{\kappa_m} \qquad (3)$$

where

$$\kappa_{m\varepsilon}=\sqrt{\varepsilon' k_0^2-k_{xm}^2},\qquad \kappa_m=\sqrt{k_0^2-k_{xm}^2} \qquad (4)$$

are the corresponding wavenumbers of the $m$th mode in the $z$-direction, $\varepsilon'$ is the relative permittivity of the dielectric, and $k_0=\omega\sqrt{\mu_0\varepsilon_0}=2\pi/\lambda$ is the free-space wavenumber. The parameter $k_{xm}$ is the wavenumber of the $m$th mode in the $x$-direction and is related to the scan angle $\theta$ by

$$k_{xm}=k_0\sin\theta+\frac{2\pi m}{B} \qquad (5)$$

where $m=0, \pm 1, \pm 2, \cdots$, is the mode index. In (2), $B_{in}$ is the susceptance representing stored power in the higher internal modes, which are below cutoff. An expression for $B_{in}$ is given in the Appendix; it is assumed that all internal modes are below cutoff except for the TEM mode. It is well known that the array input admittance $Y$ is related to the array reflection coefficient $\Gamma$ by

$$\Gamma=\frac{Y_f-Y}{Y_f+Y} \qquad (6)$$

where $Y_f$ is the characteristic admittance of the TEM mode in the feed waveguide ($Y_f=\sqrt{\varepsilon_0/\mu_0}$).

Since the aperture electric field is the same whether it is excited from the external region or from the internal region, (2) may be represented in the form of the shunt *equivalent network* shown in Fig. 4. The turns ratios of the transformers in this figure are seen from (2) to be

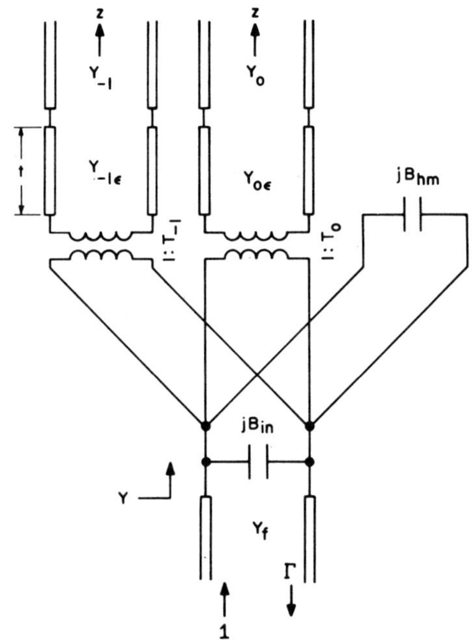


Fig. 4. Equivalent network for unit cell of array in Fig. 2.

$$T_m=\sqrt{\frac{b}{B}}\,\frac{\sin\left(k_{xm}\frac{d}{2}\right)}{k_{xm}\frac{d}{2}}. \qquad (7)$$

The $m=0$ and $m=-1$ mode transmission lines are displayed explicitly since these modes are the essential ones in the external resonance phenomenon. All the other higher modes outside the aperture are assumed to be below cutoff in both dielectric and air and are lumped together as a susceptance $B_{hm}$ (where the subscript signifies "higher modes") given by

$$jB_{hm}=\sum_{\substack{m=-\infty\\ \neq 0,-1}}^{\infty}(T_m)^2 Y_{m\varepsilon}\frac{jY_{m\varepsilon}+Y_m\cot(\kappa_{m\varepsilon}t)}{Y_{m\varepsilon}\cot(\kappa_{m\varepsilon}t)+jY_m}. \qquad (8)$$

It should be mentioned here that a similar equivalent network has been obtained by Oliner and Malech [10] except that they do not give explicit relations for $B_{hm}$ and $T_m$, since they were concerned only with the particular scan angle for $|\Gamma|=1$ and not with the variation of $\Gamma$ about this angle. Parad [13], using a different approach, has obtained an expression for $Y$ similar to (2) for a structure similar to that in Fig. 2. However, he does not consider a specific feeding arrangement and does not derive an equivalent network. In addition, he also is concerned only with the particular scan angle for which the element pattern has a null.

It is noteworthy that the *shunt form* of the equivalent network in Fig. 4 is a consequence of the *choice* of working with the tangential $E$-field at the aperture plane and the *assumption* that the aperture electric field has the same shape regardless of whether it is excited from the external region or from the internal region. The advantage of choosing a small aperture where one can assume the electric field to be constant in $x$ for all scan angles is that $B_{in}$ is independent of scan and $T_m$ is only slightly scan-dependent.

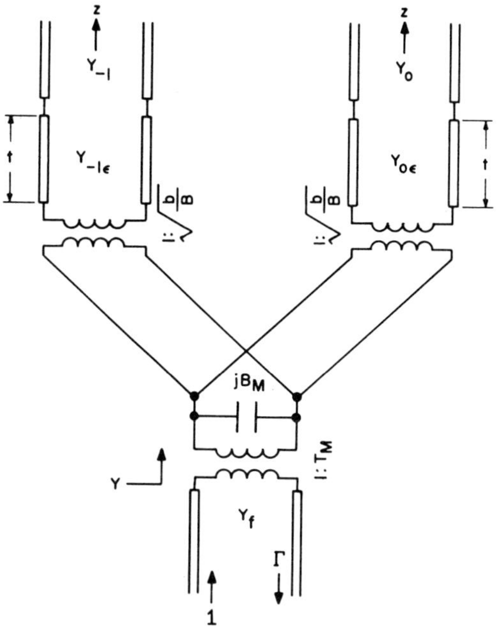


Fig. 5. Equivalent network for array in Fig. 2 after matching at broadside.

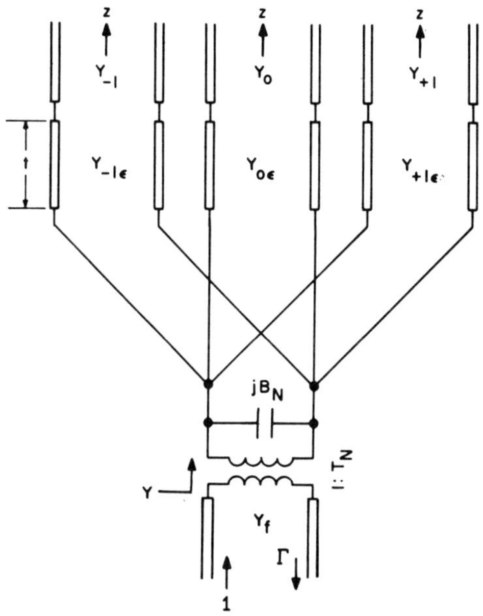


Fig. 6. Simplified equivalent network utilizing only three external modes ($B_N$, $T_N$ are independent of scan).

Thus, for a small slot the equivalent network is essentially scan-invariant, the effect of scanning being determined by the variation with $\theta$ of the modal wavenumbers and admittances. Because only two or three external modes are important in this array (to be shown later), *one can easily estimate the qualitative features* of the reflection coefficient versus scan angle from the equivalent network.

In order that the performance of the array in Fig. 2 be related to that of a practical phased array, it is necessary to include a *matching network* to match the admittance of the array to that of the feed guide at, say, broadside scan. This can be done in practice, over a narrow band of frequencies, by adding an appropriate susceptance (iris) and transformer (change in $b$-dimension of the feed guide) a half-wavelength below the aperture plane. Since this is the same, in the equivalent network, as adding the suceptance and transformer at the aperture, the latter is done as shown in Fig. 5. In this figure, $B_M$ (subscript $M$ signifying "match") is the sum of three susceptances,

$$B_M = B_{\text{in}} + B_{\text{hm}} + B_{\text{ex}} \tag{9}$$

where $B_{\text{ex}}$ is the extra susceptance added for matching. $T_M$ is the transformer added for matching, and $T_0 \approx T_{-1} \approx \sqrt{b/B}$ since $d$ is small.

It will be seen later that $B_M$ varies only slightly with $\theta$, and that this variation is due mostly to the $m=+1$ mode. If this mode is removed from $B_M$ and displayed explicitly, the variation in the resulting $B_M$ with $\theta$ is almost nil. Therefore, one may accurately obtain the performance of the array from the simplified equivalent network shown in Fig. 6 and based upon only three external modes and one internal (TEM) mode. The susceptance $B_N$ is just that which cancels the susceptance from the external mode transmission lines at $\theta=0$. The transformer $T_N$ is just that which converts the conductance from the $m=0$ transmission line at $\theta=0$ to the characteristic admittance of the feed guide.

In the following sections the equivalent network will be utilized to determine the nature of the external resonance. The network parameters which will be constant for this purpose and the values chosen for them are



$$\frac{b}{\lambda} = 0.3, \qquad \frac{d}{B} = 0.05$$

where $\lambda$ is the free-space wavelength. The remaining parameters, which are varied in what follows, are $\theta$, $t$, $B$, $\varepsilon'$. Broadside matching is assumed unless otherwise stated. A "reference" array, to which a good deal of attention is devoted, is chosen as

$$\frac{t}{\lambda} = 0.2, \qquad \frac{B}{\lambda} = 0.5, \qquad \varepsilon' = 2.56.$$

## III. Existence and Location of Null in Element Pattern

The equivalent network derived in Section II yields the array reflection coefficient $\Gamma$ when all the elements of the array are excited. It also yields the transmission coefficient $T$ for a beam radiating into space (external mode propagating in the air). It has been shown [14] that for a lossless array radiating a single beam $|\Gamma(\theta)|$ is related to the element pattern via

$$g_e = \cos\theta[1 - |\Gamma(\theta)|^2] \tag{10}$$

where $g_e$ is the realized gain (power ratio) of one element in the array with all other elements terminated in their generator admittances. Expression (10) is normalized so that $g_e=1$ for broadside match, i.e., for $|\Gamma(0)|=0$. More generally, if there are one or more radiating beams, or if the array aperture region contains lossy material, it is easy to show that (10) should be replaced by

$$g_e = \cos\theta |T(\theta)|^2 \tag{11}$$

where $|T(\theta)|^2$ is the ratio of power transmitted into the air in the unit cell by a particular external mode to power incident in the feed waveguide. The $g_e$ thus computed is that at the angle of radiation of the mode for which $T$ was taken. Defining the ideal element pattern as that for which $|\Gamma|=0$ for all scan angles leads to $g_e = \cos\theta$, $|T|=1$, for $\theta$ less than or equal to the angle for which the main lobe and first grating lobe are at equal angles from broadside, and $g_e=0$, $|T|=0$, for $\theta$ greater than this angle [15]. The computed and the ideal element patterns shown in this and the following sections are calculated from (11) or (10) after $|T|$ or $|\Gamma|$ is determined from the equivalent network.

One can easily ascertain from the equivalent network in Fig. 5 that there is a null in the element pattern and that it is not generally at the same scan angle for which an endfire grating lobe exists. Assume that the $m=0$ and $m=-1$ modes are above cutoff (propagating) in the dielectric but that only the $m=0$ mode is propagating in the air. This means that $Y_{-1}$ is a capacitive susceptance since it is the admittance of a cutoff TM (or E) mode. If $t=0$, then, from (3), $Y_{-1}$ is infinite at a scan angle for which $\kappa_{-1}=0$, i.e., a main-beam angle for which the first grating lobe is at endfire. Since the modal transmission lines are in shunt at the aperture, $Y_{-1}=\infty$ places a short circuit across the $m=0$ line and causes complete reflection ($|\Gamma|=1$) and hence a null in the element pattern.

If the dielectric slab thickness $t>0$, then $Y_{-1}=\infty$ must be moved through the length of line $t$ and therefore a finite admittance is placed across the $m=0$ line by the $m=-1$ line at this scan angle. However, at a somewhat smaller scan angle, $Y_{-1}$ is a capacitive susceptance and this finite susceptance moved through the length of line $t$ will be infinite at the aperture plane and hence will short the $m=0$ line. Thus, for $t>0$, the null in the element pattern generally occurs for scan angles smaller than (closer to broadside than) the scan angle for which an endfire grating lobe occurs in the air. As $t\to\infty$ the scan angle for which the null occurs will approach that for which there exists an endfire grating lobe in the dielectric ($\kappa_{-1\varepsilon}=0$). One perceives, therefore, that for an arbitrary slab thickness there is a scan angle for which $|\Gamma|=1$, and that it lies between the angles for which there exists an endfire grating lobe in the air and in the dielectric, respectively. And now the resonance nature of the phenomenon becomes apparent, since one sees that the element pattern null is the result of a resonance of the $m=-1$ line between the dielectric-air interface and the aperture.

The precise location of the null may be determined from the equations in the Appendix. From (35), (36), and the requirement that the admittance looking into the $m=-1$ line at the aperture be infinite, one obtains

$$\varepsilon'|\kappa_{-1}|t \cot(\kappa_{-1\varepsilon}t) - \kappa_{-1\varepsilon}t = 0 \tag{12}$$

where $\kappa_{-1} = -j|\kappa_{-1}|$, since the $m=-1$ mode is cut off in the air. Condition (12) is identical with the transverse resonance condition for a surface wave on a dielectric slab located on

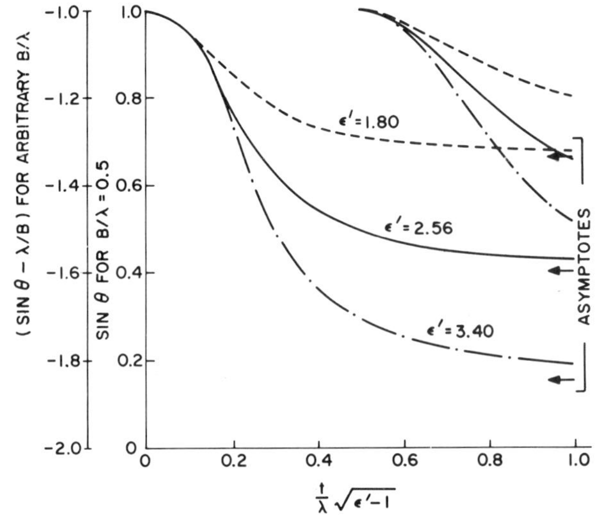


Fig. 7. Relation between $\theta$, $t/\lambda$, and $\varepsilon'$ for the location of nulls in the element pattern for the array in Fig. 2.

a perfectly conducting plane; the relevance of this surface wave to the element pattern behavior is discussed in Section V. From (4), one also has

$$(|\kappa_{-1}|t)^2 + (\kappa_{-1\varepsilon}t)^2 = (k_0t)^2(\varepsilon' - 1). \tag{13}$$

Using (4), (5), (12), and (13), one obtains the relation between $\theta$, $t/\lambda$, and $\varepsilon'$ for resonance of the $m=-1$ transmission line; this relation has been plotted in Fig. 7 in the form $\sin\theta$ versus $(t/\lambda)\sqrt{\varepsilon'-1}$ for three values of $\varepsilon'$ and for $B/\lambda=0.5$. The ordinate scale is also presented in an alternative form which is valid for arbitrary values of $B/\lambda$. As expected, the value of $\sin\theta$ for resonance migrates from that for which an endfire grating lobe exists in the air (for $t=0$) to that for which an endfire grating lobe exists in the dielectric (for $t\to\infty$). These latter values are $\sin\theta=0.66, 0.40, 0.15$ for $\varepsilon'=1.80, 2.56, 3.40$, respectively. Note that multiple resonances may occur in the $m=-1$ line. Thus, for $\varepsilon'=2.56$ and $(t/\lambda)\sqrt{\varepsilon'-1}=0.6$, one would observe two nulls in the element pattern, at $\sin\theta=0.46$ and at $\sin\theta=0.96$, whereas for no dielectric slab only the null at $\sin\theta=1.0$ occurs.

The value of $\varepsilon'=4.0$ is in a sense a "critical" value for $B/\lambda=0.5$ since for $\varepsilon'>4.0$ the resonance curves actually intersect the abscissa in Fig. 7. This means that for a suitable value of $t/\lambda$ the resonance occurs at $\sin\theta=0$ and that in such a case it would be impossible to match the array at broadside scan. Another interesting feature for $\varepsilon'>4.0$ is that the $m=+1$ mode is above cutoff in the dielectric for some scan angles and there is the possibility of resonances in the $m=+1$ transmission line as well as in the $m=-1$ transmission line. In general, each external mode (grating lobe) which is propagating in the dielectric but cut off in the air may have one or more scan angles at which it resonates.[4]

[4] If the abscissa of Fig. 7 were plotted as $(t/\lambda)(\varepsilon'-1)$ rather than $(t/\lambda)\sqrt{\varepsilon'-1}$, the curves for the first resonance for different values of $\varepsilon'$ would be much more closely bunched together for low abscissa values. This feature indicates that it is the product of $t/\lambda$ and the excess dielectric $(\varepsilon'-1)$ which essentially determines the angle of the null, for small deviations from the endfire grating lobe condition.

Since the equations from which Fig. 7 was plotted are independent of the internal portion of the array, the angle at which the element pattern has a null is not dependent on the admittance of the generators or on the angle of match of the array.

## IV. Element Patterns

This section is subdivided into three parts. In Part A, the qualitative behavior of the element pattern and reflection coefficient with scan angle is discussed as the dielectric slab thickness is varied. It is shown that a great deal of information can be obtained about the element pattern from a rather simple examination of the equivalent network. Part B presents computed element patterns, utilizing twenty-one external modes in the equivalent network. Part C shows that, for this array, three external modes permit an accurate computation of the element pattern, and therefore leads one to suspect that the performance of phased-array antennas may be computed quite accurately taking into account relatively few of the large number of modes necessary for an exact solution.

### *A. Qualitative Behavior*

With the aid of the equivalent network in Fig. 5 one may easily sketch roughly the shape of the reflection coefficient magnitude $|\Gamma|$ and the element pattern $g_e$ with scan angle as the thickness of the slab is varied. This has been done in Fig. 8 for element spacing between one-half and one wavelength for values of $\theta$ between zero and that for which an endfire grating lobe exists in the air.[5] It is assumed that the array has been matched for broadside scan. Then, from the reasoning at the beginning of Section III it is known that for $t=0$ the element pattern null (reflection coefficient peak) is at the angle for which the grating lobe is at endfire in air, and that as $t$ increases it moves toward the angle for which the grating lobe is at endfire in the dielectric. With this information, and the expectation of a more rapid variation in $|\Gamma|$ near the resonance than at broadside, one sketches the curves shown in Fig. 8.

Fig. 9 shows a similar estimate of the variation of $|\Gamma|$ and $g_e$ for element spacing less than or equal to a half-wavelength. In this case the resonance in the $m=-1$ line has the same effect upon the shape of the curves as $t$ is changed. Now, however, $|\Gamma|$ is always unity at $\sin\theta=1$. This feature of the $|\Gamma|$ curve is actually contributed by the $m=0$ line and results from the radiation conductance at the aperture plane in this line being infinite (for $t=0$) or being zero (for $t>0$).

Note that in Figs. 8 and 9 the widths of the peaks of $|\Gamma|$ have all been drawn equal, as have the widths of the nulls of $g_e$, since information concerning these features is not easily obtained from the equivalent network. As shown in Part B, these widths are actually not equal; except for this deficiency, the curves are reasonable estimates of $|\Gamma|$ and $g_e$.

[5] The equivalent network provides sufficient information to permit a determination of $|\Gamma|$ and $g_e$ when more than one beam radiates. However, since the present concern is with what one may obtain from the equivalent network by inspection, this is deferred until later.

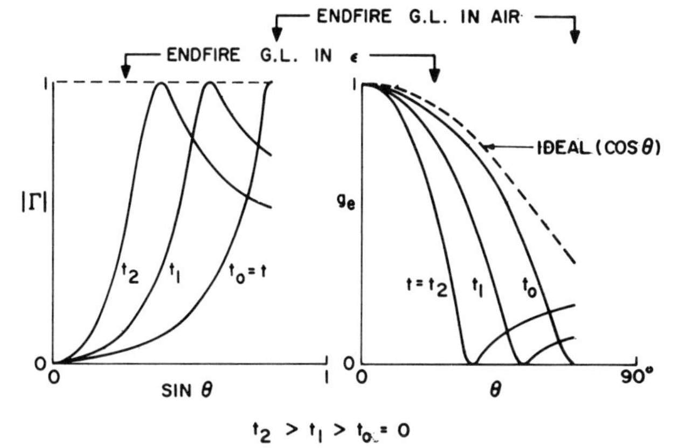


Fig. 8. Qualitative variation of $|\Gamma(\theta)|$ and $g_e(\theta)$ with scan angle $\theta$ for various slab thicknesses $t$ for the array in Fig. 2, with $0.5 < B/\lambda < 1.0$.

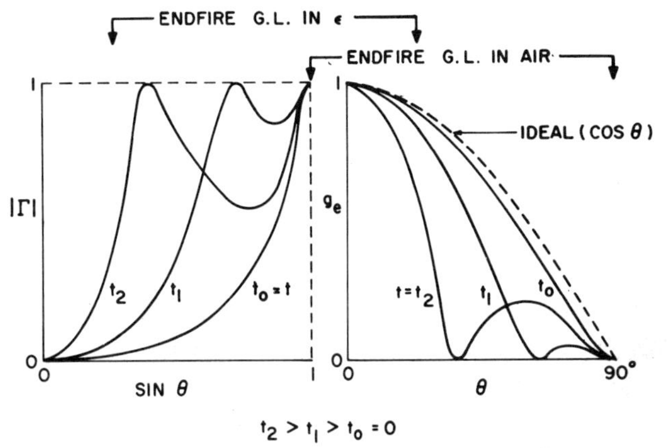


Fig. 9. Qualitative variation of $|\Gamma(\theta)|$ and $g_e(\theta)$ with scan angle $\theta$ for various slab thicknesses $t$ for the array in Fig. 2, with $B/\lambda \leq 0.5$.

### *B. Computations—Twenty-One External Modes*

Computations have been made of $\Gamma$ and $g_e$ with the aid of an IBM 360 computer utilizing the equivalent network in Fig. 5. Twenty-one modes ($m=-10$ to $m=+10$) have been taken into account outside the aperture; this is more than sufficient for computations involving the broadside matched array, and in fact far fewer modes would suffice.

Fig. 10 shows a plot of $\Gamma$ versus $\sin\theta$ for the reference array ($t/\lambda=0.2$, $B/\lambda=0.5$, $\varepsilon'=2.56$, $b/\lambda=0.3$); for comparison, a similar plot is shown for the reference array with no slab. With no slab, it is seen that the reflection coefficient is influenced appreciably by both the variation in conductance ($m=0$ line) and the variation in susceptance. However, for the array with slab the variation in $\Gamma$ is caused almost entirely by the variation in susceptance (principally from the $m=-1$ line), and the conductance only influences the curve for $\sin\theta > 0.92$ where its variation with scan becomes comparable with that of the susceptance.

The reflection coefficient magnitudes and consequent element patterns for various slab thicknesses are shown in Fig. 11 for $B/\lambda=0.5$. These curves exhibit the characteristics predicted previously, with the exception of the widths of the peaks in $|\Gamma|$. If one measures the half-width of the peak to the left of $|\Gamma|=1$, he finds that this half-width increases as $t$ increases from 0.00 to 0.20. (The same effect could be observed in $g_e$ if it were plotted versus $\sin\theta$.) However, as

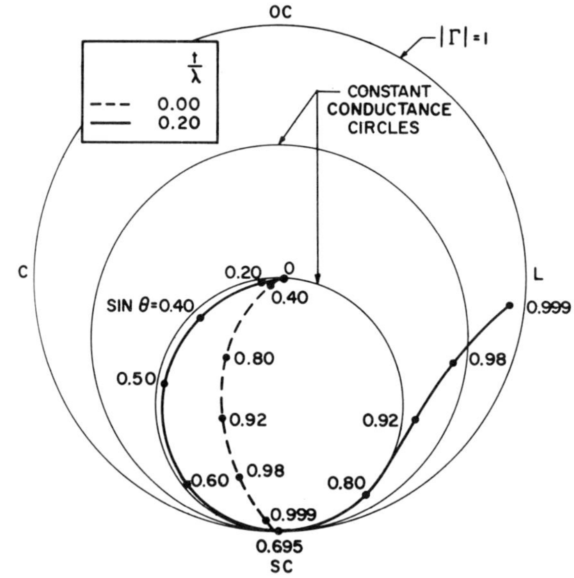


Fig. 10. Reflection chart plot as a function of scan angle $\theta$ for the array in Fig. 2, with $B/\lambda=0.5$, $b/\lambda=0.3$, $\varepsilon'=2.56$.



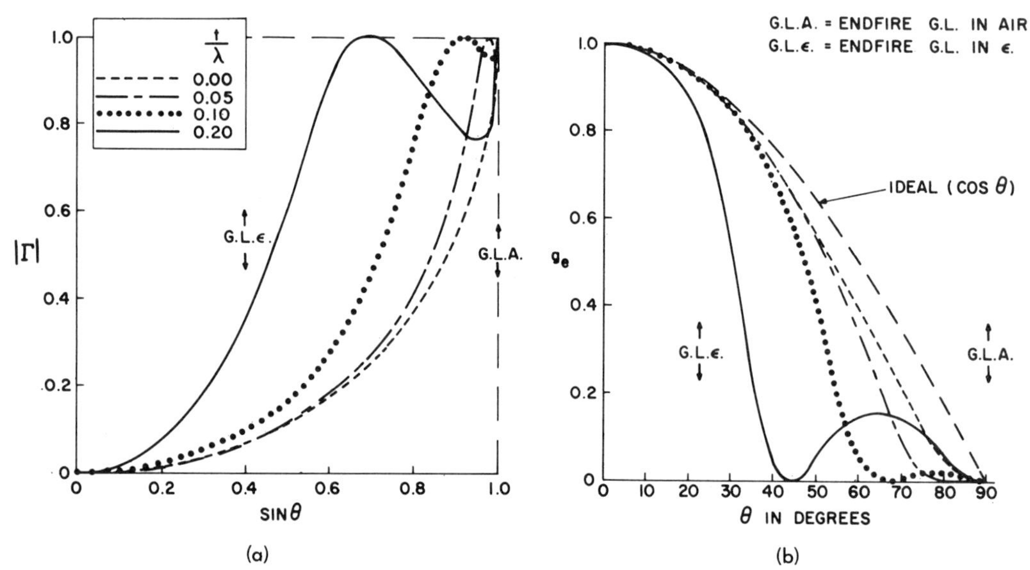


Fig. 11. Reflection coefficients and element patterns for various slab thicknesses $t$ for the array in Fig. 2, with $B/\lambda=0.5$, $\varepsilon'=2.56$, $b/\lambda=0.3$. (a) Reflection coefficient magnitudes. (b) Element patterns.

$t\to\infty$ and the angle at which $|\Gamma|=1$ approaches $\sin\theta=0.4$, the half-width must decrease since its value for a space completely filled with dielectric is comparable to that for a space completely filled with air. Thus, as $t$ increases from zero, the half-width of the $|\Gamma|$ peak and of the $g_e$ null increases and then decreases.

Fig. 12 shows $|\Gamma|$ and $g_e$ for various slab thicknesses for $B/\lambda=0.6$. The resonance has a similar effect on the shape of these curves for $\sin\theta<0.66$, where only one beam is radiating in air. For $\sin\theta>0.66$, where both the $m=0$ and $m=-1$ modes are above cutoff in air and where there can be no resonance in the $m=-1$ line, the curves are quite different from those in Fig. 11.

An example of an element pattern with a multiple (double) null is shown in Fig. 13. The double null is caused by a double resonance in the $m=-1$ transmission line, and the angles of the nulls are predictable from Fig. 7. Note that the width of the first null is clearly narrower than that in Fig. 11(b) for $t=0.20$, confirming that the widths become narrower again as $t\to\infty$.

### *C. Computations—Three External Modes*

Because the higher modes outside the aperture do not contribute much to the variation of $\Gamma$ with scan angle, it is possible to make a fairly accurate computation of $|\Gamma|$ and hence $g_e$ by taking into account the variation with scan of only two or three of the external modes. This has been done and the results are plotted in Fig. 14(a) and (b). The three-mode curves are derived from the simplified equivalent network in Fig. 6; $B_N$ and $T_N$ in this network are adjusted to give broadside match and are scan-invariant. The two-mode curves are obtained in a similar manner with the

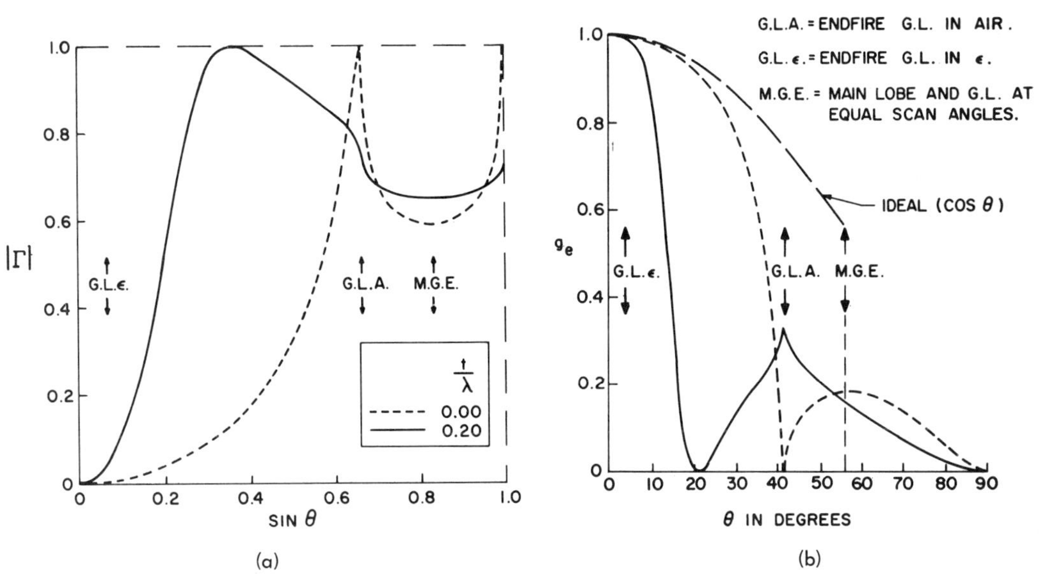


Fig. 12. Reflection coefficients and element patterns for two slab thicknesses $t$ for the array in Fig. 2, with $B/\lambda=0.6$, $\varepsilon'=2.56$, $b/\lambda=0.3$. Two radiating beams exist for $\sin\theta\geq 0.66$. (a) Reflection coefficient magnitudes. (b) Element patterns.

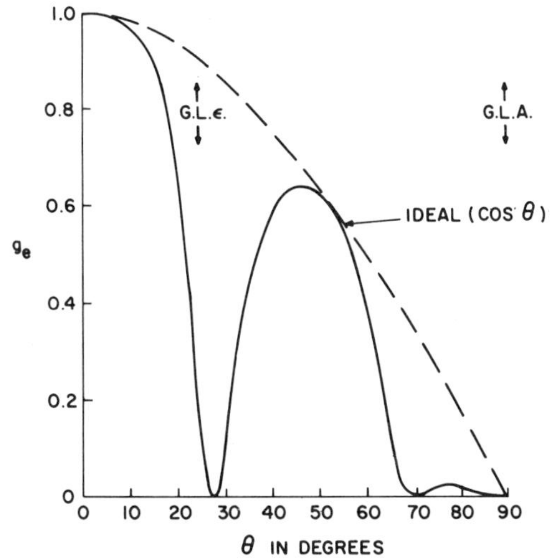


Fig. 13. Element pattern exhibiting multiple nulls for the array in Fig. 2, with a thick dielectric layer; $B/\lambda=0.5$, $t/\lambda=0.5$, $\varepsilon'=2.56$, $b/\lambda=0.3$.

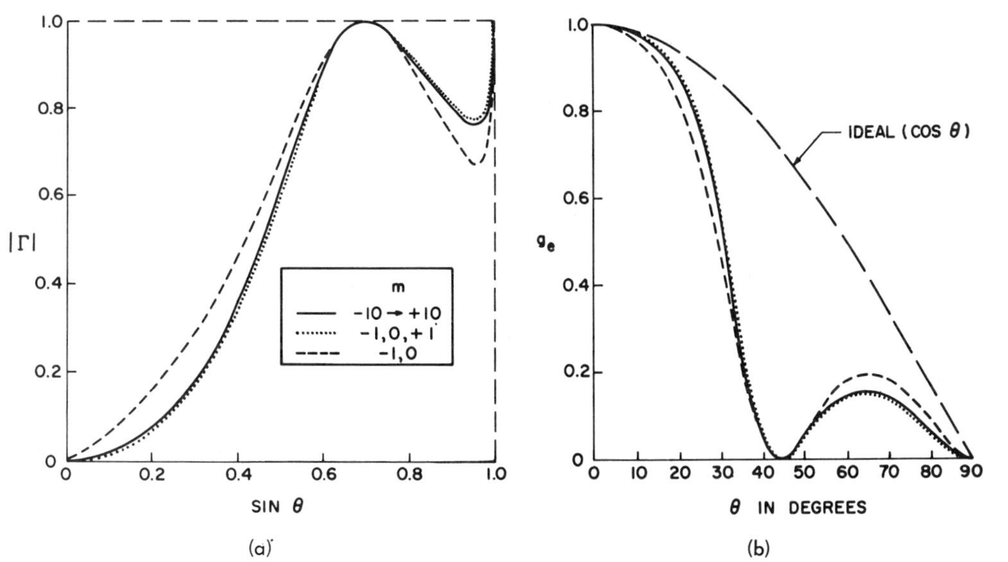


Fig. 14. Reflection coefficient magnitudes and element patterns computed under different approximations; $m$ represents the number of external modes for which the variation with scan angle is taken into account. $B/\lambda=0.5$, $t/\lambda=0.2$, $\varepsilon'=2.56$, $b/\lambda=0.3$. (a) Reflection coefficient magnitudes. (b) Element patterns.

$m = +1$ transmission line deleted. From Fig. 14 it is clear that only three external modes are important in determining the performance of the array, and that all higher modes ($B_{hm}$ in Fig. 4) contribute a susceptance which is essentially constant with scan. The agreement between the element patterns is seen to be better than that between the reflection coefficient curves because the cos $\theta$ factor suppresses differences.

## V. Relation to Surface and Leaky Waves

Although it is generally appreciated that the element pattern nulls discussed above are in some way associated with a surface wave or a surface-wave-like field which may occur on the array face, the nature of this wave and its role in the phenomenon have not been well understood. The intention of this section is to clarify the character of this surface wave, and to introduce into this dialogue another type of guided wave—the leaky wave—in order to show its influence on the properties of these element pattern nulls.

Prior to this section, the analysis of the chosen array has been carried out entirely in terms of modes of the structure propagating in the $z$-direction (transversely to the structure); this choice of modal representation was convenient for solving the driven problem (the response of the structure to the imposed phased excitation). In this section, we examine the field behavior in terms of modes propagating *along* the array face, that is, in terms of guided-wave modes of the structure propagating in the $x$-direction. These modes can be determined by solving the transverse resonance expression

$$Y + Y_f = 0 \tag{14}$$

for the normalized wavenumber $k_{x0}/k_0$. The symbols are defined in Section II, where it is also observed from (6) that (14) determines the poles of the reflection coefficient $\Gamma$. Only those wavenumbers which satisfy (14) represent guided-wave modes of the array; these modes are also *free resonances* of the array. In general, wavenumber solutions to the transverse resonance relation are of two kinds, real and complex. Real wavenumbers represent surface waves of the structure, propagating unattenuated along the structure with a velocity less than that of light, and decaying transversely away from the array face. Complex wavenumbers represent complex waves, of course; the pertinent waves here are leaky waves [16], which are guided along the array face but which also carry power away from it. These waves occur on periodic traveling-wave structures when the element spacing is of the order of $\lambda/2$ or larger [17].

We will see that the only true guided waves which can be supported by the actual array are of the leaky-wave type, and that guided surface waves on the actual array are not possible. There is, however, a surface-wave-like field present on the array face at exactly the null condition, but it corresponds to a surface wave on a modified structure, as will be shown.

### A. Relation to Surface Waves

It is well known that the reflection coefficient magnitude $|\Gamma|$ is equal to unity at the particular scan angle correspond-

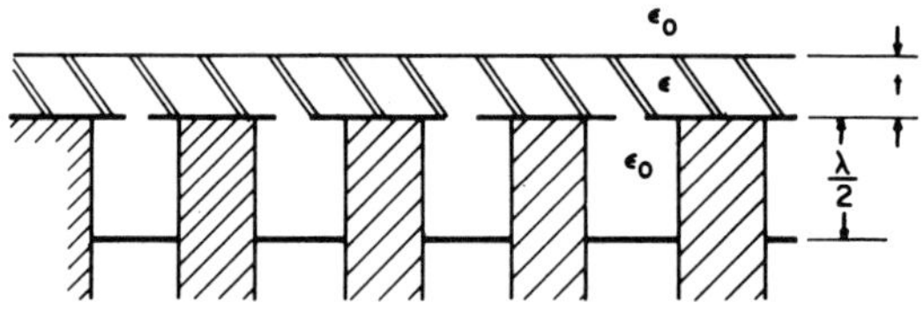


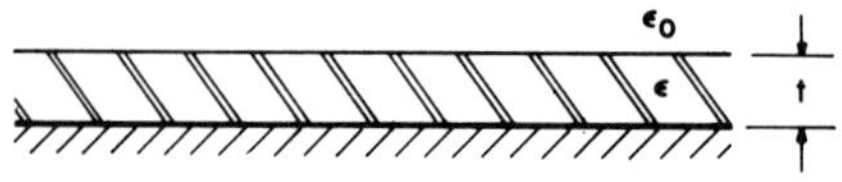


Fig. 15. Modified structure produced by inserting short-circuiting plates into the phased array at the perfect-reflection condition $|\Gamma| = 1$. (a) Dielectric-loaded corrugated structure formed by plates $\lambda/2$ below the aperture plane. (b) Dielectric slab on a perfectly conducting plane formed by closing off the apertures.

ing to the element pattern null. Thus, at this precise condition all power from the sources is reflected back from the apertures, and the dominant-mode input admittance $Y$ in a typical feed guide becomes infinite at an appropriate reference plane. According to the network of Fig. 5, this reference plane occurs at the aperture plane, so that the dominant-mode admittance is infinite there or at any reference plane $m\lambda/2$ away from it in the feed guide. At the aperture plane, higher modes are present; however, at a plane $\lambda/2$ underneath the aperture plane it is safe to assume that only the dominant mode exists, so that a short-circuiting plate may be placed there without disturbing the field. The resulting structure is shown in Fig. 15(a) to be a dielectric layer placed on a type of corrugated surface.

The approximate analysis presented herein of the array of Fig. 2 has assumed that a constant field, independent of scan angle, exists in the aperture. This field form has been given as

$$\boldsymbol{E}_{ap} = C\boldsymbol{x}_0 \tag{15}$$

so that the magnitude of the dominant-mode voltage in the aperture plane becomes

$$|V_{f0}| = |C| \frac{d}{\sqrt{b}} \tag{16}$$

as in (29) Since this dominant-mode voltage must be zero in the aperture plane because the dominant-mode admittance is infinite there, it follows that $C$ must be zero. Hence, in accordance with this approximation, the total electric field in the aperture would vanish identically and a short-circuiting plate may be placed there, actually closing off the slots. Note that if a more accurate, but still scan-invariant, functional form for the field were employed in place of (15), this same result would be obtained. The structure then becomes that shown in Fig. 15(b), a dielectric slab located on a perfectly conducting plane.

Due to the original source excitation, however, a forced field exists on this modified structure which corresponds to that of the first higher mode in the dielectric region, which can propagate in the dielectric but not in the air. It therefore decays away from the air-dielectric interface and it has a phase variation in the $x$-direction along the array face which is slow compared to light; its field form is therefore identical with that of a TM (or $E$) surface wave propagating along the dielectric slab. Relation (12) was given in Section III as the condition on the first higher mode which corresponds to $|\Gamma|=1$; this relation is consistent with the above considerations because it is also the transverse resonance condition for a TM surface wave on a dielectric slab placed on a perfectly conducting plane. The above discussion is also consistent with that presented by Oliner and Malech [10].

Because of the approximation referred to above, the continuity of magnetic field across the aperture at $|\Gamma|=1$ is not satisfied. The magnetic field is not zero at the aperture plane and its continuity across it forms the tie between the feed guide and the exterior region. On the feed guide side of the aperture a constant field is present, the TEM mode feeding the aperture. On the exterior side of the aperture the first higher-mode field exists, and it possesses a phase variation along the aperture surface. Hence, the two magnetic fields do not match each other, indicating an inconsistency. The solution is therefore incomplete, and higher-mode contributions must be present in the aperture. The difficulty lies in the assumption of an aperture field consisting of only one scan-independent modal component. It may be readily shown that another component is actually produced by the scan process, and that when this added component is included, the difficulty disappears. As a result, the form shown in Fig. 15(a) must be employed, rather than that in Fig. 15(b), and the location of the short-circuiting plane is no longer exactly $\lambda/2$ away from the aperture plane. For narrow slots this shift in the location of the short-circuiting plane is negligibly small, but it increases for larger slots. This statement is in agreement with observations made by Wu and Galindo [18], who indicate that the shift is very small for aperture width to periodic unit-cell width ratios of less than 0.1 or so, but that greater departures from $\lambda/2$ occur for very wide apertures.

The above discussion has brought out that at the particular scan angle corresponding *exactly* to the element pattern null one may place short-circuiting plates at an appropriate location in the feed guides, as shown in Fig. 15(a), except that their distance from the aperture plane will differ from $\lambda/2$ for wider apertures. Under these conditions, the *forced field* existing on the array face is identical with a *surface-wave mode of the dielectric-loaded corrugated structure* shown in Fig. 15(a). This surface wave is *not* a mode of the original array, which includes the finite conductances of the generators below the aperture plane. It is a mode of the *modified structure* which was produced by introducing short-circuiting plates at an appropriate location below the aperture plane. It must also be stressed that this modified structure has meaning only at exactly the scan angle corresponding to the element pattern null. For any deviation from this angle $|\Gamma|$ is no longer unity, and the perfectly conducting plates can no longer be inserted. The modified structure has then lost its significance.

The surface wave which is associated with the element pattern null thus corresponds to a forced field associated with the condition of perfect reflection. It has no meaning away from this condition, and it is a mode not of the actual array but of a modified version of it. If this modified version can be deduced without numerically solving the complete problem, then the location in angle of the element pattern null can be found in a rapid fashion. For narrow apertures the structure of Fig. 15(b) applies, but in more general situations the usefulness of this approach is dubious.

An additional point of interest may be indicated in connection with the surface wave on the modified (corrugated) structure. In general, on a lossless periodic traveling-wave structure with internal reactive terminations and with a period greater than a half-wavelength, a surface-wave solution of the transverse resonance relation cannot be found because one or more of the space harmonics is always radiating. Under certain special conditions, the radiating mode amplitude is zero, so that the total field becomes purely bound and the result is a surface wave. It is well known that this condition is satisfied at broadside radiation for many structures, but satisfaction off broadside is relatively rare. The surface wave which occurs on the modified phased-array face at the $|\Gamma|=1$ condition is of that type, however. Another example has recently been reported [19], associated with a strip-loaded dielectric layer on a perfectly conducting ground plane.

The preceding discussion has demonstrated that the surface wave being considered is a mode of the modified array, but not of the actual, complete array. The question remains as to whether or not the actual array, with the feed guide propagating the dominant mode, can support a true surface wave. This question may be resolved by an examination of the transverse resonance condition (14), which states that the sum of the input admittances looking both ways away from a given reference plane is zero. Let us choose a reference plane located somewhere in the feed waveguide. If a true surface wave exists on the array face, the input admittance looking into the aperture plane must be reactive; on the other hand, the input admittance looking back into the feed guide must have a conductive component. Since these two admittances can never sum to zero, we conclude that *no* true surface-wave solutions are possible on the actual array, although complex solutions may be admitted.

### *B. Relation to Leaky Waves*

It has just been shown that true surface waves cannot be supported by the actual array, but that complex solutions of the leaky-wave type may be possible. A simple physical argument involving the element pattern may also be adduced to show that the presence of a leaky wave is plausible. Let us recall that the element pattern is obtained when only one element of the array is excited and all other elements are passively terminated in their generator admittances. This situation is illustrated in Fig. 16. The single excited element radiates a space wave (in all directions) and it also excites a leaky wave (we shall assume) on the neighboring portions of the passively terminated array. The array periodic spacings relative to wavelength are such that the radiated beam associated with the leaky wave proceeds at some angle in the backward direction, as shown in Fig. 16 for the right-

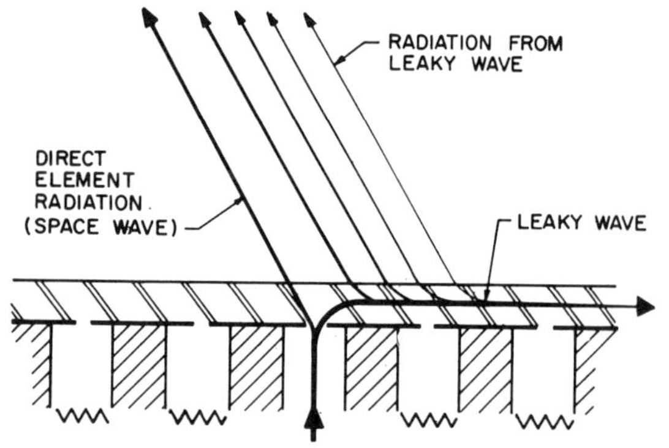


Fig. 16. Sketch showing interference between direct element radiation and that from leaky wave to produce element pattern null on left side of pattern.

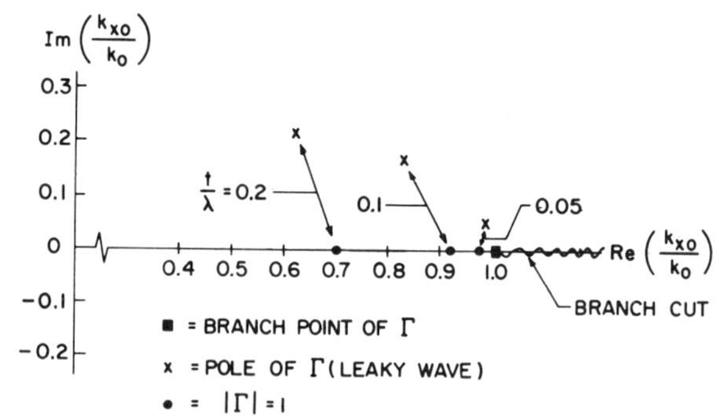


Fig. 17. Leaky-wave poles of the reflection coefficient $\Gamma$ in the complex $k_{x0}/k_0$-plane, $B/\lambda=0.5$, $\varepsilon'=2.56$. Slab thickness $t$ is a parameter.

hand portion of the array. This radiated wave then interferes with the portion of the space wave radiating in the same direction in such a way as to produce the expected null in the element radiation pattern. Precisely the same thing occurs on the other side of the array so that a symmetrical element pattern is produced. A surface wave along the array could not serve to produce the cancellation required in achieving the radiation null; the radiating leaky wave offers a simple and physically satisfying mechanism.

The above discussion assumes, but does not prove, that a leaky wave is possible. A confirmation of its existence has been established, however, by the use of a perturbation procedure; a real solution of the transverse resonance condition is known for the case for which the aperture width $d=0$ (in fact, it is plotted in Fig. 7), and a complex solution is shown to exist for small finite values of $d$ by employing this procedure. In addition, the actual transverse resonance expression with twenty-one external modes included was solved numerically using Newton's method, and the numerical values of the leaky wave's phase and attenuation constants were explicitly exhibited.

It was found that the complex solution for the leaky wave possesses a real value which is very close to the value of $k_{x0}/k_0 \equiv \sin\theta$ corresponding to the condition $|\Gamma|=1$; this closeness has a physical implication which is discussed later. The complex values for the leaky wave for different values of dielectric slab thickness $t$ are shown in Fig. 17, which presents the complex $k_{x0}/k_0$-plane; on this plane, the real axis between 0 and 1 represents $\sin\theta$ for scanning from broadside to endfire, respectively. The positions of $|\Gamma|=1$ are labeled on this axis for several values of slab thickness, as are the locations of the nearby leaky waves (which are also the poles of the reflection coefficient $\Gamma$). At $k_{x0}/k_0=1$ for $B/\lambda=0.5$ there is a double branch point of $\Gamma$ which arises from the two grating lobes which then exist at endfire ($\kappa_0=0$, $\kappa_{-1}=0$). The propagation constants $\kappa_m$ corresponding to the leaky-wave poles, and also to the points for $\Gamma$ on the Re $(k_{x0}/k_0)$-axis in Fig. 17 for the driven case, are characterized by

$$\operatorname{Re}(\kappa_0) > 0, \qquad \operatorname{Im}(\kappa_m) < 0, \qquad m \neq 0, \tag{17}$$

that is, by an outward-traveling dominant-mode wave and by decaying higher-mode waves. Note that the Re $(k_{x0}/k_0)$ for the pole is about the same as that for the point $|\Gamma|=1$, i.e., that *the leaky wave propagates at about the same angle from broadside as the angle for which* $|\Gamma|=1$.

The significance of the leaky-wave pole near $|\Gamma|=1$ is that it essentially determines the *width or the "Q" of the peak in* $\Gamma$. One can visualize this by thinking of $|\Gamma|$ as the elevation of the terrain in the complex $k_{x0}/k_0$-plane. A mountain peak exists at the position of the leaky-wave pole, the terrain sloping off to $|\Gamma|=1$ at a point nearby on the real axis. Moving along the real axis from broadside to endfire, one must traverse a shoulder of the mountain. The width of the shoulder is controlled almost entirely by the location of the mountain peak since (except for limiting values of $t$) there are no other significant terrain features (poles or zeros) nearby. If one defines the $Q$ of the peak in $|\Gamma|$ as

$$Q = \frac{P}{W} \tag{18}$$

where $P$ is the value of $k_{x0}/k_0$ for $|\Gamma|=1$, and $W$ is the width of the peak at $|\Gamma|=q$, $q\leq 1$, then for $B/\lambda=0.5$, $\varepsilon'=2.56$, $t/\lambda=0.2$, $q=0.9$, one calculates from Fig. 11 that $Q=3.33$. This $Q$ may alternatively be computed from the leaky-wave pole by expanding the denominator of $\Gamma$ in (6) in a Taylor series about the leaky-wave pole and keeping only the first two (constant and linear) terms, and approximating the numerator of $\Gamma$ as $2Y_f$ in view of (6) and (14). This procedure yields

$$Q \approx \frac{\operatorname{Re}\left(\dfrac{k_{x0}}{k_0}\right)_p}{2\operatorname{Im}\left(\dfrac{k_{x0}}{k_0}\right)_p}\sqrt{\frac{q^2}{1-q^2}} \tag{19}$$

where $(k_{x0}/k_0)_p$ refers to the location of the leaky-wave pole and it has been assumed that Re $(k_{x0}/k_0)_p \approx k_{x0}/k_0$ for $|\Gamma|=1$. Upon evaluating (19) one finds $Q=2.91$, a value which agrees quite well with that computed from the actual $|\Gamma|$ curve.

This agreement between the $Q$ computed from the leaky-wave pole and that obtained from the actual curve of $|\Gamma|$ versus $\sin\theta$ is illustrated graphically in Fig. 18. The actual $|\Gamma|$ curve is reproduced from Fig. 11, and it corresponds to an accurate calculation involving twenty-one external modes. The dotted curve which appears only near the $|\Gamma|=1$

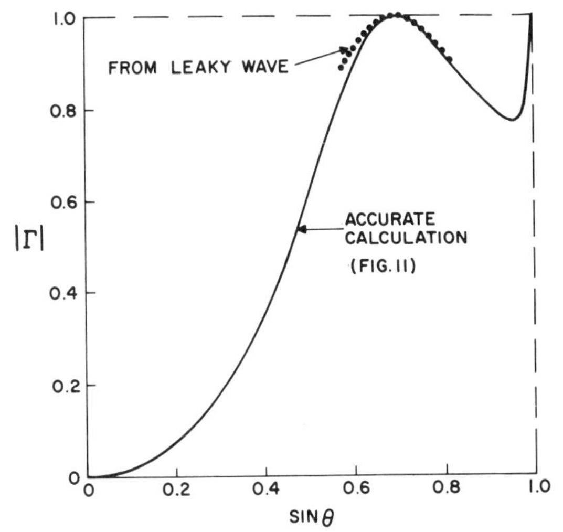


Fig. 18. Comparison of region around peak ($|\Gamma|=1$) between accurate calculation and results obtained from the leaky-wave pole via (19).

point is that obtained from the leaky-wave pole [see (19)]. Thus, whereas the earlier discussion regarding the surface wave on the modified structure relates only to the precise location of the $|\Gamma|=1$ point, and yields no information as to the neighborhood of that point, the leaky-wave result offers a measure of the width of the peak in $|\Gamma|$.

The above information about the peak of $|\Gamma|$ permits an identification of the phenomenon as a relative of the famous Wood's anomalies on optical reflection gratings [20], [21]. These were reported by Wood in 1902 as rapid variations in the intensity of various grating lobes diffracted from gratings in certain narrow frequency bands or ranges of incidence angle. They were termed anomalies by Wood since they could not then be explained by ordinary grating theory. It was only much later that the theory was developed sufficiently to explain all the anomalous results observed by Wood.

It seems that history has repeated itself again but with a slight twist. Instead of optical *reflection* gratings, anomalies have been observed on microwave *transmission* gratings. In both situations, however, the explanation of the phenomenon is the same: a nearness of the incident wavenumber for scattering to that for which there exists a free resonance (leaky wave) of the structure, and the consequent reaction of this resonance on the various diffracted waves. A detailed explanation of Wood's anomalies in these terms has been presented by Hessel and Oliner [20].

## VI. Conclusions and Discussion

The purposes of this paper have been to investigate the basic cause of element pattern nulls in phased arrays, and to explore the relationship between these nulls and possible guided waves which may be supported by the array face.

There are two types of guided waves which are of concern here, surface waves and leaky waves; the properties of these waves are summarized in Section V. It is shown that the *actual* array *cannot* support true surface waves, but that leaky waves can be guided by it. The surface wave of which everyone speaks is seen to correspond to a driven surface-wave-like field distribution which appears on the array face at precisely the null condition, and which is indistinguishable in its properties from a true surface wave on a *modified* (corrugated) structure, not on the actual array. This surface wave has meaning *only* at *precisely* the null condition, and not elsewhere. It is not an independently excitable wave but is a driven field distribution which is merely a concomitant of the total reflection produced at the null condition. If one can independently ascertain the properties of this surface-wave-like field, however, then the location in angle of the element pattern null can be obtained rapidly without solving the complete problem.

The leaky wave, which is a true mode of the actual array, is shown to be useful in a simple direct explanation (see Fig. 16) of the element pattern null. When the complete array is driven, the rôle of the leaky wave in influencing the radiation null is akin to the relation between free and forced resonances. The leaky wave is the *free resonance* which determines the *width* of the element pattern null. These aspects are examined in Section V.

The cause of element pattern nulls is identified herein with a *forced resonance of a higher mode* for structures n which the aperture field shape is essentially invariant with scan. A specific structure (an array of narrow slots covered by a dielectric slab) was chosen for detailed investigation as a typical example of a whole class of such arrays, and the behavior of this structure is illustrative of that for the wider class. A simple equivalent network was derived from which reflection coefficients and element patterns were computed. Although all of the curves shown pertain to a lossless structure, the effect of dielectric loss like that of rexolite was computed and was found to have a negligible effect on the performance, even at resonance.

An experimental confirmation of this theory as applied to *H*-plane scan resonances has been obtained from a phased-array simulator measurement. The array is a rectangular lattice of ordinary capacitive slots in a conducting plane. The slots are fed from below by $TE_{10}$ modes in rectangular waveguides, and are covered above by a dielectric slab. A simple $1\frac{1}{2}$-element simulator is used which simulates a scan angle of about 40° in the *H*-plane; the slab thickness and permittivity are chosen for resonance at this angle [10]. In the measurement, frequency is swept over a narrow band and $\Gamma$ and $T$ is measured. Good agreement has been obtained between the measured and predicted locations of the resonance. It is planned to publish these experimental results at a later date.

In the design of a practical phased-array element, where impedance matching over a range of angles and frequencies is of prime concern, it is usually best to design to avoid the resonances. The way to guarantee this is to choose an element whose external region is free of dielectric or other kinds of "slow-wave" loading and whose higher internal modes are well below cutoff. When an element is designed which deviates significantly from these criteria, whether for better wide-angle matching [22], [23] or for other reasons, the possibility of resonances must be considered and appropriate measurements made.

## Appendix

### Derivation of Input Admittance of Array

As indicated in Section II, the input admittance $Y$ at the aperture plane of the array in Fig. 2 can be derived from the unit cell in Fig. 3, assuming that the electric field in the aperture is known for all scan angles $\theta$. This is done below for the case $b/\lambda < 0.5$ so that only the TEM mode propagates in the feed waveguide.

Assuming a time dependence of $\exp(j\omega t)$, which is suppressed in the following, the electric vector mode functions for the unit-cell waveguide are given by [24]

$$\boldsymbol{e}_{um} = \boldsymbol{x}_0 \frac{1}{\sqrt{B}} \exp(jk_{xm}x) \tag{20}$$

$$k_{xm} = k_0 \sin\theta + \frac{2\pi m}{B} \tag{21}$$

where $m=0, \pm 1, \pm 2, \cdots$, is the mode index, $k_{xm}$ is the $x$-component of the $m$th mode wavenumber, $k_0 = \omega\sqrt{\mu_0\varepsilon_0}$ is the free-space wavenumber, and $\boldsymbol{x}_0$ is a unit vector in the $x$-direction. The TEM vector mode function for the feed waveguide is

$$\boldsymbol{e}_{f0} = \boldsymbol{x}_0 \frac{1}{\sqrt{b}} \tag{22}$$

where $e_{fn}$, $n \neq 0$, is not given since it will not be needed.

The modal tangential electric and magnetic fields in the unit-cell waveguide are representable in terms of the vector mode functions, and the modal voltages $V_{um}$ and modal currents $I_{um}$ by

$$\boldsymbol{E}_{um} = V_{um}(z)\boldsymbol{e}_{um}(x), \qquad \boldsymbol{H}_{um} = I_{um}(z)[\boldsymbol{z}_0 x \boldsymbol{e}_{um}(x)] \tag{23}$$

where $\boldsymbol{z}_0$ is a unit vector in the $z$-direction. Likewise, the tangential electric and magnetic fields of each mode in the feed waveguide may be represented in terms of the modal voltages $V_{fn}$ and modal currents $I_{fn}$ by

$$\boldsymbol{E}_{fn} = V_{fn}(z)\boldsymbol{e}_{fn}(x), \qquad \boldsymbol{H}_{fn} = I_{fn}(z)[\boldsymbol{z}_0 x \boldsymbol{e}_{fn}(x)]. \tag{24}$$

The total tangential electric and magnetic fields in each waveguide are then given by

$$\boldsymbol{E}_u = \sum_{m=-\infty}^{\infty} \boldsymbol{E}_{um}, \qquad \boldsymbol{H}_u = \sum_{m=-\infty}^{\infty} \boldsymbol{H}_{um} \tag{25}$$

$$\boldsymbol{E}_f = \sum_{n=0}^{\infty} \boldsymbol{E}_{fn}, \qquad \boldsymbol{H}_f = \sum_{n=0}^{\infty} \boldsymbol{H}_{fn}. \tag{26}$$

Assuming that the aperture electric field is

$$\boldsymbol{E}_{ap} = \boldsymbol{x}_0 C \tag{27}$$

as discussed in Section II, one may calculate the modal voltage magnitudes at the aperture plane in the unit-cell waveguide as

$$|V_{um}| = \left| \iint_{ap} \boldsymbol{E}_{ap} \cdot \boldsymbol{e}_{um}^{*} da_{ap} \right| = \left| \frac{2C}{\sqrt{B}} \frac{\sin\left(k_{xm}\frac{d}{2}\right)}{k_{xm}\frac{d}{2}} \right| \tag{28}$$

where * means complex conjugate and the integration is performed only over the aperture since the tangential $E$-field is zero at other points in the aperture plane. In a similar manner the TEM modal voltage magnitude at the aperture plane in the feed waveguide is computed as

$$|V_{f0}| = \frac{d}{\sqrt{b}} |C| \tag{29}$$

where $|V_{fn}|$ has not been determined for $n \neq 0$ since it will not be needed.

If now the $z$-component of the Poynting vector in each guide is integrated over the surfaces $S_U$ and $S_F$ in Fig. 3 and continuity of the tangential $E$- and $H$-fields is required across the *aperture* while the volume enclosed by each surface shrinks to zero (so that in the limit both surfaces lie in the aperture plane), the following relation is obtained:

$$V_{f0}I_{f0}^{*} = -\sum_{n=1}^{\infty} V_{fn}I_{fn}^{*} + \sum_{m=-\infty}^{\infty} V_m I_m^{*} \tag{30}$$

where the currents are defined to be positive in the positive $z$-direction. Equation (30) relates the complex power flowing into the aperture in the feed guide TEM mode to that flowing away from the aperture in all the other modes in both guides (at the plane $z=0$). The modal voltages and currents of these modes are related to each other by

$$I_{um} = V_{um}Y_{um}, \qquad -I_{fn} = V_{fn}Y_{fn}, \qquad n > 0 \tag{31}$$

where $Y_{um}$ is the admittance seen by the $m$th mode in the unit cell in the positive $z$-direction, and $Y_{fn}$ is the characteristic admittance of the $n$th mode in the feed guide. For $n=0$ (TEM) mode in the feed guide

$$I_{f0} = V_{f0}Y \tag{32}$$

where $Y$ is the admittance seen by the TEM mode in the feed guide in the positive $z$-direction; i.e., it is the *active admittance* of the array at the aperture plane.

Upon substituting (31) and (32) into (30) and solving for $Y$, one finds

$$Y = \sum_{m=-\infty}^{\infty} \frac{|V_{um}|^2}{|V_{f0}|^2} Y_{um} + \sum_{n=1}^{\infty} \frac{|V_{fn}|^2}{|V_{f0}|^2} Y_{fn}. \tag{33}$$

The second sum in (33) represents the susceptance resulting from stored power in the cutoff modes on the inside of the aperture and can be approximated for a small aperture as one-half the susceptance of a symmetrical capacitive iris of opening $d$ in a parallel-plate waveguide of width $b$. Hence

$$\sum_{n=1}^{\infty} \frac{|V_{fn}|^2}{|V_{f0}|^2} Y_{fn} = jB_{in} \approx j\frac{2b}{\lambda} Y_f \ln \csc\left(\frac{\pi d}{2b}\right) \tag{34}$$

where $B_{in}$ stands for internal susceptance and $\lambda$ is the free-space wavelength, $\lambda = 2\pi/k_0$.

Now, since the dielectric in the unit-cell waveguide is of thickness $t$, and since air fills the unit cell for $z > t$, $Y_{um}$ is given by

$$Y_{um} = Y_{m\varepsilon} \frac{jY_{m\varepsilon} + Y_m \cot(\kappa_{m\varepsilon}t)}{Y_{m\varepsilon}\cot(\kappa_{m\varepsilon}t) + jY_m} \tag{35}$$

where $Y_{m\varepsilon}$ and $Y_m$ are the modal admittances for the $m$th mode in the dielectric and in air, respectively, and are given by

$$Y_{m\varepsilon} = \frac{\omega\varepsilon'\varepsilon_0}{\kappa_{m\varepsilon}}, \qquad Y_m = \frac{\omega\varepsilon_0}{\kappa_m} \tag{36}$$

where

$$\kappa_{m\varepsilon} = \sqrt{\varepsilon' k_0^2 - k_{xm}^2}, \qquad \kappa_m = \sqrt{k_0^2 - k_{xm}^2} \tag{37}$$

are the components of the $m$th mode wavenumber in the $z$-direction.

Insertion of (28), (29), (34), and (35) into (33) yields the input admittance of the array, namely

$$Y = \sum_{m=-\infty}^{\infty} \frac{b}{B}\left[\frac{\sin\left(k_{xm}\frac{d}{2}\right)}{k_{xm}\frac{d}{2}}\right]^2 Y_{m\varepsilon}\frac{jY_{m\varepsilon} + Y_m\cot(\kappa_{m\varepsilon}t)}{Y_{m\varepsilon}\cot(\kappa_{m\varepsilon}t) + Y_m} \tag{38}$$

$$+ jB_{\text{in}}.$$

ACKNOWLEDGMENT

The authors have benefited from interesting discussions with H. A. Wheeler and P. W. Hannan of Wheeler Laboratories, and with B. L. Diamond and C. Blake of the M.I.T. Lincoln Laboratory.

REFERENCES

[1] J. L. Allen, "On surface-wave coupling between elements of large arrays," *IEEE Trans. Antennas and Propagation (Communications)*, vol. AP-13, pp. 638–639, July 1965.
[2] L. W. Lechtreck, "Effects of coupling accumulation in antenna arrays," *IEEE Trans. Antennas and Propagation*, vol. AP-16, pp. 31–37, January 1968.
[3] ——, "Cumulative coupling in antenna arrays," *IEEE G-AP Internat'l Symp. Digest*, pp. 144–149, August–September 1965.
[4] R. H. T. Bates, "Mode theory approach to arrays," *IEEE Trans. Antennas and Propagation (Communications)*, vol. AP-13, pp. 321–322, March 1965.
[5] G. F. Farrell, Jr., and D. H. Kuhn, "Mutual coupling effects of triangular-grid arrays by modal analysis," *IEEE Trans. Antennas and Propagation (Communications)*, vol. AP-14, pp. 652–654, September 1966.
[6] B. L. Diamond, "Resonance phenomena in waveguide arrays," *IEEE G-AP Internat'l Symp. Digest*, pp. 110–115, October 1967.
[7] R. F. Frazita, "Surface-wave behavior of a phased array analyzed by the grating-lobe series," *IEEE Trans. Antennas and Propagation (Communications)*, vol. AP-15, pp. 823–824, November 1967.
[8] H. A. Wheeler, "The grating-lobe series for the impedance variation in a planar phased-array antenna," *IEEE Trans. Antennas and Propagation*, vol. AP-14, pp. 707–714, November 1966.
[9] P. W. Hannan, "Discovery of an array surface wave in a simulator," *IEEE Trans. Antennas and Propagation (Communications)*, vol. AP-15, pp. 574–576, July 1967.
[10] A. A. Oliner and R. G. Malech, "Speculations on the role of surface waves" in *Microwave Scanning Antennas*, vol. 2, R. C. Hansen, Ed. New York: Academic Press, 1966, pp. 308–322.
[11] A. Hessel, G. H. Knittel, and A. A. Oliner, "On the theory of resonances in phased-array antennas," *URSI Fall Meeting Digest* (Ann Arbor, Mich.), p. 62, October 1967.
[12] A. A. Oliner and R. G. Malech, "Periodic structure approach," in *Microwave Scanning Antennas.*, vol. 2, R. C. Hansen, Ed. New York: Academic Press, 1966, pp. 216–239, 256–268.
[13] L. I. Parad, "The input admittance to a slotted array with or without a dielectric sheet," *IEEE Trans. Antennas and Propagation (Communications)*, vol. AP-15, pp. 302–304, March 1967.
[14] A. A. Oliner and R. G. Malech, "Pattern of an element in a passively terminated array environment," in *Microwave Scanning Antennas*, vol. 2, R. C. Hansen, Ed. New York: Academic Press, 1966, pp. 301–308.
[15] P. W. Hannan, "The element-gain paradox for a phased-array antenna," *IEEE Trans. Antennas and Propagation*, vol. AP-12, pp. 423–433, July 1964.
[16] T. Tamir and A. A. Oliner, "Guided complex waves, Part 1. Fields at an interface," *Proc. IEE* (London), vol. 110, pp. 310–324, February 1963.
[17] A. A. Oliner and A. Hessel, "Guided waves on sinusoidally-modulated reactance surfaces," *IEEE Trans. Antennas and Propagation*, vol. AP-7, pp. S201–S208, December 1959 (special suppl.).
[18] C. P. Wu and V. Galindo, "Effects of dielectric sheaths on an infinite phased array of rectangular waveguides," *URSI Spring Meeting Digest* (Ottawa, Canada), pp. 176–177, May 1967.
[19] J. Jacobsen, "Analytical, numerical and experimental investigation of guided waves on a periodically strip-loaded dielectric slab," Lab. of Electromagnetic Theory, Technical University of Denmark, Rept. LD 11.a, December 1967.
[20] A. Hessel and A. A. Oliner, "A new theory of Wood's anomalies on optical gratings," *Appl. Opt.*, vol. 4, no. 10, October 1965.
[21] ——, "Wood's anomalies and leaky waves," in *Proc. Symp. on Electromagnetic Theory and Antennas* (Copenhagen, Denmark). New York: Pergamon Press, 1963, pp. 887–890.
[22] E. G. Magill and H. A. Wheeler, "Wide-angle impedance matching of a planar array antenna by a dielectric sheet," *IEEE Trans. Antennas and Propagation*, vol. AP-14, pp. 49–53, January 1966.
[23] S. W. Lee, "Impedance matching of an infinite phased array by dielectric sheets," *Electronics Letters*, vol. 2, pp. 366–368, October 1966.
[24] N. Marcuvitz, *Waveguide Handbook*, M.I.T. Radiation Lab. Ser., vol. 10. New York: McGraw-Hill, 1951, pp. 5, 64, 84–89.

*The following article, which appeared after the completion of the present paper, is relevant to the material presented herein.*

C. P. Wu and V. Galindo, "Surface wave effects on dielectric sheathed phased arrays of rectangular waveguides," *Bell Sys. Tech. J.*, vol. 47, pp. 117–142, January 1968.

# Surface Waves and Anomalous Wave Radiation Nulls on Phased Arrays of TEM Waveguides with Fences

ROBERT J. MAILLOUX, MEMBER, IEEE

*Abstract*—This paper describes the behavior of an infinite array of parallel plane waveguides in a conducting ground plane with metallic fences perpendicular to the ground plane and spaced halfway between each pair of radiators. The study is mainly concerned with anomalous wave nulls and their relation to endfire surface-wave radiation, and in addition some data is presented to demonstrate the scan matching properties of the fence geometry.

An interesting new development in open periodic structure theory is reported and is shown to apply to both surface waves and anomalous waves. Certain restrictions on the phase distributions of the aperture fields and fence currents are revealed and their origin outlined briefly. These restrictions are introduced to further emphasize relationships between surface and anomalous waves. In addition to these basic relationships, a numerical study of the scanning properties of waveguide-fence arrays is presented which has indicated that structures of this type can reduce impedance variation during scan and can be especially useful when it is desirable to scan very near to, or in the limited scan case, beyond the point at which a grating lobe moves into real space.

154

## Introduction

THE DISCOVERY of radiation pattern nulls in waveguide phased arrays have stimulated researchers to study the complex phenomena associated with these effects and to seek methods of eliminating them. The nulls were initially attributed to surface waves [1]–[3], but that terminology alone was not satisfying because the nulls occur for element spacings which do not support pure surface waves, and also because the feeding waveguides were not reactively terminated and so could not support a pure surface wave. The first rigorous analysis of this behavior, performed by Farrel and Kuhn [4], included the higher order modes in the waveguide apertures, and showed that nulls can exist in the $E$ plane of a rectangular grid array of rectangular waveguides and the $H$ plane and intercardinal planes of a triangular grid array. Other authors have revealed similar phenomena for rectangular and circular waveguides [5]–[8] and have emphasized the importance of higher order modes in obtaining the cancellation necessary for resonance. Recently, Knittel *et al.* [9] have proposed an explanation on the basis of leaky waves and a driven surface-wave-like condition (see [10]) which is a normal mode for a modified corrugated structure, and is a solution for the driven array problem only at a radiation null. These authors used the leaky wave concept to obtain an estimate of the width of the null. Jacobson [11] has considered complex guided waves on a strip-loaded dielectric slab and has also observed anomalous nulls.

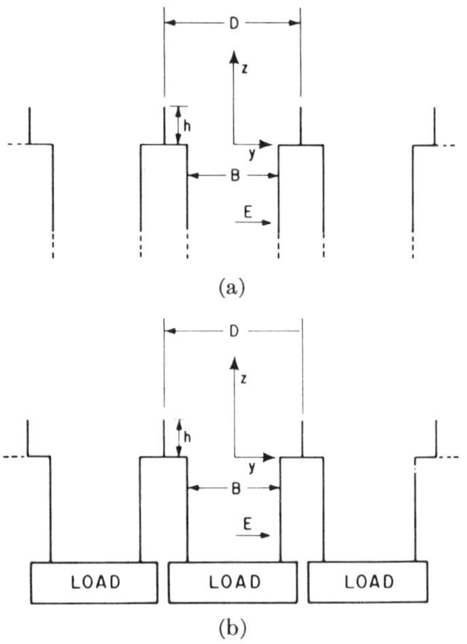


Fig. 1. Array geometry and modified corrugated structure geometry. (a) Parallel plane array structure. (b) Modified parallel plane array structure.

The present investigation concerns an infinite array with metallic fences perpendicular to the ground plane and spaced halfway between each parallel plane radiator. Similar fence structures have been proposed and used [12]–[15] to improve the scan parameters of phased arrays. A terminated structure [Fig. 1(b)] is introduced to compare two types of normal mode solutions, the endfire surface wave, for which all spacial harmonics are slow ($\beta_m^2 > k_0^2$) and the anomalous wave occurring for elements spaced greater than $\lambda/2$, for which one spacial harmonic is fast but that does not radiate. These two normal mode solutions can propagate when the terminations of Fig. 1(b) are reactive and are recognized as equivalent to the forced surface wave solution and the forced anomalous wave solution at a radiation null. In addition, some data is presented to demonstrate the scan-matching properties of the fence geometry.

Manuscript received July 13, 1971; revised September 14, 1971.
The author is with the Microwave Physics Laboratory, Air Force Cambridge Research Laboratories, L. G. Hanscom Field, Bedford, Mass. 01730.

An interesting phenomenon is reported which applies to both surface waves and anomalous waves. Certain restrictions on the phase distributions of the aperture fields and the fence currents are studied briefly and are shown to be required for both types of solution. These conditions are emphasized because they are apparently a new development in open periodic structure theory and yield insight into the energy transfer properties of these structures. They also serve as a means of comparing the surface and anomalous wave solutions.

## General Formulation

The electromagnetic field of the TEM excited structure of Fig. 1(a) can be written in terms of two scalar potential functions, one component of a Hertzian magnetic potential derived from the waveguide aperture fields and one component of the vector potential (Hertzian electric potential) derived from the fence currents. An alternative procedure which considers the space between the fences as a parallel plate waveguide section has not been followed here because this paper is concerned mainly with short fences ($0 \leq h \leq 1.5\lambda$) for which the waveguide approach requires a larger number of higher order modes. The two formulations are equivalent except for convergence properties however. The boundary value problem obtained by requiring continuity of the tangential aperture fields and from requiring the tangential electric field to vanish on the field. The following integral equations are taken directly from [16].

Aperture equation

$$k_0^2 \pi_{mx}(y,0) = k_0^2 \pi_{mx}'(y,0) + \frac{\partial A_z}{\partial y}, \qquad -\frac{B}{2} \leq y \leq \frac{B}{2}, \quad z = 0. \tag{1}$$

Fence equation

$$A_z\left(\frac{D}{2}, z\right) = A \cos k_0 z + C \sin k_0 z + k_0 \int_0^z \frac{\partial \pi_{mx}}{\partial y} \sin k_0(z - \eta)\, d\eta, \qquad 0 \leq z \leq h, \quad y = \frac{D}{2}. \tag{2}$$

The constant $C$ is zero because the boundary condition $E_y = 0$ at $z = 0$, but the constant $A$ is required so that the current can be set equal to zero at the top of the fence. The vector $A_z$ is derived from contributions from each of the fence currents, and the Hertzian potentials $\pi_{mx}'$ and $\pi_{mx}$ represent fields in regions exterior and interior to the waveguides, respectively. The exterior Hertzian potential $\pi_{mx}'$ is derived from contributions from each of the aperture fields

$$\Pi_{mx}'(y,z) = -\frac{1}{\omega D} \sum_{m=-\infty}^{\infty} \exp(-j\beta_m y) \left\{ \begin{array}{c} j\dfrac{\exp(-K_m z)}{K_m} \\[2ex] \dfrac{\exp(-jK_m' z)}{K_m'} \end{array} \right\} \int_{-B/2}^{B/2} dy'\, E_y(y') \exp(+j\beta_m y') \tag{3}$$

$$A_z(y,z) = -\frac{j}{\nu_0 D} \sum_{m=-\infty}^{\infty} \exp\left[-j\beta_m\left(y - \frac{D}{2}\right)\right] \int_0^h dz'\, I_0(z') \left\{ \begin{array}{c} \dfrac{j}{K_m}[\exp(-K_m|z - z'|) + \exp(-K_m|z + z'|)] \\[2ex] \dfrac{1}{K_m'}[\exp(-jK_m'|z - z'|) + \exp(-jK_m'|z + z'|)] \end{array} \right\} \tag{4}$$

fences can be written in the form of Fredholm integral equations, one for each aperture and each fence. These are given as [16, eqs. (7), (8)] for a 2-element array. In the present case the incident field in each waveguide varies progressively along the array with a phase constant $\beta_0 = k_0 \sin\theta$ where $\theta$ is the scan angle from broadside. The same notation is used for the slow-wave regime ($\beta_0 > k_0$) and in this case $\sin\theta$ no longer represents a real angle $\theta$, but is still the ratio $\beta_0/k_0$. The periodic excitation implies that only the integral equations at one aperture and one fence need be solved simultaneously, and that all others are satisfied identically by the periodic nature of

where

$$K_m = (\beta_m^2 - k_0^2)^{1/2}, \qquad K_m' = (k_0^2 - \beta_m^2)^{1/2},$$

$$\beta_m = \beta_0 - \frac{2\pi m}{D}, \quad \text{and} \quad \nu_0 = \frac{1}{\mu_0} = \frac{10^7}{4\pi}\frac{m}{H}.$$

These expressions were obtained by applying the Poisson summation formula to an infinite sum of the Hankel function kernels of [16]. The upper term in each bracket applies for a slow wave ($\beta_m^2 > k_0^2$) and the lower term is used for a fast wave ($\beta_m^2 < k_0^2$). The potential function $\pi_{mx}$ for the waveguide located at the origin is given as a superposition of waveguide modes

$$\pi_{mx}(y,z) = (\exp(-jk_0z) - \Gamma \exp(+jk_0z)) + \sum_{n=1}^{\infty} A_n \cos \frac{n\pi}{B}\left(y + \frac{B}{2}\right) \exp(\alpha_n z). \quad (5)$$

The fence current $I_0(z')$ is also expanded as a sum of functions; each chosen to be zero at the top of the fence ($z = h$) and to have a zero derivative at $z = 0$. The fence current must have a zero derivative at $z = 0$ or the field $E_z$ would be discontinuous, thereby implying a source function at the base of each fence. This condition is similar to that encountered for parasitic unloaded dipole antennas, where the derivative of the current is also required to be set equal to zero. The fence current is therefore normalized to the incident waveguide signal

$$I_0(z') = j2\nu_0 k_0^2 \sum_{p=1}^{P_T} I_p \cos\left(\frac{2p-1}{2h}\right)\pi z'. \quad (6)$$

After carrying out the required differentiations and integrations and truncating the aperture sum at $n = N_T$, (1) and (2) are reduced to a set of functional equations at the aperture and fence. Details are indicated in [17], as are the definitions of summations $S_1$ through $S_4$.

Aperture equation

$$1 - \frac{k_0}{D} S_3(0,y) = \Gamma\left(1 + \frac{k_0}{D} S_3(0,y)\right) + \sum_{n=1}^{N_T} A_n\left[-\cos\left(\frac{n\pi}{B}\left(y + \frac{B}{2}\right)\right) + j\frac{\alpha_n}{D} S_3(n,y)\right] - j\frac{4}{D}\sum_{p=1}^{P_T} I_p S_4(p,y), \qquad -\frac{B}{2} \le y \le \frac{B}{2}. \quad (7)$$

Fence equation

$$-j\frac{k_0^2}{D} S_1(0,z) = \Gamma\left(j\frac{k_0^2}{D} S_1(0,z)\right) - \frac{k_0}{D}\sum_{n=1}^{N_T} \alpha_n A_n S_1(n,z) + 2\frac{k_0^2}{D}\sum_{p=1}^{P_T} I_p S_2(p,z) + A\cos k_0 z, \qquad 0 \le z \le h. \quad (8)$$

With the preceding equations truncated as noted, the equations are satisfied at $P_T + 1$ points on the fence, and at $N_T + 1$ points in the aperture and the system reduces to a matrix equation which is solved numerically. The results are shown in Figs. 2 through 5 and Tables I and II. Throughout this paper reference is made to the power transmission factor (PTF), sometimes called the array element power pattern, which is that part of the input power transmitted in the main beam direction for an infinite array. It is equal to unity minus the square of the

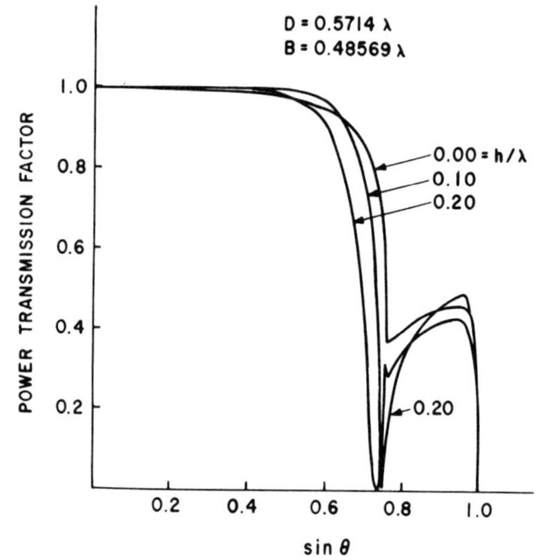


Fig. 2. Array power transmission factor ($h/\lambda = 0.0, 0.1, 0.2$).

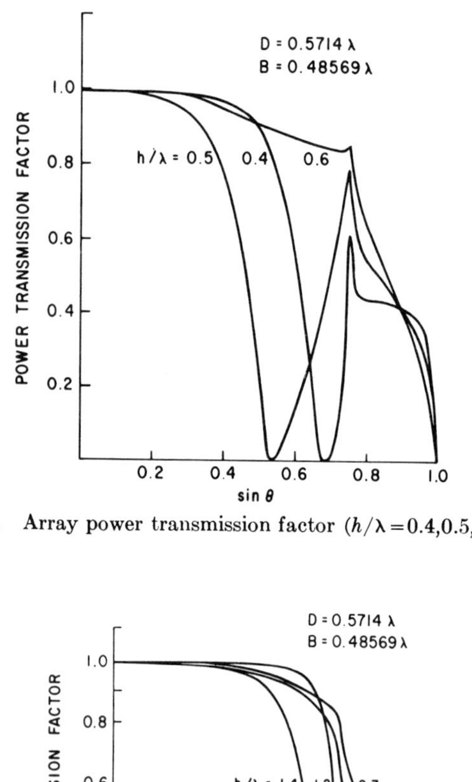


Fig. 3. Array power transmission factor ($h/\lambda = 0.4, 0.5, 0.6$).

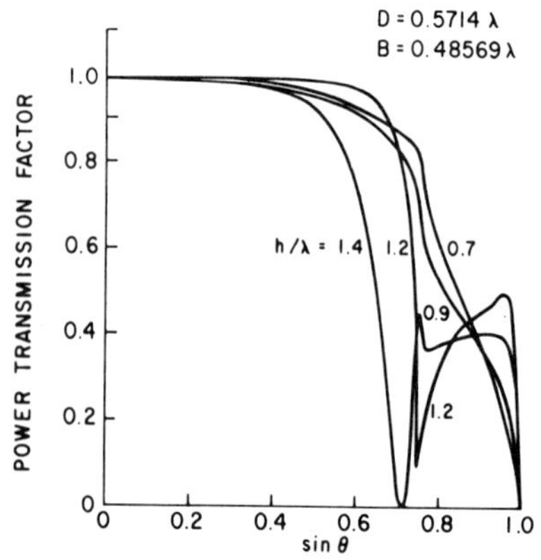


Fig. 4. Array power transmission factor ($h/\lambda = 0.7, 0.9, 1.2, 1.4$).

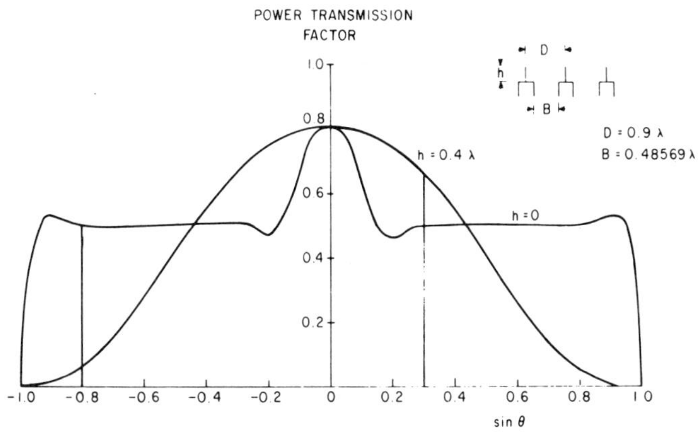


Fig. 5. Grating lobe suppression using fences.

TABLE I
FENCE CURRENT AND WAVEGUIDE MODEL COEFFICIENTS FOR SURFACE WAVE

Parameters: D = 0.400 , B = 0.100
h = 0.201 , $\sin\theta$ = 1.2

$\Gamma$ = 0.63215 + j 0.77484

$(1 + \Gamma)$ Phase = 0.443233

Fence Current:

| Position | Amplitude | Phase |
|---|---|---|
| z = 0.0 | 0.71763530E 00 | -0.26355294E 01 |
| 0.0365 | 0.70498705E 00 | -0.26355295E 01 |
| 0.0731 | 0.66493355E 00 | -0.26355295E 01 |
| 0.1096 | 0.59155061E 00 | -0.26355294E 01 |
| 0.1462 | 0.47369049E 00 | -0.26355295E 01 |
| 0.1827 | 0.27954390E 00 | -0.26355297E 01 |
| 0.2010 | 0.23485438E-06 | -0.26355299E 01 |

Waveguide Modal Coefficients:

| Coefficient | Amplitude | Phase |
|---|---|---|
| A( 1) | 0.13962592E-01 | 0.44322649E 00 |
| A( 2) | 0.50819239E-01 | -0.11275647E 01 |
| A( 3) | 0.19790671E-02 | 0.44322574E 00 |
| A( 4) | 0.19703501E-01 | -0.11275670E 01 |
| A( 5) | 0.91894874E-03 | 0.44313001E 00 |
| A( 6) | 0.12158654E-01 | -0.11275707E 01 |
| A( 7) | 0.60921106E-03 | 0.44326937E 00 |
| A( 8) | 0.92554097E-02 | -0.11275665E 01 |
| A( 9) | 0.48523916E-03 | 0.44327865E 00 |
| A(10) | 0.80245202E-02 | -0.11275713E 01 |
| A(11) | 0.43777547E-03 | 0.44315155E 00 |
| A(12) | 0.38870229E-02 | -0.11275566E 01 |

TABLE II
FENCE CURRENT AND WAVEGUIDE MODEL COEFFICIENTS FOR ANOMALOUS WAVE NULL

Parameters: D = 0.5714 , B = 0.48569
h = 0.4 , $\sin\theta$ = 0.6837

Data: $\Gamma$ = 0.92500 - j 0.37996

$(1+\Gamma)$ Phase = 2.94672 radians

Phase of $[j(1+\Gamma)\exp(-j\beta_o D/2)]$ = -2.992987 radians

PTF = $0.5117 \times 10^{-7}$

Fence Current:

| Position | Amplitude | Phase |
|---|---|---|
| z = 0 | 0.73052983E 00 | -0.29930993E 01 |
| 0.0727 | 0.92537977E 00 | -0.29930779E 01 |
| 0.1455 | 0.11811032E 01 | -0.29930575E 01 |
| 0.2182 | 0.13002224E 01 | -0.29930461E 01 |
| 0.2909 | 0.11957231E 01 | -0.29930405E 01 |
| 0.3636 | 0.76291432E 00 | -0.29930371E 01 |
| 0.4000 | 0.37017483E-06 | -0.29930365E 01 |

Waveguide Modal Coefficients

| Coefficient | Amplitude | Phase |
|---|---|---|
| A( 1) | 0.33793225E 01 | 0.29465872E 01 |
| A( 2) | 0.11466665E 00 | -0.17656664E 01 |
| A( 3) | 0.98275323E-01 | 0.29465799E 01 |
| A( 4) | 0.53254084E-01 | -0.17656733E 01 |
| A( 5) | 0.43881823E-01 | 0.29465863E 01 |
| A( 6) | 0.35194809E-01 | -0.17656759E 01 |
| A( 7) | 0.28957536E-01 | 0.29465869E 01 |
| A( 8) | 0.27870547E-01 | -0.17656730E 01 |
| A( 9) | 0.23108873E-01 | 0.29466033E 01 |
| A(10) | 0.24662994E-01 | -0.17656766E 01 |
| A(11) | 0.20883338E-01 | 0.29465876E 01 |
| A(12) | 0.11868085E-01 | -0.17656629E 01 |

element reflection coefficient amplitude when there are no grating lobes in real space, and must be computed separately otherwise.

## SYMMETRY OF SURFACE WAVES AND ANOMALOUS WAVES AS NORMAL MODES OF MODIFIED STRUCTURE

Since the reflection coefficient amplitude is unity at an array null or at endfire, there exist nonradiating normal mode solutions in each case for equivalent reactively terminated parallel plane structures. This point was demonstrated graphically by the work of Knittel *et al.* [9] and has served to clarify the relationship between the forced anomalous wave solution and its normal mode equivalent. The structure of Fig. 1(b) is introduced here to study the propagation of normal mode surface (end fire) and anomalous waves, and to describe their common symmetry conditions.

These relations will be given and are derived in detail in [17]. This derivation begins with rewriting (7) and (8) in a homogeneous form for the modified structure of Fig. 1(b). Equation (7) is then separated into two complex equations, one containing only terms even about $Y = 0$ and one whose terms are all odd about $Y = 0$. A further subdivision of each of these two complex equations and the complex equation (8) into real and imaginary parts reveals that these parts are proportional and can be made equal by multiplying each equation by a suitable constant. The end result demonstrates that if the amplitude of the reflection coefficient of the terminating admittances is unity for the structure of Fig. 1(b), the following relations exist between the model coefficients

$$-A_n{}^I R_s{}^I = A_n{}^R(1 + R_s{}^R), \qquad \text{for } n \text{ even} \tag{9}$$

$$A_n{}^R R_s{}^I = A_n{}^I(1 + R_s{}^R), \qquad \text{for } n \text{ odd} \tag{10}$$

$$C_p{}^I R_s{}^I = -C_p{}^R(1 + R_s{}^R) \tag{11}$$

where $C_p = \exp(j\beta_0 D/2) I_p = C_p{}^R + jC_p{}^I$, and the superscripts $R$ and $I$ refer to the real and imaginary parts of the variables, respectively. The constant $R_s$ is the reflection coefficient of the terminating admittance for Fig. 1(b), and it is also the complex conjugate of the reflection coefficient at each waveguide for the forced surface wave case [Fig. 1(a)]. Equations (9)–(11) are derived for arbitrary $N_T$ and $P_T$ and are not influenced by truncation errors.

The condition that the amplitude of $R_s$ be unity is obviously necessary for the normal mode solution so that there is no loss in power along the array axis. A less trivial result, however, is that each of the $P_T$ terms in the current expansion have the same phase, and so the phase of the fence current is constant everywhere apart from possible sign changes. This condition is also required for the existence of nonattenuating wave solutions on infinite Yagi arrays, but in the present case it is clearly only one of the many restrictions imposed upon the electromagnetic field. The phase of the fence current is $-\beta_0 D/2$ different from the phase of the aperture parameters $A_n$ (for $n$ even), and apart from sign changes the $A_n$ ($n$ even) are 90° out of phase with the $A_n$ ($n$ odd). Therefore, the even part of the aperture electric field has one constant phase angle, and the odd part is 90° out of phase. Moreover, each of these parameters has its phase precisely defined with respect to the termination reflection coefficient because unless $R_s = -1$, the $A_n$ ($n$ odd) are in phase with $(1 + R_s)$ and $C_p$ and $A_n$ ($n$ even) are 90° out of phase with $(1 + R_s)$.

*Surface Waves*

The special case $R_s = -1$ also corresponds to a surface wave condition, but the phase of both $A_n$ and the $(1 + R_s)$ are discontinuous.

Further implications of (9)–(11) can be seen by writing detailed expressions for the potential functions. In the outer half space

$$\pi_{mx}' = u \frac{j}{D} \sum_{m=-\infty}^{\infty} \exp(-j\beta_m y) \frac{\exp(-K_m z)}{K_m}$$

$$\cdot \Big[(1 + R_s) k_0 \Psi^R(\beta_m, 0, B) + j \sum_{n=2,\text{even}}^{N_T} \alpha_n A_n \Psi^R(\beta_m, n, B)$$

$$- \sum_{n=1,\text{odd}}^{N_T} \alpha_n A_n \Psi^I(\beta_m, n, B)\Big] \tag{12}$$

$$A_z = u \frac{2k_0{}^2}{D} \sum_{m=-\infty}^{\infty} \exp(-j\beta_m y)(-1)^m \sum_{p=1}^{P_T} C_p \Phi(\beta_m, p, h). \tag{13}$$

The functions $\Psi^R$, $\Psi^I$ are real for the surface wave case and $\Phi$ is imaginary (see [17]) and so all of the terms within the brackets of (12) have the same phase for all modes. Thus, as observed from the interior field distribution the even part of $\pi_{mx}'$ is in phase with $j(1 + R_s)$ and the odd part with $(1 + R_s)$. Moreover, the even and odd parts of $A_z$ are 90° out of phase with the even and odd parts of $\pi_{mx}$, so the even and odd parts of $E_z$ and $B_x$ are in phase and are 90° out of phase with the even and odd parts of $E_y$, and the Poynting vector is real in the $y$ direction and imaginary in the $z$ direction at every point in space. Conservation of energy requires only that the average of the Poynting vector over a unit cell be real in the $y$ direction and imaginary in the $z$ direction, but for $\beta_0 > k_0$ this average condition is automatically fulfilled and imposes no restrictions on the currents or fields. Therefore the preceding restrictions on the Poynting vector are clearly much stronger then those dictated by conservation of energy.

*Anomalous Waves*

If $0.5 \leq D/\lambda \leq 1$, one or several of the modes radiates in general for the structure of Fig. 1(a). The anomalous null usually occurs for angles less than the angle of grating lobe onset, and so only the $m = 0$ mode radiation will be considered here.

For the spacings chosen one would expect the periodic structure of Fig. 1(b) to radiate and therefore support no normal mode. Referring to the frequency or wavelength range for efficient waveguide operation this is known as a stop band of the periodic structure. Wu and Galindo [10], and later, Knittel *et al.* [9] explained in connection with similar structures than at this isolated point the amplitude of the radiating spatial harmonic mode is zero everywhere in space and so a modified sort of surface wave can propagate at this one point in the stopband region. A very similar argument, but one which differs fundamentally in one aspect, holds for the present structure. In contrast to these other published results, the mode zero amplitude for this structure is not zero everywhere in space, but only for $z \geq h$. For $z \leq h$, the fields are shown in [17] to assume a form that satisfies (9) through (11) for the anomalous wave, thereby indicating that the same symmetry conditions are maintained for this wave as for the endfire surface wave.

Table I shows aperture coefficients and fence current for $h = 0.201$, $D = 0.4$, $B = 0.1$ and $\sin\theta = 1.2$, a region where all of the spatial harmonics are in the slow-wave regime. The data are listed as they appeared in the computer printout of the solution to (7) and (8) in order to emphasize the number of significant figures to which (9)–(11) are satisfied. Table II shows the amplitude and phase distributions for an anomalous null, and confirms that the symmetry conditions are maintained to a high degree of accuracy in this situation as well as for the slow-wave case.

## Numerical Results: Scanning Properties of Arrays and Fences

The convergence of the solution with regard to the parameters $N_T$, $P_T$, and $M_T$ is discussed in [17], where it is shown that values of $N_T = 8$ and $P_T = 7$ are sufficiently large to guarantee an accurate solution everywhere except in the region of a null. In the immediate vicinity of a null, $P_T = 11$ and $N_T = 12$ were used to insure the accuracy of the solution. The value $M_T = 200$ was used for Figs. 2 through 4 and $M_T = 20$ was used for Fig. 5. Questions of relative convergence are treated in the report.

Convergence was judged by using several criteria. By conservation of energy, the PTF should be equal to the normalized power transmitted through each waveguide aperture $1 - |\Gamma|^2$ for scan angles for which there is no grating lobe in visible space. Comparing these two parameters proved to be the simplest test to monitor and it was always found consistent with the convergence of the complex waveguide reflection coefficient, aperture field and fence current, etc. The conservation of energy is a valid test for the point matching solution employed in this analysis because it is not identically satisfied as it is in the case of certain Ritz–Galerkin and method of moments solutions [13].

Figs. 2, 3, and 4 show the behavior of a parallel plane structure studied by Wu and Galindo [10] but equipped with fences between each of the waveguides. The $h = 0$ case agrees with the previously published work. The grating lobe point is at $\sin\theta = 0.75$, and here the aperture susceptance reaches a peak value and there is a corresponding reduction in gain. The first result apparent as the fence height is increased is that the loss of gain at the grating lobe angle increases and becomes a null which corresponds to zero conductance just inside of this point. As the fence height is increased to about $0.5\lambda$, the null becomes wider and moves in so as to be visible at lesser scan angles.[1] With still further increases the cusp in the data at the grating lobe angle and the null itself disappear. Fig. 4 shows that as the fence height is increased beyond one wavelength, the null returns within the visible scan space and once again dominates the element pattern.

Fig. 2 shows that before the null becomes a dominant factor, however, it is possible to achieve some improvement in VSWR by using a fence height $h/\lambda = 0.1$. This structure has very little impedance variation out to about 30° of scan.

Perhaps the most promising engineering use of fences, however, is as a means for suppressing grating lobe radiation. By comparing the $h = 0.7$ data of Fig. 4 with the $h = 0$ data of Fig. 2 one can see that the sharp dip in the power transmission factor at the grating lobe angle for $h = 0$ does not occur for $h = 0.7$. Instead, the power loss is much more gradual because the grating lobe power increase is much slower. The grating lobe power can be seen by imagining the negative half of the PTF curve and observing that the grating lobe occurs a distance $1/D$ to the left of the main beam in $\sin\theta$ space, so the relative growth of the grating lobe signal is evidenced by comparing the slope of the two curves near $\sin\theta = -1$. In this example a scan increase of between 3° and 7° can be achieved.

Fig. 5 shows an example of very effective grating lobe suppression for limited scan applications using fences. Apertures having $B = 0.48569\lambda$ are spaced at $0.9\lambda$ apart to form the array. The vertical lines represent the position of the main beam and grating lobe at the scan limit ($\sin\theta = 0.3$). Their separation remains fixed for all angles chosen. With no fence the gain falls off rapidly as the array is scanned in the positive direction from broadside because of the large grating lobe which arises very near to $\sin\theta = -1$, and at the end of scan about one half the power is in the main beam, while the other half is in the grating lobe. For a fence height $0.4\lambda$ however the grating lobe entering at $\sin\theta = -1$ is very small and so the main beam power loss is minimal out to the scan limit. Hence for limited scan applications, and over a limited range of frequencies, it may be possible to achieve considerable grating lobe suppression using this technique.

## Conclusion

This paper has discussed the relationships between the anomalous nulls and true surface waves of a driven phased array antenna. In both cases restrictions are uncovered which control the phase distributions of the resulting electromagnetic field.

In addition, a brief numerical study of the scanning properties of waveguide-fence arrays has indicated that structures of this type could be useful in cases when it is desirable to scan near to, or, in limited scan applications, beyond the angle at which a grating lobe moves into real space.

[1] A reviewer has pointed out the basic similarity between these results and those found by Amitay and Galindo [19] for varying the thickness of a dielectric plug in a waveguide array. These authors have explained the phenomenon by relating it to the existence of "trapped" higher order modes that are cutoff in the waveguides but can propagate in the plugs. The similarity is indeed revealing because the fence structure can also be considered a row of parallel plane waveguides of width sufficient, in this case, to propagate the first higher order mode. Therefore, it is clear that many of the insights of the Amitay and Galindo paper could be applied to this geometry as well.

## References


[1] J. L. Allen, "On surface-wave coupling between elements of large arrays," *IEEE Trans. Antennas Propagat.* (Commun.), vol. AP-13, pp. 638–639, July 1965.

[2] L. W. Lechtreck, "Effects of coupling accumulation in antenna arrays," *IEEE Trans. Antennas Propagat.*, vol. AP-16, pp. 31–37, Jan. 1968.

[3] ——, "Cumulative coupling in antenna arrays," *IEEE G-AP Int. Symp. Dig.*, pp. 144–149, Aug.–Sept. 1965.

[4] G. F. Farrell, Jr., and D. H. Kuhn, "Mutual coupling in infinite arrays of rectangular waveguide horns," *IEEE Trans. Antennas Propagat.*, vol. AP-16, pp. 405–414, July 1968.

[5] B. L. Diamond, "A generalized approach to the analysis of infinite planar array antennas," *IEEE Proc.*, vol. 56, pp. 1837–1851, Nov. 1968.

[6] A. G. Ehlenberger, L. Schwartzman, and L. Topper, "Design criteria for linearly polarized waveguide arrays," *IEEE Proc.*, vol. 56, pp. 1861–1872, Nov. 1968.
[7] G. V. Borgiotti, "Modal analysis of periodic planar phased arrays of apertures," *IEEE Proc.*, vol. 56, pp. 1881–1892, Nov. 1968.
[8] N. Amitay and V. Galindo, "The analysis of circular waveguide phased arrays," *Bell Syst. Tech. J.*, vol. 47, no. 11, pp. 1903–1932, Nov. 1968.
[9] G. H. Knittel, A. Hessel, and A. A. Oliner, "Element pattern nulls in phased arrays and their relation to guided waves," *IEEE Proc.*, vol. 56, pp. 1822–1836, Nov. 1968.
[10] C. P. Wu and V. Galindo, "Surface-wave effects on dielectric sheathed phased arrays of rectangular waveguides," *Bell Syst. Tech. J.*, pp. 117–142, Jan. 1968.
[11] J. Jacobsen, "Analytical, numerical, and experimental investigation of guided waves on a periodically strip-loaded dielectric slab", *IEEE Trans. Antennas Propagat.*, vol. AP-18, pp. 379–388, May 1970.
[12] S. Edelberg and A. A. Oliner, "Mutual coupling effects in large antenna arrays II: compensation effects," *IRE Trans. Antennas Propagat.*, vol. AP-8, pp. 360–367, July 1960.
[13] C. C. Allen, "Mutual coupling effects in phased array antennas," presented at 13th Annu. Symp. USAF Antenna Res. Dev. Prog., Univ. Illinois, Oct. 14, 1963.
[14] E. C. Dufort, "Design of corrugated plates for phased array matching," *IEEE Trans. Antennas Propagat.* vol. AP-16, pp. 37–46, Jan. 1968.
[15] ——, "A design procedure for matching volumetrically scanned waveguide arrays," *IEEE Proc.*, vol. 56, pp. 1851–1860, Nov. 1968.
[16] R. J. Mailloux, "Reduction of mutual coupling using perfectly conducting fences," *IEEE Trans. Antennas Propagat.*, vol. AP-19, pp. 166–173, Mar. 1971.
[17] ——, "Surface and anomalous waves on phased arrays of TEM waveguides with fences," Air Force Cambridge Res. Lab, Bedford, Mass., AFCRL-71-0060 Phys. Sci. Res. Papers **444**, Jan. 28 1971.
[18] N. Amitay and V. Galindo, "On energy conservation and the method of moments in scattering problems," *IEEE Trans. Antennas Propagat.*, vol. AP-17, pp. 747–751, Nov. 1969.
[19] ——, "Characteristics of dielectric loaded and covered circular waveguide phased arrays," *IEEE Trans. Antennas Propagat.*, vol. AP-17, pp. 722–729, Nov. 1969.



*Reprinted by permission from* IEEE TRANSACTIONS ON ANTENNAS AND PROPAGATION
Vol. AP-20, No. 2, March 1972, pp. 160-166

# MATCHING AND SIMULATION IV

Copyright © 1967 American Telephone and Telegraph Company
Reprinted from THE BELL SYSTEM TECHNICAL JOURNAL
Vol. XLVI, No. 7, September, 1967
*Printed in U.S.A.*

# Some Properties and Limitations of Electronically Steerable Phased Array Antennas*

By D. VARON and G. I. ZYSMAN

(Manuscript received April 4, 1967)

*This paper is a treatment on linear and planar phased arrays of current sources, whose amplitudes are uniform and scan-invariant. By recognition that the radiation impedance of an array element is an analytic function of a complex scan variable, a powerful mathematical tool becomes available for the investigation of some important properties of the impedance as a function of scan. For example, it is proven that in a finite array the impedance seen by such a scan-invariant current source cannot be perfectly matched over a continuous scanning range using lossless, linear, passive and time-invariant elements. This result is extended to the infinite-array case by treating the latter as a periodic structure, and assuming that the Green's function of the unit cell is analytic with respect to the scan variable. The theory includes both linear and planar arrays. Among other results it is shown that the element impedance in an infinite array must be of a specific mathematical form. It is hoped that by recognizing the limitations imposed thereby, useful guidelines will be established for achieving optimal match of an array into space.*

## I. INTRODUCTION

The class of antennas widely known as phased arrays includes essentially two types of radiators: stationary and steerable ones. The first operates at fixed amplitude and fixed relative phase between the array elements. Consequently, the antenna characteristics, such as radiation pattern, input impedance, and mutual coupling between elements, remain unchanged during the entire operational lifetime of the antenna. The steerable antenna is characterized by time varying ex-

* This work was supported by the U. S. Army under contract DA-30-069-AMC-333(Y).

citation. The relative phase between adjacent elements is varied either mechanically or electronically to bring about a variation in the orientation of the beam. In most instances scanned arrays are large in size and may contain several thousand elements. Their illumination has a linear phase taper. As a result the antenna characteristics become scan dependent. The relationship between scan angle and various parameters of interest such as gain, element impedance, and mutual coupling between elements have been the subject of intense investigation in recent years.[1,2] One particular direction has been towards improvement of the impedance match over wide scanning ranges.[3] At present the merit of a matching technique can be determined only relatively to other techniques. To the best of the authors' knowledge an absolute mathematical criterion, based on physical realizability requirements, has not been formulated. Some investigators[4,5] claim that a perfect match of an infinite array for all scan angles (at which the active impedance is not infinite, zero or purely reactive) can be achieved by an infinite set of interconnecting network elements. However, the proof is based on the assumption that the scan-dependent equivalent load impedance at the array-space interface remains unchanged after the sources have been interconnected by coupling elements. Although this assumption has been successfully applied[3,5] to improve the matching capability of an infinite array, it is incorrect to use it in a perfect matching scheme.



In this paper a new mathematical approach to phased array analysis is presented. The model for the analysis is a phased array of ideal current sources (electric or magnetic) of scan-invariant uniform amplitude. This model is further discussed in Section II. The analysis itself is based on the general laws of antenna theory and on those properties which are common to all phased arrays represented by the model.

The first part of the theory is devoted to finite arrays and is treated in Section III. The starting point of the theory is a theorem which establishes that the radiation impedance of an element in a finite array is an analytic function of the scan angle. Further, it is shown that an element in a linear or planar phased array cannot be perfectly matched over a continuous scanning range by using lossless, linear, passive and time-invariant elements. Then it is demonstrated that the directions in space of the beams' maxima are eigenvalues of a Laplacian differential operator with periodic boundary conditions which are related to the phase taper of the array, and several useful properties of those eigenvalues are derived.

The second part of the theory appears in Section IV and is devoted to infinite arrays, which play an important role in the analysis of large phased arrays. The investigation is based on a transformation between the scan angle and a complex variable $s = \alpha + j\beta$, which can be interpreted on $0 < \alpha \leqq 1, \beta = 0$ as the trigonometric sine function of the angle between the plane of the array and the direction in which a chosen grating lobe propagates. It is subsequently shown that the element impedance, as a function of $s$, is restricted to a specific mathematical form. Recognition of the limitations imposed thereby may provide new insight into the behavior of such arrays.

## II. PRELIMINARY REMARKS

The model chosen for the following treatment is a linear or planar phased array excited by a set of ideal current generators of uniform amplitude and linear phase taper. The description *ideal* implies that the sources have no internal impedance and are invariant under any loading. This means that except for the relative phasing between contiguous generators the currents are scan independent. Frequently in antenna analysis induced currents are replaced by equivalent sources by application of the equivalence principle.[6] Such currents are not part of the sources. The induced currents are accounted for automatically by fulfillment of the requirement that the tangential component of the electric field has to vanish on all conductors. In general, the source-current amplitude in each element of the array may be a function of scan. However, this dependence is generally unknown and is often neglected in theoretical work. The types of excitations commonly used are the "free excitation" and "forced excitation".* The first assumes a generator with a scan-invariant internal impedance which is capable of delivering scan-invariant incident power. In the latter a constant terminal voltage or current is maintained. As pointed out by Oliner and Malech free excitation is easier to realize in high-frequency technology than forced excitation. The latter, however, is more tractable here. The results of this study remain valid for scan-dependent excitation as well, provided the current density of the source is a smoothly varying function of scan angle and can be analytically continued into a complex scan-angle plane.

Under the assumption that the array is excited by a uniform amplitude and a linear phase taper, the current density excitation function

---

* A. A. Oliner and R. G. Malech, Ref., 1, pp. 209–211.

of an $M$-element linear array (Fig. 1) is given by

$$\mathbf{J}(x, y, z, \psi) = \begin{cases} \mathbf{J}_0(x - ma, y, z)e^{jm\psi}, & ma \leqq x \leqq (m+1)a, \\ & m = 0, 1, \cdots, M-1, \\ 0 & \text{otherwise} \end{cases} \tag{1}$$

and that of an $M \times N$ element planar array of rectangular symmetry (Fig. 2) is given by

$$\mathbf{J}(x, y, z, \psi_x, \psi_y) = \begin{cases} \mathbf{J}_0(x - ma, y - nb, z)e^{j(m\psi_x + n\psi_y)} & \\ & ma \leqq x \leqq (m+1)a, \\ & nb \leqq y \leqq (n+1)b, \\ & m = 0, 1, 2, \cdots, M-1, \\ & n = 0, 1, 2, \cdots, N-1, \\ 0 & \text{otherwise.} \end{cases} \tag{2}$$

The above currents can be either electric or magnetic the latter being regarded as equivalent to ideal electric voltage sources.

Note that the spherical coordinate systems in Fig. 1 and 2 differ from those commonly used in phased array analysis. The poles are located at endfire instead of broadside and the ranges of colatitude and azimuth are such that the upper hemisphere is spanned by $0 \leqq \theta \leqq \pi$, $0 \leqq \varphi < \pi$. This convention is chosen for reasons of mathematical convenience. The results derived in Section III are valid for linear as well as planar arrays. The inclusion of both cases in a single treatment is facilitated by a generalized notation for the current density excitation function. The steering phases $m\psi$ and $m\psi_x + n\psi_y$ are replaced by an equivalent "steering coefficient" $\sigma_{mn}(\varphi_{pq})$ in the plane of scan oriented at azimuth angle $\varphi_{pq}$. The steering coefficient

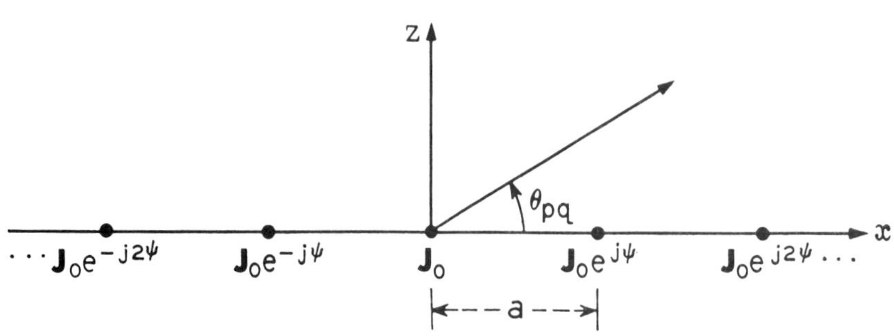


Fig. 1 — Linear phased array.

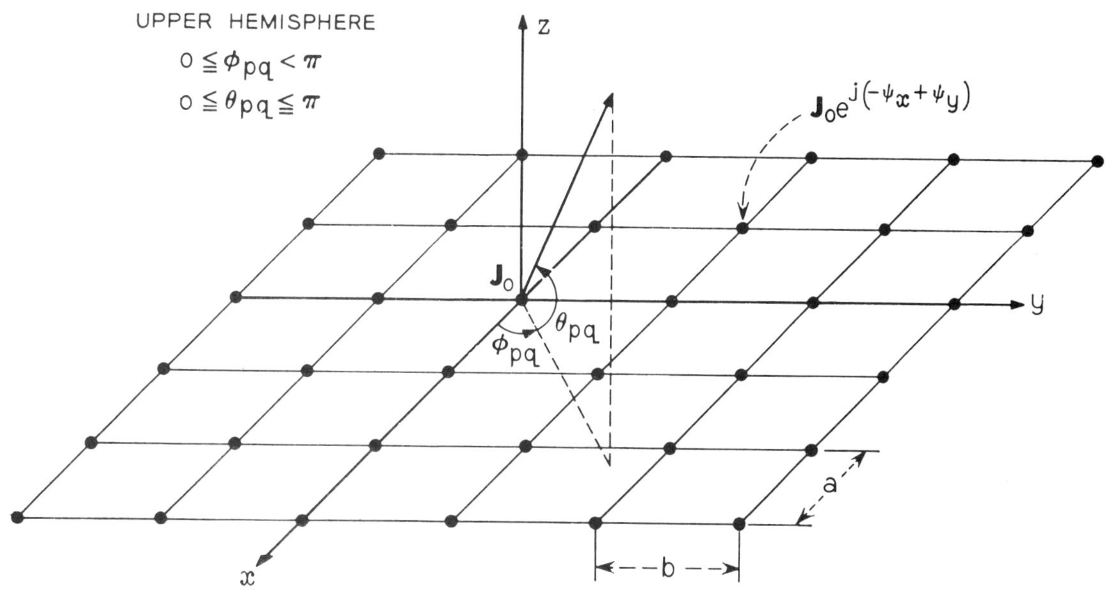


Fig. 2 — Planar phased array.

is derived by its relationship to the direction of a beam's maximum, which is determined for linear arrays by the equation

$$\psi + 2p\pi = ka \cos \theta_{p0} \qquad p = 0, \pm 1, \pm 2, \cdots \pm \infty \tag{3}$$

and for planar arrays by

$$\psi_x + 2p\pi = ka \cos \theta_{pq} \cos \varphi_{pq} \qquad p = 0, \pm 1, \pm 2, \cdots \pm \infty \tag{4a}$$

$$\psi_y + 2q\pi = kb \cos \theta_{pq} \sin \varphi_{pq} \qquad q = 0, \pm 1, \pm 2, \cdots \pm \infty, \tag{4b}$$

where $k$ is the wave number in the medium, and $\theta_{pq}$ is as shown in Fig. 1 and 2. The steering coefficient is then defined by

$$\sigma_{mn}(\varphi_{pq}) = k(ma \cos \varphi_{pq} + nb \sin \varphi_{pq})$$

$$p, q = 0, \pm 1, \pm 2, \cdots \pm \infty. \tag{5}$$

Equations (1) and (2) can now be written as

$$\mathbf{J}(x, y, z, \theta_{pq}) = \begin{cases} \mathbf{J}_0(x - ma, y - nb, z) \exp (j\sigma_{mn} \cos \theta_{pq}), \\ \qquad ma \leqq x \leqq (m + 1)a, \\ \qquad nb \leqq y \leqq (n + 1)b, \\ \qquad m = 0, 1, \cdots, M - 1, \\ \qquad n = 0, 1, \cdots, N - 1, \\ 0 \qquad \text{otherwise} \end{cases} \tag{6}$$

at $\varphi_{pq} = \text{const.}$

167

Under the above generalization the excitation function for the linear array becomes a special case, $q = 0$, $N = 1$, $\varphi_{pq} = 0$, and the period in the $y$-direction extends from $-\infty$ to $+\infty$; or alternatively $p = 0$, $M = 1$, $\varphi_{pq} = \pi/2$ and the period in the $x$-direction extending from $-\infty$ to $+\infty$. Since the phase constant $\exp \{j\sigma_{mn}(\varphi_{pq}) \cos \theta_{pq}\}$ is independent of $(p, q)$, any $\theta_{pq}$ may be chosen as the independent variable of scan. The subscript $pq$ will henceforth be omitted whenever the mathematical expressions are independent of $(p, q)$.

The time dependence $e^{j\omega t}$ is assumed throughout the analysis. In a steerable array the phase taper is time dependent. However, it is understood that the rate of change of the phase taper is very small in comparison to the angular frequency, i.e., $d\psi/dt \ll \omega$, since only under that condition do the classical concepts of directivity and radiation impedance remain meaningful. If $\psi(t)$ is a step function it is assumed that the time interval is long enough to allow all transients to reach negligible values before a new step is initiated.

The formal solution of the array problem is obtained from Maxwell's Equations via a vector potential $\mathbf{A}(x, y, z, \theta)$ which is a solution of the inhomogeneous reduced wave equation

$$\nabla^2\mathbf{A} + k^2\mathbf{A} = -\mu\mathbf{J}(x, y, z, \theta), \tag{7}$$



where $\mu$ is the permeability of the medium. The magnetic field is given by

$$\mathbf{H} = \frac{1}{\mu} \nabla \times \mathbf{A}, \tag{8a}$$

and the electric field (under Lorentz gauge) by

$$\mathbf{E} = -j\omega\left(\mathbf{A} + \frac{1}{k^2} \nabla\nabla \cdot \mathbf{A}\right). \tag{8b}$$

The solution to (7) over infinite space $V$ can be written in closed form in terms of a dyadic Green's function[7]

$$\mathbf{A}(x, y, z, \theta) = \mu \int_V \bar{\bar{\mathfrak{G}}}(x, y, z \mid \xi, \eta, \zeta) \cdot \mathbf{J}(\xi, \eta, \zeta, \theta)\, d\xi\, d\eta\, d\zeta, \tag{9}$$

where $\bar{\bar{\mathfrak{G}}}(x, y, z \mid \xi, \eta, \zeta)$ is a solution of

$$\frac{\partial^2\bar{\bar{\mathfrak{G}}}}{\partial x^2} + \frac{\partial^2\bar{\bar{\mathfrak{G}}}}{\partial y^2} + \frac{\partial^2\bar{\bar{\mathfrak{G}}}}{\partial z^2} + k^2\bar{\bar{\mathfrak{G}}} = -\bar{I}\,\delta(x - \xi)\,\delta(y - \eta)\,\delta(z - \zeta), \tag{10}$$

$\bar{I}$ being the unit dyadic $\mathbf{a}_x\mathbf{a}_x + \mathbf{a}_y\mathbf{a}_y + \mathbf{a}_z\mathbf{a}_z$. The boundary conditions which $\bar{\bar{\mathfrak{G}}}$ has to satisfy are derivable via the Vector Green's Theorem*

* P. M. Morse and H. Feshbach, Ref. 7, p. 1767.

by imposition of the requirement that the tangential component of the electric field has to vanish on all conductors. This guarantees that all induced currents are accurately determined.

It can be shown[8] that the average complex power delivered by the $mn$th element in the array is

$$P_{mn} = -\frac{1}{2}\int_{V_{mn}} \mathbf{E}\cdot\mathbf{J}_{mn}^{*}\,dv, \tag{11}$$

where

$$\mathbf{J}_{mn}(x, y, z, \theta) = \mathbf{J}(x, y, z, \theta),$$
$$ma \leqq x \leqq (m+1)a, \qquad nb \leqq y \leqq (n+1)b \tag{12}$$

the asterisk (*) denotes complex conjugate, and $V_{mn}$ is a simply connected volume occupied by $\mathbf{J}_{mn}$. If $S_{mn}$ is a surface obtained by taking a cross section through $V_{mn}$, the total current, $I_{mn}$, flowing through the cross section $S_{mn}$ is

$$I_{mn} = \iint_{S_{mn}} \mathbf{J}\cdot d\mathbf{s}. \tag{13}$$

The element radiation impedance, $Z_{mn}$, is defined in terms of the complex power by

$$P_{mn} = \tfrac{1}{2}\,|\,I_{mn}\,|^2\,Z_{mn}\,. \tag{14}$$

By (10) and (13) via (8b) and (9), the element radiation impedance can be defined directly in terms of the array geometry and the excitation:

$$Z_{mn}(\theta) = \frac{1}{|\,I_{mn}\,|^2}\int_{V_{mn}}\int_{V} \mathbf{J}_{mn}^{*}(x, y, z, \theta)\cdot\bar{\bar{G}}(x, y, z \mid \xi, \eta, \zeta)$$
$$\cdot\mathbf{J}(\xi, \eta, \zeta, \theta)\,d\tau\,dv, \tag{15a}$$

where $d\tau = d\xi d\eta d\zeta$, $dv = dxdydz$, and

$$\bar{\bar{G}}(x, y, z \mid \xi, \eta, \zeta) = j\omega\mu\left(\bar{\bar{I}} + \frac{1}{k^2}\nabla\nabla\right)\cdot\bar{\bar{\mathcal{G}}}(x, y, z \mid \xi, \eta, \zeta). \tag{15b}$$

Operator $\nabla$ operates on $(x, y, z)$. The quantity $|I_{mn}|^2$ is introduced for the purpose of normalization, and may depend on the choice of the cross section $S_{mn}$.

The definition of the impedance includes both linear and planar array elements. It is consistent with the commonly known definition of impedance[9] if the latter is viewed as a relation between the average complex power delivered by the generator and the rms current flowing into the load. The definition given by (15) is necessary in view of the fact that in a system excited by distributed currents, a terminal voltage in the time domain is not always uniquely defined. In a system excited by magnetic currents, (15) defines the element admittance if the permeability $\mu$ is replaced by the permittivity $\epsilon$ and the electric currents by their magnetic counterparts.

In the following theoretical discussion, it is assumed that the phased arrays are excited by a uniform amplitude and a linear phase taper.

## III. FINITE ARRAYS

*Theorem 1: The element radiation impedance in a finite, steerable, linear or planar phased array of scan-invariant current sources, radiating into a linear, lossless, passive and time-invariant system, is an entire function*[10] *of the scan angle $\theta$ in any given plane of scan, with an essential singularity* at $\theta \to \infty$.*

*Proof:* By (15a)

$$|I_{rs}|^2 Z_{rs} = \int_{V_{rs}} \int_V \mathbf{J}_{rs}^*(x, y, z, \theta) \cdot \bar{\bar{G}}(x, y, z \mid \xi, \eta, \zeta) \cdot \mathbf{J}(\xi, \eta, \zeta, \theta)\, d\tau\, dv. \tag{16}$$

On expanding (16) in a double sum of integrals over all cells $\{(m, n)\}$, $m = 0, 1, \cdots, M-1$; $n = 0, 1, \cdots, N-1$, and using the relationships of (6) followed by a change of variable in each term of the sum, one obtains

$$Z_{rs}(\theta) = \sum_{m=0}^{M-1} \sum_{n=0}^{N-1} \tilde{z}_{mnrs} \exp\,[j(\sigma_{mn} - \sigma_{rs}) \cos\theta], \tag{17a}$$

where

$$\tilde{z}_{mnrs} = \frac{1}{|I_{rs}|^2} \int_{V_{00}} \int_{V_{00}} \mathbf{J}_0^*(x, y, z) \cdot \bar{\bar{G}}(x + ra, y + sb, z \mid \xi + ma, \eta + nb, \zeta) \cdot \mathbf{J}_0(\xi, \eta, \zeta)\, d\tau\, dv. \tag{17b}$$

In any given plane of scan $\varphi$ is constant, so that

$$\sigma_{mn} - \sigma_{rs} = k[(m - r)a \cos\varphi + (n - s)b \sin\varphi] = \sigma_{m-r,n-s} \tag{18}$$

is independent of $\theta$. Both $\cos\theta$ and the exponential function are entire functions.† Consequently, the exponential function appearing in (17a)

* R. V. Churchill, Ref. 10, Sec. 68, p. 157; Sec. 112, p. 270.
† R. V. Churchill, Ref. 10, Sec. 21, p. 47; Sec. 23, p. 50.

is an entire function of an entire function, which is likewise entire[11] (entire functions are also called integral functions). $Z_{rs}(\theta)$ is a finite sum of entire functions and is also entire.

The nature of the essential singularity at $\theta \to \infty$ is obtained by first expanding $\cos \theta$ in the complex $\theta$-plane

$$\cos(\theta_r + j\theta_i) = \cos\theta_r \cosh\theta_i - j\sin\theta_r \sinh\theta_i \; . \tag{19}$$

Then, if $|\theta_i| \to \infty$ in such a way that $(\sigma_{mn} - \sigma_{rs})\theta_i > 0$, the *mn*th term behaves as $\exp\{|\sigma_{mn} - \sigma_{rs}| \sin\theta_r \exp[|\theta_i|]\}$ Q.E.D. Note that even when $\mathbf{J}_0(x, y, z, \theta)$ is scan dependent, $Z_{rs}(\theta)$ is analytic provided $\mathbf{J}_0(x, y, z, \theta)$ is analytic. However, other isolated singularities may exist.

*Corollary 1a: Re* $\{Z_{rs}\}$ *and* $\frac{\partial}{\partial\theta}$ *Re*$\{Z_{rs}\}$ *are entire functions of* $\theta$ *each with an essential singularity at* $\theta \to \infty$. Proof appears in Appendix A.

*Theorem 2: The power radiated by an element in a finite, steerable, linear or planar phased array of scan-invariant current sources, radiating into a lossless, linear, passive and time-invariant system cannot be kept constant over a continuous scanning range with lossless, linear, passive and time-invariant network elements and scatterers only.*

*Proof:* Let $\bar{\bar{G}}(x, y, z \mid \xi, \eta, \zeta)$ be the dyadic Green's function of the entire system including all equalizing elements. The radiation impedance of the *mn*th element of the array is given by (15a) for a lossless, linear, passive, time-invariant system. If the array is radiating constant power over a continuous scanning range, the real part of the radiation impedance, $R_{rs}(\theta) = \text{Re}\{Z_{rs}\}$, must remain constant in that range and

$$\frac{\partial}{\partial\theta}[R_{rs}(\theta)] = 0, \qquad \theta_1 \leqq \theta_r \leqq \theta_2 \, , \qquad \theta_i = 0 \tag{20}$$

where $\theta = \theta_r + j\theta_i$. By Corollary 1a, $\frac{\partial}{\partial\theta}[R_{rs}(\theta)]$ is analytic in the closed $\theta$-plane and has an essential singularity at $\theta \to \infty$. However, if the derivative vanishes along the line $\theta_1 \leqq \theta_r \leqq \theta_2$ it must vanish everywhere in the $\theta$-plane*. Hence, it cannot have an essential singularity at infinity. The contradiction implies that $R_{rs}(\theta)$ cannot be constant over a continuous scanning range. Q.E.D.

Equations (3) and (4) specify the directions of the beams' maxima, however, not all of them correspond to real directions in space. Whereas $\varphi_{pq}$ is real for all $(p, q)$, $\theta_{pq}$ can be either real or imaginary, as may be

* P. M. Morse and H. Feshbach, Ref. 7, Vol. I, p. 390.

seen from the solution of (4):

$$\varphi_{pq} = \tan^{-1} \frac{(\psi_y + 2q\pi)a}{(\psi_x + 2p\pi)b}, \qquad 0 \leqq \varphi_{pq} < \pi \quad (21a)$$

$$\theta_{pq} = \cos^{-1} \frac{\psi_x + 2p\pi}{ka \cos \varphi_{pq}} = \cos^{-1} \frac{\psi_y + 2q\pi}{kb \sin \varphi_{pq}}, 0 \leqq \theta_{pq} \leqq \pi. \quad (21b)$$

If $\theta_{pq}$ is real it is said that the beam is in real space. By way of mathematical generalization it is said that all those beams having an imaginary $\theta_{pq}$ are in "imaginary space". If $\theta_{pq} = 0$, or $\theta_{pq} = \pi$, it is said that the beam is in a grazing position between real and imaginary space. It can easily be verified from (21) that for a given phasing $(\psi_x, \psi_y)$ every pair $(p, q)$ corresponds to a unique direction $(\varphi_{pq}, \theta_{pq})$ in the complex domain $0 \leqq \varphi < \pi$, $0 \leqq \mathrm{Re}\{\theta\} \leqq \pi$. These directions are the characteristic directions of the system. They are directly related, through (4), to the eigenvalues of

$$\frac{\partial^2 F}{\partial x^2} + \frac{\partial^2 F}{\partial y^2} + \Gamma^2 F(x, y) = 0 \quad (22)$$

with the following periodic boundary conditions

$$F(x, y) = F(x + a, y) \exp(-j\psi_x), \quad (23a)$$

$$\frac{\partial F}{\partial x}(x, y) = \frac{\partial F}{\partial x}(x + a, y) \exp(-j\psi_x), \quad (23b)$$

$$F(x, y) = F(x, y + b) \exp(-j\psi_y), \quad (23c)$$

$$\frac{\partial F}{\partial y}(x, y) = \frac{\partial F}{\partial y}(x, y + b) \exp(-j\psi_y). \quad (23d)$$

The eigenfunctions, which form a complete orthogonal set in the interval $0 \leqq x \leqq a$, $0 \leqq y \leqq b$ are

$$F_{pq}(x, y) = \exp\left[j(\psi_x + 2p\pi)\frac{x}{a}\right] \exp\left[j(\psi_y + 2q\pi)\frac{y}{b}\right],$$

$$p, q = 0, \pm 1, \pm 2, \cdots \pm \infty. \quad (24)$$

By (4) they can also be written as

$$F_{pq}(x, y) = \exp\{jk \cos \theta_{pq}(x \cos \varphi_{pq} + y \sin \varphi_{pq})\}. \quad (25)$$

The eigenvalues $\{\Gamma_{pq}\}$ are

$$\Gamma_{pq} = k \cos \theta_{pq} \qquad p, q = 0, \pm 1, \pm 2, \cdots \pm \infty. \quad (26)$$

The results thus derived lead to several interesting conclusions which are summarized in the following lemmas.

*Lemma 1: Every steerable linear or planar phased array with a linear phase taper has only a finite number of beams in real space.* Proof appears in Appendix *B*.

For every pair of phasing $(\psi_x, \psi_y)$ there exists an infinite set of characteristic directions $\{\theta_{pq}, \varphi_{pq}\}$. As the array is scanned by varying the values of $(\psi_x, \psi_y)$ in the intervals $-\pi \leqq \psi_x \leqq \pi$, $-\pi \leqq \psi_y \leqq \pi$ some characteristic directions will go through a grazing position going from imaginary to real space or vice versa. We shall call such characteristic directions "transitive characteristic directions".* Since the condition for a grazing position is $|\cos \theta_{pq}| = 1$, it follows from Lemma 1 that the number of transitive characteristic directions is finite.

*Lemma 2: The radiation impedance of an element in a linear or planar phased array can be expanded by an infinite series over all characteristic directions of the system.* Proof appears in Appendix C.

## IV. INFINITE ARRAYS

In analyzing large arrays it has been found useful to approximate the behavior of the center elements by the behavior of identical elements in an infinite array of the same geometry.[12] This approximation is motivated by the fact that the performance of the center elements is strongly affected through mutual coupling by contiguous elements, but very weakly by elements far away.[13]

The formulation of the infinite array problem may be obtained from the results derived for finite-size arrays by letting the number of elements $M$ and $N$ approach infinity. The infinite array problem can also be treated as a periodic structure by application of Floquet's theorem. In the following, the latter approach is adopted, but first it is demonstrated that both methods are consistent.

The electric field of an infinite array as given by (8b) must satisfy the same periodicity conditions as the source function, i.e.,

$$\mathbf{E}(x + ma, y + nb, z) = \mathbf{E}(x, y, z) \exp [j\sigma_{mn}(\varphi) \cos \theta]. \qquad (27)$$

* Note the distinction made between "grazing position" and "transitive characteristic direction". A beam associated with a transitive characteristic direction may attain a grazing position for a particular phasing, but may also point in other directions.

On the other hand, the electric field

$$\mathbf{E}(x, y, z, \theta) = -\int_V \bar{\bar{G}}(x, y, z \mid \xi, \eta, \zeta) \cdot \mathbf{J}(\xi, \eta, \zeta, \theta)\, d\tau \tag{28}$$

can be expanded in an infinite sum of integrals using the relationships of (6):

$$\mathbf{E}(x, y, z, \theta) = -\sum_{m=-\infty}^{\infty} \sum_{n=-\infty}^{\infty} \exp(j\sigma_{mn} \cos\theta) \cdot \int_{V_{00}} \bar{\bar{G}}(x, y, z \mid \xi + ma, \eta + nb, \zeta) \cdot \mathbf{J}_0(\xi, \eta, \zeta)\, dv, \tag{29}$$

where $V_{00}$ is the volume occupied by $\mathbf{J}_0$. Define a new Green's function

$$\bar{\bar{G}}_0(x, y, z \mid \xi, \eta, \zeta) = \sum_{m=-\infty}^{\infty} \sum_{n=-\infty}^{\infty} \exp(j\sigma_{mn} \cos\theta) \bar{\bar{G}}(x, y, z \mid \xi + ma, \eta + nb, \zeta) \tag{30}$$

and notice that

$$\bar{\bar{G}}_0(x, y, z \mid \xi + Ma, \eta + Nb, \zeta) = \exp(-j\sigma_{MN} \cos\theta) \bar{\bar{G}}_0(x, y, z \mid \xi, \eta, \zeta) \tag{31}$$

since by (5)

$$\sigma_{m+M, n+N} = \sigma_{mn} + \sigma_{MN}. \tag{32}$$

From (27) and (31) it follows that $\bar{\bar{G}}_0(x, y, z \mid \xi, \eta, \zeta)$ can be expanded by the eigenfunctions (25) as

$$\bar{\bar{G}}_0 = \sum_{p=-\infty}^{\infty} \sum_{q=-\infty}^{\infty} \bar{g}_{pq}(z \mid \zeta) F_{pq}(x, y) F_{pq}^*(\xi, \eta), \tag{33a}$$

where

$$\bar{g}_{pq} = \frac{1}{ab} \int_0^a \int_0^b \bar{\bar{G}}_0 F_{pq}(\xi, \eta) F_{pq}^*(x, y)\, dx\, dy\, d\xi\, d\eta. \tag{33b}$$

Substituting (30) via (33a) into (15a) for the center element, $m = n = 0$, one obtains

$$Z_{00} = \sum_{p=-\infty}^{\infty} \sum_{q=-\infty}^{\infty} z_{pq}, \tag{34a}$$

where

$$z_{pq} = \frac{1}{|I_{00}|^2} \int_{V_{00}} \int_{V_{00}} \mathbf{J}_0^*(x, y, z) \cdot \bar{g}_{pq}(z \mid \zeta) F_{pq}^*(\xi, \eta) F_{pq}(x, y) \cdot \mathbf{J}_0(\xi, \eta, \zeta)\, d\tau\, dv. \tag{34b}$$

Equation (34) is an alternate representation to (86) for the radiation impedance of the infinite array element and it demonstrates that Lemma 2 is valid for infinite arrays as well.

By substituting the new representation for $\bar{\bar{G}}_0$, (30), (33), into (29) and noting that the electric field satisfies the homogeneous reduced wave equation in the source-free region, one obtains for the unbounded space

$$\mathbf{E}(x, y, z) = \sum_{p=-\infty}^{\infty} \sum_{q=-\infty}^{\infty} \mathcal{E}_{pq} F_{pq}(x, y) \exp(-\gamma_{pq} |z|) \qquad |z| > d_{\max}, \quad (35a)$$

where

$$\gamma_{pq} = \sqrt{\Gamma_{pq}^2 - k^2} = jk \sin \theta_{pq} \quad (35b)$$

$$\mathcal{E}_{pq} e^{-\gamma_{pq}|z|} = -\int_{V_{00}} \bar{\bar{g}}_{pq}(z \mid \zeta) \cdot \mathbf{J}_0(\xi, \eta, \zeta) F_{pq}^*(\xi, \eta) \, d\tau, \quad (35c)$$

and $d_{\max}$ is the projection on the $z$-axis of the largest distance between two points on the surface enclosing $V_{00}$. It can be seen that the electric field in the source-free region, above the central area of a large array may be approximated by a finite number of homogeneous plane waves propagating in the real characteristic directions, and an infinite number of nonhomogeneous plane waves, exhibiting exponential decay in the direction perpendicular to the plane of the array. The latter are interpreted as waves propagating in the imaginary characteristic directions.

In an infinite array all elements are embedded in an identical environment, and therefore the power radiated by each element is the same. There is no net power flow into a unit cell through the "side walls". Consequently, the quantity $\mathrm{Re}\{|I_{00}|^2 z_{pq}\}$ of (34b) is equal to the power propagated by the plane wave $(p, q)$ within a unit cell in the direction perpendicular to the plane of the array. By Lemma 1 there is only a finite number of plane waves with transitive characteristic directions (see footnote p. 1571). Let them be distinguished from all other plane waves by assignment of the subscript $(p, q) = (\tau, \nu)$.

$$\mathbf{E}_{\tau\nu} = \mathcal{E}_{\tau\nu} F_{\tau\nu}(x, y) \exp(-jk |z| \sin \theta_{\tau\nu}), \qquad |z| > d_{\max} \quad (36)$$

$$\mathbf{H}_{\tau\nu} = \mathcal{H}_{\tau\nu} F_{\tau\nu}(x, y) \exp(-jk |z| \sin \theta_{\tau\nu}), \qquad |z| > d_{\max}, \quad (37)$$

where $F_{\tau\nu}(x, y)$ is given by (26), and $\mathcal{E}_{\tau\nu}$ by (35c). If

$$\mathfrak{D}_{\tau\nu} = jk[\cos \theta_{\tau\nu} \cos \varphi_{\tau\nu} \mathbf{a}_x + \cos \theta_{\tau\nu} \sin \varphi_{\tau\nu} \mathbf{a}_y - \sin \theta_{\tau\nu} \mathbf{a}_z] \quad (38)$$

175

then

$$\mathcal{H}_{\tau\nu} = \frac{j}{\omega\mu} \mathfrak{D}_{\tau\nu} \times \mathcal{E}_{\tau\nu} \, . \tag{39}$$

The power radiated by a $(\tau, \nu)$ plane wave per unit cell into the upper hemisphere is

$$P_{\tau\nu} = \frac{1}{2} \text{Re} \int_0^a \int_0^b (\mathbf{E}_{\tau\nu} \times \mathbf{H}_{\tau\nu}^*) \cdot \mathbf{a}_z \, dx \, dy. \tag{40}$$

Substitution of (36) through (39) into (40) gives

$$P_{\tau\nu} = \frac{ab}{\eta_0} \sin \theta_{\tau\nu}^* \left[ | \mathcal{E}_{\tau\nu} \cdot \mathbf{a}_x |^2 + | \mathcal{E}_{\tau\nu} \cdot \mathbf{a}_y |^2 + \frac{\sin \theta_{\tau\nu}}{\sin \theta_{\tau\nu}^*} | \mathcal{E}_{\tau\nu} \cdot \mathbf{a}_z |^2 \right] \tag{41}$$

where $\eta_0 = (\mu/\epsilon)^{\frac{1}{2}}$. From (41) a radiation resistance per wave is defined as

$$R_{\tau\nu} = \frac{P_{\tau\nu}}{| I_{00} |^2}. \tag{42}$$

Since the entire system is passive and lossless, then by conservation of energy, the power $P_{\tau\nu}$ must originate from the element itself. Hence,

$$R_{\tau\nu} = \text{Re} \, \{z_{\tau\nu}\}, \qquad \theta_{\tau\nu} \ \text{real}, \tag{43}$$



where $z_{\tau\nu}$ is given by (34b).

From (41) it follows that when a wave $(\tau, \nu)$ is in real space $R_{\tau\nu}$ is real, and when it is in imaginary space $R_{\tau\nu}$ is imaginary (in which case $\text{Re}\{z_{\tau\nu}\} = 0$). Hence, of all the elements comprising the source's load, $R_{\tau\nu}$ appears either resistive or reactive, depending upon the scan angle. Such properties of a load, which are unknown in lumped network theory, are a consequence of the losslessness postulate. When propagation is possible power is carried away from the source. When propagation is inhibited there is no net loss of power and the load must be reactive. By Lemma 1 only $\text{Re}\{z_{\tau\nu}\}$ has those properties. All other $z_{pq}$, $(p, q) \neq (\tau, \nu)$ and $\text{Im}\{z_{\tau\nu}\}$ always retain their dissipative or reactive characteristics. Further, there is only a finite number of terms having $\text{Re}\{z_{pq}\} > 0$. In practical phased arrays the spacing between the elements and the scanning range are such that only one such term exists at a time.

The following two definitions summarize the properties described above:

*Definition 1:* An O-type network function is a scan-dependent immittance (impedance or admittance) which is seen by the source as resistive when the beam is in real space and as reactive when the beam is in imaginary space, and it behaves like an open circuit for impedance and like a short circuit for admittance in the grazing position.

*Definition 2:* An E-type network function is a scan-dependent immittance (impedance or admittance) which remains either resistive or reactive when the beam passes through the grazing position.

The motivation behind the nomenclature introduced by the two definitions will become clear later, in Theorems 4 and 5. The O-type and E-type immittances are of distinct mathematical form. To arrive at it consider first the following transformation:*

$$s = \sin \theta_{mn} \tag{44}$$

$$\chi = \cos \varphi_{mn} , \tag{45}$$

where $(m, n)$ is one particular transitive characteristic direction out of all $(\tau, \nu)$. Given $s$ and $\chi$ all other characteristic directions are uniquely determined. By (4)

$$\psi_x = ka\chi\sqrt{1 - s^2} - 2m\pi \tag{46}$$

$$\psi_y = kb\sqrt{1 - \chi^2}\,\sqrt{1 - s^2} - 2n\pi, \tag{47}$$

where $(1 - \chi^2)^{\frac{1}{2}} \geqq 0$ for all possible $\chi$ and $(1 - s^2)^{\frac{1}{2}} > 0$ if $0 \leqq \theta_{mn} < \pi/2$, and $(1 - s^2)^{\frac{1}{2}} < 0$ if $\pi/2 < \theta_{mn} \leqq \pi$. Then by substitution of (47) into (22) all other characteristic directions are found:

$$\tan \varphi_{pq} = \frac{kab(1 - \chi^2)^{\frac{1}{2}}(1 - s^2)^{\frac{1}{2}} + 2(q - n)\pi a}{kab\chi(1 - s^2)^{\frac{1}{2}} + 2(p - m)\pi b} \tag{48a}$$

$$\cos \theta_{pq} = f_{pq}(s), \tag{48b}$$

where

$$f_{pq}(s) = \frac{ka\chi(1 - s^2)^{\frac{1}{2}} + 2(p - m)\pi}{ka \cos \varphi_{pq}}. \tag{48c}$$

This suggests that when characteristic direction $(m, n)$ is scanned in a plane $\chi =$ const, each of the components $z_{pq}$ of the total input impedance as given by (34) can be expressed as a function of the same variable $s$. The conformal mapping between the $\theta_{mn}$-plane and the $s$-plane is shown in Fig. 3. In view of the branch cut $-1 \leqq \alpha \leqq 1$ it will be understood that $s = \alpha$ denotes $s = \alpha - j0$ if $0 \leqq \theta_{mn} \leqq \pi/2$ and $s = \alpha + j0$ if $\pi/2 \leqq \theta_{mn} \leqq \pi$. Let $s = s_{\tau\nu}$ be the value at which characteristic direc-

* Recall that $\theta_{mn}$ and $\varphi_{mn}$ are not in the conventional spherical coordinate system (see p. 1564).

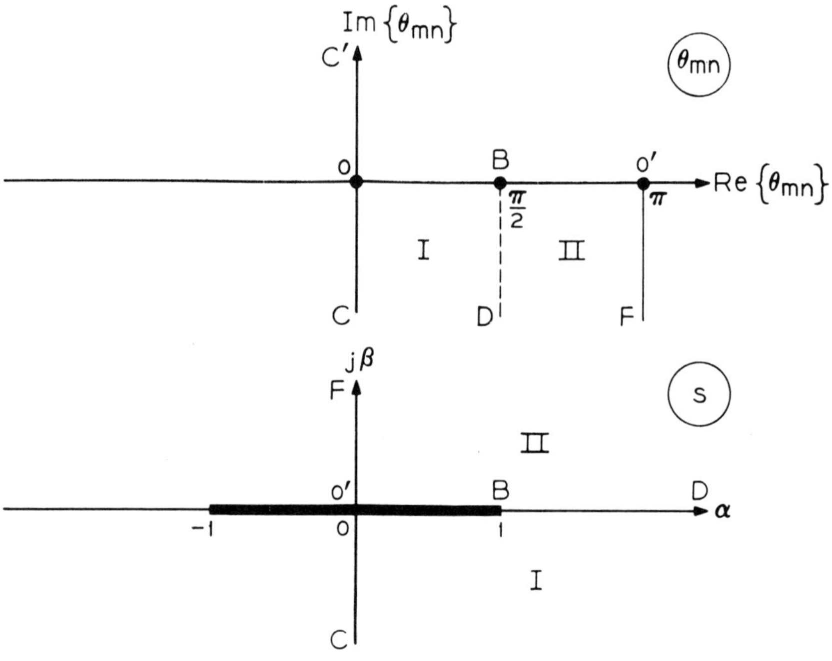


Fig. 3 — Conformal mapping $s = \sin \theta_{mn}$.

tion $(\tau, \nu)$ is in grazing position. At this value

$$f_{\tau\nu}^2(s_{\tau\nu}) = 1. \tag{49}$$



Of all values $\{s_{\tau\nu}\}$ there is at least one which satisfies (49) for $s_{\tau\nu} = 0$. From (48) it is obvious that $f_{mn}^2(0) = 1$, and there may be other transitive characteristic directions $(\tau, \nu) \neq (m, n)$ which attain their grazing positions at $s_{\tau\nu} = 0$.

*Theorem 3: In an obstacle-free space, the impedance function $z_{pq}(s)$, associated with the characteristic direction $(p, q)$, is an analytic function of the complex variable $s = \alpha + j\beta$, with branch points at $s = s_{\tau\nu}$ and an essential singularity at $|s| \to \infty$. If $(p, q) = (m, n)$ then $z_{mn}(s)$ may have a simple pole at $s = s_{mn} = 0$.*

*Proof:* The general definition of $z_{pq}$ is given by (34b) in which the $\theta_{pq}$ dependence is contained in the Green's function component $\bar{g}_{pq}(z \mid \zeta)F_{pq}^*(\xi, \eta)F_{pq}(x, y)$. The Green's function is derived from (10) via (15b). Green's function $\bar{\mathcal{G}}(x, y, z \mid \xi, \eta, \zeta)$ satisfies the same periodic boundary conditions as $\bar{G}_0(x, y, z \mid \xi, \eta, \zeta)$ and can be expanded in a series similar to (33a):

$$\bar{\mathcal{G}} = \sum_{p=-\infty}^{\infty} \sum_{q=-\infty}^{\infty} \bar{C}_{pq}(z \mid \zeta)F_{pq}(x, y)F_{pq}^*(\xi, \eta). \tag{50}$$

By substitution of (50) into (10) and use of the orthogonality property of $F_{pq}(x, y)$ one obtains a differential equation for $\bar{\bar{C}}_{pq}(z \mid \zeta)$

$$\frac{d^2\bar{\bar{C}}_{pq}}{dz^2} - \gamma_{pq}^2\bar{\bar{C}}_{pq}(z \mid \zeta) = -\bar{\bar{I}}\,\frac{\delta(z - \zeta)}{ab}$$

$$\gamma_{pq} = jk \sin \theta_{pq} \tag{51}$$

with the additional requirement that as $|z| \to \infty$, $\bar{\bar{C}}_{pq}$ behaves as an outgoing or evanescent wave. The solution of (51) for free space is

$$\bar{\bar{C}}_{pq}(z \mid \zeta) = \bar{\bar{I}}\,\frac{1}{2jabk \sin \theta_{pq}} \exp\{-jk\,|z - \zeta|\sin \theta_{pq}\} \tag{52}$$

$\bar{\bar{g}}_{pq}(z \mid \zeta)$ is obtained from $\bar{\bar{C}}_{pq}(z \mid \zeta)$ through an operator $\bar{\bar{\mathcal{R}}}_{pq}$:

$$\bar{\bar{g}}_{pq}(z \mid \zeta) = j\omega\mu\bar{\bar{\mathcal{R}}}_{pq}\cdot\bar{\bar{C}}_{pq}(z \mid \zeta), \tag{53a}$$

where

$$\bar{\bar{\mathcal{R}}}_{pq} = \bar{\bar{I}} + \frac{1}{k^2}\,\mathfrak{D}_{pq}\mathfrak{D}_{pq} \tag{53b}$$

$\mathfrak{D}_{pq}$ being given by (38). Substitution of (52) into (53a) followed by substitution into (34b) gives

$$z_{pq} = \frac{\omega\mu}{2abk\,|I_{00}|^2 \sin \theta_{pq}} \int_{V_{00}} \int_{V_{00}} \mathbf{J}_0^*(x, y, z)$$
$$\cdot\bar{\bar{\mathcal{R}}}_{pq}\cdot\mathbf{J}_0(\xi, \eta, \zeta) \exp\{jk \cos \theta_{pq}[(x - \xi)\cos \varphi_{pq}$$
$$+ (y - \eta)\sin \varphi_{pq}] - jk\,|z - \zeta|\sin \theta_{pq}\}\,d\tau\,dv. \tag{54}$$

The integrand is an entire function of $\theta_{pq}$ with an essential singularity at $|\operatorname{Im}\{\theta_{pq}\}| \to \infty$. Hence,* if $\mathbf{J}_0(x, y, z)$ is piecewise continuous, the integral is also an entire function with the same essential singularity. By (48)

$$\sin \theta_{pq} = [1 - f_{pq}^2(s)]^{\frac{1}{2}}. \tag{55}$$

By Lemma 1,

$$f_{pq}^2(s) \neq 1 \quad \text{if} \quad (p, q) \neq (\tau, \nu). \tag{56}$$

From Fig. 3 it is readily seen that $|s| < \infty$ when $|\theta_{pq}| < \infty$ which implies, via (48), (55) that $|\cos \theta_{pq}| < \infty$ and $|\sin \theta_{pq}| < \infty$ as long as $|s| < \infty$. Thus, the singularities introduced by the transformation (44), (45) are the branch points at $s = s_{\tau\nu}$. Also if $(p, q) = (m, n)$†

* E. J. Copson, Ref. 11, Sec. 5.5, pp. 107–109.

† Recall that (m, n) is the characteristic direction which defines the transformation from $(\psi_x, \psi_y)$ into $(s, \chi)$, (44)–(47).

$z_{pq}$ may have a simple pole at $s = 0$. (Note, for example, that for horizontal polarization, $\mathbf{J}_0 = \mathbf{a}_x J_0$, there is a simple zero in the plane of scan corresponding to $\varphi_{mn} = 0$, at $s = 0$.) Q.E.D.

The above proof can be applied separately to the real and imaginary parts of the right-hand side of (54). If $z_{pq} = R_{pq}(s) + jX_{pq}(s)$, then $R_{pq}(s)$ and $X_{pq}(s)$ are analytic functions of $s$, real on the real axis of $s$, with an essential singularity at $|s| \to \infty$, branch points at $s = s_{\tau\nu}$, and possibly simple poles at $s = 0$.

In systems other than obstacle-free space, the normalized complex power $z_{pq}(s)$ has different forms. Except for isolated values of $s$, the radiated power and the stored energy per unit cell are bounded and continuous functions of $s$ over those portions of the real and imaginary axes of the $s$-plane which have physical meaning. Hence, it is reasonable to postulate that an analytic continuation of $z_{pq}$ as a function of scan can be made into a region of the complex $s$-plane which includes portions of both the real and imaginary axes. It may be of interest to note that the impedance function $z_{pq}(s)$ derived by L. Stark[14] for the planar dipole array over a ground plane is analytic. The regularity of $z_{pq}(s)$ depends directly on the regularity of $\bar{g}_{pq}(z \mid \zeta; s)$. The singularities of $z_{pq}$ in the $s$-plane are determined by the boundary conditions which $\bar{g}_{pq}(z \mid \zeta; s)$ satisfies.



*Theorem 4: An E-type immittance function $V(s)$ is an even function of $s$.*

*Proof:* Let the complex variable $s$ be defined with respect to the transitive characteristic direction $(m, n)$. Once $(m, n)$ is chosen, the proper branch of $(1 - s^2)^{\frac{1}{2}}$ in (48) is uniquely determined. Let $(k, l)$ denote all other transitive characteristic directions which reach their transitive position simultaneously with $(m, n)$. Formally, this implies

$$f_{\tau\nu}^2(0) = 1 \qquad (\tau, \nu) = (m, n), (k, l). \tag{57}$$

As a consequence of Definition 2 and Lemma 1, $V(s)$ is recognizable as

$$V(s) = \begin{cases} R_{pq}(s) & (p, q) \neq (m, n), (k, l), \\ X_{pq}(s) & \text{all } (p, q), \end{cases} \tag{58}$$

where $R_{pq}(s) + jX_{pq}(s) = z_{pq}(s)$, $z_{pq}$ given by (54). Thus, (58) establishes the connection between the defined E-type function and physical quantities corresponding to $R_{pq}(s)$ and $X_{pq}(s)$. Consider Definition 2 which states

$$V(s) - V^*(s) = 0 \qquad s = \alpha \qquad 0 < \alpha < 1, \tag{59a}$$

$$V(s) - V^*(s) = 0 \qquad s = j\beta. \tag{59b}$$

Since $V(s)$ is analytic and also real on the real axis of $s$, (59) may be rewritten as*

$$V(s) - V(s^*) = 0 \qquad s = \alpha \qquad 0 < \alpha < 1, \tag{60a}$$

$$V(s) - V(s^*) = 0 \qquad s = j\beta. \tag{60b}$$

On the real axis

$$V(\alpha) - V(\alpha) = 0. \tag{61a}$$

On the imaginary axis

$$V(j\beta) - V(-j\beta) = 0. \tag{61b}$$

By analytic continuation† of (61b) from the imaginary axis to a point $s$ in the complex plane one obtains

$$V(s) - V(-s) = 0. \tag{62}$$

Hence, $V(s)$ is an even function of $s$. Q.E.D.

*Theorem 5: An O-type immittance $W(s)$ is an odd function of $s$.* The proof is similar to that of Theorem 4 and it appears in Appendix D.

It has been shown in Theorem 2 that a finite phased array cannot be perfectly matched over a continuous scanning range. The proof is limited to finite arrays and cannot be directly extended to infinite arrays since the representation of the element impedance by (17a) does not guarantee convergence in the complex $\theta$-plane if the limits of the summations are extended to infinity. In treating the infinite array, the element impedance is derived by symmetry considerations from which it is concluded that the net complex power radiated from each element is conserved entirely within the unit cell of that element. It has been shown that the two definitions are consistent. Although the problem of whether an infinite array can be perfectly matched is of academic interest only, it is worthwhile noting that as for finite arrays, the answer in this case is *also* negative. To show this the reader may recall that the impedance has been defined as normalized power and postulated to be an analytic function of the scan variable $s = \alpha + j\beta$. The normalization constant is $|I_{00}|^2$ given by (13). If the complex power as a function of scan is represented by

$$\hat{P}(s) = |I_{00}|^2 [R(s) + jX(s)], \tag{63}$$

* P. M. Morse and H. Feshbach, Ref. 7, Vol. I, p. 393.
† Morse and Feshbach, *Op. Cit.*, p. 389.

then by Lemma 1, the term $R(s)$ is a finite sum of analytic functions of the complex variable $s$. Consequently, $R(s)$ is an analytic function of $s$. In general, it may be represented as

$$R(s) = E(s) + \mathcal{O}(s), \tag{64}$$

where $E(s)$ is an even function of $s$ and $\mathcal{O}(s)$ is an odd function of $s$. Under conditions of perfect match over a continuous range, constant power, $P_r$, is radiated over that range. Since $R(s)$ is analytic it implies $R(s) = P_r \mid I_{00} \mid^{-2}$ everywhere in the $s$-plane. Since a constant is even, $\mathcal{O}(s) = 0$. Further, $E(s)$ must have a branch cut on the real axis of the $s$-plane in the interval $[-1,1]$. But the branch cut does not exist if $E(s) = P_r \mid I_{00} \mid^{-2}$. The contradiction implies that $\hat{P}(s)$ in (63) cannot equal a constant over a continuous range of $s$.

*Theorem 6: The resistance and reactance functions of an element, or their derivatives, in an infinite linear or planar phased array of current sources are discontinuous when a grating lobe is in a grazing position.*

*Proof:* In an infinite array the grating lobes are plane waves propagating in the characteristic directions. By Theorems 4 and 5 the element impedance $Z(s)$ in an obstacle-free space can be written as



$$Z(s) = P(s) + \frac{Q(s)}{s}. \tag{65}$$

For real values of $s$, $P(s)$ is an even complex function of $s$ bounded at $s = 0$, and $Q(s)$ is an even real function of $s$ nonzero at $s = 0$. On the real axis of $s$

$$Z(\alpha) = P(\alpha) + \frac{Q(\alpha)}{\alpha}. \tag{66a}$$

On the imaginary axis of $s$

$$Z(j\beta) = P(j\beta) - j\frac{Q(j\beta)}{\beta}. \tag{66b}$$

A grating lobe is in its transitive position at $s = 0$. The pole discontinuities are established by showing that

$$\operatorname{Re}\{\lim_{\alpha\to 0} Z(\alpha) - \lim_{\beta\to 0} Z(j\beta)\} = \lim_{\alpha\to 0}\frac{Q(\alpha)}{\alpha} = \infty \tag{67a}$$

$$\operatorname{Im}\{\lim_{\alpha\to 0} Z(\alpha) - \lim_{\beta\to 0} Z(j\beta)\} = \lim_{\beta\to 0}\frac{Q(j\beta)}{\beta} = \infty. \tag{67b}$$

The pole discontinuity has to be interpreted as an invalid mathematical solution at the transitive position. It is a result of the idealization introduced by the concept of an "infinite array." If $R_{mn}(s)$ has a simple zero at $s = 0$, as is the case when a horizontally polarized array is placed above a ground plane, then the active impedance in the neighborhood of $s = 0$ can be written as

$$Z(s) = R(s) + jX(s), \tag{68a}$$

where $R(s)$ and $X(s)$ are real functions of $s$ (real for s real).

$$R(s) = \sum_{i=0}^{\infty} a_i s^i \tag{68b}$$

$$X(s) = \sum_{i=0}^{\infty} b_{2i} s^{2i}. \tag{68c}$$

When the beam whose transitive characteristic direction is in real space, $s = \alpha$

$$R_\alpha \triangleq R(\alpha) = \sum_{i=0}^{\infty} a_i \alpha^i \tag{69a}$$

and when it is in imaginary space, $s = j\beta$

$$R_\beta \triangleq \text{Re}\,\{R(j\beta)\} = \sum_{i=0}^{\infty} (-1)^i a_{2i} \beta^{2i}. \tag{69b}$$



The discontinuity in the derivative of the resistance is

$$\lim_{\alpha \to 0} \frac{dR_\alpha}{d\alpha} - \lim_{\beta \to 0} \frac{dR_\beta}{d\beta} = a_1. \tag{70}$$

Similarly, the reactance

$$X_\alpha \triangleq X(\alpha) = \sum_{i=0}^{\infty} b_{2i} \alpha^{2i} \tag{71a}$$

$$X_\beta \triangleq \text{Im}\,\{Z(j\beta)\} = \sum_{i=0}^{\infty} (-1)^i [b_{2i}\beta^{2i} + a_{2i+1}\beta^{2i+1}] \tag{71b}$$

$$\lim_{\alpha \to 0} \frac{dX_\alpha}{d\alpha} - \lim_{\beta \to 0} \frac{dX_\beta}{d\beta} = -a_1\,. \tag{72}$$

The proof can be generalized for any order algebraic singularity or zero at $s = 0$. For example, if there is a zero of multiplicity $N$ the discontinuity will be in the $N$th derivatives of the resistance and reactance. A noninteger order zero yields a discontinuity after a sufficient number of differentiations. Q.E.D.

## V. SUMMARY AND CONCLUSIONS

A new mathematical approach to phased arrays has been adopted to investigate and discover various properties of the radiation impedance of an array element as a function of scan angle. The underlying idea of the method is the treatment of the impedance as an analytic function of a complex scan variable, which enables one to prove that an array element subject to the model chosen cannot be perfectly matched over a continuous scanning range by using lossless, linear, passive and time-invariant elements.

The first half of the theory is devoted to finite arrays. It is shown that the directions (in space) of the beams' maxima are eigenvalues of a Laplacian differential operator with periodic boundary conditions, which are related to the phase taper of the array. It is proven that there exists only a finite number of real eigenvalues. The known concept of imaginary space is then adopted to accommodate the imaginary eigenvalues. Furthermore, it is demonstrated that all beams except a finite number are completely confined either to real space or to imaginary space, and that only a finite number of beams may attain a grazing position. The unique properties of the latter beams have been found to play an important role in the investigation of infinite arrays, to which the second half of the theory is devoted.



The interest in infinite arrays, apart from its academic aspect, stems from the good approximation it provides for the behavior of the center portion of a large finite array. It has been found that the infinite array element impedance as a function of scan is restricted to a specific mathematical form. It is the authors' hope that recognition of the limitations imposed by that form may provide useful guidelines in achieving optimal match of an array to space.

## VI. ACKNOWLEDGMENT

The authors wish to thank N. Amitay, E. R. Nagelberg, and R. G. Pecina for their critical reading of the manuscript and for valuable comments and suggestions regarding this work.

## APPENDIX A

### *Proof of Corollary 1a*

*Corollary 1a:* $Re\{Z_{rs}\}$ *and* $\frac{\partial}{\partial \theta} Re\{Z_{rs}\}$ *are entire functions of* $\theta$ *each with an essential singularity at* $\theta \rightarrow \infty$.

*Proof:* Denoting

$$\tilde{z}_{mnrs} = \rho_{mnrs} + j\chi_{mnrs} \tag{73}$$

one obtains from (17a)

$$\begin{aligned} \text{Re}\{Z_{rs}(\theta)\} &\triangleq R_{rs}(\theta) \\ &= \sum_{m=0}^{M-1} \sum_{n=0}^{N-1} \{\rho_{mnrs} \cos [B_{mnrs}(\theta)] - \chi_{mnrs} \sin [B_{mnrs}(\theta)]\}, \end{aligned} \tag{74}$$

where

$$B_{mnrs}(\theta) = (\sigma_{mn} - \sigma_{rs}) \cos \theta \tag{75}$$

and

$$\begin{aligned} \frac{\partial}{\partial \theta} R_{rs}(\theta) = - \sum_{m=0}^{M-1} \sum_{n=0}^{N-1} B'_{mnrs}(\theta) \{ \rho_{mnrs} \sin [B_{mnrs}(\theta)] \\ + \chi_{mnrs} \cos [B_{mnrs}(\theta)]\}. \end{aligned} \tag{76}$$

Since $\cos \theta$ is an entire function of $\theta$, $\cos[B_{mnrs}(\theta)]$ and $\sin[B_{mnrs}(\theta)]$ are entire functions of an entire function, and are therefore entire. The existence of the essential singularity can be demonstrated in a similar fashion to that in Theorem 1. Q.E.D.

## APPENDIX B



### *Proof of Lemma 1*

*Lemma 1: Every steerable linear or planar phased array with a linear phase taper has only a finite number of beams in real space.*

*Proof*: A beam $(p, q)$ is in real space if $|\cos \theta_{pq}| \leqq 1$. Dividing (4a) by $ka$ and (4b) by $kb$, squaring and adding, one obtains

$$\left(\frac{\psi_x + 2p\pi}{ka}\right)^2 + \left(\frac{\psi_y + 2q\pi}{kb}\right)^2 \leqq 1 \tag{77}$$

or

$$\left(\frac{\psi_x}{2\pi} + p\right)^2 \left(\frac{\lambda}{a}\right)^2 + \left(\frac{\psi_y}{2\pi} + q\right)^2 \left(\frac{\lambda}{b}\right)^2 \leqq 1. \tag{78}$$

Necessary conditions for the above inequality to be satisfied are

$$\left|\frac{\psi_x}{2\pi} + p\right| \frac{\lambda}{a} \leqq 1 \tag{79}$$

$$\left|\frac{\psi_y}{2\pi} + q\right| \frac{\lambda}{b} \leqq 1. \tag{80}$$

Since

$$-\frac{1}{2} \leqq \frac{\psi_x}{2\pi} \leqq \frac{1}{2} \quad \text{and} \quad -\frac{1}{2} \leqq \frac{\psi_y}{2\pi} \leqq \frac{1}{2},$$

$$|p| \leqq \frac{a}{\lambda} + \frac{1}{2} \tag{81}$$

$$|q| \leqq \frac{b}{\lambda} + \frac{1}{2}. \tag{82}$$

Hence, both $p$ and $q$ are bounded. Q.E.D.

## APPENDIX C

### *Proof of Lemma 2*

*Lemma 2: The radiation impedance of an element in a linear or planar phased array can be expanded by an infinite series over all characteristic directions of the system.*

*Proof:* The current density excitation function of a finite-size array given by (1), (2) satisfies the periodic boundary conditions (23) in the finite domain occupied by the array. Let this domain be denoted by $D$. The current density can, therefore, be uniquely expanded in $D$ in terms of the eigenfunctions (25):



$$\mathbf{J}(x, y, z) = U(x, y, D) \sum_{p=-\infty}^{\infty} \sum_{q=-\infty}^{\infty} \mathbf{j}_{pq}(z) F_{pq}(x, y), \tag{83}$$

where

$$\mathbf{j}_{pq}(z) = \frac{1}{ab} \int_0^a \int_0^b \mathbf{J}_0(x, y, z) F_{pq}^*(x, y)\, dx\, dy \tag{84}$$

and $U(x, y, D)$ is a two-dimensional unit step function

$$U(x, y, D) = \begin{cases} 1 & (x, y) \text{ in } D, \\ 0 & \text{otherwise.} \end{cases} \tag{85}$$

Substitution of (31a) into (15a) yields

$$Z_{mn} = \sum_{p=-\infty}^{\infty} \sum_{q=-\infty}^{\infty} \hat{z}_{mnpq}, \tag{86a}$$

where

$$\hat{z}_{mnpq} = \frac{1}{|I_{mn}|^2} \int_{V_{mn}} \int_V \mathbf{J}_{mn}^*(x, y, z) \cdot \bar{\bar{G}}(x, y, z \mid \xi, \eta, \zeta) \cdot \mathbf{j}_{pq}(\zeta) F_{pq}(\xi, \eta) U(\xi, \eta, D)\, d\tau\, dv. \tag{86b}$$

Q.E.D.

APPENDIX D

## *Proof of Theorem 5*

*Theorem 5: An O-type immittance $W(s)$ is an odd function of $s$.*

*Proof:* Let the complex variable $s$ be defined with respect to the transitive characteristic direction $(m, n)$. Let $(k, l)$ be all other transitive characteristic directions which reach their transitive position simultaneously with $(m, n)$. Then as a consequence of Definition 1 and Lemma 1

$$W(s) = R_{pq}(s) \qquad (p, q) = (m, n), (k, l), \tag{87}$$

where $R_{pq}(s) = \mathrm{Re}\{z_{pq}\}$, $z_{pq}$ given by (54). Thus, (87) establishes the connection between the defined O-type function and a physical quantity corresponding to $R_{pq}(s)$. From Definition 1

$$W(s) - W^*(s) = 0 \qquad s = \alpha, \qquad 0 < \alpha < 1 \tag{88}$$

$$W(s) + W^*(s) = 0 \qquad s = j\beta. \tag{89}$$

Since $W(s)$ is real on the real axis of $s$, (88), (89) may be rewritten as

$$W(s) - W(s^*) = 0 \qquad s = \alpha \tag{90}$$

$$W(s) + W(s^*) = 0 \qquad s = j\beta. \tag{91}$$

On the real axis

$$W(\alpha) - W(\alpha) = 0. \tag{92}$$

On the imaginary axis

$$W(j\beta) + W(-j\beta) = 0. \tag{93}$$

By analytic continuation of (93) from the imaginary axis to a point $s$ in the complex plane one obtains

$$W(s) + W(-s) = 0. \tag{94}$$

Hence, $W(s)$ is an odd function of $s$. Q.E.D.

REFERENCES

1. Hansen, R. C., Editor, *Microwave Scanning Antennas,* Vol. II, Academic Press, 1964.
2. Wu, C. P., private communication.
3. Amitay, N., Improvement of Planar Array Match by Compensation through Contiguous Element Coupling, IEEE Trans. Antennas Prop., *AP-14,* No. 5, September, 1966, pp. 580–586.
4. Hannan, P. W., Proof that a Phased-Array Antenna Can Be Impedance

Matched for all Scan Angles, Radio Science, *2* (new series), No. 3, March, 1967, pp. 361–369.
5. Hannan, P. W., Lerner, D. S., and Knittel, G. H., Impedance Matching a Phased-Array Antenna Over Wide Scan Angles by Connecting Circuits, IEEE Trans. Antennas Prop., *AP-13,* No. 1, January, 1965, pp. 28–34.
6. Schelkunoff, S. A., Some Equivalence Theorems of Electromagnetics and their Application to Radiation Problems, B.S.T.J., *15,* January, 1936, pp. 92–112.
7. Morse, P. M., and Feshbach, H., *Methods of Theoretical Physics,* McGraw-Hill Book Co., Inc., 1953, Vol. II, p. 1769.
8. Stratton, J. A., *Electromagnetic Theory,* McGraw-Hill Book Co., Inc., 1941, pp. 132–133.
9. Guillemin, E. A., *Communication Networks,* John Wiley and Sons, Inc., 1935, Vol. II, pp. 474–476.
10. Churchill, R. V., *Complex Variables and Applications,* McGraw-Hill Book Co., Inc., 1960, Sec. 19, p. 40.
11. Copson, E. J., *Theory of Functions of a Complex Variable,* Oxford University Press, 1935, Sec. 7.1, p. 159.
12. Wheeler, H. A., The Radiation Resistance of an Antenna in an Infinite Array or Waveguide, Proc. IRE, *36,* April, 1948, pp. 478–488.
13. Galindo, V. and Wu, C. P., Asymptotic Behavior of the Coupling Coefficients for an Array of Thin-Walled Rectangular Waveguides, IEEE Trans. Antennas Prop., *AP-14,* No. 2, March, 1966, pp. 248–249.
14. Stark, L., Radiation Impedance of a Dipole in an Infinite Planar Phased Array, Radio Science, *1* (new series), No. 3, March, 1966, pp. 361–377.

# Aperture Matching for an Infinite Circular Polarized Array of Rectangular Waveguides

SHUNG W. LEE, MEMBER, IEEE

*Abstract*—This paper studies the polarization of the main beam radiated from an infinite array of rectangular waveguides, and in particular, illustrates the matching procedure for the design of a circularly polarized array. It is demonstrated that the use of resonant irises at the waveguide openings together with a conventional matching junction inside the waveguides can improve both the power and the axial ratio of the main beam in a circularly polarized array. In general, a good matching design is the one which gives a proper compromise between the power and the polarization, and is not necessarily the one minimizing the reflection.

## I. Introduction

RECENT years have witnessed considerable progress in the aperture design of a large phased array of rectangular waveguides. Most of the work has been devoted to the case of linearly polarized radiating elements, where the waveguide width-to-height ratio is small and supports only the dominant $TE_{10}$ modes [1]. In many practical applications, however, an array which radiates circularly polarized beams is needed, and systematic design information about such arrays is apparently lacking. Some scattered computations concerning polarization can be found in [2] and [3], but none of them have considered the matching design.

While the method of aperture analysis is the same for the linearly and circularly polarized arrays, the design considerations used for the aperture matching can be quite different and may be explained as follows.

1) In a linearly polarized array, the radiating beam is essentially linearly polarized regardless of the type of matching devices (see [8]). Therefore, as long as the impedance of an array has small variation with scan, it can be well matched by using the conventional matching networks inside the waveguide elements. On the other hand, in a circularly polarized array, the important characteristics of the main beam are not only its power but also its polarization. A matching device which minimizes the reflection often times does not produce a beam with good axial ratio. Therefore, in the matching design of a circularly polarized array, a tradeoff between the power and polarization is generally necessary.

2) The waveguide spacing along the $E$-plane direction is usually quite small in a linearly polarized array. Consequently, the impedance behavior in the $E$-plane scan is generally satisfactory even without any matching device. Taking the iris matching technique [4] as an example, it is only necessary to introduce the iris along the $H$ plane (inductive iris) in order to obtain a good impedance behavior in all planes of scan. However, in a circularly polarized array, the width-to-height ratio of the waveguide element is generally close to unity (in order to support two orthogonal modes). Thus the impedance behavior may be erratic in all planes of scan, which makes the matching much more difficult.

3) In a linearly polarized array, there is only one propagating mode ($TE_{10}$) at the waveguide terminal. Therefore, the conventional internal matching by using shunt susceptance can be easily analyzed with a Smith chart. However, the waveguide element used in a circularly polarized array supports two or even more propagating modes. Due to the coupling among these modes, the internal matching is no longer a simple task.

In this paper we will present an iris matching technique for a circularly polarized array. The use of an inductive iris has been very successful in a linearly polarized array [4]. The success of the iris matching technique may be qualitatively attributed to the following two factors. First, the inductive iris effectively reduces the waveguide opening along the $H$ plane to a proper degree but does not alter the internal dimension of the waveguide. Secondly, the iris discontinuity is placed right at the array aperture which eliminates the critical transmission distance usually existing in other matching devices between the matching junction and the array aperture, and consequently, it is relatively insensitive to the frequency variation. In the matching of a circularly polarized array, it is apparent that we need to use a resonant iris to reduce the opening of the nearly square waveguide element. As will be shown by numerical results, a proper design of a resonant iris together with two shunt susceptances inside each waveguide can indeed enhance the power transmission as well as improve the axial ratio of the main beam.

This paper is divided into five parts. In Section II, we will outline the statement of problem and define some terminologies to be used later. In Section III, the numerical results for a circular polarized array with and without iris will be presented, while the design of its further internal matching will be discussed in Section IV. Finally, we will give a short conclusion in Section V. In presenting this paper, we have to make frequent reference to the polarization characteristics of a linearly polarized array. That subject is discussed in [8].

Manuscript received August 27, 1970; revised November 12, 1970. This work was supported by U. S. Air Force Cambridge Research Laboratory under Contract F19628-68-C-0185, Mod. No. P001.

The author was with the Ground Systems Group, Hughes Aircraft Company, Fullerton, Calif. He is now with the Department of Electrical Engineering, University of Illinois, Urbana, Ill. 61801.

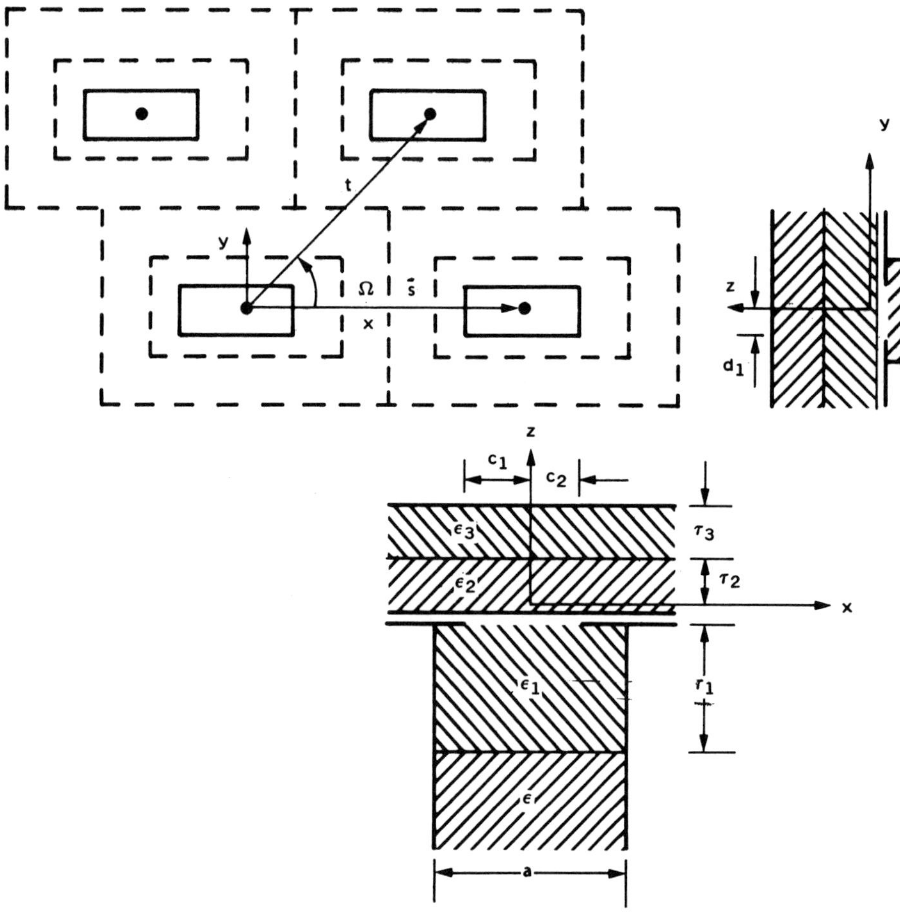


Fig. 1. Geometry of infinite array of rectangular waveguides.



## II. Statement of Problem

The configuration of an infinite array of rectangular waveguides is shown in Fig. 1. The parameters of the array are: 1) the array lattice $(s,t,\Omega)$; 2) the waveguide dimension $(a,b)$; 3) the iris $(c,d)$; and 4) the dielectric loading $\epsilon$ and $(\epsilon_1,\tau_1)$ inside the waveguides and the dielectric sheets $(\epsilon_2,\tau_2)$ and $(\epsilon_3,\tau_3)$ outside the waveguides. Each waveguide in the array supports (at least) two propagating modes, namely $TE_{10}$ and $TE_{01}$, and is excited in a cophasal manner. To be explicit, the incident electric field in the $(p,q)$th element is given by

$$\boldsymbol{E}^{(i)} = \left\{\hat{a}_x A_{01}\left(\frac{2}{\pi a\beta_{01}}\right)^{1/2} \sin\frac{\pi}{b}\left(y+\frac{b}{2}\right)\exp\left(-j\beta_{01}z\right) - \hat{a}_y A_{10}\left(\frac{2}{\pi b\beta_{10}}\right)^{1/2} \sin\frac{\pi}{a}\left(x+\frac{a}{2}\right)\exp\left(-j\beta_{10}z\right)\right\} \cdot \frac{\exp\left[-jk_0[(ps+qt\cos\Omega)u+(qt\sin\Omega)v]\right]}{\left[(|A_{10}|^2+|A_{01}|^2)/120\pi\right]^{1/2}}. \tag{1}$$

Here $\beta_{mn} = [k_0^2\epsilon - (m\pi/a)^2 - (n\pi/b)^2]^{1/2}$ is the propagation constant of the $TE_{mn}$ mode at the feed section in the waveguide. The last exponential factor in (1) is the interelement phase shift with $(u = \sin\theta_0\cos\varphi_0,\ v = \sin\theta_0\sin\varphi_0)$, where $(\theta_0,\varphi_0)$ is the desired direction of the radiating main beam. The amplitude of the incident electric field in (1) is normalized such that the total incident power from each waveguide is unity.

The problem posed is then to determine the scattered field due to the incidence in (1). In particular, we are interested in the reflection coefficients $(R_{10},R_{01})$ for the $(TE_{10},TE_{01})$ modes defined by

$$\boldsymbol{E}^{(r)} = \left\{\hat{a}_x R_{01}\left(\frac{2}{\pi a\beta_{01}}\right)^{1/2} \sin\frac{\pi}{b}\left(y+\frac{b}{2}\right)\exp\left(+j\beta_{01}z\right) - \hat{a}_y R_{10}\left(\frac{2}{\pi a\beta_{10}}\right)^{1/2} \sin\frac{\pi}{a}\left(x+\frac{a}{2}\right)\exp\left(+j\beta_{10}z\right)\right\} \cdot \frac{\exp\left[-jk_0[(ps+qt\cos\Omega)u+(qt\sin\Omega)v]\right]}{\left[(|A_{10}|^2+|A_{01}|^2)/120\pi\right]^{1/2}} \tag{2}$$

and the transmission coefficients $(T_\theta,T_\varphi)$ of the radiating main beam defined by

$$\boldsymbol{E}^{(t)} = (\hat{a}_\theta T_\theta - \hat{a}_\varphi T_\varphi)\frac{(jk_0)\exp\left(-j\boldsymbol{k}_0\cdot\hat{r}\right)}{\left[st\sin\Omega\cos\theta_0/120\pi\right]^{1/2}} \tag{3}$$

where $(\hat{a}_r,\hat{a}_\theta,\hat{a}_\varphi)$ are the unit vectors in the spherical coordinate system and $\boldsymbol{k}_0$ is a vector with components $(k_0,\theta_0,\varphi_0)$ in the spherical coordinate. The total power transmitted in the main beam is given by

$$p^{(t)} = |T_\theta|^2 + |T_\varphi|^2. \tag{4}$$

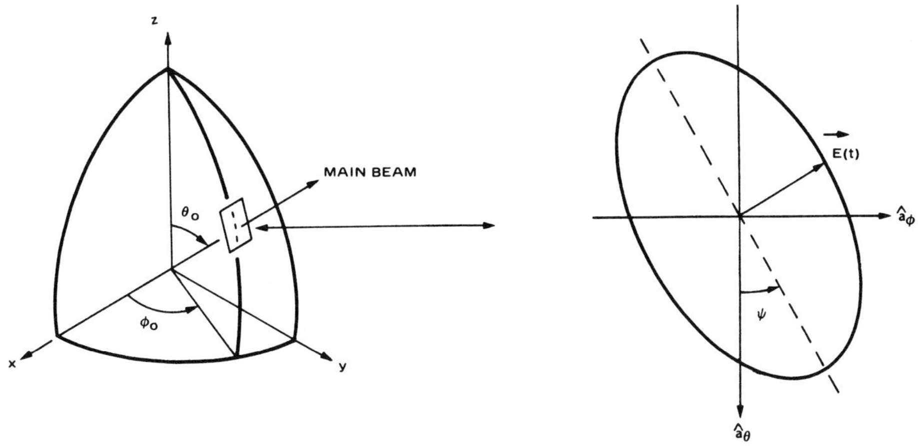


Fig. 2. Polarization ellipse of main beam.

The polarization of the main beam can be described by the familiar polarization ellipse with two parameters, namely the axial ratio of the ellipse

$$\eta = \frac{\text{minor axis}}{\text{major axis}} = \frac{\| T_\theta + jT_\varphi | - | T_\theta - jT_\varphi \|}{| T_\theta + jT_\varphi | + | T_\theta - jT_\varphi |} \tag{5}$$

and by the inclination angle of the major axis with respect to $\theta$ axis

$$\psi = \frac{\arg (T_\theta - jT_\varphi) - \arg (T_\theta + jT_\varphi)}{2} \tag{6}$$

as indicated in Fig. 2.

The scattered problem described previously can be formulated in a first-order Fredholm integral equation with the aperture electric field as the unknown function. Solving this integral equation by the moment method transforms the integral equation into an infinite system of linear equations, which can then be truncated and solved with the aid of a computer. Since such a procedure has been repeatedly described in [4], [5] we will give no further detail except the following facts concerning the numerical computations: 1) the number of the space harmonics used in the free space region is roughly 100 and that of the normal modes used in the waveguide region is roughly 25; 2) the numerical solution of this problem exhibits a relative convergence phenomenon, which has been studied in detail elsewhere [6]. Thus, to obtain the correct solution most efficiently, it should be specified that the number of normal modes used in the aperture and the number used in the waveguide have a ratio that is proportional to the physical dimensions of the aperture and the waveguide, as described in [6].

## III. Iris Matching of a Circularly Polarized Array

We have considered the matching of a circularly polarized array for several different array lattice configurations. However, for the sake of brevity we will present only a typical array which is to be scanned over the range $0 \leq \theta_0 \leq 60°$ in all planes. The parameters of the array are as follows.

*Lattice*: $s = t = 0.605\lambda_0$ and $\Omega = 60°$. This is an equilateral triangular lattice with the grating lobe angles in $\varphi_0 = (0°,45°,90°)$ planes[1] [9] at $\theta_0 = $ (none,66°,76°). It can be shown that this lattice is optimum in the sense of minimum elements for a given aperture and scan range (i.e., $0 \leq \theta_0 \leq 65°$).

*Waveguide*: $a = b = 0.51\lambda_0$, $\epsilon_1 = \epsilon = 1.5$. The dielectric loading lifts the $TE_{10}$ and $TE_{01}$ modes further above cut-off, but still prevents other modes to propagate.

*Dielectric Sheet and Iris*: None.

*Excitation*: The amplitude ratio of $TE_{10}$ and $TE_{01}$ modes are $A_{10}:A_{01} = 1:j$.

The reflection coefficients for $TE_{10}$ mode ($R_{10}$) and $TE_{01}$ mode ($R_{01}$) for the preceding array are presented in Fig. 3 for scan in the planes $\varphi_0 = 0°$, 45°, 90°. Note that they vary quite erratically as a function of scan. In particular, at $\varphi_0 = 90°$, $\theta_0 = 60°$, the transmitted power $P^{(t)}$ is $-4.1$ dB, corresponding to a 61 percent reflected power. The axial ratio $\eta$ at this scan angle is 0.023, and the inclination angle $\psi$ is $-75.5°$, implying a linearly polarized beam with $\boldsymbol{E}$ vector roughly pointing at $x$ direction.

The first step in the matching design process is to reduce the variation of the reflections. This may be achieved by introducing a resonant iris at the array aperture with (Fig. 1)

$$\frac{c_1}{a} = \frac{c_2}{a} = \frac{d_1}{b} = \frac{d_1}{b} = 0.4 \tag{7}$$

which will be referred to as (20 by 20) percent (symmetrical) iris. After such a modification, the reflection coefficients are shown in Fig. 4, and they are much more localized. The power and polarization of its radiating main beam are given in Figs. 5 and 6. The power transmission loss is between 0.9 to 1.7 dB up to $\theta_0 = 60°$, and beyond that there is a sharp dip in the plane $\varphi_0 = 90°$ immediately before the grating lobe angle ($\theta_0 = 66°$). The polarization is nearly perfect around the broadside

[1] In a circularly polarized array it is no longer convenient to use the $H$, $D$, $E$ symbols to refer to planes defined by $\varphi_0 = (0°, 45°, 90°)$ since we excite two orthogonal modes.

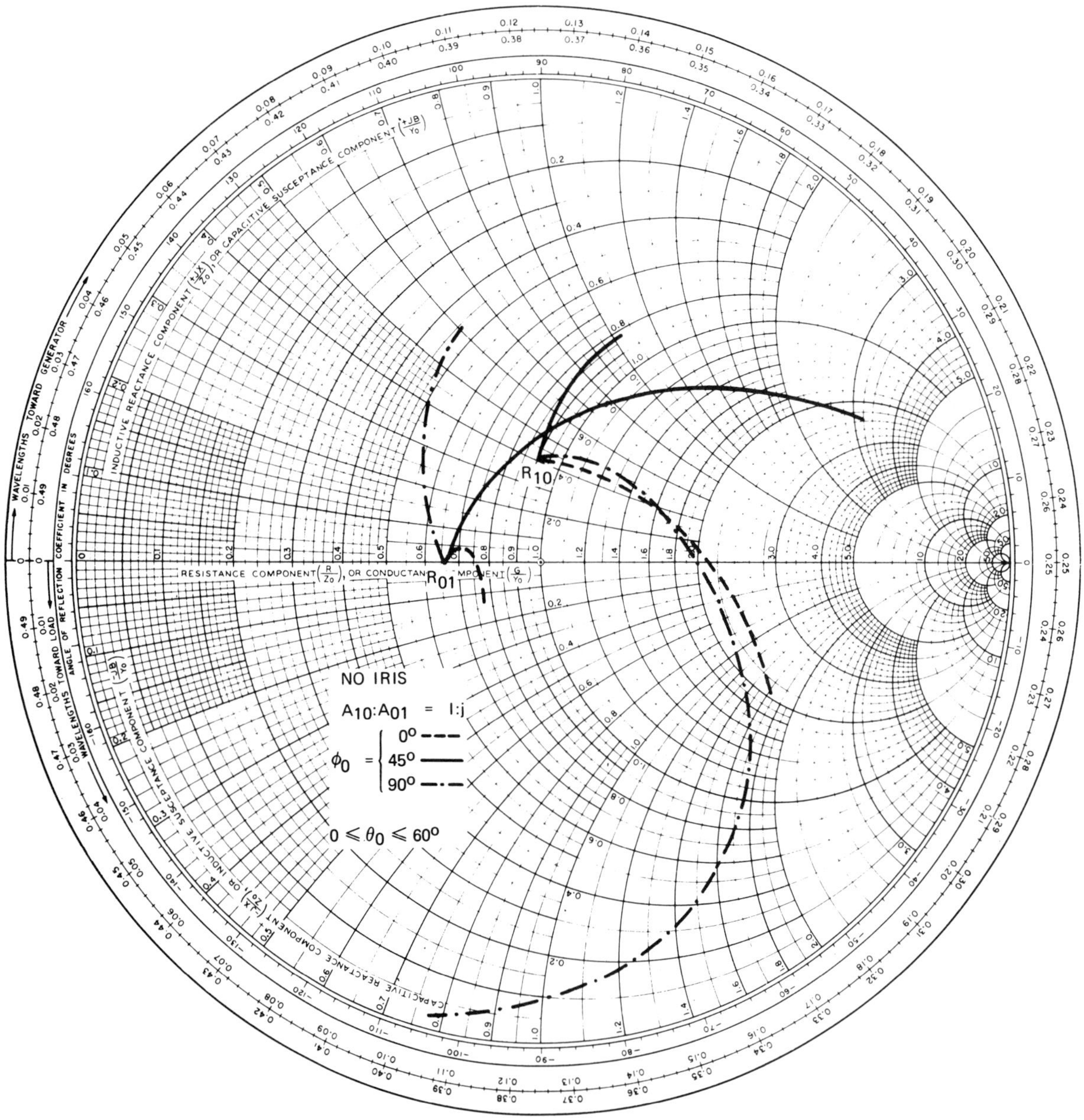


**Fig. 3.** Reflection coefficients $R_{10}$ and $R_{01}$ for $TE_{10}$ and $TE_{01}$ modes, respectively, of circularly polarized array with excitation $TE_{10}:TE_{01} = 1:j$. Parameters of array are $s = t = 0.605\lambda_0$, $\Omega = 60°$, $a = b = 0.51\lambda_0$, $\epsilon_1 = \epsilon = 1.5$; no iris and no dielectric sheets outside waveguides.

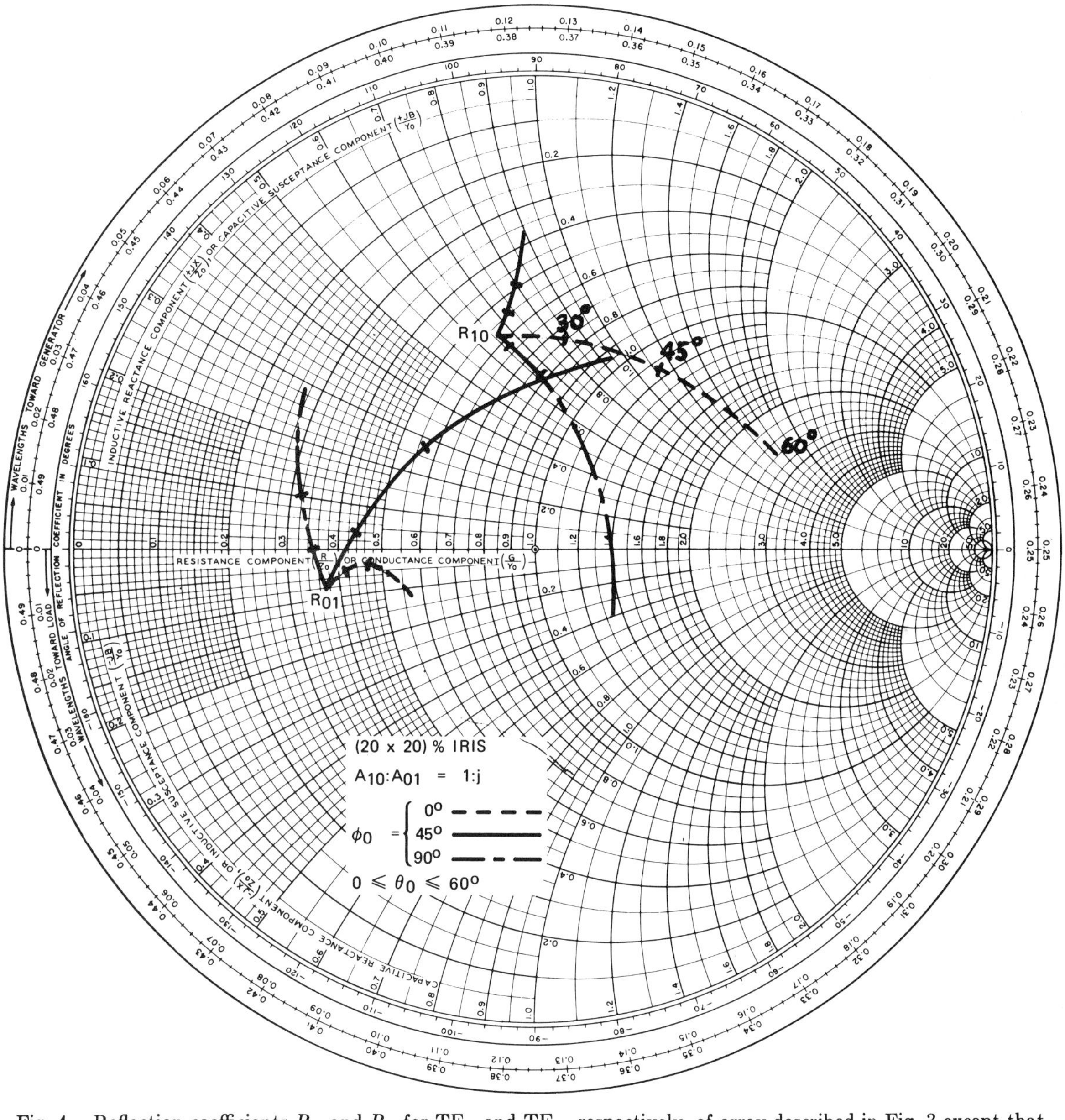


Fig. 4. Reflection coefficients $R_{10}$ and $R_{01}$ for $TE_{10}$ and $TE_{01}$, respectively, of array described in Fig. 3 except that a (20 by 20) percent iris is added ($c_1 = c_2 = d_1 = d_2 = 0.4a$).

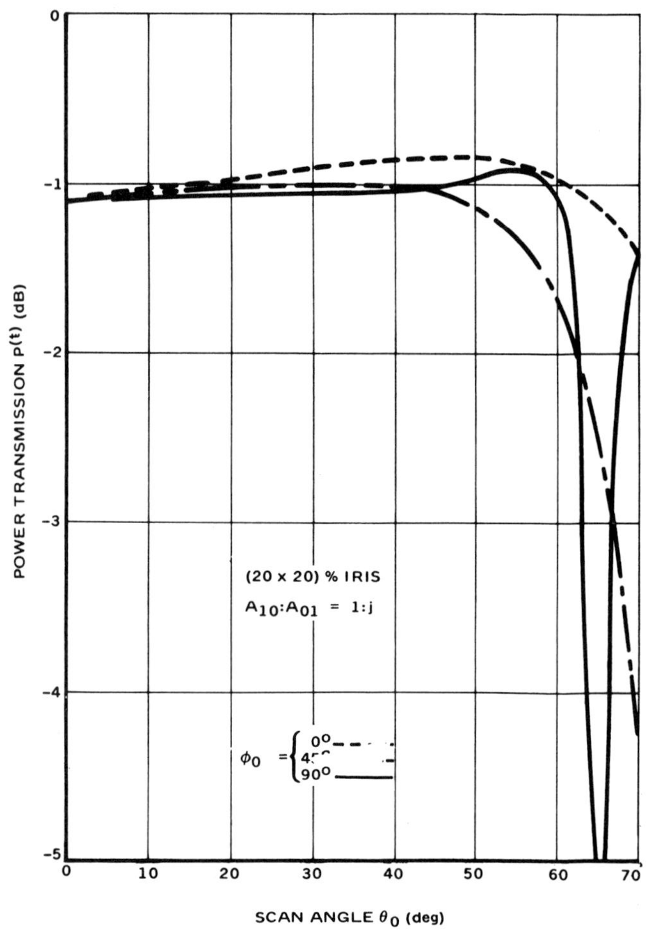


Fig. 5. Power transmission in main beam of array described in Fig. 4.

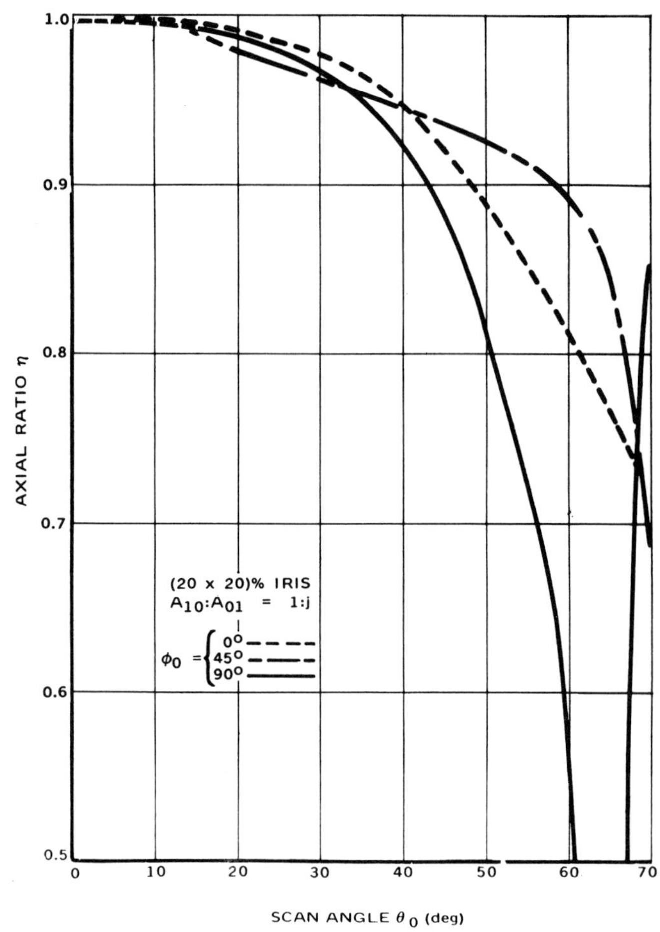


Fig. 6. Axial ratio of main beam of array described in Fig. 4.

direction, and deteriorates at large scan angles. Within the scan range of interest ($0 \leq \theta_0 \leq 60°$), the worst axial ratio occurs at ($\varphi_0 = 90°$, $\theta_0 = 60°$) where it takes the value of 0.55 (or 2.6 dB). This is indeed a great improvement over the case without iris loading.

With the variation of the reflection coefficients being localized as shown in Fig. 4, the next step is to use a conventional matching device inside the waveguides for further improvement.

Note that the results shown in Figs. 4–6 are due to a superposition of two excitations, namely a $TE_{10}$ with unit amplitude and a $TE_{01}$ mode with amplitude ($j$). However, for the present purpose, we first need to know the scattering coefficients at the array aperture due to $TE_{10}$ and $TE_{01}$ separately. For convenience in notation, let us represent the scattering at the aperture by a four-port (or six-terminal) network, as sketched in Fig. 7, and introduce the (normalized) scattering coefficient: $s_{mn}$ is the scattering coefficient at the $m$th terminal due to a unit power incidence at terminal $n$. Its amplitude is so normalized that its square represents power, and its phase is referred to that of the electric field. For example, $s_{42}$ is the transmission coefficient of $TM_{00}$ mode in the free space, i.e., $T_\theta$ in (3) due to an incident $TE_{01}$ mode from the waveguides. We have computed all the scattering

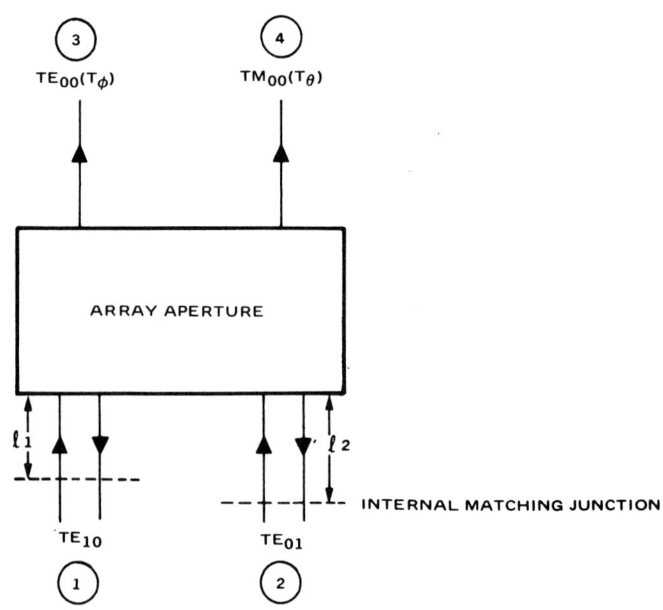


Fig. 7. Four-port network representation for scattering at array aperture.

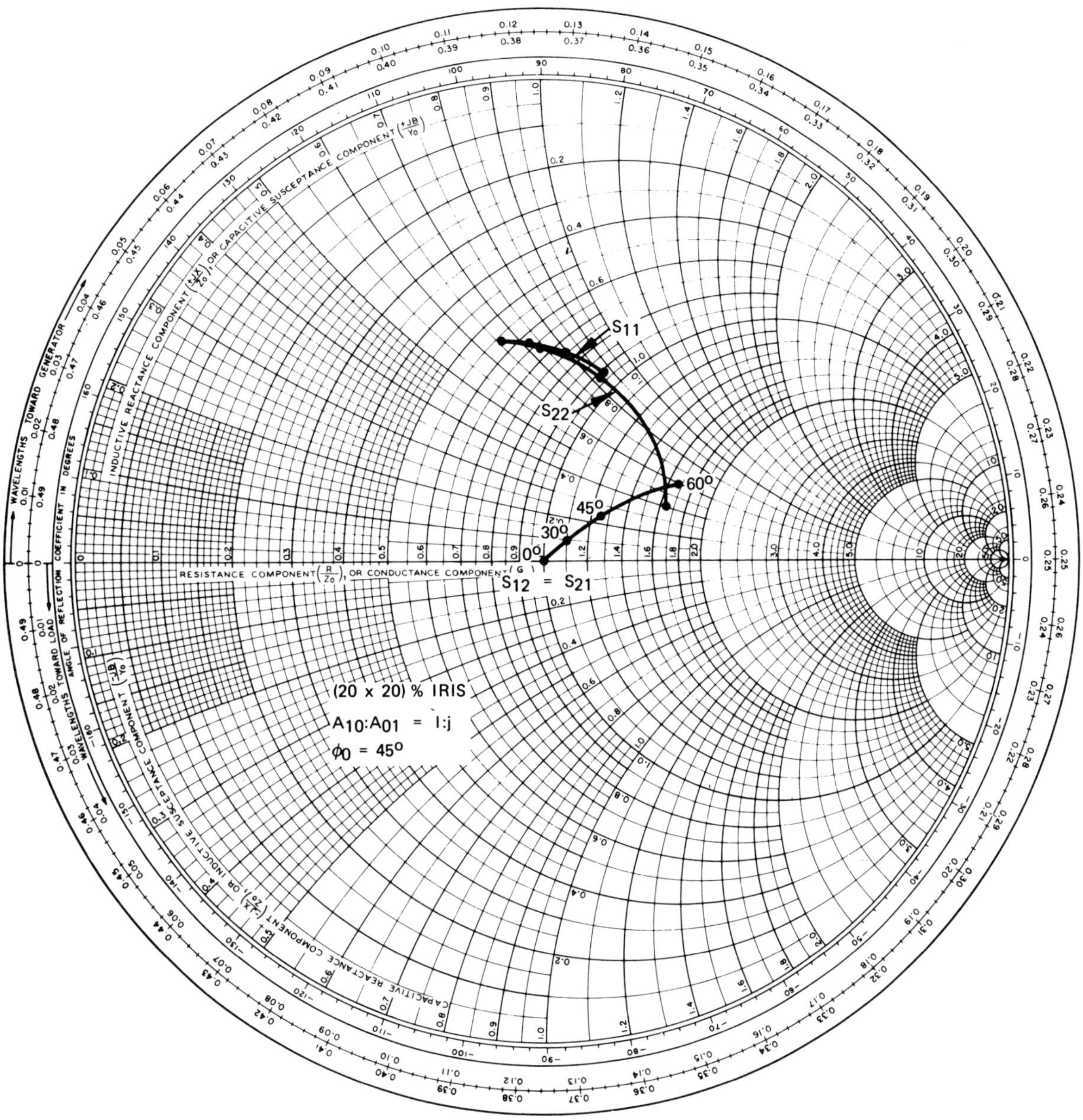


Fig. 8. Scattering coefficients of reflected field of array described in Fig. 4 with scan in plane $\varphi_0 = 45°$.

coefficients for the array under consideration and the results are summarized in the following.

1) *Reflection for Scan in the Planes* $\varphi = 0°$ *and* $90°$: The coupling to the cross polarization in the waveguide, i.e., $s_{21}$ and $s_{12}$ are negligibly small. A typical example is that at $(\varphi_0 = 0°, \theta_0 = 60°)$, $s_{11} = 0.73 \exp(-j83.5°)$ and $s_{21} = 0.0014 \exp(-j136.8°)$. Therefore, for these two planes of scan, the curves presented in Fig. 4 can be taken as the self-reflection coefficients, i.e., $R_{10} = s_{11}$ and $R_{01} = js_{22}$.

2) *Reflection for Scan in the Plane* $\varphi_0 = 45°$: In this case, the coupling between $TE_{10}$ and $TE_{01}$ increases as the scan angle $\theta_0$ as shown in Fig. 8. For example, at $(\varphi_0 = 45°, \theta_0 = 60°)$, $s_{11} = 0.4 \exp(j71.8°)$, $s_{22} = 0.28 \exp(j23°)$, and $s_{12} = s_{21} = 0.32 \exp(j27.5°)$. The magnitudes of these four quantities are comparable. Note also the relations that $R_{10} = s_{11} + js_{12}$, and $R_{01} = s_{21} + js_{22}$ between Figs. 4 and 8.

3) *Transmission for Scan in the Plane* $\varphi = 0°$ *and* $\varphi = 90°$: Similar to the case of a linearly polarized array described in [8], the polarization of the transmitted field is essentially linear, and its electric field is in the same direction as the projection of the incident $\boldsymbol{E}$ vector in the waveguide on the plane transverse to the main beam. Explicitly, for scan in the plane $\varphi = 0°$, $|s_{31}| \gg |s_{41}|$ and $|s_{42}| \gg |s_{32}|$, which implies that the incident $TE_{10}$ mode in the waveguide accounts for the $\varphi$ component of the transmitted electric vector, while incident $TE_{01}$ accounts for the $\theta$ component. Thus, to obtain a good circularly polarized beam, $s_{31}$ should be approximately equal to $s_{42}$. At $(\varphi_0 = 0°, \theta_0 = 30°)$, for example, $s_{31} = 0.89 \exp(-j67.8°)$ and $s_{42} = 0.895 \exp(-j66.7°)$, and hence the axial ratio is close to unity (Fig. 6). At $(\varphi_0 = 0°, \theta = 60°)$, $s_{31} = 0.32 \exp(-j83.4°)$ and $s_{42} = 0.95 \exp(-j74.5°)$, which explains the relatively bad axial ratio for this scan angle (Fig. 6). The same comments apply to the case $\varphi_0 = 90°$, expect that, in that case, $|s_{41}| \gg |s_{31}|$ and $|s_{32}| \gg |s_{42}|$.

4) *Transmission for Scan in the Plane* $\varphi_0 = 45°$: At large scan angles the magnitudes of $(s_{31}, s_{41}, s_{32}, s_{42})$ become comparable. Only when the relation $s_{31} + js_{32} = s_{41} + js_{42}$ is approximately true can a good circularly polarized beam be assured.

## IV. Internal Matching

With the various scattering coefficients found in Section III for the array with (20 by 20) percent iris, one may now proceed to introduce the conventional shunt admittances inside the waveguides for further improvement. As mentioned in the Introduction, this step is not as simple as the case of a linearly polarized array because of the coupling between the $TE_{10}$ and $TE_{01}$ modes.

Since there is no longer any apparent advantage of working with self- and mutual impedance concepts, we will consider only the scattering coefficients. For matching purposes, let us introduce two transverse discontinuities in each waveguide. The first discontinuity, situated at a distance $l_1$ from the aperture, presents a (normalized) shunt susceptance $B_{11}$ to the $TE_{10}$ mode, while it does not disturb $TE_{01}$ mode (i.e., $B_{21} \rightarrow \infty$, a simple example is a thin conducting strip oriented along the $y$ direction) as indicated in Fig. 7. The second discontinuity, situated at a distance $l_2$ from the aperture, presents a shunt susceptance $B_{22}$ to $TE_{01}$ mode while it does not disturb $TE_{10}$ mode (i.e., $B_{12} \rightarrow \infty$). Then the scattering property of two such discontinuities can be uniquely described by a matrix for the reflected field:

$$U = \begin{bmatrix} -jB_{11}/(2 + jB_{11}) & 0 \\ 0 & -jB_{22}/(2 + jB_{22}) \end{bmatrix}. \tag{8}$$

Note that in writing (8) we have assumed that the reference plane for $TE_{10}$ mode is at $z = -l_1$ and that for $TE_{01}$ mode is at $z = -l_2$. Such a convention will be used throughout this section.

Now, let us consider the multiple scattering problem between the aperture junction described in Section III and the junction of shunt susceptances described by (8). Let the incident field from the waveguide (below the shunt susceptances of course) be represented by the vector $\varphi^{(i)} = [A_{10}, A_{01}]$. Then, by tracing the multiple scattering between these two junctions, one may show that the reflected field is given by [7]

$$\varphi^{(r)} = \begin{bmatrix} R_{10} \\ R_{01} \end{bmatrix} = \{U + (I + U)S^{(r)}(I - US^{(r)})^{-1}(I + U)\}\varphi^{(i)} \tag{9}$$

and the transmitted field by

$$\varphi^{(t)} = \begin{bmatrix} T_\varphi \\ T_\theta \end{bmatrix} = S^{(t)}[1 - US^{(r)}]^{-1}[I + U]\varphi^{(i)}. \tag{10}$$

In deriving the preceding equations, the following three facts have been used.

1) For the junction of shunt susceptances, the matrix $[I + U]$ describes its transmission property where $I$ is an identity matrix.

2) $S^{(r)}$ describes the reflection property of the aperture junction and is explicitly given by

$$S^{(r)} = \begin{bmatrix} \exp(-j\beta_{10}l_1) & 0 \\ 0 & \exp(-j\beta_{01}l_2) \end{bmatrix} \begin{bmatrix} s_{11} & s_{12} \\ s_{21} & s_{22} \end{bmatrix} \cdot \begin{bmatrix} \exp(-j\beta_{10}l_1) & 0 \\ 0 & \exp(-j\beta_{01}l_2) \end{bmatrix} \tag{11}$$

where the first and the third matrices in (11) are introduced to shift the reference plane for $TE_{10}$ mode to $z = -l_1$, and that for $TE_{01}$ mode to $z = -l_2$.

3) $S^{(t)}$ describes the transmission property of the aperture junction and is given by

$$S^{(t)} = \begin{bmatrix} s_{31} & s_{32} \\ s_{41} & s_{42} \end{bmatrix} \begin{bmatrix} \exp(-j\beta_{01}l_1) & 0 \\ 0 & \exp(-j\beta_{01}l_2) \end{bmatrix}. \tag{12}$$

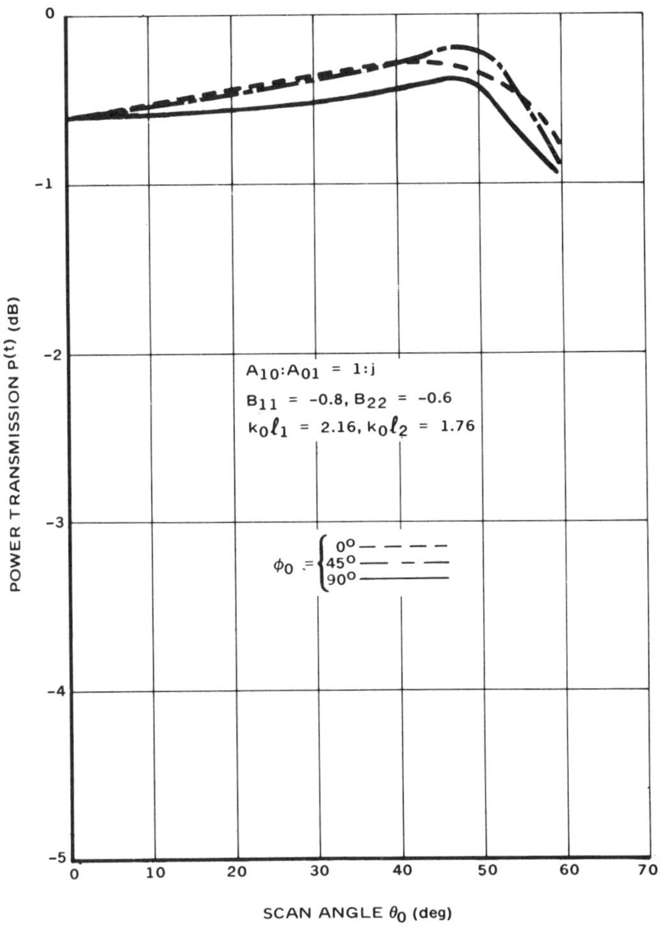


Fig. 9. Power transmission in main beam of array described in Fig. 4 with additional internal matching.

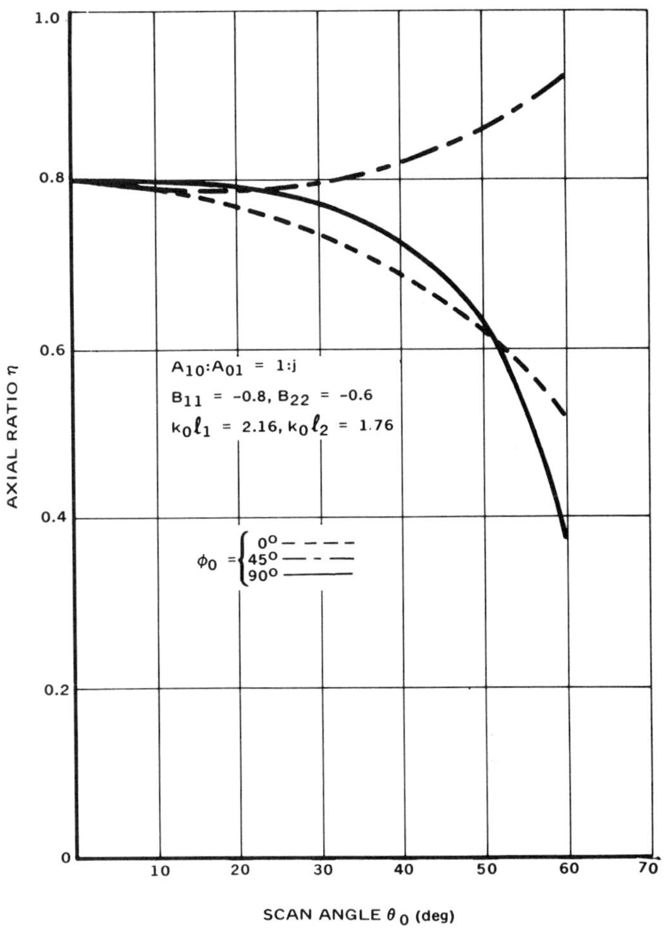


Fig. 10. Axial ratio of main beam of array described in Fig. 9.



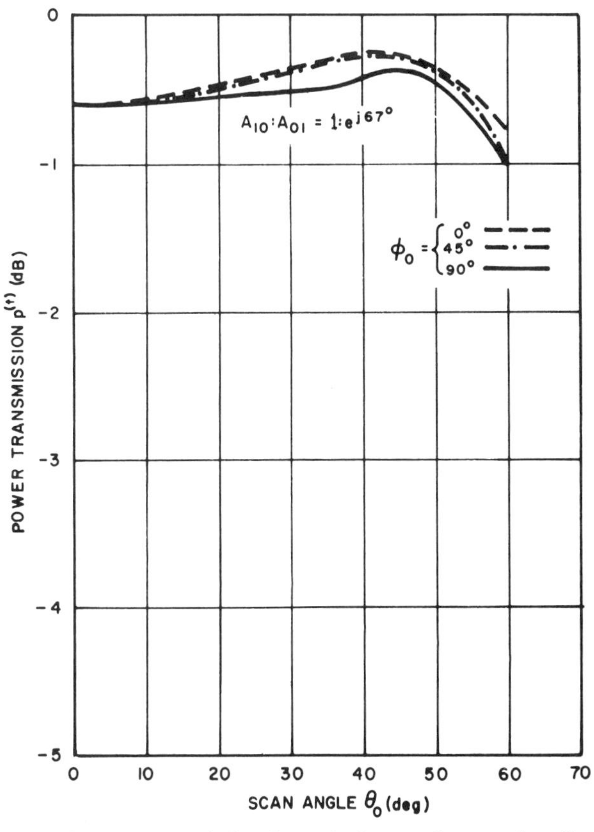


Fig. 11. Power transmission in main beam of array described in Fig. 9 with different excitation.

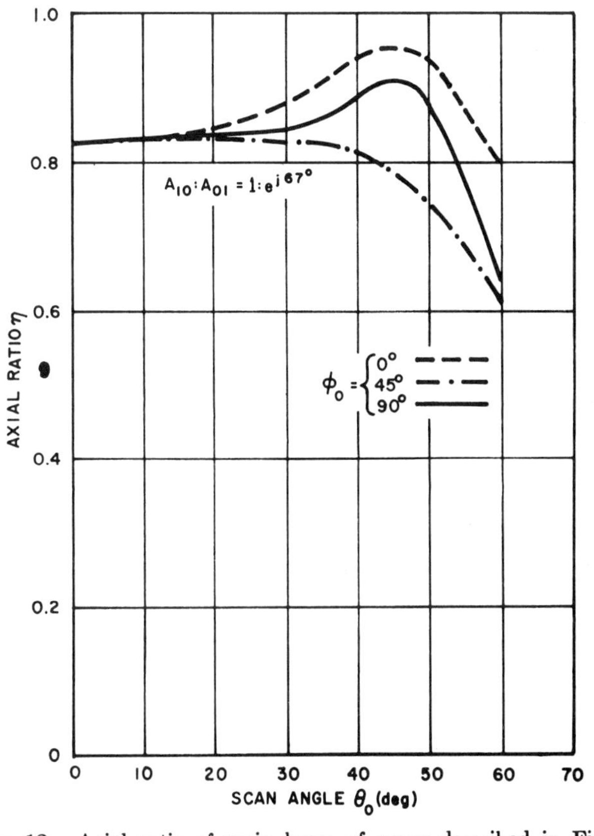


Fig. 12. Axial ratio of main beam of array described in Fig. 9 with different excitation.

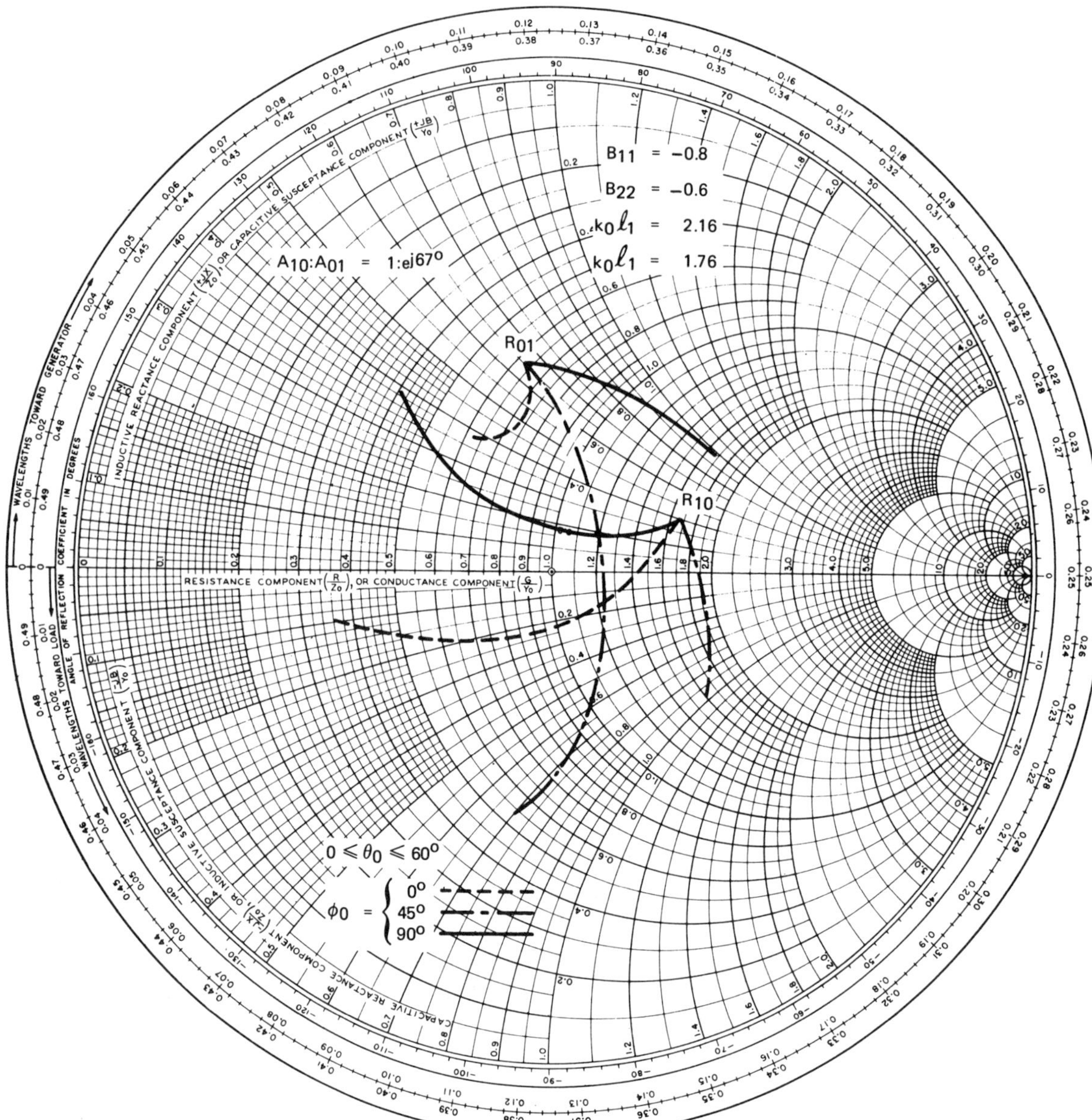


Fig. 13. Reflection coefficient of main beam of array described in Fig. 9 with different excitation.



Since the scattering coefficients $(s_{11},s_{12},\cdots,s_{42})$ are known from the computation in Section III, our present task is to select a suitable set of parameters, namely $(B_{11},B_{22},l_1,l_2)$ and $(A_{10},A_{01})$, for the computation in (9) and (10) in order to obtain better matching.

At this point, we should emphasize again the difficulty in the design of a circularly polarized array. There seems no analytical or systematic way of selecting these parameters, and one has to depend upon trial and error.[2] Furthermore, an optimum power transmission of the main beam usually is not accompanied by a good axial ratio. Thus, in a practical design, one often has to select a set of parameters which compromises these two characteristics.

We have computed a large number of cases by using (9) and (10) for different combinations of $(B_{11},B_{22},l_1,l_2)$, but with fixed $(A_{10}:A_{01}) = (1:j)$. A relatively good case is shown in Figs. 9 and 10. A comparison of Fig. 9 with Fig. 5 reveals that the use of suitable internal matching can indeed reduce the transmission power loss at most scan angles. However, this is done at an expense of the axial ratio, as may be seen by comparing Figs. 6 and 10.

One remedy for the poor axial ratio in Fig. 10 is to readjust the excitation $(A_{10},A_{01})$. We note that, at $(\varphi_0 = 90°,\ \theta_0 = 60°)$, the transmission coefficients are $T_\varphi = 0.64 \exp(-j34.5°)$ and $T_\theta = 0.62 \exp(j6.3°)$. Even though $|T_\varphi| \approx |T_\theta|$, their phase difference is far from being 90°, which is the main reason for the poor axial ratio $(\eta = 0.37)$. Recalling from Section III that, in the plane $\varphi_0 = 90°$, $T_\varphi$ is mainly due to incident $TE_{01}$ and $T_\theta$ due to $TE_{10}$, an apparent solution is to decrease the phase of incident $TE_{01}$ from 90°. After several attempts, we found that the excitation $A_{10}:A_{01} = 1:\exp(j67°)$ is a

[2] For the special case when $s_{12} = s_{21} = 0$ ($TE_{10}$ and $TE_{01}$ uncoupled), the formula in (9) is reduced to the conventional form equivalent to the parallel combination of two admittances, and it then can be worked out directly on the Smith chart. Unfortunately, in a circularly polarized array, $s_{12}$ and $s_{21}$ are not negligibly small for all planes of scan.

good choice. The corresponding main beam characteristics for such an excitation are shown in Figs. 11 and 12, and the reflection coefficients are given in Fig. 13. Note that the reflection coefficients are quite centralized as compared with those obtained before the internal matching and the adjusting in excitation (Fig. 4). The power transmission loss is roughly half a dB up to $\theta_0 = 55°$, and drops to about one dB at $\theta_0 = 60°$. The axial ratio is greater than 0.6 in all scan ranges ($0 \leq \theta_0 \leq 60°$), which is adequate in many practical applications.

## V. Conclusions

Generally speaking, the variation of the reflection coefficients with scan in a circularly polarized array is more pronounced than that in a linearly polarized array. This is mainly due to the relatively larger dimensions and/or the heavier dielectric loading of the waveguide elements. Such a variation usually can be reduced by introducing a proper resonant iris at each waveguide opening. Once the reflection coefficients are localized, one may design a conventional matching junction inside the waveguide for a further adjustment of the reflections. However, since the important characteristics of a circularly polarized array are both its power and its polarization, a good design for the internal junction is one which gives a proper compromise between the power and the polarization, and is not necessarily the one minimizing the reflection.

Finally, we mention that in the design of a circularly polarized array, the use of circular waveguides is sometimes favorable to some practical workers. We have also analyzed the iris matching technique for arrays of circular guides, and the results will be reported in the near future. It is found that from the impedance point of view there is no apparent advantage for using the circular guides over the rectangular ones. In the design of an array, there are so many parameters in the lattice and the loading. A well-matched array lies heavily on the proper choice of those parameters, while it is relatively less important whether we start out with rectangular or circular guides.

## Acknowledgment

Useful comments from W. R. Jones, N. Wong, and R. Tang are appreciated.

## References

[1] G. H. Knittel, "Wide-angle impedance matching of phased-array antennas—a survey of theory and practice," in *Dig. Phased Array Antenna Symp.*, 1970, pp. 62–65.

[2] N. Amitay and V. Galindo, "The analysis of circular waveguide phased arrays," *Bell Syst. Tech. J.*, vol. 47, 1968, pp. 1903–1931.

[3] S. W. Lee, N. Wong, and R. Tang, "Analysis of infinite planar array of rectangular waveguides by generalized scattering matrix approach," in *Dig. Phased Array Antenna Symp.*, 1970, pp. 40–42.

[4] S. W. Lee and W. R. Jones, "On the suppression of radiation nulls and broadband and impedance matching of rectangular waveguide phased arrays," *IEEE Trans. Antennas Propagat.*, vol. AP-19, Jan. 1971, pp. 41–51.

[5] N. Amitay, C. P. Wu, and V. Galindo, "Methods of phased array analysis," in *Dig. Phased Array Antenna Symp.*, 1970, pp. 33–37.

[6] S. W. Lee, W. R. Jones, and J. J. Campbell, "Convergence of numerical solutions of iris-type discontinuity problems," presented at the 1970 Int. IEEE G-AP Symp., Columbus, Ohio, Sept. 13–17; also, *IEEE Trans. Microwave Theory Tech.*, vol. MTT-19, June 1971, to be published.

[7] J. Pace and R. Mittra, "Generalized scattering matrix analysis of waveguide discontinuity problems," in *Quasi-Optics*, vol. 24. New York: Polytechnic Institute of Brooklyn Press, 1964, pp. 177–197.

[8] S. W. Lee, "Aperture matching for an infinite circular-polarized array of rectangular waveguides," Hughes Aircraft Co., Fullerton, Calif., Tech. Rep. FR70-14-574, Aug. 1970.

[9] Y. T. Lo and S. W. Lee, "Affine transform and its application to antenna arrays," *IEEE Trans. Antennas Propagat.*, vol. AP-13, Nov. 1965, pp. 890–896.

*Reprinted from* IEEE TRANSACTIONS
ON *ANTENNAS AND PROPAGATION*
Volume AP-15, Number 2, March, 1967
pp. 313-314


# ACTIVE IMPEDANCE OF A PHASED-ARRAY ANTENNA ELEMENT SIMULATED BY A SINGLE ELEMENT IN WAVEGUIDE

200

BY

M. A. BALFOUR

**Active Impedance of a Phased-Array Antenna Element Simulated by a Single Element in Waveguide**

The use of simulators in the design of a phased-array antenna was described in a previous publication [1]. The array elements were dielectric-loaded circular waveguides in a triangular arrangement. Conducting planes were placed coincident with the symmetry planes of the array so as to image a small number of elements into an infinite array. The conducting planes were arranged to form rectangular or triangular waveguides. Each of the simulators sampled the performance of the array at a given set of scan conditions. Impedance measurements in the simulator waveguides yielded the impedance of the array as seen by a plane wave impinging upon the array face from free space. This impedance is not, however, the *active impedance* of the array; the active impedance is defined as that impedance seen by a generator connected to one array element when all other array elements are active. The simulator measurements described above can, however, be related to the active impedance by an equivalent circuit procedure [2].

Manuscript received October 3, 1966.

A novel simulator cross section has been developed which facilitates a *direct* measurement of the active impedance of a phased-array element at two different scan angles, with a *single* active element in waveguide. Figure 1(a) is a cross-section view of this simulator. As shown, the cross section comprises one complete element and four quarter elements. The quarter elements are not propagating since they contain two orthogonal conducting image planes. Hence, the simulator shown in Fig. 1(a) contains only one active element and will be referred to as a *one-port* simulator. The simulator is operated in the TE-10 mode and the scan angle that is simulated is given by the relation

$$\sin\theta = (1/\sqrt{3})(\lambda/2d). \qquad (1)$$

This is derived from the general relationship [1], [2]

$$\sin\theta = \lambda/\lambda_c, \qquad (2)$$

where $\theta$ is the scan angle from broadside, $\lambda$ is free-space wavelength, and $\lambda_c$ is cutoff wavelength of the simulator waveguide. The polarization of the element is indicated as well as the direction of scan (projected into the transverse plane of the array). Note, for example, when the interelement spacing $d/\lambda$ is $1\sqrt{3}$, the scan angle simulated is 30°.

Figure 1(b) shows the same one-port cross section turned on its side. The element polarization is indicated. When the simulator is operated with this polarization in the TE-10 mode, a scan angle given by

$$\sin\theta = \lambda/2d \qquad (3)$$

is simulated. If, for example, the array spacing is $d/\lambda = 1\sqrt{3}$ the angle simulated is $\theta = 60°$. Thus, the simulator indicated in Fig. 1 is capable of simulating the performance of an infinite array at two significantly different scan angles, while employing only a single active array element.

Figure 2(a) is a photograph of such a one-port simulator fixture constructed at Wheeler Laboratories. In the foreground is a sample of the array face, shown in the cross section of Fig. 1(a). The single active element is shown, partially disassembled, extending behind the face of the array. The nonpropagating quarter elements extend only a short distance behind the array face to satisfy the reactive fields near the aperture

Fig. 2(b) shows this simulator fixture in a measurement configuration which permits a direct measurement of active impedance. As shown, the element is fed at its input port, through a slotted line, causing it to radiate into the simulator waveguide. The simulator waveguide is terminated in a reflectionless load, shown to the right of the simulator waveguide. This load simulates free space in front of the infinite array. Under these conditions, the single element radiating into the terminated waveguide exactly simulates one radiating element in the environment of an infinite number of radiating elements, all phased to radiate at a scan angle $\theta$. The scan angle $\theta$ is determined by the cross-section dimensions of the simulator waveguide, as in (2). With the simulator oriented as shown in Fig. 2(b), $\theta = 30°$ is being simulated. The impedance measured in the simulator at the element input port, under these conditions, is precisely the *active impedance* of a single element in an infinite array.

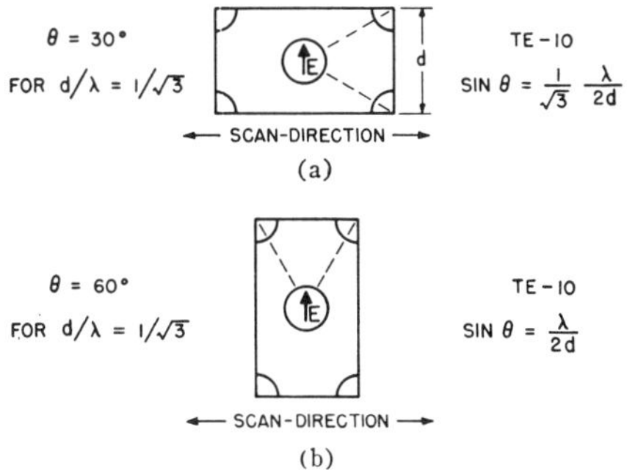


Fig. 1. One-port simulator cross section for two angles of scan. (a) Low-angle scan; (b) high-angle scan.

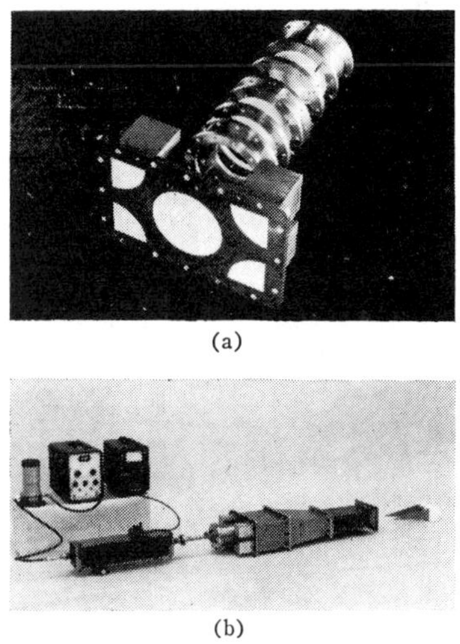


Fig. 2. Configurations of (a) one-port simulator for two angles of scan; (b) measurement of active impedance.

It is interesting to note that a direct measurement of active impedance in simulators is also feasible when the array sample encompasses several active elements. If the array sample contains several active elements, it would be necessary to connect individual generators to the input port of each element of the array sample and adjust the amplitudes of these generators such that the sample excites the simulator waveguide in the waveguide mode necessary to simulate $\theta$, as given by (2).

Figure 3 summarizes possible one-port simulators for both triangular and square arrangements of circular elements. In Fig. 3(a), for triangular arrays, cross sections I and II represent the simulator that was discussed in detail in this paper. Cross section III is a one-port simulator in triangular waveguide; it is operated in the dominant TE-mode of triangular waveguide [3] and will simulate $\theta = 42°$ when $d/\lambda = 1/\sqrt{3}$. Fig. 3(b) summarizes possible single-element simulators for a square arrangement of circular elements. For $d/\lambda = 1/2$, for example, $\theta = 30°$, 42°, 45°, for cross sections IV, V, and VI respectively.

To summarize the advantages of one-port simulators, they provide a simple technique for evaluating the performance of a typical active array element in an infinite array of active elements under actual scan conditions. In addition to serving as an effective design tool, they may be employed as a production

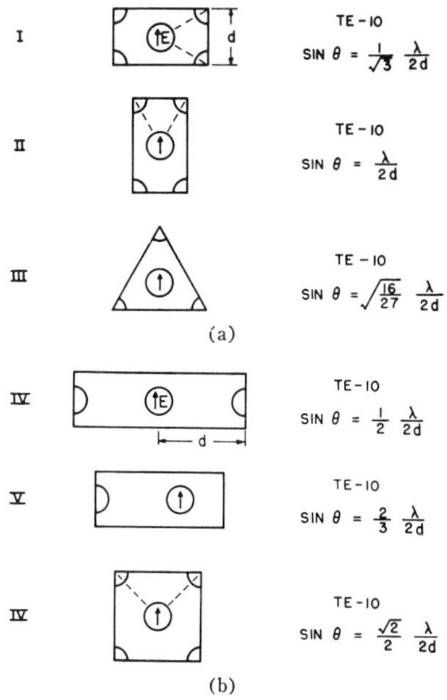


Fig. 3. Summary of one-port cross sections for (a) triangular and (b) square arrays.

test fixture; mass-produced array elements may be tested under *actual operating conditions* at both low and high power levels.

ACKNOWLEDGMENT

The technique described in this communication was devised for the design and testing of a phased-array antenna element for the Bell Telephone Laboratories under a prime contract from the U. S. Army.

The project is under the direction of R. L. Mattingly and P. H. Smith, at the Bell Telephone Laboratores, and G. E. Vaupel at Wheeler Laboratores. Consultation was provided by H. A. Wheeler and V. Mazzola.

M. A. BALFOUR
Wheeler Laboratories, Inc.
Great Neck, N. Y.

REFERENCES

[1] M. A. Balfour, "Phased-array simulators in waveguide for a triangular arrangement of elements," *IEEE Trans. on Antennas and Propagation (Communications)*, vol. AP-13, pp. 475–476, May 1965.
[2] P. W. Hannan and M. A. Balfour, "Simulation of a phased-array antenna in waveguide," *IEEE Trans. on Antennas and Propagation*, vol. AP-13, pp. 342–353, May 1965.
[3] S. A. Shelkunoff, "Modes in triangular waveguide," in *Electromagnetic Waves*. Princeton, N. J.: Van Nostrand, 1943, pp. 392–397.

IEEE TRANSACTIONS ON ANTENNAS AND PROPAGATION, VOL. AP-20, NO. 5, SEPTEMBER 1972

# The Determination of Active Array Impedance with Multielement Waveguide Simulators

JACOB J. GUSTINCIC, MEMBER, IEEE

***Abstract*—The basic principles behind a multielement waveguide simulator are presented for the case of general symmetric elements in a rectangular or triangular spaced infinite array. The techniques are applicable to any polarization and are valid for array spacings which allow grating lobes in real space. It is shown that a single simulator containing $N \times M$ elements can be used to measure the reflection coefficient for the infinite array at $N \times M$ scan angles. The measurement involves the determination of the $N \times M$ transmission coefficients between one element and each of the other elements in the simulator when the simulator waveguide is terminated in a matched load. It is shown that the $N \times M$ reflection coefficients so obtained are actually the eigenvalues of the scattering matrix of the network formed by the $N \times M$ input ports to the simulator. An interpolation scheme is presented which allows the infinite array impedance to be approximated over a large range of scan angles by means of an expansion in terms of the eigenvalues of the simulator. The physical significance of the interpolation is discussed, and experimental results for a 25-element simulator utilizing stripline slots are presented.**

202

## Introduction

THE FIRST significant report of the application of waveguide simulators for the measurement of infinite array properties can be found in the work of Hannan *et al.* [1] in 1963. In a later more comprehensive paper Hannan and Balfour [2] present many of the fundamental ideas behind the use of waveguide simulators, and a detailed discussion based on their treatment can be found in the text of Oliner and Malech [3]. This basic technique utilizes the concept of a "looking in" type simulator in which waveguide modes are launched at a small number of loaded elements and the mode reflections are noted to provide the impedance a plane wave would see looking into the infinite array. An equivalent circuit is obtained which then allows the computation of the impedance the array would see "looking out" into free space [4]. A single polarization and incidence angle is simulated for each waveguide mode used. By means of several simulators with various modes, the array behavior at most of the significant angles and polarizations can then be sampled. The technique works very well and has been used in a number of applications [5]–[9] with rectangular and triangular waveguides.

A series of simulators utilizing a single active circular waveguide radiating element to launch a single mode in rectangular and triangular guide has been presented by Balfour [10]. Each of these simulators provides a direct measurement of the looking out impedance for the incidence angle and polarization corresponding to the simulator mode launched. These simulators have the advantage that a single element alone can be easily evaluated under a given discrete scan condition. Recently such a single-element simulator has been used in the study of array blindness effects [11].

The schemes discussed in the preceding have two major disadvantages. A separate simulator must be fabricated for each scan angle simulated. In addition, the looking in type simulator requires special high-order mode launchers and reflection measuring devices. The measurement of impedances when grating lobes are present becomes impractical in this type of simulator as it requires the simultaneous launching and reflection measurement of two waveguide modes. The second disadvantage lies in the fact that the array impedance can be obtained only at discrete scan angles. No straightforward general scheme exists to extrapolate these values to other angles.

In the following the principles behind a very general multielement looking out type simulator are presented. It is shown that by making a limited number of measurements on a single simulator of reasonable size, the input reflection coefficient of the infinite array can be obtained simultaneously at a considerable number of scan angles. An interpolation to the reflection coefficient for more general scan angles is then given by an expansion in terms of these discrete values. The technique has an advantage over the computational method for elements which are symmetrical but not easily computed. The practicality of the theory is demonstrated with experimental results from the simulation of an infinite stripline slot array.

## Basic Considerations

Consider an infinite two-dimensional array of uniformly spaced identical radiating elements. Let the elements be capable of radiating either vertical or horizontal polarization. For the present, let the elements be arrayed in the $x$–$y$ plane with center positions as indicated in Fig. 1 where the center of the coordinate system is taken to be halfway between the centers of four adjacent elements. The element positions are then given by

$$[x_r, y_s] = [(r - \tfrac{1}{2})d_x, (s - \tfrac{1}{2})d_y], \qquad r = 0, \pm 1, \cdots, \quad s = 0, \pm 1, \cdots \quad (1)$$

Manuscript received January 13, 1972; revised March 27, 1972. This work was supported by the Swedish Board for Technical Development and the LM Ericsson Telephone Company, MI Division, Mölndal, Sweden.

The author is with the Microwave Antenna Laboratory, Division of Network Theory, Chalmers University of Technology, Gothenburg, Sweden.

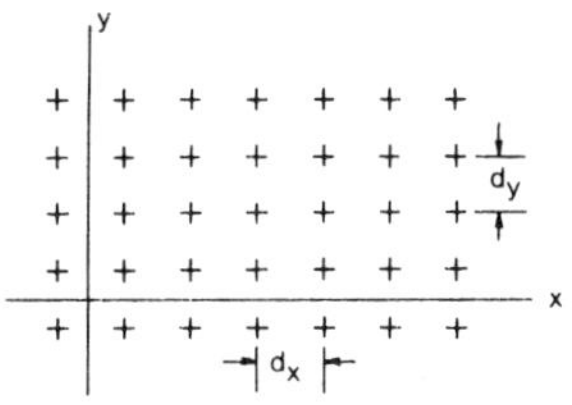


Fig. 1. Element positions in infinite array.

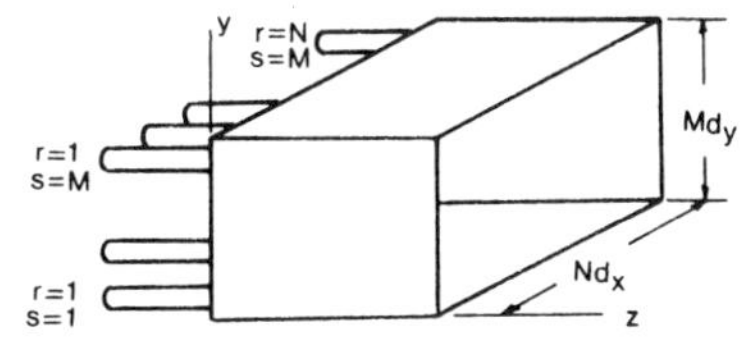


Fig. 2. Multielement simulator with $N \times M$ input lines.

where $d_x$ and $d_y$ are the element spacings along the $x$ and $y$ axes, respectively. Let the elements in the array be fed by identical uniform transmission lines. With no loss in generality, it will be assumed that only the vertically polarized portion of the elements are excited. The results of an analysis for horizontal polarization can be obtained directly by interchanging the $x$ and $y$ coordinates and their respective associated parameters.

Let the transmission line feeding the vertically polarized portion of the $r,s$ element be excited by a voltage wave of amplitude

$$a(x_r,y_s) = \exp\left[-j\left(\frac{n\pi}{a}x_r + \frac{m\pi}{b}y_s\right)\right]. \tag{2}$$

It is well known that the modes of a rectangular waveguide can be considered to be made up of four plane wave components. The excitation (2) has been chosen so that the radiated field corresponds to one of the plane wave components of an $n,m$th-order waveguide mode in a rectangular waveguide of width $a$ and height $b$. The principal plane wave radiated by the excitation (2) has a scan angle off broadside given by

$$\theta_{nm} = \sin^{-1}\frac{1}{k_0}\left[\left(\frac{n\pi}{a}\right)^2 + \left(\frac{m\pi}{b}\right)^2\right]^{1/2} \tag{3}$$

in a scan plane which makes an angle

$$\varphi_{nm} = \tan^{-1}\frac{ma}{nb} \tag{4}$$

with the $x$–$z$ plane. In general, a reflected wave will occur when the excitation is applied and the amplitude of the reflected wave at the $r,s$ element is

$$b(x_r,y_s) = R_{nm}a(x_r,y_s) \tag{5}$$

where $R_{nm}$ is the reflection coefficient measured at a suitable reference plane common to all the input lines. Two required conditions of symmetry and their consequences are now noted.

1) *Array Symmetry*: Since all the elements in the array have the same environment with the exception of a constant phase gradient in the excitation, the same reflection coefficient must occur at each element, and the $R_{nm}$ must be independent of the choice of $r,s$ and only dependent on the excitation numbers $n,m$.

2) *Element Symmetry*: If the radiating element is symmetric about both the $x$ and $y$ axes, then the reflection coefficient will be the same for all the plane wave components which make up the $n,m$th-order waveguide mode. Specifically, the element symmetry implies

$$b(\pm x_r,\pm y_s) = R_{nm}a(\pm x_r,\pm y_s), \qquad r = 0,\pm 1,\cdots, \quad s = 0,\pm 1,\cdots \tag{6}$$

where the sign of $x_r$ and $y_s$ can be chosen independently. From (2) it is seen that by choosing the proper sign of $x_r$ and $y_s$, the scan plane angle can be changed from $\varphi$ to $-\varphi$, $\pi - \varphi$, and $\pi + \varphi$ to generate the four component waves which make up the waveguide mode.

Now let the following linear superposition of the component excitations be applied to the $y$ polarized $r,s$ radiating element:

$$A_{rs}^{(nm)} = \left[\frac{a(-x_r,y_s) - a(x_r,y_s)}{4j} + \frac{a(-x_r,-y_s) - a(x_r,-y_s)}{4j}\right] = \sin\frac{n\pi}{a}x_r \cos\frac{m\pi}{b}y_s. \tag{7}$$

Clearly this excitation is an odd function about the plane $x = 0$ and an even function about the plane $y = 0$. Also, provided the width $a$ and the height $b$ are multiples of the respective $x$ and $y$ element spacings, i.e.,

$$a = Nd_x \tag{8}$$

$$b = Md_y \tag{9}$$

the excitation will be an odd function about the plane $x = a$ and an even function about the plane $y = b$ with the element positions symmetrically disposed about these planes. Since the elements have been assumed to be vertically polarized radiators, it then follows that the tangential component of electric field must vanish on the aforementioned planes. This is true for both the reactive and radiative fields. Note that if the element spacing and excitation (2) are such that a grating lobe exists in real space, the superimposed grating lobe fields must also vanish on the null planes.

When conductors are placed on the null planes and the portion of the array exterior to the enclosed volume is removed, a rectangular waveguide is formed. The waveguide contains $N \times M$ array elements which radiate fields into the guide as shown in Fig. 2. From (6) and (7) it is seen that the reflection coefficient for the elements radiating into the waveguide with the excitation (7) is the same as when the elements radiate with the single-component excitation (2) in the infinite array. Thus the waveguide with its $N \times M$ mode-launching elements can be used as an infinite array simulator.

### The Multielement Simulator

The simulator in Fig. 2 will be shown to exhibit a number of interesting properties when its characteristics are interpreted in terms of the multiterminal network formed by the $N \times M$ input lines. Let these lines be numbered in the same way as the element positions in the infinite array as indicated in Fig. 2. Since vertical polarization has been assumed, the radiating elements can be excited to launch $N \times M$ longitudinal-section electric ($LSE_y$) modes into the waveguide. The excitation at the $r,s$th line for the $LSE_{ynm}$ modes is obtained from (1), (7)–(9) as

$$A_{rs}^{(nm)} = \sin \frac{n\pi}{N}\left(r - \frac{1}{2}\right) \cos \frac{m\pi}{M}\left(s - \frac{1}{2}\right),$$

$$n = 1,2,\cdots,N, \quad m = 0,1,\cdots,M-1. \quad (10)$$

The resulting reflected wave amplitude at the $r,s$th line due to this excitation is given from (6) and (7) as

$$B_{rs}^{(nm)} = R_{nm} \sin \frac{n\pi}{N}\left(r - \frac{1}{2}\right) \cos \frac{m\pi}{M}\left(s - \frac{1}{2}\right) \quad (11)$$

where $R_{nm}$ is the same reflection coefficient as in the infinite array when the elements are excited with vertical polarization and scanned off broadside through an angle

$$\theta_{nm} = \sin^{-1}\left[\left(\frac{n\lambda_0}{2Nd_x}\right)^2 + \left(\frac{m\lambda_0}{2Md_y}\right)^2\right]^{1/2} \quad (12)$$

in a scan plane which makes an angle



$$\varphi_{nm} = \tan^{-1} \frac{mNd_x}{nMd_y} \quad (13)$$

with the $x$–$z$ plane.

There will be some choices of $m$ and $n$ for which the square root in (12) exceeds unity and $\theta_{nm}$ does not represent an angle in visible space. In these cases the $LSE_{ynm}$ mode is beyond cutoff and is evanescent in the simulator guide. The scan plane which contains the largest number of real scan angles is the $H$ plane where $m = 0$. As seen from (12), when the element spacing along the $x$ axis is at least a half wavelength, there are at least $N$ real scan angles in the $H$ plane corresponding to propagating $LSE_{yn0}$ modes. As $m$ is increased from zero and the scan plane is tilted up from the $x$–$z$ plane, the number of values of $n$ for which (12) has a real solution is seen to decrease. The scan plane closest to the $E$ plane is the plane for which $m = M - 1$, and this plane contains the fewest number of real scan angles.

When the element spacing in either plane exceeds a half wavelength, at least one grating lobe will exist in real space for certain of the scan angles $\theta_{nm},\varphi_{nm}$. Under these conditions the infinite array will radiate secondary grating waves at scan angles $\theta_{gnm},\varphi_{gnm}$. The grating lobe conditions which determine these angles are given by

$$k_0 d_x \sin\theta_{nm} \cos\varphi_{nm} - k_0 d_x \sin\theta_{gnm} \cos\varphi_{gnm} = 2\pi q$$

$$k_0 d_y \sin\theta_{nm} \sin\varphi_{nm} - k_0 d_y \sin\theta_{gnm} \sin\varphi_{gnm} = 2\pi p$$

where $p$ and $q$ are any integers not simultaneously zero (main beam condition). By making use of (12) and (13), these conditions can be solved to yield

$$\theta_{gnm} = \theta_{n'm'} \quad (14)$$

$$\varphi_{gnm} = \varphi_{n'm'} - \pi \quad (15)$$

where $\theta_{n'm'},\varphi_{n'm'}$ are the expressions (12) and (13) with $n$ and $m$ replaced by the primed quantities and

$$n' = |2Nq - n|, \qquad q = 0,1,\cdots$$

$$m' = |2Mp - m|, \qquad p = 0,1,\cdots. \quad (16)$$

The scan angle off broadside has been defined as a positive angle so that the 180° rotation of the scan plane represented by the factor $-\pi$ in (15) rotates the grating lobes to the other side of broadside in the scan plane. As can be seen in (14) and (15), the grating lobe fields must have the same scan angles as certain components of the $N \times M$ waveguide modes, which can exist in the simulator. In particular, the superimposed grating lobe fields exhibit themselves as $LSE_{yn'm'}$ modes in the simulator, and the grating lobes exist in real space only for those values of $p$ and $q$ for which the $LSE_{yn'm'}$ mode propagates.

When the simulator waveguide is terminated in a matched load, the $N \times M$ input transmission lines form a lossy reciprocal $N \times M$ port microwave network. These $N \times M$ ports are completely characterized by a symmetric $(N \times M) \times (N \times M)$ scattering matrix $\boldsymbol{S}$. In order to be consistent with the double-index notation already in use, the element of $\boldsymbol{S}$ which represents the scattering coefficient between the $r,s$ input line and the $j,k$ line will be written $S_{rs,jk}$. The exact form of $\boldsymbol{S}$ is determined by the detailed construction of the elements and their associated feed lines. In a practical situation the various components of $\boldsymbol{S}$ are easily measured, and from these measured values the reflection coefficients for the infinite array can be determined as follows.

Letting $\boldsymbol{A}^{(nm)}$ and $\boldsymbol{B}^{(nm)}$ be $1 \times (N \times M)$ column matrices with elements given by the respective amplitudes in (10) and (11), it is clear from the definition of the scattering matrix that

$$\boldsymbol{B}^{(nm)} = \boldsymbol{S}\boldsymbol{A}^{(nm)} = R_{nm}\boldsymbol{A}^{(nm)} \quad (17)$$

so that

$$(\boldsymbol{S} - R_{nm}\boldsymbol{I})\boldsymbol{A}^{(nm)} = \boldsymbol{O},$$

$$n = 1,2,\cdots,N, \quad m = 0,1,\cdots,M-1 \quad (18)$$

where $\boldsymbol{I}$ is the unit matrix. From (18) it is seen that the reflection coefficient for the $n,m$ scan angle $R_{nm}$ is simply the $n,m$ eigenvalue of the scattering matrix $\boldsymbol{S}$. The amplitudes $\boldsymbol{A}^{(nm)}$ are the corresponding eigenvectors. Since the $R_{nm}$ are the eigenvalues of $\boldsymbol{S}$, they can be determined from $\boldsymbol{S}$ in a number of ways. One of the simplest methods is to make use of the fact that the eigenvectors are known exactly. The $R_{nm}$ can be obtained directly from any one of the rows of the matrix equation (17). In particular, making use of (10) and the $p,q$th row of (17), the reflection coefficients can be determined by the expression

$$R_{nm} = \sum_{r=1}^{N} \sum_{s=1}^{M} S_{pq,rs} \frac{\sin (n\pi/N)(r - \frac{1}{2}) \cos (m\pi/M)(s - \frac{1}{2})}{\sin (n\pi/N)(p - \frac{1}{2}) \cos (m\pi/M)(q - \frac{1}{2})} \quad (19)$$

valid for $p = 1,2,\cdots,N$ and $q = 0,1,\cdots,M - 1$, except for those isolated choices of $p$ and $q$ for which the denominator vanishes. It is most important to note that (19) only involves the scattering coefficients between the $p,q$ input line and every line in the simulator. Thus only one row of the scattering matrix needs to be measured, and only $N \times M$ measurements determine the $N \times M$ values of $R_{nm}$.

## Interpolation to Other Angles

An interpolation to give results for more general angles will now be considered. Although the method is more involved than a direct linear interpolation between the eigenvalues, the results have a definite physical interpretation in terms of the impedance of a zoned infinite array. It is easily demonstrated that the eigenvectors of a symmetric matrix are orthogonal to one another provided that their eigenvalues are distinct. The eigenvector excitations $\boldsymbol{A}^{(nm)}$ must therefore have this property. This can be shown directly by expanding the trigonometric functions in (10) into exponentials and summing the geometric series which result when the $n,m$ and $k,j$ eigenvectors are scalar multiplied to yield

$$\boldsymbol{A}^{(nm)T}\boldsymbol{A}^{(kj)} = \frac{NM}{4} \delta_{nk}\delta_{mj}\epsilon_{nN}\epsilon_{m0}. \quad (20)$$

The superscript $T$ indicates the transposed matrix, $\delta_{nk}$ is the Kronecker delta, and $\epsilon_{pq} = 2$ for $p = q$ and unity otherwise. This result is seen to be true even if the eigenvectors happen to be equal and $R_{nm} = R_{kj}$. Any orthogonal set of vectors which have the properties given by (20) forms a complete set spanning the vector space formed by the $N \times M$ possible excitations of the simulator lines. It is then possible to expand *any* excitation of the simulator in terms of a linear combination of these vectors [13].

Let the desired excitation of the lines of the simulator be given by the $1 \times (N \times M)$ column matrix $\boldsymbol{E}$ with elements $E_{rs}$. This excitation can then be written as the following linear combination of the eigenvectors:

$$\boldsymbol{E} = \sum_{n=1}^{N} \sum_{m=0}^{M-1} \frac{\boldsymbol{E}^T\boldsymbol{A}^{(nm)}}{\boldsymbol{A}^{(nm)T}\boldsymbol{A}^{(nm)}} \boldsymbol{A}^{(nm)}. \quad (21)$$

As each of the eigenvectors experiences a reflection with a reflection coefficient $R_{nm}$, the ratio of the reflected to the incident wave amplitude at the $r,s$ line is given by

$$\boldsymbol{\Gamma}_{rs} = \frac{1}{E_{rs}} \sum_{n=1}^{N} \sum_{m=0}^{M-1} \frac{\boldsymbol{E}^T\boldsymbol{A}^{(nm)}}{\boldsymbol{A}^{(nm)T}\boldsymbol{A}^{(nm)}} R_{nm}\boldsymbol{A}^{(nm)}. \quad (22)$$

Because each of the eigenvectors is reflected in a different manner, the reflected wave amplitudes in the input lines will not necessarily form a vector proportional to the vector excitation $\boldsymbol{E}$. Some input lines will reflect energy into other lines, and the value of the reflection coefficient given by (22) can take on values exceeding unity. It is now interesting to go back and determine the significance of $\boldsymbol{\Gamma}$ and $\boldsymbol{E}$ in terms of the infinite array.

The extension of the eigenvectors into the infinite $x$–$y$ plane given by (7) are all odd functions about the plane $x = 0$ and even functions about the plane $y = 0$. In addition, these functions are periodic in $x$ with period $2a$ and periodic in $y$ with period $2b$. The expansion of the general excitation in terms of the eigenvectors of the simulator is equivalent to an excitation of the infinite array which is the same as the simulator excitation only over the region occupied by the simulator from $x = 0$ to $x = a$ and from $y = 0$ to $y = b$. In the region from $x = -a$ to $x = 0$ the array excitation is an odd reflection about the $x = 0$ plane, and in the region from $y = -b$ to $y = 0$ the excitation is an even reflection about the $y = 0$ plane. This excitation then repeats itself periodically throughout the infinite $x$–$y$ plane.

Consider now the situation when the four component excitation of plane waves radiating at angles $\theta,\varphi$ is applied to the simulator as follows:

$$E_{rs}(\theta,\varphi) = \sin (k_0 \sin \theta \cos \varphi \, x_r) \cos (k_0 \sin \theta \sin \varphi \, y_s),$$
$$r = 1,2,\cdots,N, \quad s = 1,2,\cdots,M. \quad (23)$$

A discontinuity in the excitation results when (23) is periodically continued. When $\theta,\varphi$ are not one of the $\theta_{nm},\varphi_{nm}$ pairs, the excitation at the planes $x = -a$ and $x = +a$ does not repeat itself since

$$E_{N,s}(\theta,\varphi) = -E_{1-N,s}(\theta,\varphi), \qquad s = 1 - M,\cdots,M. \quad (24)$$

Also, the normal derivative of the excitation is discontinuous at the planes $y = -b$ and $y = +b$:

$$\frac{dE_{rM}(\theta,\varphi)}{dy_M} = -\frac{dE_{r,1-M}(\theta,\varphi)}{dy_{1-M}}, \qquad r = 1 - N,\cdots,N. \quad (25)$$

Because of the divergence condition of the electric field, (25) represents a discontinuity in the excitation of the $z$ component of the electric field.

Under these conditions the $x$–$y$ plane is filled with $2N \times 2M$ element subarrays, each excited identically, but forming a total excitation with amplitude discontinuities at the boundaries where the arrays meet. When $\theta,\varphi$ are one of the $\theta_{nm},\varphi_{nm}$ pairs, (24) and (25) vanish, and the total excitation becomes exactly continuous with identical reflection coefficients for all the elements.

It has been demonstrated [12] that in certain cases reflection coefficient values within a few percent of infinite array values can be observed as near as four or five elements away from the edge of a truncated infinite array. As there are $2N \times 2M$ elements in the four quadrants of the basic subarray, the $r = 1$, $s = 1$ element of the basic simulator region is at least $(N - 1),(M - 1)$ elements from the boundaries of the subarray. It is then seen that the infinite array criteria can be achieved with only a moderate number of elements in the simulator. This observation immediately suggests that if the simulator contains enough elements, the impedance at one of the center elements in a subarray with excitation (23) can be used to approximate the input impedance to any element in the infinite array when it is scanned to the general angles $\theta,\varphi$.

Making use of (10), (20), (22), and (23), the following explicit expression for the reflection coefficient at the $r,s$ input line to the simulator under the excitation (23) can be obtained:

$$\Gamma_{rs}(\theta,\varphi) = \sum_{n=1}^{N}\sum_{m=0}^{M-1} \frac{(F_n - F_{-n})(G_m + G_{-m}) \sin (n\pi/N)(r - \frac{1}{2}) \cos (m\pi/M)(s - \frac{1}{2})}{\epsilon_n N \epsilon_{m0} \sin u(r - \frac{1}{2}) \cos v(s - \frac{1}{2})} R_{nm} \tag{26}$$

where

$$F_n(\theta,\varphi) = \frac{1}{2N}\frac{\sin N(u - n\pi/N)}{\sin \frac{1}{2}(u - n\pi/N)} \tag{27}$$

$$G_m(\theta,\varphi) = \frac{1}{2M}\frac{\sin M(v - m\pi/M)}{\sin \frac{1}{2}(v - m\pi/M)} \tag{28}$$

and

$$u = k_0 d_x \sin\theta \cos\varphi$$

$$v = k_0 d_y \sin\theta \sin\varphi. \tag{29}$$

At the angles $\theta_{nm},\varphi_{nm}$ all the coefficients in (26) but one vanish identically and

$$\Gamma_{rs}(\theta_{nm},\varphi_{nm}) = R_{nm}.$$

For the intermediate angles (26) gives an interpolation with the physical significance previously discussed.

When the excitation of the simulator is chosen such that the scan angles $\theta,\varphi$ are not near any of the eigenvector angles, mutual coupling effects dominate. Since no particular eigenvector is excited, the reflected amplitudes in the input lines no longer need be proportional to the vector excitation, and the infinite array situation is not approximated. Such is always the case for $E$-plane scan conditions. By taking the limit in (26) as $\varphi$ approaches $\pi/2$, it can be shown that the resulting formula for $\Gamma_{11}(\theta,\pi/2)$ has coefficients in the sum over $n$ with equal magnitudes and alternating sign.

### The Experimental Simulator

In order to provide an actual experimental demonstration of the simulator concepts presented here, an $N = 5$, $M = 5$ simulator was constructed for a stripline slot array. The simulator is shown in Fig. 3. The radiating elements are dumbell shaped off-center fed slots with each element surrounded by six parallel plate mode suppression rivets. The element spacing is $0.475\lambda_0$ in the $E$ plane and $0.538\lambda_0$ in the $H$ plane at a design frequency of 9.5 GHz. The $H$-plane spacing allows a grating lobe to enter real space at an azimuth scan angle of 60°. Twenty-three of the elements are terminated in matched stripline pill terminations, which are shown exposed in the figure. The two remaining elements are provided with connectors for use in the transmission measurements. The simulator was constructed so that the unit cells containing the various elements are moveable. Thus by alternately interchanging the position of one of the active elements with each of the loaded elements, 24 separate transmission measurements can be made to provide all but one of the elements in one row of the scattering matrix. A reflection measurement at the other active element then completes the procedure. The output waveguide was provided with a flat moveable absorbing termination. Sliding load techniques were used to eliminate measurement errors due to load reflections. A paper dealing with the experimental details of the study is available [14].

Direct amplitude and phase measurements were made with a network analyzer, and the eigenvalues of the scattering matrix were calculated from (19) with the use of a digital computer. Contour plots of the amplitude and phase of the input reflection coefficient for the $r = 1, s = 1$ subarray element as computed from the interpolation equation (26) are shown in Figs. 4 and 5. In these figures the amplitude and phase contours are shown plotted as a function of the direction cosines of the scan ray, $\sin\theta\cos\varphi$ and $\sin\theta\sin\varphi$. The eigenvalues are indicated by black dots which are arranged in a rectangular grid in the direction cosine space. For reasons previously discussed, the interpolation formula (26) gives unrealistic reflection coefficient values which exceed unity along the $E$-plane scan axis so that the contours have been truncated along the line, $\sin\theta\cos\varphi = \sin\theta_{10}$, for viewing purposes. Also, the contours of Fig. 4 have been truncated at unity magnitude.

Several interesting features can be observed in the contour plots. Of foremost importance is the fact that the interpolation formula is seen to generate a surface which passes through the eigenvalues in a rather smooth fashion throughout most of the region over which the eigenvalues are distributed. The curves are particularly smooth for $H$-plane scan conditions, and appreciable fluctuations between eigenvalues are only observed for scan angles near endfire with scan planes near the $E$ plane. As the array is scanned to endfire in the $H$ plane, the reflection coefficient magnitude does not go to unity due to the presence of a grating lobe. As the angle $\varphi$ is increased from zero, the grating lobe vanishes, and total reflection is seen to occur with five of the eigenvalues representing the complete reflection experienced when the simulator elements are phased to launch evanescent modes. For scan into imaginary space beyond these eigenvalue locations, the inter-

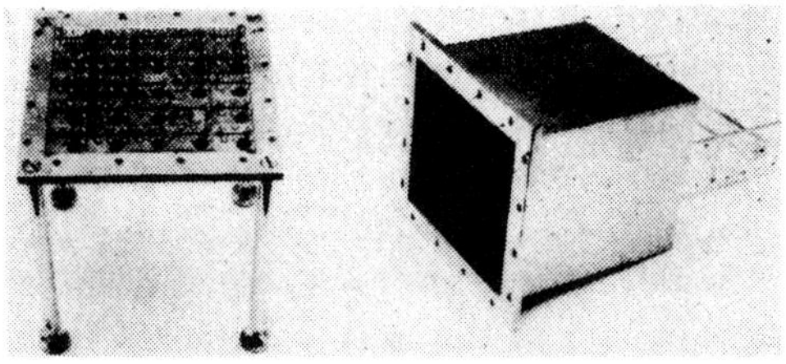

Fig. 3. Twenty-five-element simulator.

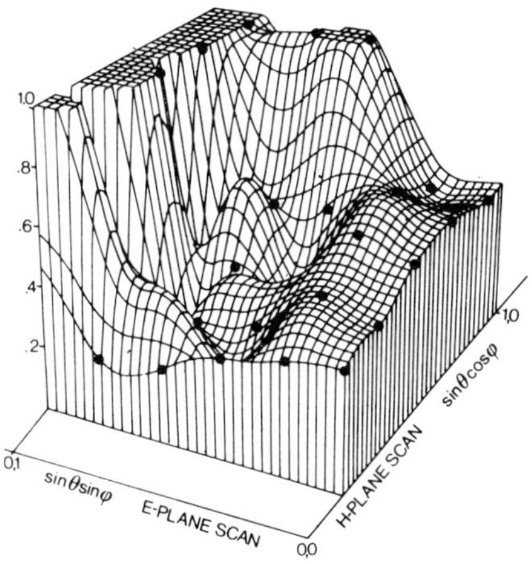


Fig. 4. Magnitude of $\Gamma_{11}(\theta,\varphi)$.

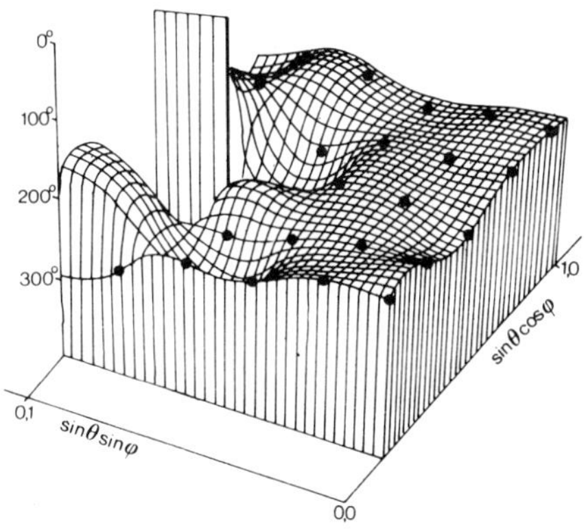


Fig. 5. Relative phase of $\Gamma_{11}(\theta,\varphi)$.

polation scheme breaks down again, and the magnitude of the reflection coefficient exceeds unity.

An interesting resonance can be seen in the phase contours of Fig. 5. A computer generated branch cut can be observed, indicating that the reflection coefficient has a zero near endfire in the $\varphi = 50°$ scan plane. The edge of a "hole" in the magnitude contour can also be observed in the truncated region between the 3, 4 and 4, 4 eigenvalues at the very back of the figure. At the resonant scan angle, the 1, 1 element of the subarray is perfectly matched. The intriguing possibility of using the interpolation scheme to search for blindness resonances is readily apparent and is the subject of further investigation.

## Half Elements and Triangular Spacing

In the considerations leading to the simulator configuration of Fig. 2, it is possible to choose the excitations so that some or all of the null planes actually bisect elements instead of falling exactly halfway between them. When this is done the elements are still symmetrically disposed about these planes. The elements which are bisected by planes parallel to the electric field receive zero incident excitation, and the mutual coupling from the excited elements of the array induces only a higher order evanescent field in the feed structure behind the elements. If the feed structure has the same symmetry as the radiating elements, the simulator conducting planes can be extended into the feed region to bisect the evanescent fields. Half elements are then constructed in the simulator to exactly account for the reactive effects of the bisected elements [2]. These half elements always have zero excitation and may be disregarded as far as the simulator is concerned. Enough of the bisected input line should be provided to contain the evanescent fields.

Elements bisected by the planes perpendicular to the electric field, on the other hand, receive full excitation in the infinite array. Their transmitted and received fields are even functions about the bisecting plane. In this case, when a half element for the simulator is constructed by extending the conducting plane into the symmetric feed structure, the input line to the half element must be treated as one of the active ports of the simulator.

When the simulator is constructed with half elements, the eigenvectors are given by (7) with $x_r$ and $y_s$ replaced by the active element positions and $a$ and $b$ replaced by the appropriate simulator dimensions. With these substitutions, (3), (4), (18), and (22) are valid, and the scan angles and input reflection coefficients can be calculated as before. When bisected elements exist, the subarrays of the continued excitations have elements exactly on the array boundaries.

The preceding statements apply directly to the important case of the triangular spaced array. The triangular spaced array is the special case of the previous analysis in which the elements in Fig. 1 are removed for $|r + s|$ an odd integer. Such a removal of elements does not destroy the array symmetry, and waveguide simulation is still possible [3]. For the triangular spaced array, the null planes must generally be chosen to bisect radiating elements. Once again, only the active elements are used as input ports in the simulator, and the aforementioned formulas are applicable with the indicated substitutions.

## Conclusions

It has been demonstrated that it is possible to obtain extensive knowledge of the active infinite array impedance under scan conditions with the use of a single multielement waveguide simulator. The measurements involved are the straightforward determination of transmission coefficients between simulator elements. Although the measured data requires some small computer processing, the amount of information generated for the evaluation of the array should justify the procedure. This is especially true in light of the great expense and effort involved in the construction of a scanning array of appreciable size.

The number of exact reflection coefficient values given by the eigenvalues of the simulator and the accuracy of the interpolation scheme are determined by the size of the simulator. It has been shown that a comprehensive picture of the array behavior can be obtained in a simulator with as few as 24 elements. Swept frequency measurements on larger simulators are quite feasible and well within the capabilities of presently available microwave instrumentation.

## Acknowledgment

The efforts of A. Derneryd and P. I. Jonsson in the computations and experimental work are gratefully acknowledged. The contour plotting subroutine was provided by R. Lagerlöf.

## References

[1] P. W. Hannan, P. J. Meier, and M. A. Balfour, "Simulation of phased array antenna impedance in waveguide," *IEEE Trans. Antennas Propagat.* (Commun.), vol. AP-11, pp. 715–716, Nov 1963.
[2] P. W. Hannan and M. A. Balfour, "Simulation of a phased-array antenna in waveguide," *IEEE Trans. Antennas Propagat.*, vol. AP-13, pp. 342–353, May 1965.
[3] A. A. Oliner and R. G. Malech, "Simulation of infinite arrays by waveguide," *Microwave Scanning Antennas*, vol. 2. R. C. Hansen, Ed. New York: Academic Press, 1966, pp. 322–335.
[4] G. J. Laughlin, "ARRAY 1: calculation of the active transmit impedance for a phased-array antenna of lossless elements," *IEEE Trans. Microwave Theory Tech.* (Comput Progr. Descrip.), vol. MTT-18, pp. 666–667, Sept. 1970.
[5] M. A. Balfour, "Phased array simulators in waveguide for a triangular arrangement of elements," *IEEE Trans. Antennas Propagat.* (Commun.), vol. AP-13, pp. 475–476, May 1965.
[6] E. V. Byron and J. Frank, "On the correlation between wideband arrays and array simulators," *IEEE Trans. Antennas Propagat.* (Commun.), vol. AP-16, pp. 601–603, Sept. 1968.
[7] P. W. Hannan, "Discovery of an array surface wave in a simulator," *IEEE Trans. Antennas Propagat.* (Commun.), vol. AP-15, pp. 574–576, July 1967.
[8] H. A. Wheeler, "A survey of the simulator technique for designing a radiating element," in *Symp. Dig., 1970 Phased Array Antenna Symp.*, pp. 56–59.
[9] ——, "A systematic approach to the design of a radiator element for a phased-array antenna," *Proc. IEEE*, vol. 56, pp. 1940–1951, Nov. 1968.
[10] M. A. Balfour, "Active impedance of a phased-array antenna element simulated by a single element in waveguide," *IEEE Trans. Antennas Propagat.* (Commun.), vol. AP-15, pp. 313–314, Mar. 1967.
[11] S. S. Gregorwich, A. Hessel, and G. H. Knittel, "A waveguide simulator study of a blindness effect in a phased array," *Microwave J.*, vol. 14, pp. 37–41, Sept. 1971.
[12] C. P. Wu, "Analysis of finite parallel-plate waveguide arrays," *IEEE Trans. Antennas Propagat.*, vol. AP-18, pp. 328–334, May 1970.
[13] R. Courant and D. Hilbert, *Methods of Mathematical Physics*, vol. 1. New York: Interscience, 1965, ch. 1.
[14] J. J. Gustincic, "A new simulator technique for phased arrays," in *Proc. 1972 Naval Electronics Lab. Cent. Array Antenna Conf.*, pp. 11-1–11-20.

# CONFORMAL ARRAYS V

*Reprinted from the* PROCEEDINGS OF THE IEEE
VOL. 56, NO. 11, NOVEMBER, 1968
pp. 2016-2027


# A Matrix-Fed Circular Array for Continuous Scanning



BORIS SHELEG

*Abstract*—**The Butler-matrix-fed circular array will form a focussed radiation pattern when the proper current distribution is established on the inputs to the matrix. Further, this beam can be scanned through 360° by changing only the phases of the matrix input currents, just as with a linear array scanning is accomplished by varying the phases of the element currents. This operation was experimentally demonstrated with a 32-dipole circular array and reasonable agreement was obtained between the measured and calculated patterns. Finally, a synthesis procedure is described for determining the matrix input currents required to attain a prescribed current distribution on the array.**

Manuscript received April 1, 1968; revised July 31, 1968.
The author is with the Microwave Antennas and Components Branch, Electronics Division, Naval Research Laboratory, Washington, D. C. 20390

## INTRODUCTION

ANTENNAS consisting of radiating elements arrayed on a circle have been studied and have been used for many years, but recent developments in switching and phase shifting have led to a renewed interest in them. The appeal of the circular array is that, because of its symmetry, it can be used to scan a beam in discrete steps through a full 360° without the variations in gain and pattern shape that occur when four linear arrays are used, each scanning through a single quadrant. The purpose of this study was to determine some of the possibilities and also the limitations of scanning with circular arrays and, in particular, to demonstrate the use of the Butler matrix in feeding the elements of the array. The idea of using a Butler matrix for this purpose is due to Shelton [1], who showed that it permitted the formation of a narrow radiated beam that could then be scanned essentially like the beam from a linear array, by the operation of phase shifters alone.

The operation of a Butler-matrix-fed circular array (multimode array) is first described heuristically in terms of "modes" and then, more satisfactorily, by considering the distribution of currents impressed on the radiating elements by the matrix. In addition, calculations were made to show how the radiation pattern of the multimode array varies as it is scanned continuously, rather than in discrete steps.

The experimental portion of this program was performed at *L*-band with a circular array of 32 dipoles around a conducting cylinder. Sidelobe level control was shown by using different amplitude tapers over the illuminated portion of the array.

## Theory of Operation

The principles involved in scanning a multimode array are most easily seen by considering not an array, but a continuous distribution of current. When this distribution is expressed as a Fourier series, in general infinite, each term represents a current mode uniform in amplitude but having a phase varying linearly with angle. The radiation pattern of each mode has the same form as the current mode itself, and these pattern modes are the Fourier components of the radiation pattern of the original distribution. The expression of the radiation pattern as the sum of modes of this form is then seen to be analogous to the summation of the contribution made to the pattern of a linear array by its elements, so the operation of a multimode array can be explained by referring to an equivalent linear array.

Referring to Fig. 1, consider a current distribution $I(\alpha)$ to be the sum of a finite number of continuous current modes $I_n e^{jn\alpha}$ with $-N \le n \le N$. The radiation pattern for $\theta = \pi/2$, is then given by

$$E(\phi) = \sum_{n=-N}^{N} C_n e^{jn\phi} \qquad (1)$$

where the $C_n$ are complex constants given by

$$C_n = 2\pi K j^n I_n J_n\left(\frac{2\pi a}{\lambda}\right) \qquad (2)$$

with $K$ a constant [2]. There is a one-to-one correspondence between the current modes $e^{jn\alpha}$ and the far-field pattern modes $e^{jn\phi}$, but note that their relative phases are not necessarily the same. Another property peculiar to circular arrays with isotropic radiators is that some modes can be made to give zero contribution in the plane of the circle by the selection of a proper diameter. However, for practical

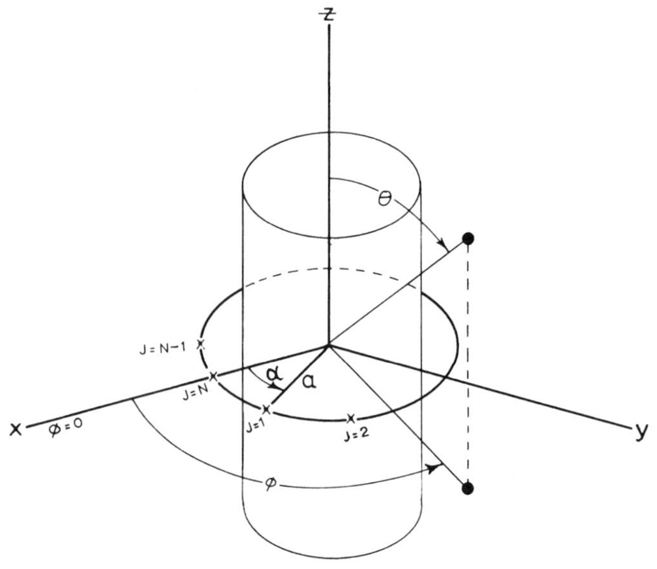


Fig. 1. Coordinates for a continuous cylindrical sheet of vertical current elements.

antennas of interest (e.g., dipoles approximately one-quarter wavelength over a reflecting cylinder) all the modes make contributions in the plane of the array.

Equations (1) and (2) demonstrate that a change in relative amplitude and phase of each current mode results in a corresponding change in the corresponding pattern mode (this can be done by controlling $I_n$). This is nearly identical to the formulation for linear arrays. A linear array of $2N+1$ isotropic elements with interelement spacing $a$ has a radiation pattern given by

$$E(u) = \sum_{n=-N}^{N} A_n e^{jnu} \qquad (3)$$

where $u = ka \sin \phi$, $\phi$ is the angle off-broadside, and $A_n$ is the current on the $n$th element. Equations (1) and (3) show the similarity of the patterns of the circular current sheet and the linear array, with the role of the current mode in the circular array taken by the element in the linear array. One difference is that for the circular array the argument is $\phi$, and for the linear array it is $ka \sin \phi$. A second difference is that equally excited elements in a linear array make contributions of equal magnitude to the radiation pattern, but equally excited current modes do not contribute equally, because their elevation patterns are not identical. This results in differences in their strength of contribution in the plane of the antenna. For example, if in the antenna being considered (Fig. 1) it is desired that the pattern modes be equal in magnitude and be in phase at $\phi = 0$, the excitations of the current modes must be [from (2)]

$$I_n = \frac{1}{2\pi K j^n J_n\left(\frac{2\pi a}{\lambda}\right)}. \qquad (4)$$

Its radiation pattern is then given by

$$E(\phi) = \sum_{n=-N}^{N} e^{jn\phi}, \qquad (5)$$

which may be summed to give the pattern characteristic of a uniform array

$$E(\phi) = \frac{\sin\left(\dfrac{2N+1}{2}\phi\right)}{\sin\dfrac{\phi}{2}}. \tag{6}$$

The beam can be scanned (in theory) by a linear variation of the phases of the mode excitations, just as the beam from a linear array is scanned by a linear variation of the element phases. If the phase difference between adjacent modes is $\phi_0$ radians [multiplying $I_n$ of (4) by $e^{-jn\phi_0}$], the resultant pattern is expressed as

$$E(\phi) = \frac{\sin\left[\dfrac{2N+1}{2}(\phi - \phi_0)\right]}{\sin\left(\dfrac{\phi - \phi_0}{2}\right)}, \tag{7}$$

which is the original pattern scanned $\phi_0$ radians. Although the foregoing analysis was based on the particular example of a cylindrical sheet of infinitesimal current elements, the same reasoning applies to any circular antenna having similar pattern modes. The only difference in the analysis would be the particular relationship between the phases and amplitudes of the pattern modes and their respective current modes.

Various antennas have made use of the multimode principle in their operation [3]–[5] but, as has been mentioned, Shelton discovered that it was possible to excite simultaneously and independently all the modes, both positive and negative, from zero to $N/2$ by connecting a single ring of $N$ elements to the outputs of a Butler matrix. That this is true is evident from the definition of the Butler matrix [6]. This matrix is a lossless, passive network having $N$ inputs and $N$ outputs, where $N$ usually is some power of 2. The inputs are isolated from one another, and a signal into any input results in currents of equal amplitude on all the outputs with phase varying linearly across the elements. Specifically, if $N$ is even and the $K$th input port is energized ($K=0, \pm 1, \pm 2, \cdots, \pm(N-2)/2, N/2$), the difference in phase between adjacent ports is $2\pi K/N$ and the total phase variation around a circular array connected to the Butler matrix would be $2\pi K$, which is the $K$th mode. Hence, with the Butler matrix we may establish on the array the N current modes corresponding to $K=0, \pm 1, \cdots, \pm(N-2)/2$, $N/2$, and, because the input ports are isolated, the modes are independent. It should be noted here that there are Butler matrices that do not satisfy the definition given above and that cannot be used unmodified in a multimode array. These networks establish, across their outputs, linear phase progressions whose total variations are odd multiples of $\pi$ radians. A matrix of this type can be changed to one having the proper modes by adding fixed phase shifts to all the output ports; if the $N$ outputs are labeled $J=1, 2, \cdots, N$, the phase shift applied to the $J$th output is $J\pi/N$.

A schematic diagram of a scanning multimode array is shown in Fig. 2. The desired phase and amplitude distribution is established over the inputs to the Butler matrix by fixed phase shifters and a corporate structure. Once the pencil beam pattern is formed at some azimuth angle, it is scanned just as in a linear array; the mode amplitudes are held fixed, and a linear phase progression is set up on the mode inputs by operating the variable phase shifters.

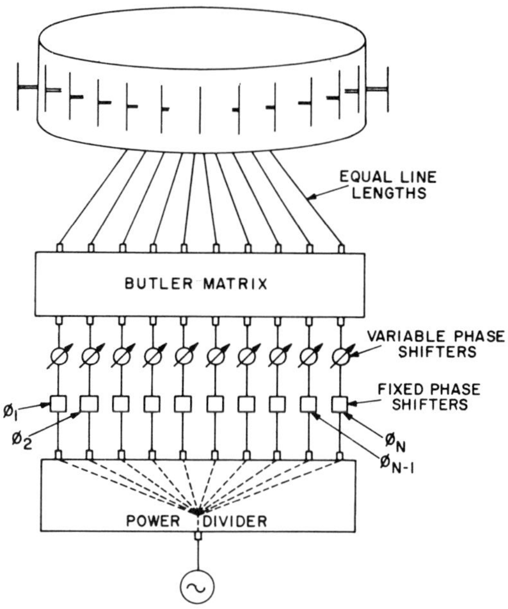


Fig. 2. Schematic diagram of a scanning multimode array.

## Pattern Calculation

Thus far, the explanation of a multimode array has been based on the summation of pattern modes of the form $e^{jK\phi}$. The summation could be exactly achieved with a continuous current sheet, or it could be approximated arbitrarily well with a ring array having a sufficient number of elements. For arrays fed from a Butler matrix, as many current modes can be established as there are elements, and it is not obvious how many of these modes have far-field patterns that fit the $e^{jK\phi}$ form sufficiently well. For example, in an $N$-element array the highest order mode ($K=N/2$) has an element-to-element phase variation of $\pi$ radians. By symmetry, its pattern must be scalloped, with $N$ nulls and $N$ peaks; therefore, it obviously cannot be used as a uniform mode.

To determine the quality of the modes established by the Butler matrix, a series of calculations was made, both of mode patterns and of pencil beam patterns obtained by summing different numbers of modes. The array consisted of 32 elements, and the interelement spacing was varied from 0.4 to 0.6$\lambda$. Two different element patterns were used, one was an approximation to the measured pattern of the elements that were actually used (dipoles around a cylinder) and the other was the exact pattern of an infinitesimal vertical current element in front of a conducting cylinder [7].[1]

[1] Also, for a more detailed description, see R. H. DuHamel, "Pattern synthesis for antenna arrays on circular, elliptical, and spherical surfaces," Elec. Engrg. Research Lab., University of Illinois, Urbana, Contract N6-ORI-71; Task 15, Tech. Rept. 16, May 1952.

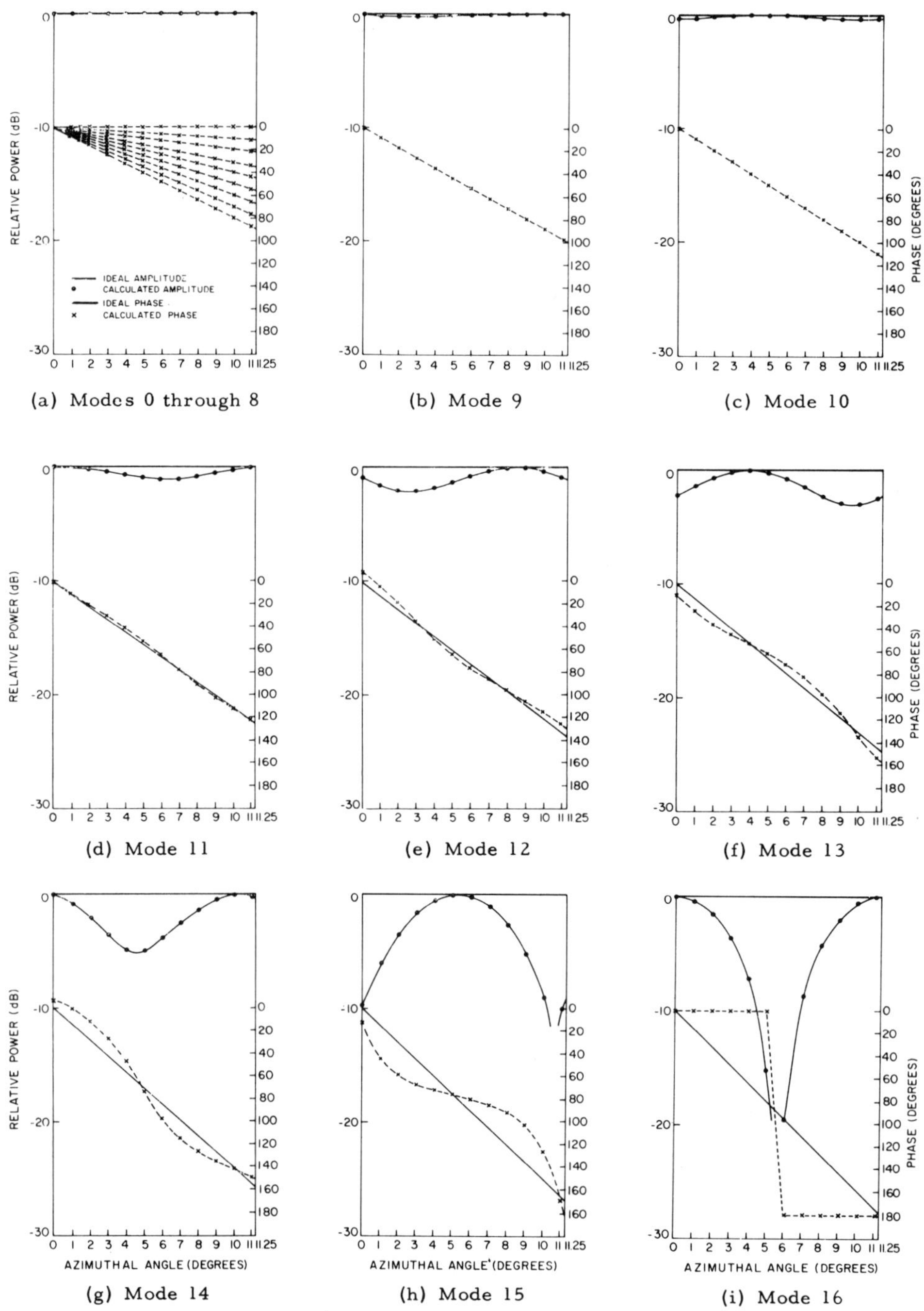


Fig. 3. Calculated mode patterns for a 32-dipole circular array compared with the ideal mode patterns.

The approximate pattern of the dipole in front of a cylinder is given by

$$A(\phi) = \tfrac{1}{2}(1 + \cos\phi), \tag{8}$$

where the phase was assumed constant in azimuth when referred to a point one-third the distance from the cylinder to the dipole. This assumption is reasonably good, at least in the unshadowed region. Mode patterns and pencil beam patterns computed using (8) were in good agreement with those obtained using the exact pattern of the vertical current element, and no results for the latter have been included.

Consider, as in Fig. 1, a circular array of radius $R$ with $N$ elements equally spaced at $\alpha_J = J2\pi/N$, where $J = 1, 2, \cdots, N$. Referred to the center of the circle, the relative space phase of the $J$th element is $(2\pi R/\lambda)\cos(\phi - \alpha_J)$, where only the plane of the array is considered. If the element pattern is $A(\phi - \alpha_J)$ and the current on the element is $A_J e^{j\psi_J}$, the radiation pattern of the array is given by

$$E(\phi) = \sum_{J=1}^{N} A_J e^{j\psi_J} A(\phi - \alpha_J) e^{j(2\pi R/\lambda)\cos(\phi - \alpha_J)}. \tag{9}$$

Mode patterns were calculated from this equation with the element pattern given by (8) and, for the $K$th mode, a current distribution given by $A_J = 1$, $\psi_J = 2\pi KJ/N$. Results are shown for a 32-element array, for which the modes correspond to $K = 0, \pm 1, \pm 2, \cdots, \pm 15, 16$. The phase and amplitude of computed mode patterns for a 32-element circular array ($0.5\lambda$ spacing) are compared with ideal modes in Fig. 3. It is seen that all modes up to $\pm 10$ are in sub-

TABLE I
THE RELATIVE GAINS AND PHASES OF THE PATTERN MODES OF A 32-ELEMENT ARRAY

| Mode | Relative Gain (dB) | Relative Phase at $\phi=0$ (degrees) |
|---|---|---|
| 0 | 0.0 | 0.00 |
| 1 | 0.2 | 0.74 |
| 2 | 0.1 | 7.51 |
| 3 | 0.2 | 14.63 |
| 4 | 0.2 | 29.27 |
| 5 | 0.0 | 44.63 |
| 6 | 0.5 | 63.06 |
| 7 | 0.2 | 91.93 |
| 8 | −0.5 | 115.35 |
| 9 | 0.8 | 144.56 |
| 10 | 1.0 | −168.45 |
| 11 | −0.5 | −128.07 |
| 12 | 0.7 | −105.74 |
| 13 | 3.5 | −35.18 |
| 14 | 5.1 | 20.91 |
| 15 | 5.6 | 120.92 |
| 16 | 5.6 | −172.05 |

stantial agreement with the ideal patterns, having at most a $\pm 0.25$-dB difference from a uniform pattern, with a maximum phase error of 3°. Modes $\pm 14$, $\pm 15$, and 16 are poor approximations. Mode patterns were also computed for arrays with element spacings of 0.4 and $0.6\lambda$. With $0.4\lambda$ element spacing, modes up to $\pm 12$ are close to ideal, and modes $\pm 15$ deviate about as much as did modes $\pm 14$ for the array with 0.5-$\lambda$ spacing. For 0.6-$\lambda$ spacing, modes up to $\pm 9$ are good, and mode 12 corresponds to mode 14 for 0.5-$\lambda$ spacing. Table I lists the relative gains of all the pattern modes for 0.5-$\lambda$ spacing and also their relative phases at $\phi=0$ when the current modes are in phase at $\alpha=0$ (i.e., at element 32). To form a narrow beam with its peak at $\phi=0$, the pattern modes must be in phase in that direction; therefore, the phase differences between the modes must be accounted for, as shown in Fig. 2, by fixed phase shifts at the inputs to the Butler matrix.

The analysis of a circular array in terms of ideal modes is adequate for a qualitative description of its operation and does predict reasonably well the position and shape of the main lobe but not the structure of the sidelobes. For a more accurate estimate, the current distribution on the array must be determined, and the pattern must then be calculated from (9). First assume that the $N$-element Butler matrix has all the current modes in phase at the $N$th element if the mode inputs are fed in phase; this means that for the $K$th mode the $J$th element has phase $2\pi KJ/N$. Then, if $B_K e^{j\beta_K}$ is the current applied to the $K$th input port, the resultant $A_J e^{j\psi_J}$ on the $J$th radiating element is given by

$$A_J e^{j\psi_J} = \frac{1}{\sqrt{N}} B_K e^{j\beta_K} e^{jKJ(2\pi/N)}. \tag{10}$$

If many input ports are simultaneously excited, the output currents may be added to give

$$A_J e^{j\psi_J} = \frac{1}{\sqrt{N}} \sum_K B_K e^{j\beta_K} e^{jKJ(2\pi/N)}. \tag{11}$$

This current distribution may then be substituted into (9) to give the radiation pattern expressed as

$$E(\phi) = \sum_{J=1}^{N} \left[\sum_K B_K e^{j\beta_K} e^{jJK(2\pi/N)}\right] \cdot A(\phi - \alpha_J) e^{j(2\pi R/\lambda)\cos(\phi - \alpha_J)}. \tag{12}$$

It is perhaps evident from the symmetry of the multimode array that the pattern shape does not change if the beam is scanned by some multiple of $2\pi/N$, the angle between elements. If for example, the beam is to be scanned $M(2\pi/N)$ radians, where $M$ is an integer, the mode amplitudes $B_K$ are held constant but there must be a mode-to-mode phase difference of $M(2\pi/N)$; hence, the phase of the $K$th mode becomes $\beta_K + KM(2\pi/N)$. When this is substituted into (11), the current on the $J$th radiating element is found to be

$$A_J e^{j\psi_J} = \frac{1}{\sqrt{N}} \sum_K B_K e^{j\beta_K} e^{j(M+J)K(2\pi/N)}, \tag{13}$$

showing that the original current distribution has been moved, intact, $M$ elements around the array.

To show the formation of the radiation patterns as the modes are superimposed, patterns of a 32-element array with 0.5-$\lambda$ spacing were computed from (12) using the cardioid element pattern of (8). With uniform excitation of the Butler-matrix inputs (all $B_K=1$) and with phases corresponding to those in Table I, the modes were successively excited and at each stage the radiation pattern was calculated, giving the series of patterns shown in Fig. 4. It can be seen that the beam narrows as more modes are added and that, until modes $\pm 11$ are reached, the far-out sidelobes decrease as they would if the modes were perfect. This improvement ceases as the higher modes are added; when all the modes but the 16th are included, the pattern is noticeably worse than when just the modes up to $\pm 10$ are used. The two patterns using modes up to $\pm 10$ and up to $\pm 15$ are compared in Fig. 5 with the patterns that would result if the corresponding numbers of ideal modes were summed. The agreement, when all the modes are used, is particularly poor in the region of the far-out lobes; when the less uniform modes (11)–(15) were not used, the patterns for the discrete and continuous cases agreed everywhere to within 0.5 dB. It is interesting to note that about 95 percent of the total power is radiated by the 13 elements nearest the direction of the beam and that the currents on 11 of these differed in phase from the cophasal condition by less than 20°.

The sidelobes from a linear array can be lowered by tapering the amplitude distribution over the array, and, by analogy, the same should be true for a circular array if the mode amplitude distribution is tapered. With the phases again as in Table I, a cosine taper was applied to the mode inputs [i.e., $B_K = \cos(K\pi/32)$], and the patterns were computed as modes were added successively. Fig. 6 is a comparison of the final pattern, when all the modes were used, with the pattern that would result if the modes were ideal. This shows that the sidelobe level has indeed been reduced but not to the level that would be obtained with perfect modes. In this

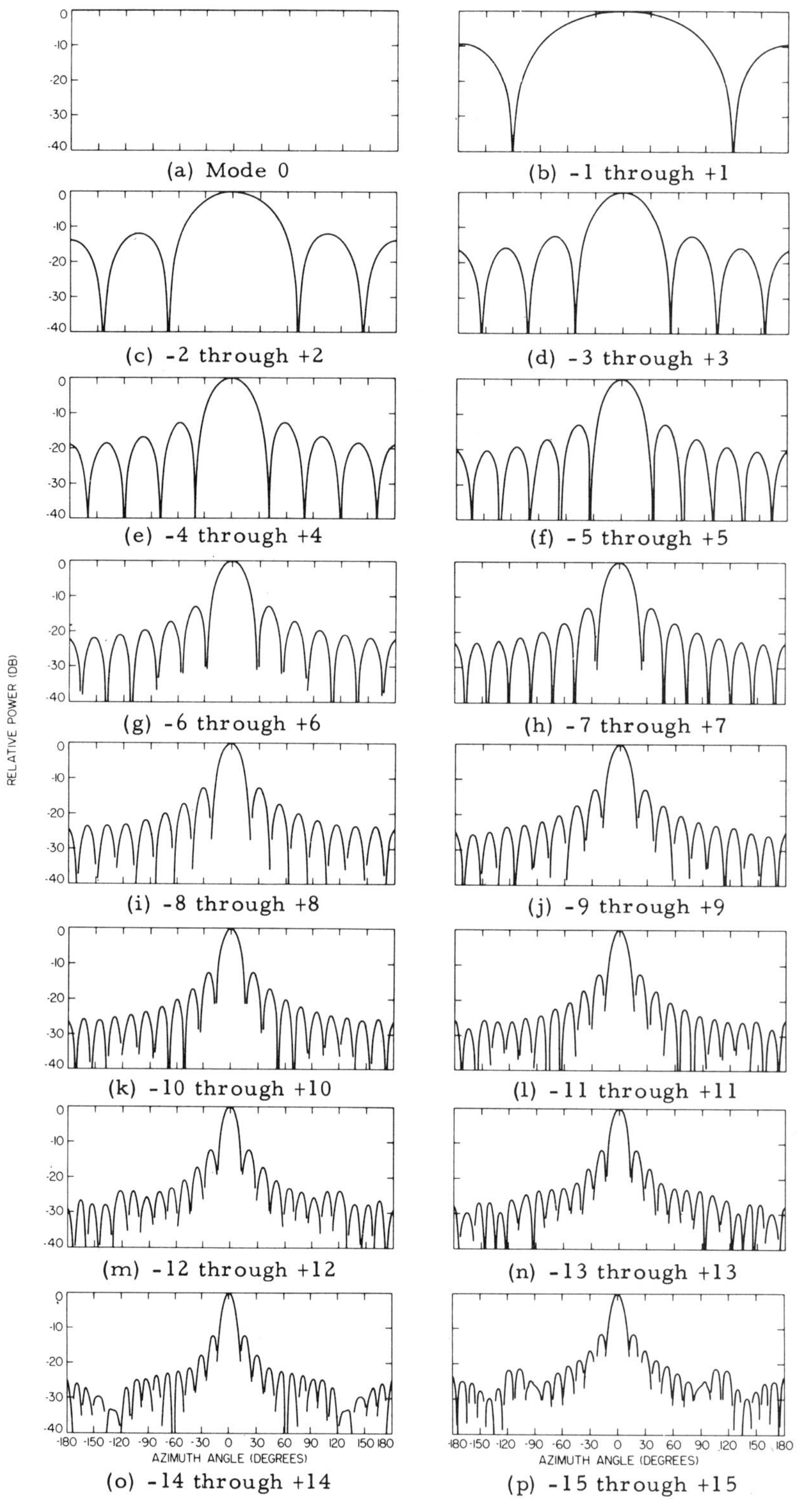


Fig. 4. Mode-by-mode buildup of the pattern of a 32-element array with uniform excitation of the modes.

case more than 95 percent of the total power is radiated from the 9 elements closest to the beam and on these elements the currents differ from cophasal by at most 5°.

One of the distributions used in the experimental program was $B_K = \cos^2 (\pi K/40)$, which provided a 17-dB taper over the 31-mode inputs. To indicate how much the pattern shape could be expected to change as the beam was scanned, patterns were computed for various beam positions. Fig. 7 shows three patterns, one phased so that its peak is in the direction of element 32 ($\phi = 0$), the other two having the same amplitude distribution over the modes but phased to scan the beam one quarter and one half, respectively, of the angle between elements. It may be seen that, at least for this distribution, the pattern changes only slightly as the beam is scanned.

Patterns were also calculated for different element spacings, element patterns, and amplitude distributions, but those shown satisfactorily illustrate the beam formation

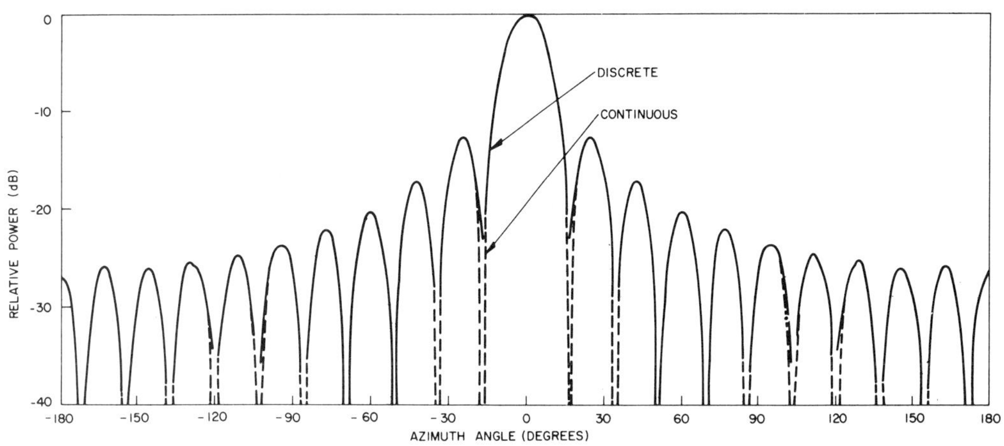


(a) Sum of modes -10 through +10

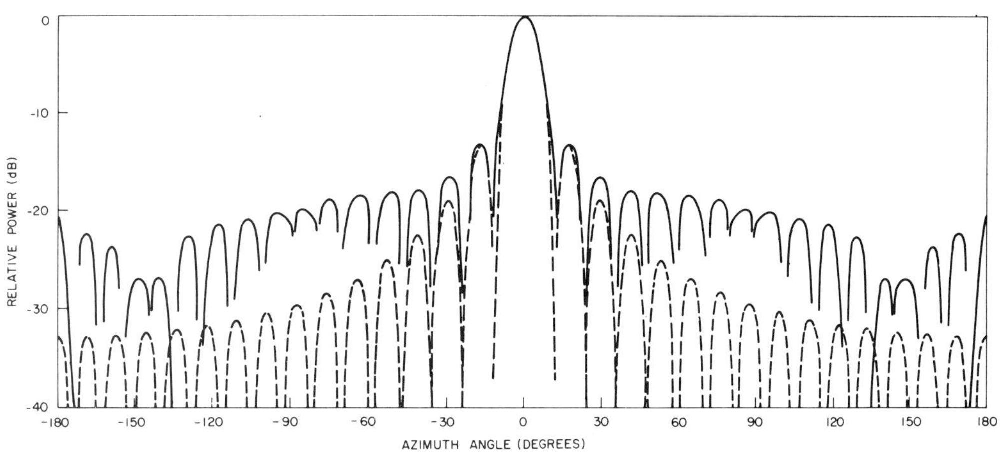


(b) Sum of modes -15 through +15

Fig. 5. Comparison of the patterns of a 32-element array and of a continuous current sheet using 21 and then 31 uniformly excited modes.



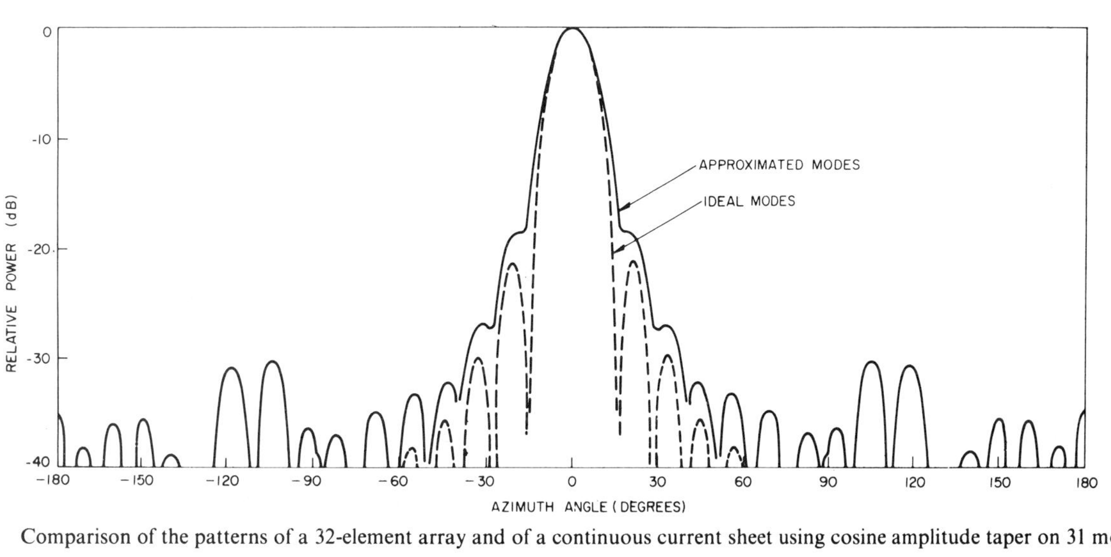


Fig. 6. Comparison of the patterns of a 32-element array and of a continuous current sheet using cosine amplitude taper on 31 modes.

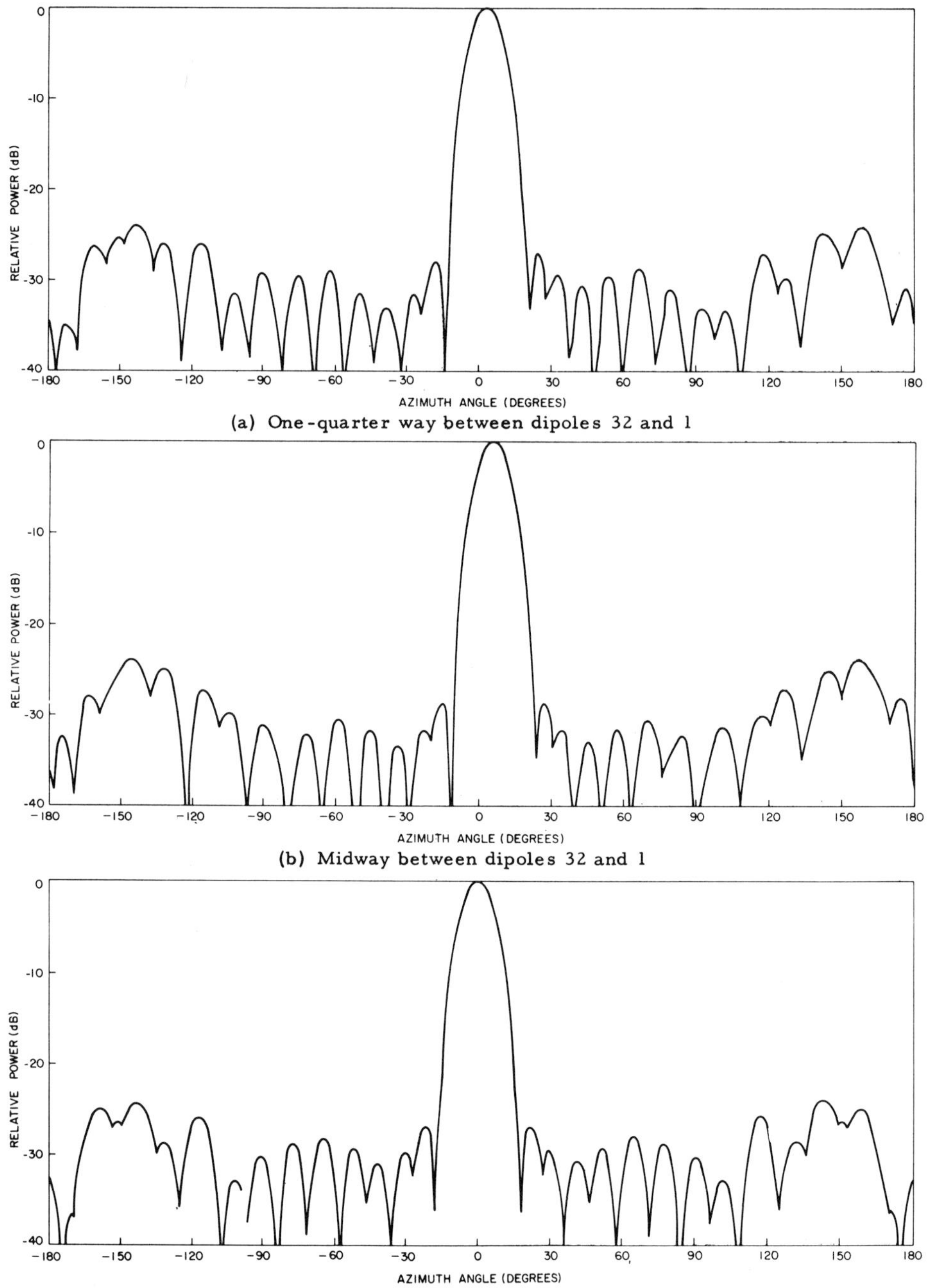


Fig. 7. Patterns and the corresponding current distributions on a 32-element array for beams at 0°, 5.625°, and 2.813°. The amplitude taper on the modes is $B_K = \cos^2(\pi K/40)$ with $K = 0, +1, \pm 2, \cdots, \pm 15$.

and scanning and also indicate how the pattern differs from one based on the existence of perfect pattern modes.

## Synthesis of Aperture Distributions

It should now be evident that the radiation pattern of a circular array computed on the assumption that the pattern modes are perfect is not the same as that computed from the actual current distribution, and that a certain amount of cut-and-try is involved in determining the number of modes to use and in adjusting the phases of the modes to form a beam in a particular direction. Instead of picking the mode excitations, only to find that the corresponding current distribution results in a poor radiation pattern, it would be preferable first to pick a current distribution having an acceptable pattern and then to find the mode excitations which will give these currents. That this is always possible was discovered by Davies [8], who showed that any prescribed output currents can be achieved with a Butler matrix by properly exciting the matrix inputs.

Consider an $N \times N$ Butler matrix with input and output ports labeled $K$ and $J$, respectively. If the prescribed currents $A_J e^{j\psi_J}$, where $J = 1, 2, \cdots, N$ are to be set up on the array, the $N$ currents that must be applied to the inputs of the matrix are

TABLE II

COPHASAL DISTRIBUTION FOR THE BEAM AT 5.625° WITH REQUIRED INPUT CURRENTS AND THE CURRENT DISTRIBUTION ON THE ARRAY FOR BEAMS SCANNED TO 8.438° AND 11.25°

| Input to Modes | | | Beam Position | | | | | | | | | | | |
|---|---|---|---|---|---|---|---|---|---|---|---|---|---|---|
| | | | Between Elements 32 and 1 | | | | On Element 1 | | | | One-Quarter Way Between Elements 1 and 32 | | | |
| Mode | Amplitude | Phase (rad) | Element | dB | Phase | Cophasal Deviations | Element | dB | Phase | Cophasal Deviations | Element | dB | Phase | Cophasal Deviations |
| −15 | 0.05014 | −2.98853 | 1 | 0.00000 | 4.41432 | 0.0 | 1 | 0.00000 | 1.30606 | 1.3 | 1 | 0.00000 | 1.98974 | 0.1 |
| −14 | 0.16982 | −2.80781 | 2 | −0.51926 | 39.47402 | 0.0 | 2 | −2.01691 | 17.80973 | 0.2 | 2 | −1.64249 | 28.35642 | 0.6 |
| −13 | 0.31358 | −2.60148 | 3 | −1.43457 | 108.24662 | 0.0 | 3 | −1.10324 | 64.83482 | 5.0 | 3 | −1.45862 | 83.40640 | 5.6 |
| −12 | 0.43767 | −2.37450 | 4 | −2.84185 | −151.91152 | 0.0 | 4 | −2.07344 | 160.91891 | 6.4 | 4 | −2.17622 | −176.63541 | 0.1 |
| −11 | 0.51999 | −2.09735 | 5 | −4.93675 | −24.83651 | 0.0 | 5 | −5.79880 | −80.13033 | 11.4 | 5 | −5.09892 | −48.91991 | 8.0 |
| −10 | 0.59752 | −1.76797 | 6 | −8.17489 | 124.58780 | 0.0 | 6 | −10.95202 | 43.41646 | 3.0 | 6 | −10.31497 | 94.83752 | 5.0 |
| −9 | 0.68904 | −1.42198 | 7 | −14.02687 | −69.38130 | 0.0 | 7 | −13.44224 | 164.97679 | 10.9 | 7 | −18.95701 | −132.54333 | 26.00 |
| −8 | 0.73786 | −1.03866 | 8 | −110.95878 | 9.40060 | — | 8 | −13.49060 | −44.26421 | 62.2 | 8 | −18.54252 | −39.69405 | — |
| −7 | 0.70416 | −0.53875 | 9 | −110.29151 | −127.37128 | — | 9 | −19.78507 | 147.62327 | — | 9 | −23.49840 | 149.10857 | — |
| −6 | 0.63277 | 0.07777 | 10 | −108.42737 | −80.12002 | — | 10 | −22.97828 | −27.73955 | — | 10 | −26.36628 | −26.97267 | — |
| −5 | 0.49759 | 0.68746 | 11 | −112.01176 | 103.52423 | — | 11 | −25.16028 | 154.81247 | — | 11 | −28.37765 | 155.27371 | — |
| −4 | 0.23592 | 1.46408 | 12 | −111.17657 | −75.03709 | — | 12 | −26.75804 | −23.60801 | — | 12 | −29.86092 | −23.31109 | — |
| −3 | 0.17912 | −2.74105 | 13 | −107.88319 | −119.27306 | — | 13 | −27.94438 | 157.42682 | — | 13 | −30.96016 | 157.62730 | — |
| −2 | 0.25706 | −1.78156 | 14 | −117.77163 | 34.95047 | — | 14 | −28.80676 | −21.90153 | — | 14 | −31.75041 | −21.77114 | — |
| −1 | 0.25264 | −0.54328 | 15 | −112.87704 | 65.89978 | — | 15 | −29.39541 | 158.52920 | — | 15 | −32.27577 | 158.60405 | — |
| 0 | 0.41310 | 0.25664 | 16 | −117.61885 | −131.95122 | — | 16 | −29.73847 | −21.22499 | — | 16 | −32.56037 | −21.19276 | — |
| 1 | 0.41310 | 0.35481 | 17 | −118.16084 | −136.67705 | — | 17 | −29.85107 | 158.85470 | — | 17 | −32.61636 | 158.85068 | — |
| 2 | 0.25264 | −0.24785 | 18 | −111.86473 | 60.67050 | — | 18 | −29.73843 | −21.22460 | — | 18 | −32.44693 | −21.26605 | — |
| 3 | 0.25706 | −1.29069 | 19 | −122.53436 | 43.74770 | — | 19 | −29.39536 | 158.52921 | — | 19 | −32.04448 | 158.44071 | — |
| 4 | 0.17912 | −2.05383 | 20 | −107.81624 | −112.96098 | — | 20 | −28.80673 | −21.90223 | — | 20 | −31.39070 | −22.04947 | — |
| 5 | 0.23592 | 2.34765 | 21 | −114.45265 | −75.31601 | — | 21 | −27.94435 | 157.42611 | — | 21 | −30.45327 | 157.20708 | — |
| 6 | 0.49759 | 1.76738 | 22 | −116.41499 | 94.23270 | — | 22 | −26.75802 | −23.60789 | — | 22 | −29.17439 | −23.93888 | — |
| 7 | 0.63277 | 1.35404 | 23 | −109.30689 | −95.61509 | — | 23 | −25.16029 | 154.81323 | — | 23 | −27.45203 | 154.29411 | — |
| 8 | 0.70416 | 0.93386 | 24 | −113.19501 | −108.60654 | — | 24 | −22.97834 | −28.73904 | — | 24 | −25.07565 | −28.62743 | — |
| 9 | 0.73786 | 0.63030 | 25 | −113.92446 | 8.94868 | — | 25 | −19.78514 | 147.62316 | — | 25 | −21.47069 | 145.78916 | — |
| 10 | 0.68904 | 0.44333 | 26 | −14.02684 | −69.38122 | 0.0 | 26 | −13.49063 | −44.26440 | 62.2 | 26 | −12.17991 | −51.94113 | — |
| 11 | 0.59752 | 0.29369 | 27 | −8.17489 | 124.58789 | 0.0 | 27 | −13.44226 | 164.97683 | 10.9 | 27 | −9.45066 | 143.16679 | 28.0 |
| 12 | 0.51999 | 0.16066 | 28 | −4.93675 | −24.83631 | 0.0 | 28 | −10.95202 | 43.41665 | 3.0 | 28 | −7.73339 | 1.07559 | 13.00 |
| 13 | 0.43767 | 0.07986 | 29 | −2.84186 | −151.91129 | 0.0 | 29 | −5.79880 | −80.13017 | 11.4 | 29 | −5.12229 | −118.84582 | 0.2 |
| 14 | 0.31358 | 0.04923 | 30 | −1.43458 | 108.24684 | 0.0 | 30 | −2.07344 | 160.91905 | 6.4 | 30 | −2.07351 | 135.92040 | 5.0 |
| 15 | 0.16982 | 0.03925 | 31 | −0.51925 | 39.47413 | 0.0 | 31 | −1.10324 | 64.83485 | 5.0 | 31 | −0.76244 | 51.05834 | 4.0 |
| 16 | 0.05014 | 0.05488 | 32 | −0.00002 | 4.41439 | 0.0 | 32 | −2.01692 | 17.80984 | 0.2 | 32 | −1.35198 | 9.49976 | 3.5 |

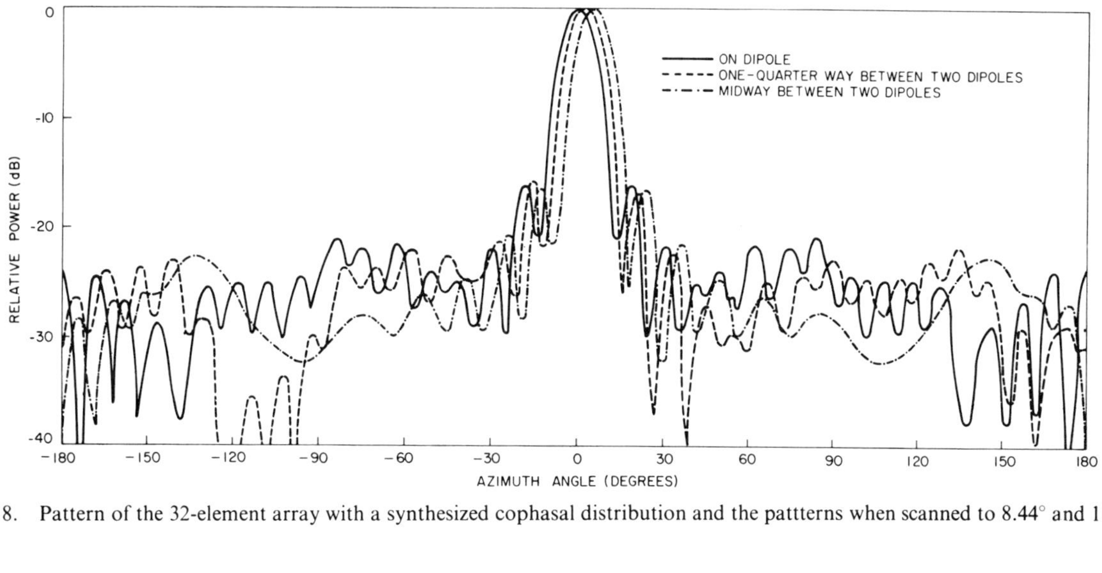


Fig. 8. Pattern of the 32-element array with a synthesized cophasal distribution and the pattterns when scanned to 8.44° and 11.25°

$$B_K e^{j\beta_K} = \sum_{J=1}^{N} A_J e^{j\psi_J} e^{-jKJ(2\pi/N)} \tag{14}$$

These currents are conjugate to those that would appear at the inputs if currents $A_J e^{-j\psi_J}$ were fed into the outputs of the matrix. The correctness of (14) may be verified by substitution into (11). As has been shown previously, since the matrix has a zero mode (looking from input to output), the current distribution that has been set up can be moved, intact, $M$ elements around the array by holding the amplitudes $B_K$ of the input currents fixed but changing their phases by applying a linear phase progression with a mode-to-mode phase difference of $M(2\pi/N)$.

It has been shown that the current distribution and the radiation pattern of a multimode array are invariant (aside from rotation) if the beam is scanned in steps equal to the angle between elements. For any other angle of scan, the current distribution will be changed. If, for example, all inputs are fed in phase with equal amplitude, only element $N$ will be excited. If the linear phase progression $e^{-jK(2\pi/N)}$

is then applied to the inputs, the excitation is switched to element 1. If, however, the linear phase progression were only half this (i.e., $e^{-jK\pi/N}$), two elements, $N$ and 1, would be strongly excited, but there would be currents on all the elements of the array. As a practical example, consider a 32-element array with a cophasal distribution on the 14-element sector which includes elements 26–7. The desired amplitude distribution is cos $[(K-1/2)\pi/16]$, which is symmetrical about a point midway between elements 32 and 1, and the elements are to be phased to form a beam in this direction. All other elements are to be inert. To show how the current distribution varies as the beam is scanned in small steps, the input currents required to achieve this distribution are first determined from (14), then their phases are changed to scan the pattern and the new distribution on the array is computed from (11). Table II gives the original distribution, phased for a peak at $\phi=5.625°$, and the corresponding input currents to the Butler matrix. Also in Table II is the distribution on the array when the beam is scanned to 11.25° (the direction of element 1) and the distribution when the beam is scanned to the angle midway between the first two. It is seen that, for the scanned beams, the currents are no longer confined to a sector; all elements are illuminated, with those on the rear of the array about 30 dB down. The stronger currents are on 15 or 16 elements, and over this sector there are only minor amplitude ripples with the currents differing from the cophasal condition by about 20°. The two scanned patterns (Fig. 8) do not differ significantly from the original one. Their beamwidths, near-in sidelobes, and the general level of their far-out lobes are comparable. If this distribution had been designed for very low sidelobes, it is likely that the pattern changes would have been more significant.

Fig. 9. Circular array of 32 dipoles ($\lambda/2$ spacing, 900 MHz).

Fig. 10. Beam-forming and scanning network for the 32-dipole array.

### Experimental Program



The circular array used in the experimental program had 32 elements and was operated at 900 MHz. Various radiating elements were used: dipoles, short back-fire elements, and Yagis (the latter two to reduce the elevation beamwidth without increasing the height of the antenna), but the only array that will be described is a 32-element array of slot-fed dipoles, vertically polarized, spaced $0.5\lambda$ apart and $0.25\lambda$ from a conducting cylinder. This antenna is shown in Fig. 9 and the associated beamforming and scanning network is shown in Fig. 10. Since 3-dB quadrature couplers were used in the matrix, it had no zero mode; therefore, the coaxial cables connecting the matrix to the dipoles had to be cut to the proper lengths to correct for this. Corporate structures made in triplate line were used to establish the various amplitude distributions over the inputs to the Butler matrix.

The measured mode patterns for this array (Fig. 11) do not compare favorably with the computed patterns in Fig. 3. The deviations are attributable primarily to phase and amplitude errors in the matrix. All the current modes were fed so as to have the same phase at element 32, and the relative phases of the pattern modes were determined by comparing the phase of each mode with that of the zero mode in the far field at $\phi=0$.

Fig. 12 shows the pattern of the array when a corporate structure was used providing currents of equal amplitude to all the mode inputs but number 16. For comparison, the corresponding calculated pattern (from Fig. 5) is shown solid. The two patterns agree reasonably well; both have beamwidths of about 10°, the measured first sidelobes are 1.5 dB higher than those calculated, and the general level of the far-out sidelobes is about 21 dB down for both.

The next series of patterns was taken with a tapered amplitude distribution over the modes. By dividing the outputs with tees, 31 modes were fed from a 16-element corporate structure. This resulted in a stepped distribution (since pairs of adjacent modes had equal amplitudes) with a 17-dB taper. The measured beamwidth (11.5°) and the first side-

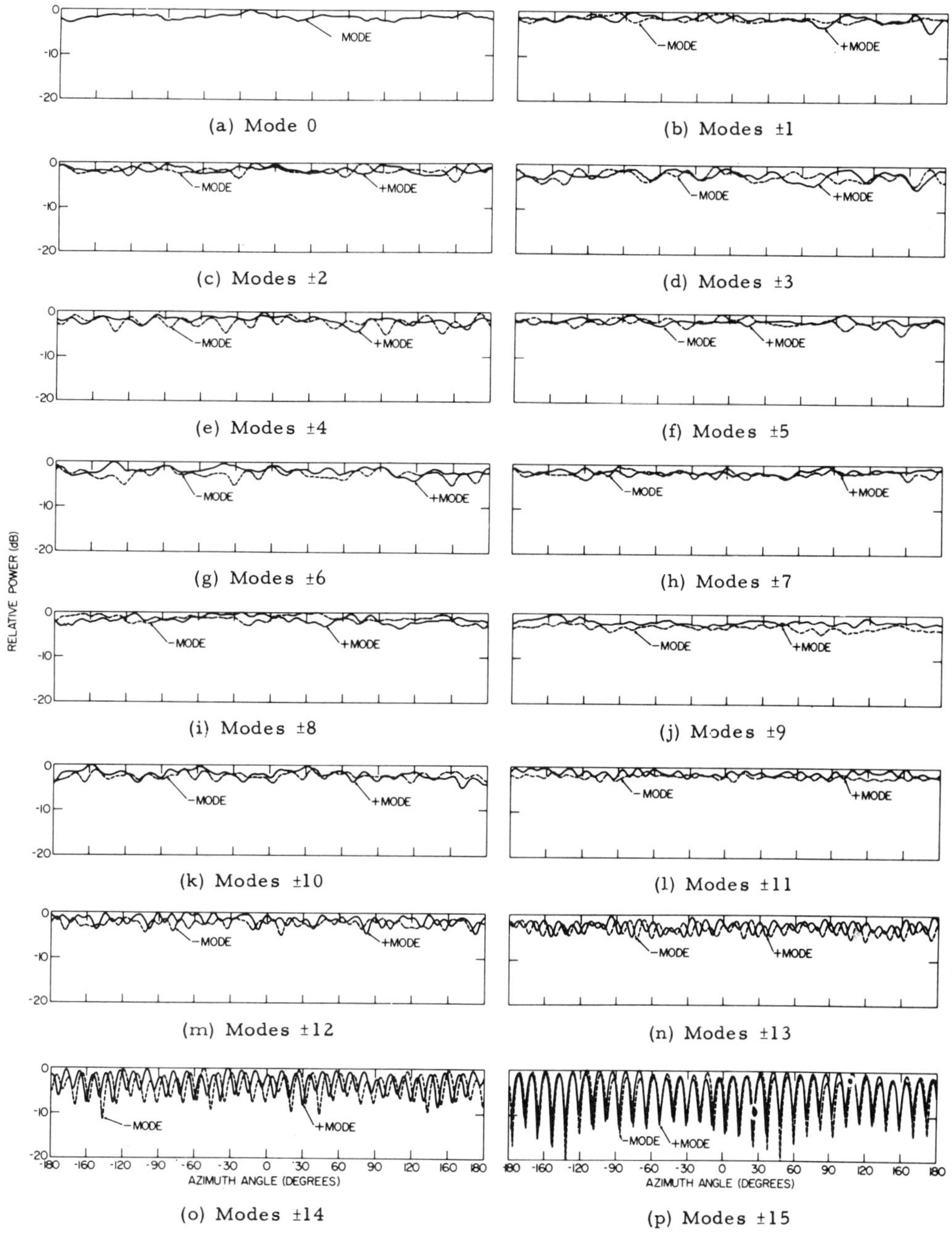


Fig. 11. Measured mode patterns for the 32-dipole array.

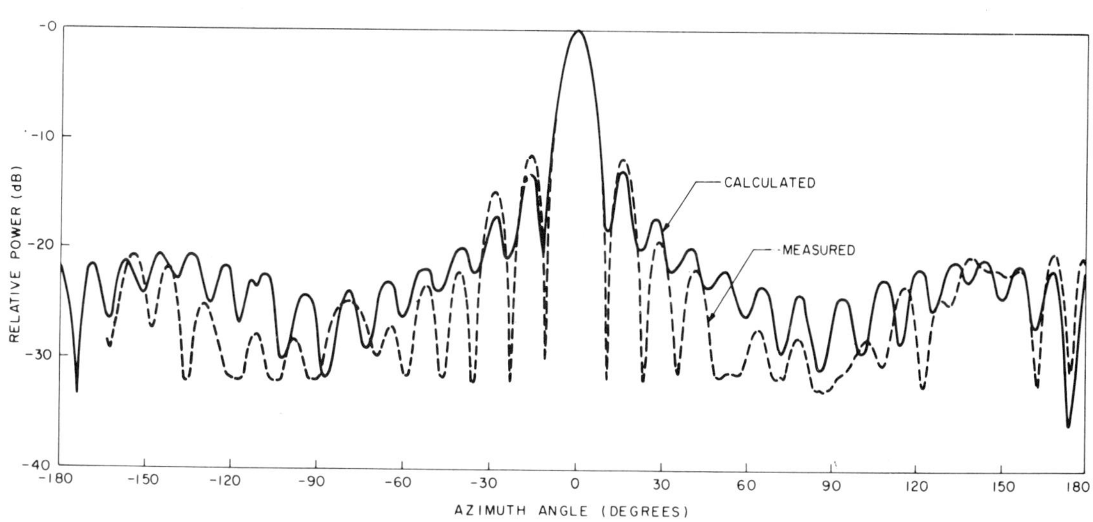


Fig. 12. Measured and calculated patterns of the 32-element array with uniformly excited modes.

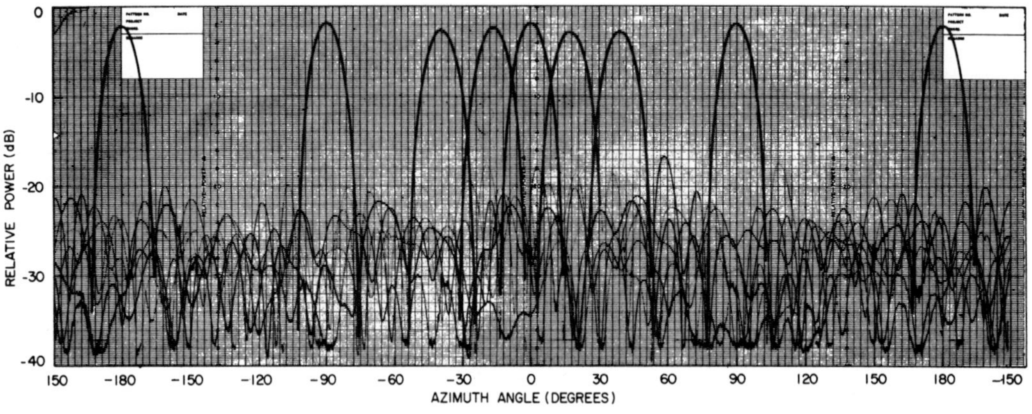


Fig. 13. Scanned patterns for the 32-element array with a tapered stepped distribution on the modes.

lobe (19 dB down) agree well with those calculated, but the level of the far-out lobes was somewhat worse than for the calculated pattern. The beam was then scanned by operating the phase shifters, and some of the patterns are shown in Fig. 13. It was found that the beamwidth and sidelobe level changed only slightly, and the gain varied by about 1 dB as the beam was scanned.

## Conclusions

It has been shown that a Butler matrix can be used to feed a circular array to form a narrow pattern that can be scanned through 360° in azimuth by the operation of phase shifters alone. One explanation of this, based on the assumption that the radiation pattern could be written as the sum of a finite number of uniform pattern modes, was found to work only qualitatively in that it could not be used to predict the structure of the sidelobes. A 32-element array of dipoles was used to demonstrate experimentally how a beam was formed by superposition of the pattern modes (even though imperfect) and how the scanning was performed. Finally, the synthesis procedure of Davies was described, and as an example, the inputs to the Butler matrix required to achieve a prescribed cophasal sector distribution on the array were determined and the change in the current distribution for other beam positions were shown.

## Acknowledgment

The author acknowledges the help given by R. M. Brown through numerous discussions and consultations, the contributions of F. W. Lashway, who was responsible for the mechanical design, and of R. J. Wiegand, who helped with the measurement program. In addition, the author expresses his gratitude to J. Tyszkiewicz of the Naval Air Systems Command, who sponsored and supported this work.

## References

[1] G. C. Chadwick and J. C. Glass, "Investigation of a multiple beam scanning circular array," Scientific Rept. 1 to USAF Cambridge Research Lab., Cambridge, Mass., Contract AF19(628)367, December 31, 1962.

[2] W. R. LePage, C. S. Roys, and S. Seely, "Radiation from circular current sheets," *Proc. IRE*, vol. 38, pp. 1069–1072, September 1950.

[3] R. C. Honey and E. M. T. Jones, "A versatile multiport biconical antenna," *Proc. IRE*, vol. 45, pp. 1374–1383, October 1957.

[4] A. C. Schell and E. L. Bouché, "A concentric loop array," *IRE WESCON Conv. Rec.*, vol. 2, pt. 1, pp. 212–215, August 1958.

[5] C. P. Clasen, J. B. Rankin, and O. M. Woodward, Jr., "A radial-waveguide antenna and multiple amplifier system for electronic scanning," *RCA Rev.*, vol. 22, no. 3, pp. 543–554, 1961.

[6] J. L. Butler, "Digital, matrix, and intermediate-frequency scanning," in *Microwave Scanning Antennas*, vol. 3, R. C. Hansen, Ed. New York: Academic Press, 1966, ch. 3.

[7] P. S. Carter, "Antenna arrays around cylinders," *Proc. IRE*, vol. 31, pp. 671–693, December 1943.

[8] D. E. N. Davies, "A transformation between the phasing techniques required for linear and circular aerial arrays," *Proc. IEE* (London), vol. 112, pp. 2041–2045, November 1965.

IEEE TRANSACTIONS ON ANTENNAS AND PROPAGATION, VOL. AP-18, NO. 5, SEPTEMBER 1970

# Step-Scanned Circular-Array Antenna

JERRY E. BOYNS, MEMBER, IEEE, C. W. GORHAM, A. DAVID MUNGER, JOSEPH H. PROVENCHER, MEMBER, IEEE, JOHN REINDEL, MEMBER, IEEE AND B. I. SMALL MEMBER, IEEE

NELC | A CONTRIBUTION OF THE NAVAL ELECTRONICS LABORATORY CENTER, SAN DIEGO, CALIFORNIA 92152

***Abstract*—A circular-array radar antenna designed for operation over a 20-percent bandwidth has demonstrated improved performance over that of linear arrays. A major advantage is that the beam pointing angle is independent of frequency. The array developed and tested provides 128 beam positions with a beam crossover of about −2 dB, a half-power beamwidth on the order of 4 or 5 degrees, and a sidelobe level of −25 dB. Step increments of one element per beam position allow practical implementation of step scanning, i.e., simple commutation of the current distribution.**

**Two approaches are possible for feeding and scanning the ring array: 1) a vector-transfer system involving 3-bit phase and amplitude boards which are capable of amplitude attenuation (VT scan) and 2) a lens feed system employing an $R$–$2R$ parallel-plate region, with the spacing between the plates less than $\frac{1}{2}$ a free-space wavelength, so that only the electric-field component perpendicular to the plates is propagated. Energy is launched and extracted from the lens by means of monopoles mounted $\frac{1}{4}$ wavelength in front of the circumferential ground plane enclosing the lens.**

**Computations involved in determining current distribution are based on [1] and [2].**

Fig. 1. Ring array.

## Introduction

It is well known that the radiation pattern produced by a planar or linear array deteriorates with increasing scan angle. To obtain 360° azimuth coverage from an electronically scanned array aperture, there are numerous advantages to the use of a circular array. When compared to a configuration of four linear arrays, for example, it can be shown that for the circular array the overall performance can be improved. Since the symmetry of the array is the same for all beam positions, the beam pointing angle is independent of the frequency. Numerous other advantages accrue depending upon the particular application.

Some of the problems involved with the design and implementation of the circular array are 1) increased complexity in the mathematical analysis of the circular array as compared with the linear array and 2) increased complexity in developing a feed system that both provides proper phase correction to a planar phase front and commutes the phase and amplitude distributions to step the beam. This paper describes some particular solutions to these problems.

The design goals of the array to be discussed here are 1) 128 beam positions with a beam crossover of about −2 dB, 2) a half-power beamwidth on the order of 4 or 5 degrees, and 3) a sidelobe level of −25 dB. A step increment of one element per beam position will meet these goals and allow practical implementation of step scanning, i.e., simple commutation of the current distribution. The project was undertaken in phases: 1) synthesis of current distributions and analysis of array performance with the aid of a high-speed computer; 2) development of a 128-element array; 3) development of a vector-transfer feed system and, independently, of a parallel-plate lens feed system; and 4) evaluation of the final experimental systems.

Manuscript received August 25, 1969; revised March 23, 1970. The authors are with the Naval Electronics Laboratory Center, San Diego, Calif. 92152.

## The Experimental Systems

### *Ring Array*

The ring array designed and constructed for the experimental circular-array program is shown in Fig. 1. The array consists of 128 radiating elements equispaced at 0.65 wavelength and embedded in a ground plane of 13.2-wavelength radius at the center frequency $f_0$. The radiating elements are sectoral horns with a length of approximately 3 wavelengths and an aperture of about $\frac{1}{2}$ wavelength in the $E$ plane and $1\frac{1}{2}$ wavelengths in the $H$ plane. The elements have a bandwidth of $\pm 10$ percent of $f_0$ with a voltage standing wave ratio (VSWR) of less than 1.5:1. The minimum half-power beamwidth (HPBW) in the $H$ plane is 40°. The gain of the element is 10.4 dB at $f_0$. Mutual coupling between adjacent elements in the array was measured at −22 dB.

### *Vector-Transfer Feed System*

Two approaches are presented to solve the problem of feeding and scanning the ring array. The first method

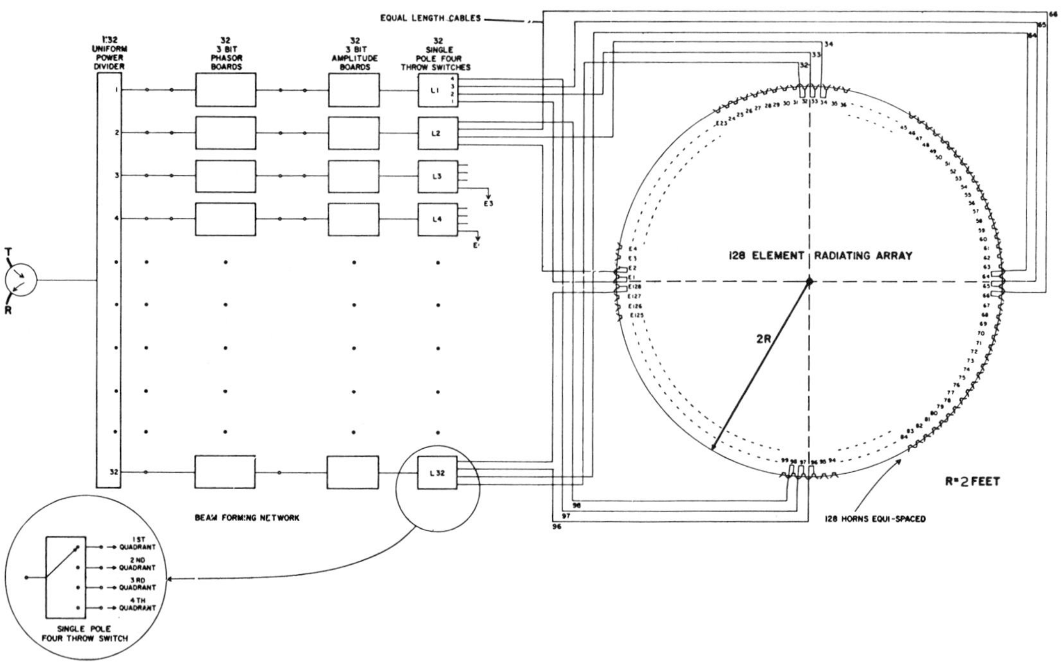


Fig. 2. Vector-transfer-fed circular array.

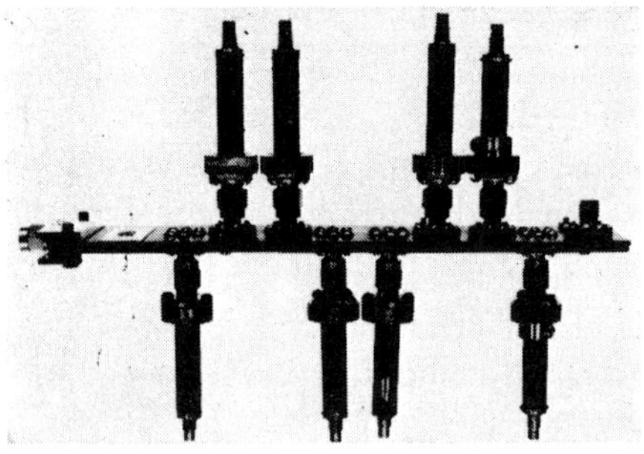

Fig. 3. Vector switch with externally mounted diodes.

Fig. 4. Single-pole 4-throw switch.



involves 3-bit phase and amplitude boards which are capable of amplitude attenuation (VT scan). On transmit, the power is equally divided 32 ways and fed to 32 vector boards. Each is connected to a single-pole four-throw (SP4T) switch with equal-length cables running to four radiating elements separated by 90°, as shown in Fig. 2. Any 32 adjacent elements can be fed at the same time since no two elements of a 90° (32-element) sector of the ring array are connected to the same vector board. The proper phase and amplitude are provided by the vector boards shown in Fig. 3.

The microwave SP4T transfer switches are built with shunt-mounted p-i-n diodes on stripline circuits. A SP4T switch is shown in Fig. 4. The circuit has the form of a corporate divider with shunt diodes placed at the junctions. The forward-biased diodes prevent transmission; the reverse-biased diodes permit transmission. Typically, a switch has a loss of 0.8 dB and a VSWR of 1.2:1 over the 20-percent band.

The vector boards consist of a 3-bit phasor and an attenuator controlled by a 4-bit logic driver. The stripline circuit board has four cascaded hybrid couplers to which are connected four pairs of matched switches, three pairs for the phase switch and one pair for the attenuator.

The phasor circuit has been called the hybrid-coupled transformed phase shifter. It is widely used and has been fully described [3]. It is preferred over other circuits because it requires only two diodes per bit, has high-power handling capacity, is relatively simple, and has low loss.

The attenuator switch consists of a hybrid coupler that is terminated in a parallel circuit composed of a 50-ohm resistor and a diode switch. At low diode currents, the attenuation decreases as the diode current is increased in steps by a logic drive circuit. It is possible to eliminate the microwave resistor from the circuit and have p-i-n diodes absorb the power. However, it is more difficult to tune the circuit over the band without the resistor. The vector boards and transfer switches are controlled by two shift registers. When the two registers are moved $n$ places, the beam is stepped to the $n$th beam position relative to the original.

Since amplitude taper is accomplished by attenuation, the amount of amplitude taper that can be practically used is limited. A $\cos^6 \alpha$ taper gives about 3.5-dB loss in addition to the system loss. Computed patterns indicate that the 3-bit phase approximation (45° steps) yields sidelobes of −20 dB at best, when no errors are taken into account, and a HPBW of 3.7°. A slower taper with an overall attenuation of about 2 dB is shown in Fig. 5. For this taper, computed patterns indicate that −19-dB sidelobes are available with HPBW of about 3.4°. Furthermore, approximation of the taper with eight discrete steps has negligible effect on the pattern but greatly simplifies implementation of the beam switching. Further computer studies indicate that the composite system errors are equivalent to random errors with a standard deviation of approximately 20°.

The aforementioned vector boards could also be built with 4 bits at an increase in cost. Some patterns were computed on this assumption, with errors also considered, and the indications are that a 4-bit phase quantization and a faster taper would significantly reduce the sidelobe level.

In summary, this method of feeding the ring array appears to have several desirable characteristics. First, in terms of power-handling capacity, the system allows the power to be divided into 32 equal parts; therefore, each component is required to handle only the power that is radiated by a single element. Second, a high degree of flexibility is achieved by having independent control of the phase and amplitude of each element in the excited sector. For example, the attenuators can be switched out during transmission to minimize transmission losses and later switched to form a low-sidelobe beam for reception only; or the phasing of all elements can be advanced an equal amount between transmission and reception (or from pulse to pulse) to redistribute the random errors and thus reduce the total apparent sidelobe level. Sum and difference can easily be obtained by dividing the input power dividers.

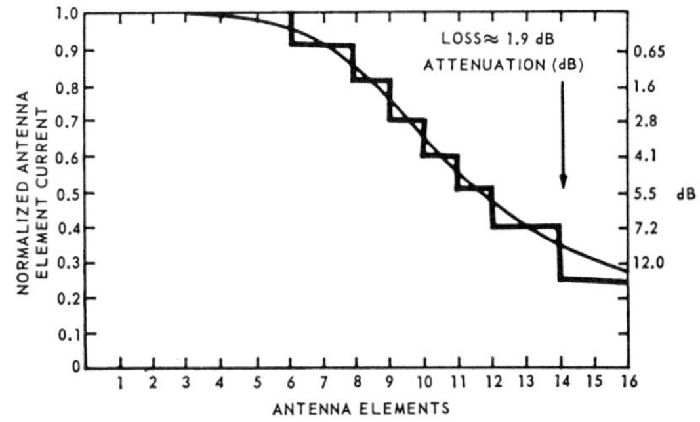


Fig. 5. Amplitude taper of vector-transfer system.

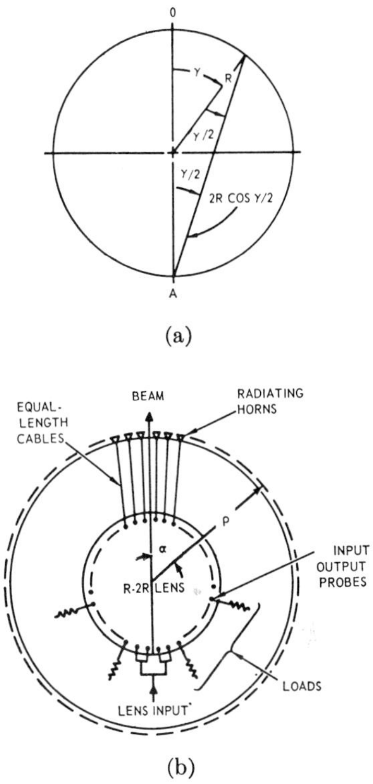


Fig. 6. Parallel-plate lens. (a) Geometry. (b) Schematic of completed array.

## *R-2R Parallel-Plate Lens Feed System*

Another method of feeding and scanning the ring array can be referred to as an $R$–$2R$ parallel-plate lens feed system [4]. The lens is a parallel-plate region of radius $R$, with the spacing between parallel plates less than $\frac{1}{2}$ a free-space wavelength in order to propagate only the electric-field component perpendicular to the plates. Energy is launched and extracted from the lens by means of monopoles mounted $\frac{1}{4}$ wavelength in front of the circumferential ground plane enclosing the lens. From Fig. 6, it can be seen that energy introduced at a point $A$ travels a distance $2R \cos \gamma/2$ when received by a pickoff probe at $\gamma$. For a beam-cophasal distribution on a ring array of radius $\rho$, the energy must be delayed a distance $\rho \cos \alpha$ for an antenna element located at $\alpha$ measured from the beam direction. Thus the proper phase distribution will be provided by the lens if $R = \frac{1}{2}\rho$ and $\gamma = 2\alpha$. The lens must be one half the radius of the ring array as shown above. The lens angle will be twice the array angle if all the lens ports are used to illuminate one half of the array aperture (180°). The illumination of the ring is accomplished with 64 equispaced probes on the lens, each switchable to one of two elements on the array separated 180°, as shown in Fig. 7. Adjacent probes using equal-length cables must feed adjacent elements on the ring array.

The use of a single horn or probe is the usual method of feeding a parallel-plate lens, but this method provides little control over the amplitude distribution at the output terminals of the lens. A small linear array of probes, therefore, is fed to give a more directional pattern within the lens and, thereby, a faster taper at the output probes. Good control of the amplitude at the output ports can be realized by varying the number of probes fed and by using an amplitude taper on their excitation. The probes are situated about $\frac{1}{4}$ wavelength in front of the enclosing ground plane so the pattern of a single probe at $\alpha$ can be approximated by $\sin \{\pi/2 \cos [(\gamma + \alpha)/2]\}$, where $\alpha$ is measured from the center of the excited probes. The output

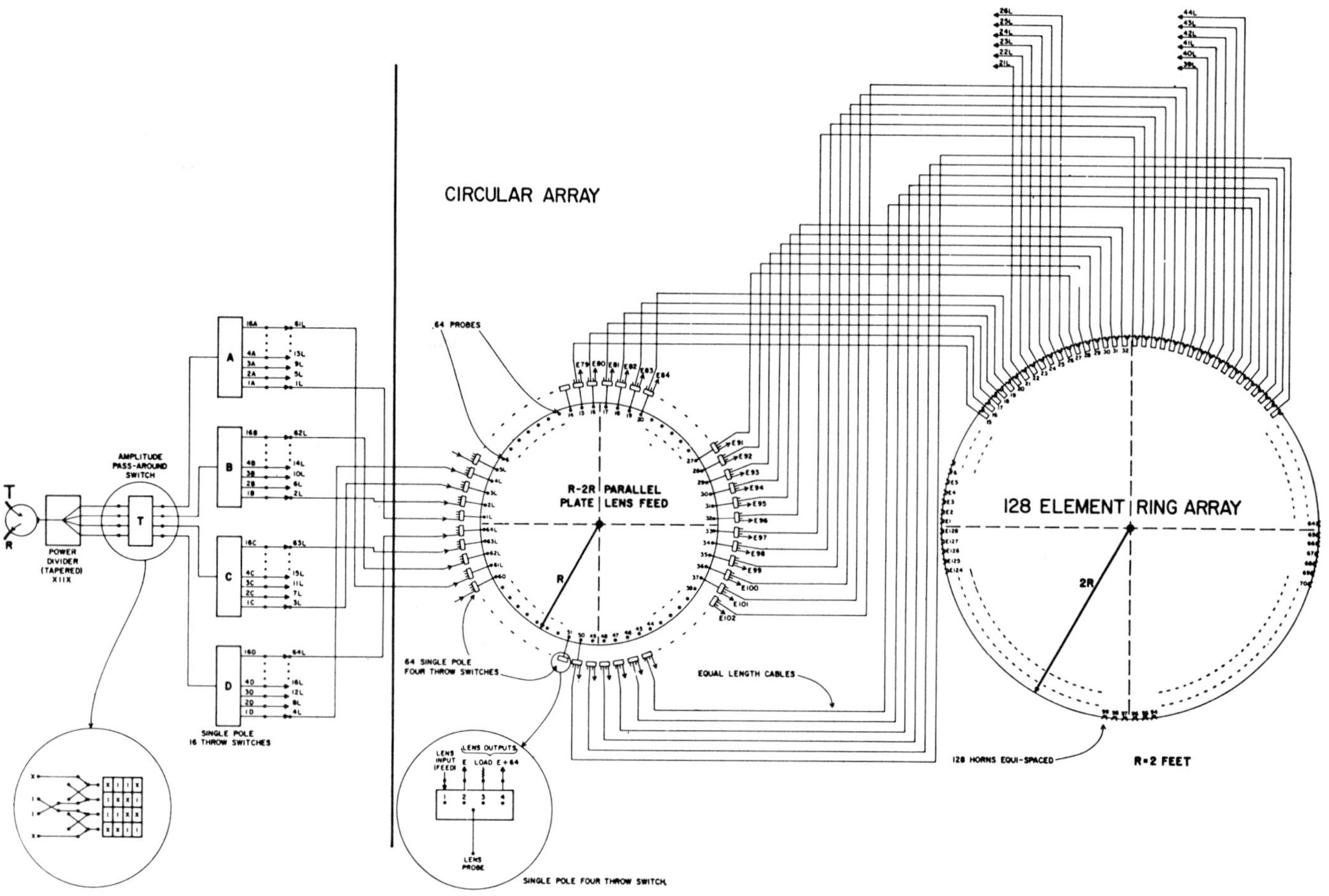


Fig. 7. *R*–2*R* parallel-plate lens-fed circular array.

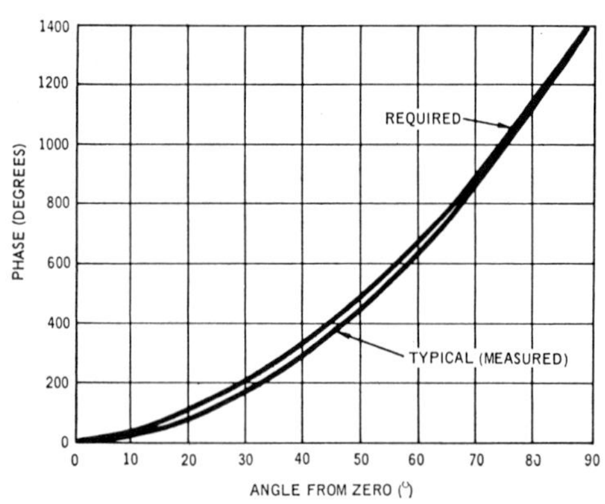


Fig. 8. Measured and predicted phase distributions versus angle from beam direction.

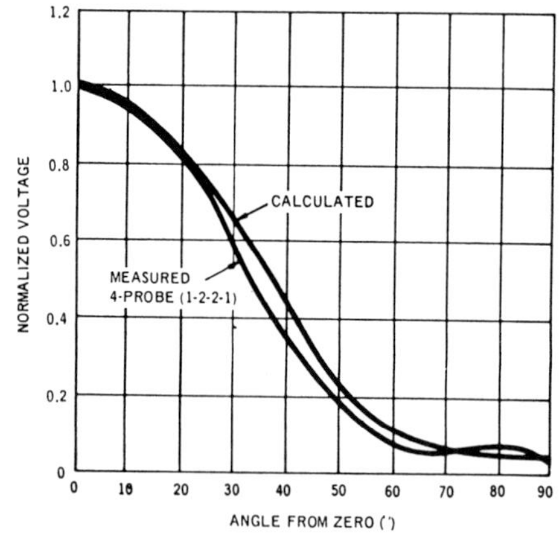


Fig. 9. Measured and predicted amplitude distributions versus angle from beam direction for 1-2-2-1 input configuration.

of a probe at $\gamma$ can be roughly computed from

$$E(\gamma) = \sum_{\alpha} I(\alpha) \frac{\sin^2 \{\pi/2 \cos [(\gamma + \alpha)/2]\}}{\{2R \cos [(\gamma + \alpha)/2]\}^{1/2}} \cdot \exp \{-j2\pi R \cos [(\gamma + \alpha)/2]\}.$$

Fig. 8 shows measured and computed phase distributions, and Fig. 9 shows measured and computed amplitude distributions for feeding with $I(\alpha) = $ 1-2-2-1.

For the lens to properly function, it is necessary that internal reflections be minimized. Internal reflections can destroy the optical properties of the lens; thus, it is necessary to terminate nonactive ports in matched loads. Since each probe has a directional element pattern, it can only absorb all incident energy at normal incidence. At high angles of incidence, that is, for elements near the excited probes, most energy will be reflected. Therefore, the lens behavior is optimum when the primary-probe array pattern is highly directional and has low sidelobes. The 1-2-2-1 input amplitude distribution satisfies this criterion and gives near-optimum results. A discrepancy of the phase from the beam-cophasal curve resulted and can be interpreted as due to an apparent radius of the lens which is electrically smaller than the actual radius. It appears that the actual phase center of the probe radiators is slightly inside the lens rather than on the enclosing ground plane.

Three-hundred-sixty-degree azimuth beam steering is implemented by means of an SP4T switch at each input–output port of the lens. This switch selects one of the two radiating elements to which it is connected, a matched load when not active, or an input port when the probe is used to feed the lens. The SP4T switch has been previously described. A discussion of the double-pole double-throw (DPDT) switch and the single-pole 16-throw (SP16T) switch is given in detail in the following section.

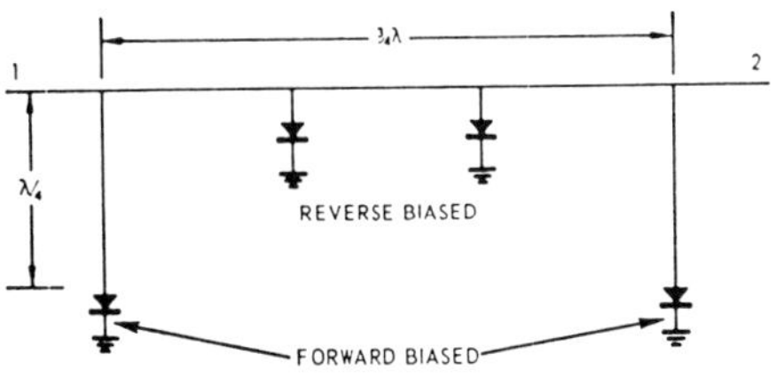


Fig. 10. Double-pole double-throw switch, effective circuit path.

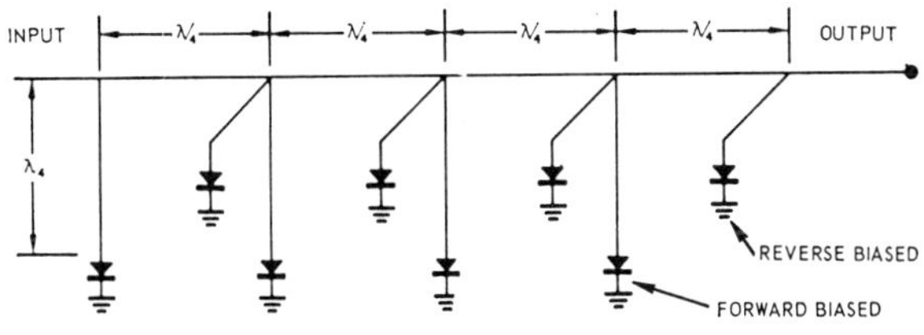


Fig. 11. Equivalent circuit of SP16T switch.

## Scanning Switches

### *DPDT Switches*

The DPDT switches used in the parallel-plate fed system consist of a four-port transmission-line circuit and eight shunt diodes. The effective circuit path for a single line is shown in Fig. 10.

Isolation between the poles is obtained by two cascaded shunt diodes and is approximately 35 dB. The forward-biased diodes reflect open circuits at the circuit ports. The VSWR of the circuit was measured at less than 1.3:1. The insertion loss was measured at 1 dB.

### *SP16T Switches*

The SP16T switches consist of five SP4T switches combined on a single stripline circuit board. The effective circuit path to any one of the 16 output ports is as shown in Fig. 11. The bandwidth of the switch is enhanced by spacing the diodes by $\frac{1}{4}$ wavelength.

The VSWR was measured at less than 1.5:1. The insertion loss was about 2 dB and varied by $\frac{1}{2}$ dB through the individual paths because of variations in diode characteristics. The circuit requires 30 diodes.

In summary, very-low-sidelobe patterns are achievable with the $R$-$2R$ lens feed system. This is primarily a result of the optical characteristics and the amplitude-tapering techniques of the lens. The power-handling capability of this system is limited by the type of coaxial fitting used for the input–output ports on the $R$–$2R$ lens. If a 1:4 divider is used as an input to the lens to obtain the proper taper on the primary pattern of the lens input, each of the four probes must withstand one quarter of the total input power. In principle the lens is an optical system, but an actual lens is bandwidth-limited because the probe must be maintained at about $\frac{1}{4}$ wavelength from the enclosing ground plane in order to provide a good match and to maintain a directional pattern in the lens. The use of ridged-waveguide techniques in the probe design will increase the effective bandwidth.

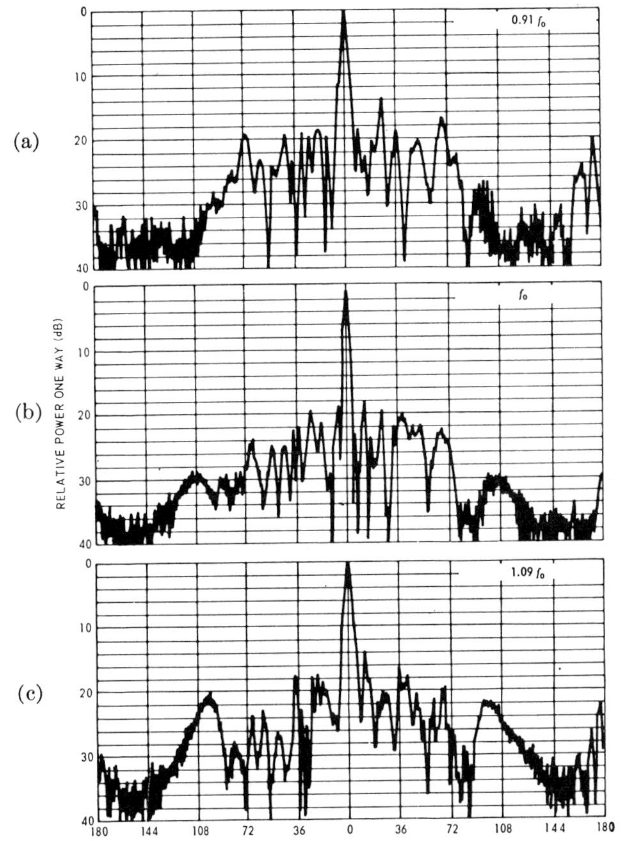


Fig. 12. Measured radiation patterns for vector-transfer-fed circular array.

TABLE I

Predicted and Measured Sidelobe Levels and Beamwidths Using 90° Aperture and Frequency = $f_0$

| | Parallel-Plate Feed | Vector-Transfer Feed |
|---|---|---|
| Predicted (theoretical) Sidelobe Level | −28 dB | −19 dB |
| Measured Sidelobe Level | −24 dB | −18 dB |
| Predicted HPBW | 5.98° | 3.4° |
| Measured HPBW | 5.2° | 3.6° |

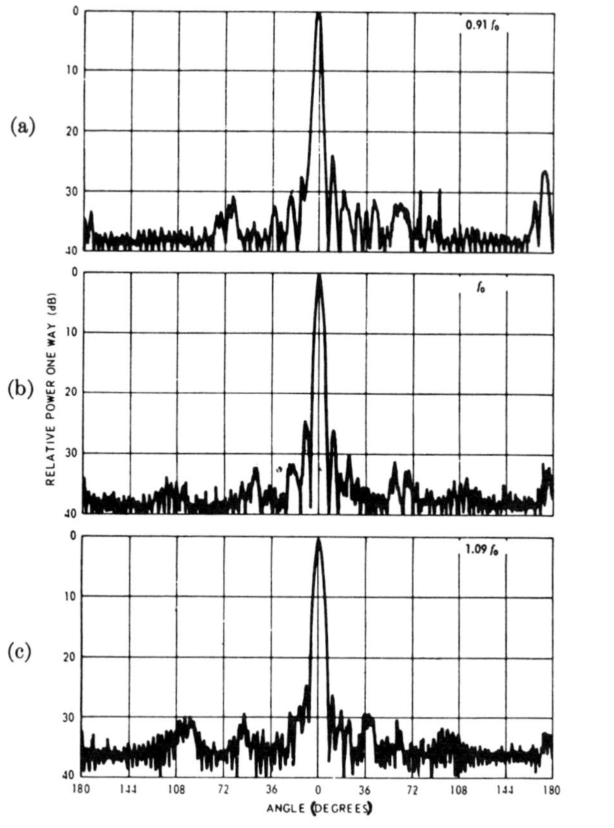


Fig. 13. Measured radiation patterns for parallel-plate lens-fed circular array.

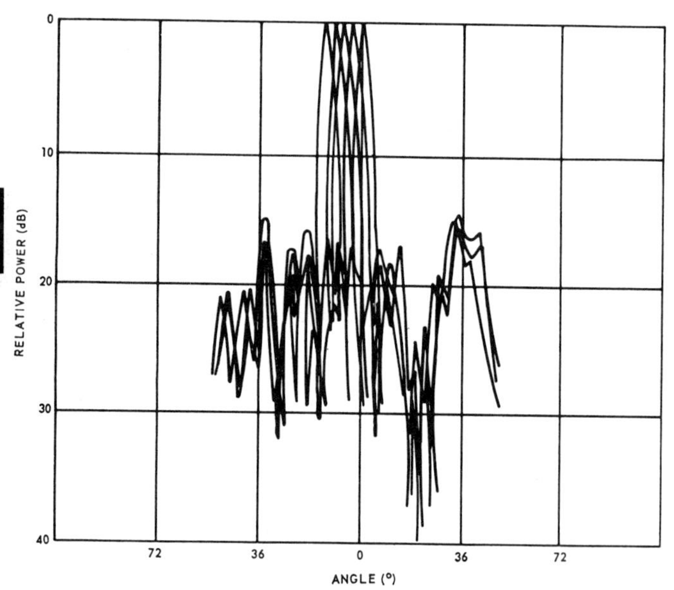


Fig. 14. Measured scan patterns for vector-transfer system, showing five beam positions.

Steering of the parallel-plate feed system was accomplished by manual switching. Computer control would be necessary for agile rapid beam steering.

## Radiation Patterns

Prediction versus measured sidelobe levels and beamwidths for the vector-transfer and parallel-plate feed systems with 32 elements illuminated are given in Table I.

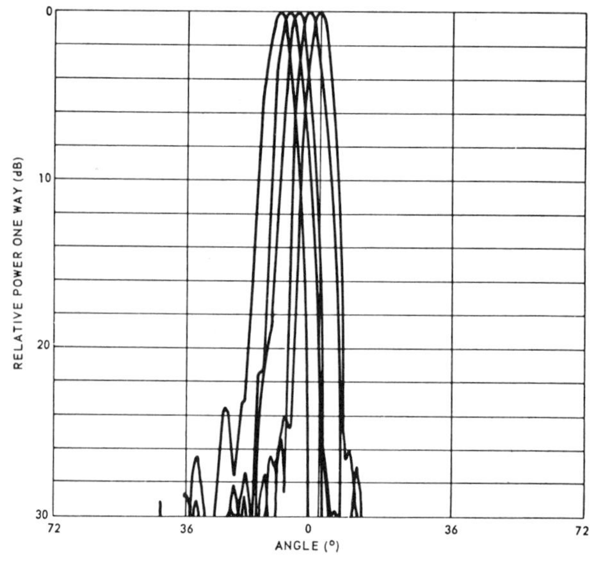


Fig. 15. Measured scan patterns for parallel-plate system, showing five beam positions.

Fig. 12 shows measured radiation patterns for the vector-transfer system, Fig. 12(a) being taken at $0.91f_0$, Fig. 12(b) at $f_0$, and Fig. 12(c) at $1.09f_0$. Fig. 13(a) shows a measured radiation pattern taken at $0.91f_0$ for the parallel-plate feed system. Fig. 13(b) is a measured pattern at $f_0$, and Fig. 13(c) is a measured pattern at $1.09f_0$.

Measured scan patterns for both systems are shown in Figs. 14 and 15. Each pattern shows five of the possible 128 beam positions. Because each beam is formed at a position normal to the radiating surface, each beam, theoretically, should be the same.

## Conclusions

Measured performance of both the vector-transfer and the parallel-plate feed systems compared favorably with computer predictions. The largest source of error in both systems appears to be in the phase-distribution scheme. The parallel-plate system gave the lowest phase errors and in addition is essentially true time delay. As a result it gave lower sidelobe patterns and has a larger bandwidth capability. The vector-transfer system is to be preferred for high power-handling capabilities and lends itself to compact packaging. It also has a high degree of versatility for providing amplitude and phase distributions.

## References

[1] R. H. DuHamel, "Pattern synthesis for antenna arrays on circular, elliptical, and spherical surfaces," Elec. Eng. Lab., University of Illinois, Urbana, Contract N6-ori-71, Tech. Rept. 16, May 1952.
[2] J. R. Wait, "Electromagnetic radiation from sources on and near cylindrical surfaces," National Bureau of Standards, Rept. 5553, January 15, 1958.
[3] R. W. Burns, "PIN diodes advance high-power phase shifting," *Microwaves*, vol. 4, pp. 38–48, November 1965.
[4] J. E. Boyns, A. D. Munger, J. H. Provencher, J. Reindel, and B. I. Small, "A lens feed for a ring array," *IEEE Trans. Antennas Propagat.* (Communications), vol. AP-16, pp. 264–267, March 1968.

*Reprinted by permission from* IEEE TRANSACTIONS ON ANTENNAS AND PROPAGATION
Vol. AP-19, No. 1, January 1971
pp. 131-134


## Mutual Coupling on a Cylindrical Array of Waveguide Elements

***Abstract*—This communication presents a modal approach for investigating mutual coupling on a cylindrical array of waveguide elements. Computed results are given in the form of element patterns and are compared with measured patterns.**

### Introduction

Attempts to predict the performance of large phased arrays by summation of individual element contributions will usually fail if the effects of mutual coupling are neglected. Such effects have been extensively investigated for infinite planar arrays. Recently, there has been increased interest in conformal array geometries, in particular the circular and cylindrical arrays. Sureau and Hessel [1] and Borgiotti and Balzano [2] have analyzed the two-dimensional problem of an array of infinite axial slots on a conducting cylinder. In the present communication, the three-dimensional problem is investigated.

### Theory

A cylindrical array of circumferentially polarized waveguide elements is considered. The axial dimension is assumed to be infinite. The array and all the space exterior to it is split into identical unit cells. Such cells are illustrated in Fig. 1 along with the coordinate system. Basic modal excitations are considered which place all unit cells in identical environments and the fields in all unit cells are identical except for a constant phase shift from cell to cell along rows and columns.

Along rings the cell-to-cell phase shifts must be such that a complete transversal of a ring produces a total phase change that is a multiple of $2\pi$. If each ring contains $M$ unit cells, the $m$th mode refers to a cell-to-cell phase shift of $2\pi m/M$. Only $M$ modes are nonredundant, chosen to be modes $-M/2+1$ to $M/2$. Specification of mode $m_0$ in this range will lead to azimuth error modes given by

$$m_l = m_0 + lM, \quad -\infty < l < \infty, \quad l \text{ integer.} \tag{1}$$

Only the primary modes $m_0$ can be specified on the aperture due to the finite number of elements per ring, but all modes can, in principle, be radiated into space.

In elevation, a cell-to-cell phase shift of $\gamma_z$ corresponds to a beam elevation angle of arcsin $(\gamma_z/d)$, where $d$ is the vertical cell-to-cell spacing in units of $\lambda/2\pi$. Error modes are given by

$$k_{zn}d = \gamma_z + 2\pi n, \quad -\infty < n < \infty, \quad n \text{ integer.} \tag{2}$$

The error modes correspond to linear array grating lobes.

Solutions in space for $\phi$ polarization are superpositions of all allowed modes. In cylindrical coordinates, they are of the form

$$H_z = \sum_{l=-\infty}^{\infty} \sum_{n=-\infty}^{\infty} A_{ln} k_{tn}^2 \begin{Bmatrix} H_{m_l}{}^2(k_{tn}\rho) \\ K_{m_l}(ik_{tn}\rho) \end{Bmatrix} \exp(ik_{zn}z) \exp(im_l\phi) \tag{3}$$

$$E_\phi = \sum_{l=-\infty}^{\infty} \sum_{n=-\infty}^{\infty} A_{ln} k_{tn} \begin{Bmatrix} H_{m_l}{}^{2\prime}(k_{tn}\rho) \\ iK_{m_l}{}'(ik_{tn}\rho) \end{Bmatrix} \exp(ik_{zn}z) \exp(im_l\phi) \tag{4}$$

where $k_{tn}^2 = 1 - k_{zn}^2$, $\rho$ is the radius, and the upper or lower solutions are taken if $k_{tn}$ is real or imaginary, respectively. $H_{m_l}{}^2(x)$ is a Hankel function of the second kind of order $m_l$, and $K_{m_l}(x)$ is the modified Bessel function. The prime denotes differentiation with respect to the argument.

Waveguide elements of uniform cross section are assumed. For convenience, LSE modes (TE to the $z$ axis) are used. In matching

Manuscript received May 27, 1970; revised July 31, 1970.

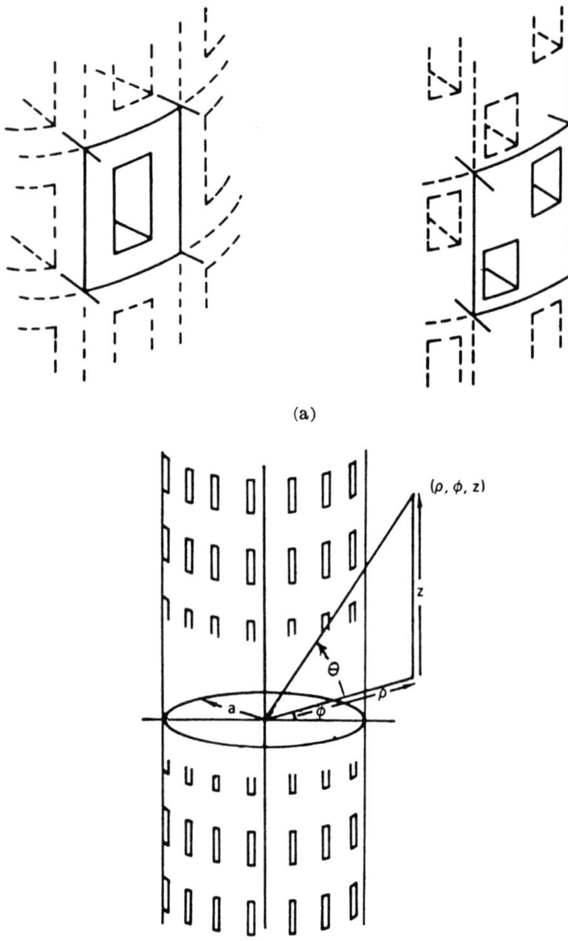


(a)

(b)

Fig. 1. (a) Examples of unit cells. (b) Coordinate system.



the boundary over the waveguide aperture, the small difference between the planar waveguide aperture and the curved radiating aperture is neglected. With this approximation, the local rectangular coordinates of the waveguide are easily aligned to the cylindrical coordinate system. For the $j$th waveguide of the unit cell, the $E_z$ and $H_z$ fields in the aperture can be expressed as

$$H_{zj} = \sum_{p=0}^{\infty}\sum_{q=1}^{\infty}[B_{pqj} + C_{pqj}][\epsilon_j - (q\pi/2c_j)^2] \cdot \cos[p\pi(\phi - \phi_j + \alpha_j)/2\alpha_j] \sin[q\pi(z - z_j + c_j)/2c_j] \quad (5)$$

$$E_{\phi j} = \sum_{p=0}^{\infty}\sum_{q=1}^{\infty}[B_{pqj} - C_{pqj}]k_{pqj} \cos[p\pi(\phi - \phi_j + \alpha_j)/2\alpha_j] \cdot \sin[q\pi(z - z_j + c_j)/2c_j] \quad (6)$$

$$k_{pqj}^2 = \epsilon_j - (p\pi/2b_j)^2 - (q\pi/2c_j)^2 \quad (7)$$

where $2c_j$ is the dimension of waveguide in $z$ direction, $2b_j$ the dimension of waveguide in $\phi$ direction, $2\alpha_j$ the angular extent of waveguide in $\phi$ direction, $\phi_j,z_j$ the coordinates of the center of the $j$th waveguide, and $\epsilon_j$ the dielectric constant of filling of $j$th waveguide. $k_{pq}$ is taken to be positive real or negative imaginary. $B_{pqj}$ is the amplitude of the $p,q$th mode traveling toward the aperture, and is zero for all modes but those exciting the array (normally the $p = 0$, $q = 1$ mode only). $C_{pqj}$ is the amplitude of the wave traveling away from the aperture. When $k_{pqj}$ is positive real, the $C_{pqj}$ term is the reflected field. When $k_{pqj}$ is negative imaginary, the $C_{pqj}$ represents evanescent waveguide fields.

Over each waveguide aperture, the fields described by (3) and (4) for $\rho = a$ must match those described by (5) and (6), where $a$ is the cylinder radius. In addition, the tangential electric field $E_\phi$ given by (4) must vanish over the surface of the cylinder not occupied by waveguide elements. Equation (4) may be equated to (7) over the waveguide aperture, and to zero over the rest of the unit cell. If both sides are multiplied by $\exp(-ik_{zs}z)\exp(-im_r\phi)$ and integrated over the entire unit cell, the following equation results:

$$\frac{2\pi d}{M} ik_{ts}A_{rs}\begin{Bmatrix} H_{m_r}^{2\prime}(k_{ts}a) \\ iK_{m_r}'(ik_{ts}a) \end{Bmatrix} = \sum_j \sum_{p=0}^{\infty}\sum_{q=1}^{\infty} k_{pqj}(B_{pqj} - C_{pqj}) \cdot 2\alpha_j \exp[-i(m_r\phi_j - p\pi/2)V_1(m_r\alpha_j, -p)] \cdot \{-2ic_j \exp[-i(k_{zs}z_j - q\pi/2]V_2(k_{zs}c_j, -q)\} \quad (8)$$

where

$$\left.\begin{matrix} V_1(x,n) \\ V_2(x,n) \end{matrix}\right\} = \frac{1}{2}\left\{\frac{\sin(x + n\pi/2)}{x + n\pi/2} \pm (-1)^n \frac{\sin(x - n\pi/2)}{x - n\pi/2}\right\}. \quad (9)$$

The subscripts 1 and 2 indicate addition or subtraction, respectively.

In a similar manner, (3) may be equated to (5), both sides multiplied by $\cos[P\pi(\phi - \phi_J + \alpha_J)/2\alpha_J] \sin[Q\pi(z - z_J + c_J)/2c_J]$, and the resulting equation integrated over the aperture of waveguide $J$. Finally, substituting for $A_{ln}$ using (8) gives, after reorganization,

$$\sum_j \sum_{p=0}^{\infty}\sum_{q=1}^{\infty} \{S(p,q,j;P,Q,J) + D(p,q,j;P,Q,J)\}F_{pqj} = G_{PQJ},$$
$$0 \le P \le \infty, \quad 1 \le Q \le \infty, \quad J \text{ over all waveguides} \quad (10)$$

where

$$S(p,q,j;P,Q,J) = M\sum_{l=-\infty}^{\infty}\sum_{n=-\infty}^{\infty}\begin{Bmatrix} k_{tn}aH_{m_l}^{2}(k_{tn}a)/H_{m_l}^{2\prime}(k_{tn}a) \\ -ik_{tn}aK_{m_l}(ik_{tn}a)/K_{m_l}'(ik_{tn}a) \end{Bmatrix} \cdot \exp[-im_l(\phi_j - \phi_J)] \exp[-ik_{zn}(z_j - z_J)] \cdot V_1(m_l\alpha_J,p)V_1(m_l\alpha_J,P)V_2(k_{zn}c_J,q)V_2(k_{zn}c_J,Q) \quad (11)$$

$$D(p,q,j;P,Q,J) = \frac{i\pi ad(1 - \delta_{0q})(1 + \delta_{0p})\epsilon_J - (q\pi/2c_J)^2}{8c_J\alpha_J k_{pqj}}\delta_{pP}\delta_{pQ}\delta_{jJ} \quad (12)$$

$$F_{pqj} = 4c_j\alpha_j k_{pqj}(-i)^{p+q}(B_{pqj} - C_{pqj}) \quad (13)$$

$$G_{pqj} = i\pi ad(1 - \delta_{pq})(1 + \delta_{0p})[\epsilon_j - (q\pi/2c_j)^2](-i)^{p+q}B_{pqj}. \quad (14)$$

The above is an infinite matrix equation for the amplitudes of all waveguide modes in terms of the incident mode amplitudes. In practice, the matrix is truncated (as is the sum over $l$ and $n$) and the resultant set of equations is solved for the lower modes on a digital computer. Spatial mode amplitudes $A_{ln}$ can be obtained from (8).

## Computations and Results

It has been assumed that all elements are identical and uniformly spaced. Two configurations, the regular (rectangular element lattice) and the staggered (triangular element lattice), are possible. In the first case, a single element per unit cell is used. For the staggered configuration, two identical elements per unit cell are used. Behavior for the two cases is nearly identical for elevation angles which exhibit no grating lobe in visible space ($k_{zn} > 1$, for $n \neq 0$). For elevation angles greater than this angle, the behavior is quite different.

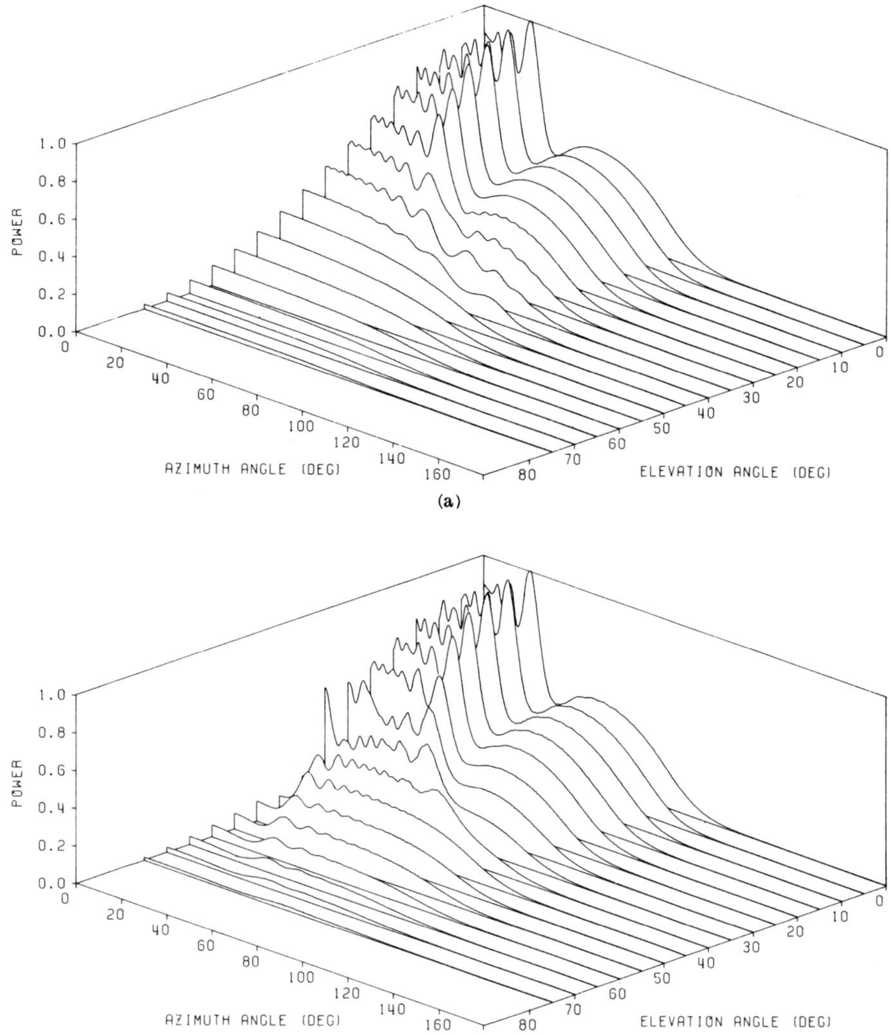


Fig. 2. Element patterns for 256-column cylindrical array. Radius 26.3λ, vertical element spacing 0.72λ. (a) Regular configuration. (b) Staggered configuration.



Equation (8) was solved for both cases, assuming that all waveguide modes except $p = 0$, $q = 1$ are nonpropagating. The incident field amplitude of mode $m_0$ is $B_{01}$, which is designated $B(m_0)$, and the modal reflection coefficient is $C_{01}$, designated $R(m_0)$. For any azimuth excitation, the $B(m_0)$ may be determined by a Fourier transform. The incident and reflected fields of column number $s$ then are

$$\text{Inc}\ (s) = \sum_{m_0=(-M/2)+1}^{M/2} B(m_0) \exp\left(\frac{2\pi i m_0 s}{M}\right), \qquad 0 \leq s \leq M-1 \tag{15}$$

$$\text{Ref}\ (s) = \sum_{m_0=(-M/2)+1}^{M/2} B(m_0) R(m_0) \exp\left(\frac{2\pi i m_0 s}{M}\right), \quad 0 \leq s \leq M-1. \tag{16}$$

The far field for such an excitation is a superposition of the radiated modes. At each elevation angle given by (2) which is in real space, an azimuth pattern is defined

$$E(\phi,\theta_n) = \sum_{m_0=(-M/2)+1}^{M/2} B(m_0) \sum_{l=-\infty}^{\infty} A_{ln} \exp\left[i(m_0 + lM)\phi\right]. \tag{17}$$

The pattern of a singly excited column of elements can be determined by letting all $B(m_0) = 1/M$. Since the space factor and

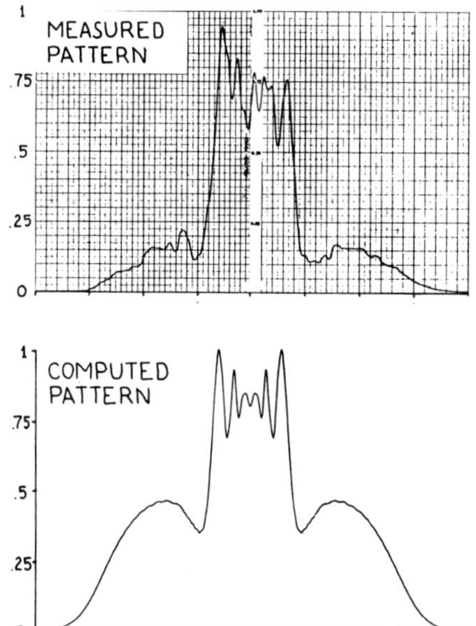


Fig. 3. Computed and measured element patterns at broadside. Azimuth spacing 0.67λ.

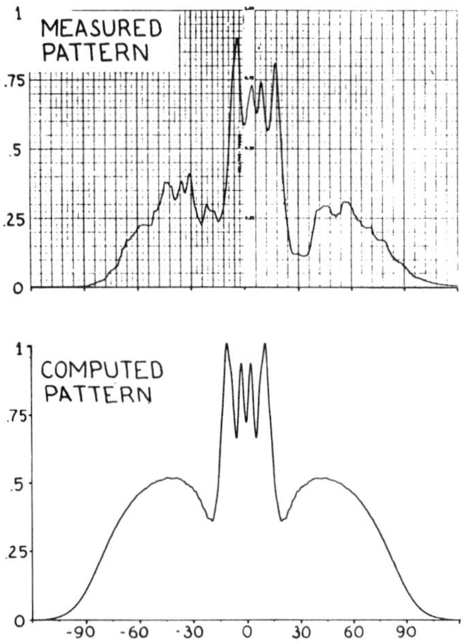


Fig. 4. Computed and measured element patterns at broadside. Azimuth spacing 0.73λ.

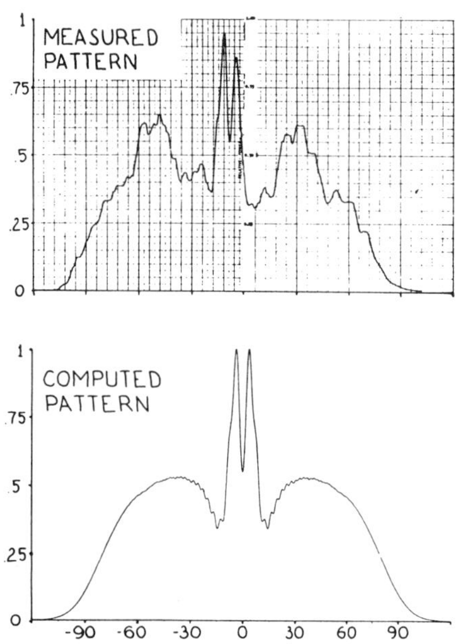


Fig. 5. Computed and measured element patterns at broadside. Azimuth spacing 0.79λ.

element pattern for a linear array are separable, the pattern through the beam of the column is equivalent to a pattern cut at the same elevation angle for an individually excited element.

Fig. 2 shows element patterns of a singly excited element for the staggered and regular configurations. The array radius is 26.3λ, and a ring has 256 elements. The waveguide aperture is 0.13λ by 0.68λ, and vertical spacing is 0.72λ. For this spacing, a grating lobe occurs at an elevation angle of about 23 degrees from broadside. The primary difference between the two cases occurs at elevation angles greater than 23 degrees. Although results are not shown here, this is also true for the active element impedances.

A 60-degree sector of a cylindrical array having the dimensions described has been developed at Naval Electronics Laboratory Center. The array has 32 elements in a column and is staggered. The total number of elements is 1344. An element near the center was excited with all other elements terminated, and azimuth patterns were measured through broadside. Measured and computed patterns for circumferential element spacings of 0.67λ, 0.73λ, and 0.79λ are compared in Figs. 3–5, using a linear power scale. Asymmetry of the measured patterns reflects errors in the experimental array. The center part of the patterns corresponds to the predicted patterns within this error. The outer skirts are perhaps affected to a greater degree by the fact that the experimental antenna is a small sector of the full array.

## Discussion

The approach presented in this communication is used to compute the behavior of a cylindrical array for basic modal excitations. One advantage in considering the modal behavior is that any arbitrary array excitation can be written as a superposition of modal excitations, and the resulting reflected and radiated fields are superpositions of the modal reflected and radiated fields. Thus, for a given array excitation, both the active impedance of every element in the array and the radiated fields at every point in space can be determined.

In this communication, element patterns are shown for two reasons: 1) they are readily measurable from an experimental array, and 2) they are useful for some aspects of array design and for predicting array performance. In azimuth, for example, the useful active sector of the array is equal to that portion of the element pattern which has significant gain, since active elements outside this sector contribute little to the main beam. In elevation, the element patterns determine scan angles achievable with minimum loss of gain.

A. D. Munger
J. H. Provencher
Naval Electronics Laboratory Center
San Diego, Calif. 92152

B. R. Gladman
Admiralty Surface Weapons Establishment
Portsmouth, Hants, England

## References

[1] J. C. Sureau and A. Hessel, "Element pattern for circular arrays of axial slits on large conducting cylinders," *IEEE Trans. Antennas Propagat.* (Communications), vol. AP-17, pp. 799–803, November 1969.

[2] G. V. Borgiotti and Q. Balzano, "Mutual coupling analysis of a conformal array of elements on a cylindrical surface," *IEEE Trans. Antennas Propagat.*, vol. AP-18, pp. 55–63, January 1970.

IEEE TRANSACTIONS ON ANTENNAS AND PROPAGATION, VOL. AP-19, NO. 1, JANUARY 1971

# Element Pattern for Circular Arrays of Waveguide-Fed Axial Slits on Large Conducting Cylinders

JEAN-CLAUDE SUREAU, MEMBER, IEEE, AND ALEXANDER HESSEL, SENIOR MEMBER, IEEE

*Abstract*—Analytical and numerical results are presented for the array element pattern of axial slits in conducting cylinders. The slits are fed by parallel-plate guides in the TEM mode and are matched to free space. Both harmonic series and integral representation for the far-field patterns are obtained and numerically evaluated as a function of $kr_0$ and spacing. The integral representation is treated asymptotically for large values of $kr_0$. The element patterns exhibit two characteristic features: a dip off broadside and a ripple in the main beam near broadside, the level of both being $kr_0$ dependent. While the harmonic series offers no basis for an explanation of these effects, the asymptotics clearly demonstrates that the pattern can be decomposed into two parts:

1) a space wave contribution which can be directly obtained by substituting the circular array "admittance crater" into the appropriate planar array formulas and exhibits a dip analogous to a grating lobe effect;

2) a set of creeping waves which, unlike the more common type, contribute significantly in the forward region and, by interfering with the space wave, cause the ripple near broadside. These creeping waves are the fast (usually $n = -1$) space harmonics (grating lobes) of the fundamental, slightly slow set, and their presence does not seem to have been recognized in the literature.

## I. Introduction

FOR a proper design and evaluation of circular arrays it is desirable to know the electrical performance of an element in the array environment. Since, at radio frequencies, the sources are of the constant incident power type, the most useful definition of the element pattern is that resulting from the excitation of a single-feed waveguide, all others being "match terminated." When any number of elements is excited, the overall radiation pattern and the realized gain may be computed by a direct superposition of the array element pattern. In this fashion the mutual coupling effects are automatically accounted for.

In a large planar array the element pattern is essentially the same for all elements. As a result, a fully excited large planar array pattern is a product of an array factor and an element pattern, and consequently the phase variations of the latter do not play a role in the power pattern of the array. Such is not the case for a circular array. Here, for a given observation point, each element contributes differently in accordance with the element pattern evaluated in the physical direction of interest. Therefore, the fully excited array pattern does not factor, and both amplitude and phase variations of the element pattern affect the gain and the sidelobes.

The isolated $E$-plane element pattern, i.e., pattern of a single slit, does not include mutual coupling and consequently is nearly isotropic with a practically constant phase in the lit region. In contrast, the circular array element pattern for large radii is expected to approach that of a planar array, and since the latter, for spacing between $\lambda/2$ and $\lambda$, exhibits large amplitude and phase variations, it is of considerable importance to investigate quantitatively the properties of the former.

Previous related theoretical results for aperture antennas on conducting cylinders have been limited to single radiating elements [1]–[3] for which the aperture field was prescribed (forced excitation). This is in contrast with the method of excitation considered in this paper and referred to as free excitation [4, pp. 295–296], whereby not the aperture field but the incident fields, in the transmission line feeding the excited elements, are specified. The significance of the difference between the radiation due to an isolated element and that of an element in the array environment has already received much attention in the planar phased array literature [4, pp. 297–308]. Related papers on circular arrays of dipoles in free space or monopoles above a ground plane focused on solutions of integral equations for the current distribution which take into account the effects of mutual coupling [5], [6]. The resulting driving point and mutual impedances were also computed. However, this work was limited to a forced-gap excitation of dipoles in moderate-size arrays and no element patterns in matched array environment were considered. Some studies of the performance of electronically scanned circular arrays have been confined to analysis and synthesis based on assumed element patterns [7]–[9] and in the calculated examples, because of lack of more accurate information, have employed idealized elements. Recently, the analysis and some numerical results for the element pattern of a circular array of

Manuscript received March 18, 1970; revised July 24, 1970. This work was supported in part by the U. S. Army Advanced Ballistic Missile Defense Agency under Contract DAHC 60-69-C-110. It is part of a dissertation submitted by J. C. Sureau to the Department of Electrophysics, Polytechnic Institute of Brooklyn, Farmingdale, N. Y., in partial fulfillment of the requirements for the Ph.D. degree.

J. C. Sureau is with Grumman Aerospace Corporation, Bethpage, N. Y.

A. Hessel is with the Department of Electrophysics, Polytechnic Institute of Brooklyn, Farmingdale, N. Y.

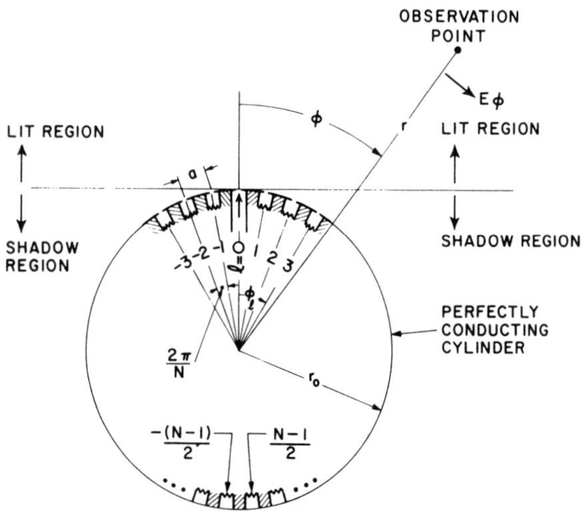


Fig. 1. Circular array geometry.

parallel-plate waveguides via harmonic series have been presented [10].

As a first step toward understanding the mutual coupling effects in circular arrays of waveguide-fed apertures, this paper presents the derivation of the $E$-plane element pattern of a two-dimensional array of parallel-plate guide-fed axial slits in a large conducting cylinder (Fig. 1). The main results of this work, which include both harmonic series and the integral representation, were previously summarized [11]. Although the present study is limited to the $H$-type (LSE polarization), it is simply extended to the $E$-type case.

The analysis employs a modal approach in which the angular periodicity of the array is directly exploited. Formal expressions are derived in terms of the aperture electric field, the knowledge of which is required for any specific element. In the special case of a narrow slit, the usual constant field assumption, with an unknown and to be determined complex amplitude, leads to simple explicit field expressions and illuminating results. The analysis proceeds along two parallel paths resulting in the harmonic series or an integral representation for the radiated field. Formal exact expressions in terms of the aperture distribution are derived in Section II. For large cylinders, the asymptotic evaluation for $kr_0 \gg 1$ of the integral representation of the far field along with the physical interpretation of the results are given in Section III. In Section IV numerical data for the case of thin slits are presented.

## II. Formal Expressions

### A. Representation of Single-Element Excitation

Fig. 1 illustrates schematically a circular array of $N$ parallel-plate waveguide-fed infinite axial slits in a perfectly conducting cylinder. The waveguides are assumed to propagate solely the TEM mode. Only the $l = 0$ element is fed, the others are match terminated. The system may be considered as an $N$-port network, the excitation of which, i.e., the incident complex wave amplitude vector $\boldsymbol{a}$, is described (e.g., for $N$ odd) by the set of its components:

$$\boldsymbol{a} = \{a_l\} = \{a(\Phi_l)\} = \{\delta_l^0\}, \qquad l = 0,\pm 1,\cdots,\pm(N-1)/2 \quad (1)$$

where $\delta_l^0$ is the Kronecker delta and $\Phi_l = 2\pi l/N$ is the angular element location.

There are two alternatives to represent the single-element excitation and the fields in an equispaced circular array. In the first, the extent of the angular region is confined to the physical space $-\pi \leq \Phi \leq \pi$ in which the $2\pi$ periodicity of the fields is explicitly imposed. The single-element excitation is expressed as a finite discrete spectrum of angular harmonics which, for $N$ odd, takes the form

$$\boldsymbol{a} = \{\delta_l^0\} = \frac{1}{N} \sum_{n=-(N-1)/2}^{+(N-1)/2} \{\exp[-jn\Phi_l]\}, \qquad \frac{-(N-1)}{2} \leq l \leq \frac{N-1}{2}. \quad (2)$$

In an alternative representation, the angular coordinate is regarded as an infinitely extended angular space [12] $-\infty < \Phi < \infty$ in which the $2\pi$ periodicity is enforced by image elements and periodic image sources, the latter placed an angular distance $2\pi$ apart. In this space the actual element excitation is given as a superposition of the excitation of individual sources $\boldsymbol{a} = \sum_{s=-\infty}^{+\infty} \boldsymbol{a}_s$, where $\boldsymbol{a}_s = \{\delta_l^{sN}\}$. Each source $\boldsymbol{a}_s$ is now represented as a band spectrum of angular wavenumbers resulting in

$$\boldsymbol{a} = \frac{\beta}{2\pi} \sum_{s=-\infty}^{+\infty} \int_{-\pi/\beta}^{+\pi/\beta} \{\exp[-j\nu(\Phi_l - 2s\pi)]\}\, d\nu, \qquad l = 0,\pm 1,\cdots \quad (3)$$

where $\beta = 2\pi/N$ and $\Phi_l = l\beta$.

For any $\nu$ in (3) or $n$ in (2), the set $\{\exp[-j\nu\Phi_l]\}$ represents a phase sequence excitation (linear phase taper), where the amplitude coefficient of the incident wave in the $l$th waveguide is given by $\exp[-j\nu\Phi_l]$.

Denoting by $E_\Phi(r,\Phi;\nu)$ the angular electric field component resulting from a phase sequence excitation, the field due to a singly excited $l = 0$ element follows from superposition, according to (2):

$$E_\Phi(r,\Phi) = \frac{1}{N} \sum_{n=-(N-1)/2}^{+(N-1)/2} E_\Phi(r,\Phi;n) \quad (4)$$

and, in view of (3), the continuous spectrum representation for the same field takes the form

$$E_\Phi(r,\Phi) = \sum_{s=-\infty}^{+\infty} E_\Phi^{(0)}(r,\Phi - 2s\pi) \quad (5)$$

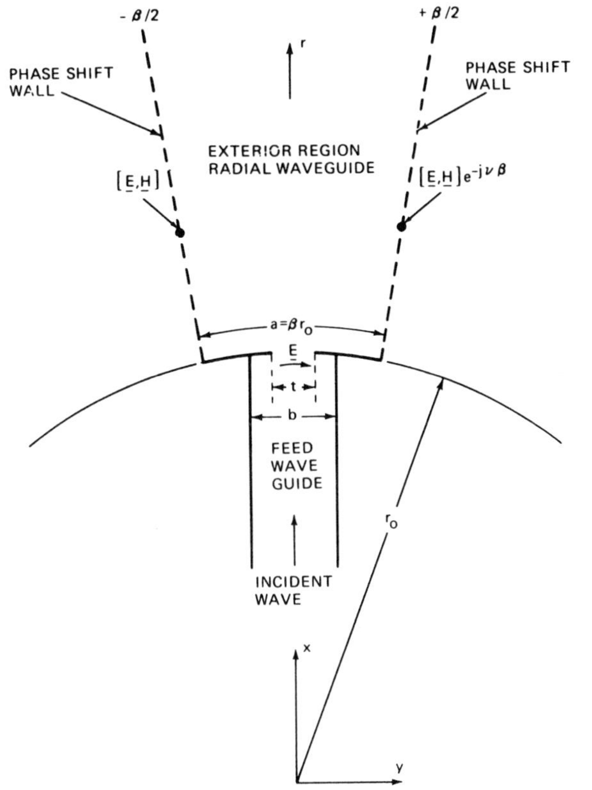


Fig. 2. Unit cell in circular array.

where

$$E_\Phi^{(0)}(r,\Phi) = \frac{\beta}{2\pi}\int_{-\pi/\beta}^{+\pi/\beta} E_\Phi(r,\Phi;\nu)\, d\nu \tag{6}$$

is the source term ($s = 0$).

### B. Phase Sequence Excitation of a Circular Array

Due to the periodicity of the structure and its extension, the electromagnetic problem for a phase sequence excitation reduces, by Floquet's theorem, to the determination of the fields within a unit cell. Such a unit cell consists of a junction of the feed waveguide and an effective radial waveguide (Fig. 2) whose walls are characterized by a phase-shift boundary condition [13].

In the feed waveguide, the modal voltages $V_m{}^i$ and currents $I_m{}^i$ are defined in the usual manner. The field in the radial waveguide can be expressed in terms of the aperture field as follows:

$$E_\Phi(r,\Phi;\nu) = \frac{1}{\beta r_0}\sum_{m=-\infty}^{+\infty} V_m{}^e(r_0;\nu)\frac{H_{\nu_m}{}^{(2)\prime}(kr)}{H_{\nu_m}{}^{(2)\prime}(kr_0)}\exp\left[-j\nu_m\Phi\right], \quad -\infty < \Phi < \infty \tag{7}$$

with $\nu_m = \nu + 2\pi m/\beta$, and the modal voltage

$$V_m{}^e(r_0;\nu) = r_0\int_{-\beta/2}^{+\beta/2} E_\Phi(r_0,\Phi;\nu) \cdot \exp\left[+j(\nu + 2\pi m/\beta)\Phi\right] d\Phi. \tag{8}$$

The effect on the dominant mode of the feed waveguide of the slit radiating into the radial waveguide is represented in the aperture plane by a $\nu$-dependent "active" admittance $Y_a(\nu) = G_a(\nu) + jB_a(\nu)$. For an efficient power transfer, a network, identical for all elements, is introduced in each waveguide to yield a "match" for some real value $\nu = \nu_0$. The equivalent circuit "looking out" of the feed waveguide is shown in Fig. 3 from which

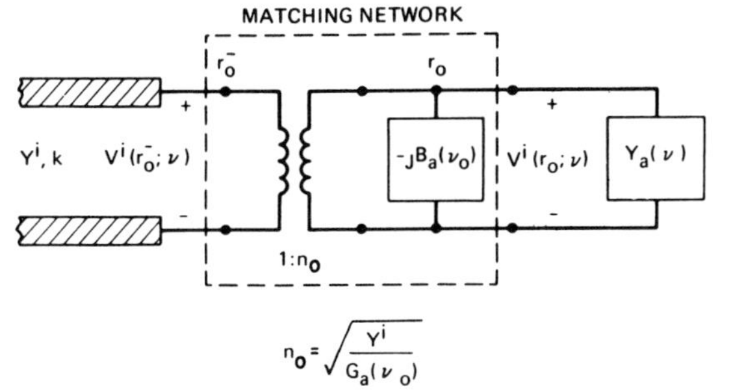


Fig. 3. Equivalent network representation for unit cell.

$$V_m{}^e(r_0;\nu) = V_0{}^{\rm inc} n_0 T(\nu,\nu_0)\frac{V_m{}^e(r_0;\nu)}{V_0{}^i(r_0;\nu)} \tag{9}$$

and the transmission coefficient $T(\nu,\nu_0)$ is given by

$$T(\nu,\nu_0) = \frac{V_0{}^i(r_0{}^-;\nu)}{V_0{}^{\rm inc}} = \frac{2G_a(\nu_0)}{Y_a{}^*(\nu_0) + Y_a(\nu)}. \tag{10}$$

Substituting (9) into (7) gives

$$E_\Phi(r,\Phi;\nu) = \frac{V_0{}^{\rm inc}}{\beta r_0} n_0 T(\nu,\nu_0) \cdot \sum_{m=-\infty}^{+\infty}\frac{V_m{}^e(r_0;\nu)}{V_0{}^i(r_0;\nu)}\frac{H_{\nu_m}{}^{(2)\prime}(kr)}{H_{\nu_m}{}^{(2)\prime}(kr_0)}\exp\left[-j\nu_m\Phi\right]. \tag{11}$$

Since all Floquet field quantities must be periodic in $\nu$ with the period $2\pi/\beta$,

$$V_m{}^e(r_0;\nu) = V_0{}^e(r_0;\nu_m) \equiv V^e(r_0;\nu_m) \tag{12}$$

$$V_0{}^i(r_0;\nu) = V_0{}^i(r_0;\nu_m) \equiv V^i(r_0;\nu_m). \tag{13}$$

Hence

$$E_\Phi(r,\Phi;\nu) = \frac{V^{\rm inc}}{\beta r_0} n_0 T(\nu,\nu_0) \cdot \sum_{m=-\infty}^{+\infty}\frac{V^e(r_0;\nu_m)}{V^i(r_0;\nu_m)}\frac{H_{\nu_m}{}^{(2)\prime}(kr)}{H_{\nu_m}{}^{(2)\prime}(kr_0)}\exp\left[-j\nu_m\Phi\right]. \tag{14}$$

Expression (14) relates the external field resulting from a phase sequence excitation to the incident voltage as well as to $T(\nu,\nu_0)$ and $V^e/V^i$, both of which are dependent only on the form of the aperture electric field.

### C. Element Pattern

Expressions for the field resulting from a single-element excitation can be obtained with the help of (14) using superposition in accordance with (4) or (6). Since $T(\nu,\nu_0) = T(\nu_m,\nu_0)$, these expressions can be contracted into a single integral and a single summation yielding, respectively,

$$E_\Phi^{(0)}(r,\Phi) = \frac{V^{\text{inc}}n_0}{2\pi r_0}\int_{-\infty}^{+\infty} T(\nu,\nu_0)\frac{V^e(r_0;\nu)}{V^i(r_0;\nu)} \cdot \frac{H_\nu^{(2)\prime}(kr)}{H_\nu^{(2)\prime}(kr_0)}\exp[-j\nu\Phi]\,d\nu \quad (15)$$

$$E_\Phi(r,\Phi) = \frac{V^{\text{inc}}n_0}{2\pi r_0}\sum_{n=-\infty}^{+\infty} T(n,\nu_0)\frac{V^e(r_0;n)}{V^i(r_0;n)} \cdot \frac{H_n^{(2)\prime}(kr)}{H_n^{(2)\prime}(kr_0)}\exp[-jn\Phi]. \quad (16)$$

In the far field (15) becomes

$$E_\Phi^{(0)}(r,\Phi) \sim -j\frac{V^{\text{inc}}n_0}{2\pi r_0}\left(\frac{2}{\pi kr}\right)^{1/2}\exp[-j(kr-\pi/4)] \cdot \int_{-\infty}^{+\infty}\frac{V^e(r_0;\nu)}{V^i(r_0;\nu)}T(\nu,\nu_0)\frac{\exp[-j\nu(\Phi-\pi/2)]}{H_\nu^{(2)\prime}(kr_0)}d\nu \quad (17)$$

and the corresponding harmonic series expression is given in [11].

### D. Transmission Coefficient $T(\nu,\nu_0)$

The main distinction between similar series and integral expressions obtained by previous authors for a forced excitation of a single element on a smooth cylinder and those obtained here, is the presence of the transmission coefficient $T(\nu,\nu_0)$ as in (17). Its appearance is a direct consequence of the free excitation of the array.

From (10) it is seen that the active admittance $Y_a(\nu)$ uniquely specifies $T(\nu,\nu_0)$ for a given match point $\nu_0$. An approximate expression for $Y_a(\nu)$ can be obtained as in planar arrays from the requirement of continuity of complex power at the aperture in a unit cell[1]

$$Y_a(\nu) = \sum_{m=-\infty}^{+\infty}\left|\frac{V^e(r_0;\nu_m)}{V^i(r_0;\nu_m)}\right|^2 Y^e(r_0;\nu_m) + \sum_{m=1}^{\infty}\left|\frac{V_m^i(r_0;\nu)}{V^i(r_0;\nu)}\right|^2 Y_m^i \quad (18)$$

where $Y^e$ and $Y^i$ are the modal admittances of the radial and TEM guides, respectively.

Since the voltage ratios in (18) depend on the form of the aperture electric field $\boldsymbol{E}_{\text{ap}}$, $Y_a(\nu)$ may be calculated only provided this form is known. For the case of a narrow slit to be treated as an example, a uniform field is a good and widely used approximation. For wider apertures, variational procedures for the determination of $\boldsymbol{E}_{\text{ap}}$ may be employed.

The first term in the right-hand side of (18) represents the contribution of the exterior region to the input admittance and is strongly $\nu$ dependent because of the form of the radial mode characteristic admittance

$$Y^e(r_0;\nu) = G^e(r_0;\nu) + jB^e(r_0;\nu) = \frac{1}{\zeta_0\beta r_0}\left[-j\frac{H_\nu^{(2)}(kr_0)}{H_\nu^{(2)\prime}(kr_0)}\right]. \quad (19)$$

The second term in (18) corresponds to the contribution of the feed waveguide to the total susceptance for a single-mode waveguide. To the extent that the aperture field shape may be assumed to be independent of $\nu$, this contribution is also independent of $\nu$ and will be canceled by the matching susceptance for all values of $\nu$.

The behavior of $Y^{e\prime}(r_0;\nu) = -jH_\nu^{(2)}(kr_0)/H_\nu^{(2)\prime}(kr_0)$ for real values of $\nu$ and for large values of $kr_0$ is given to the first order by the following:

1) for $|kr_0 - \nu| > 0(kr_0^{1/3})$, it follows from the Debye approximation [14] that

$$Y_e' \sim \pm\frac{1}{\sin\gamma} \quad (20)$$

where $\cos\gamma = \nu/kr_0$ with $0 \leq \text{Re}\,\gamma < \pi$ and $\gamma = j|\gamma|$ for $|\nu| > kr_0$;

2) for $|kr_0 - \nu| \leq 0(kr_0^{1/3})$, it follows from the first-order Airy approximation [14] that

$$Y_e' \sim j\exp\left[j\frac{2\pi}{3}\right]\left(\frac{kr_0}{2}\right)^{1/3}\frac{\text{Ai}(2^{1/3}\tau\exp[-j2\pi/3])}{\text{Ai}'(2^{1/3}\tau\exp[-j2\pi/3])} \quad (21)$$

where $\nu = kr_0 + kr_0^{1/3}\tau$.

The plots in Fig. 4 show that the conductance $G'^e$ and the susceptance $B'^e$ never exactly go to zero and are bounded. In the limit $kr_0 \to \infty$ the plots become identical with the modal admittance crater of an $E$-plane scanned planar array [15].[2] The level of the pronounced peak in $G'^e$ and the $B'^e$ near $|\nu/kr_0| = 1$ is (from (21)) proportional to $(kr_0)^{1/3}$.

The expression for the active input admittance represents, as in the planar case, a weighted sum of periodically sampled values of the circular array admittance crater. In general, $|Y_a(\nu)|$ will peak at values of $\nu$ for which one of the $|Y^{e\prime}(r_0,\nu_m)|$ will be large, i.e., for $|\nu_m/kr_0| \approx 1$.

In particular, for $\lambda/2 < a < \lambda$ the smallest values of $|\nu|$ for which $|Y_a(\nu)|$ will possess a peak are near

$$\nu/kr_0 = \pm(\lambda/a - 1). \quad (22)$$

This peak in turn causes a dip in $T(\nu,\nu_0)$, the extent of which is thus proportional to $(kr_0)^{1/3}$. This effect may be attributed to the transition of a higher $H$-type mode of the phase-shift-wall waveguide (or a higher angular-space harmonic) from below cut-off condition ($G'^e \ll 1$) to a

[1] The effect of curvature over the extent of the aperture has been neglected in arriving at this expression.

[2] The importance and the shape of the admittance crater was first recognized by Rhodes.

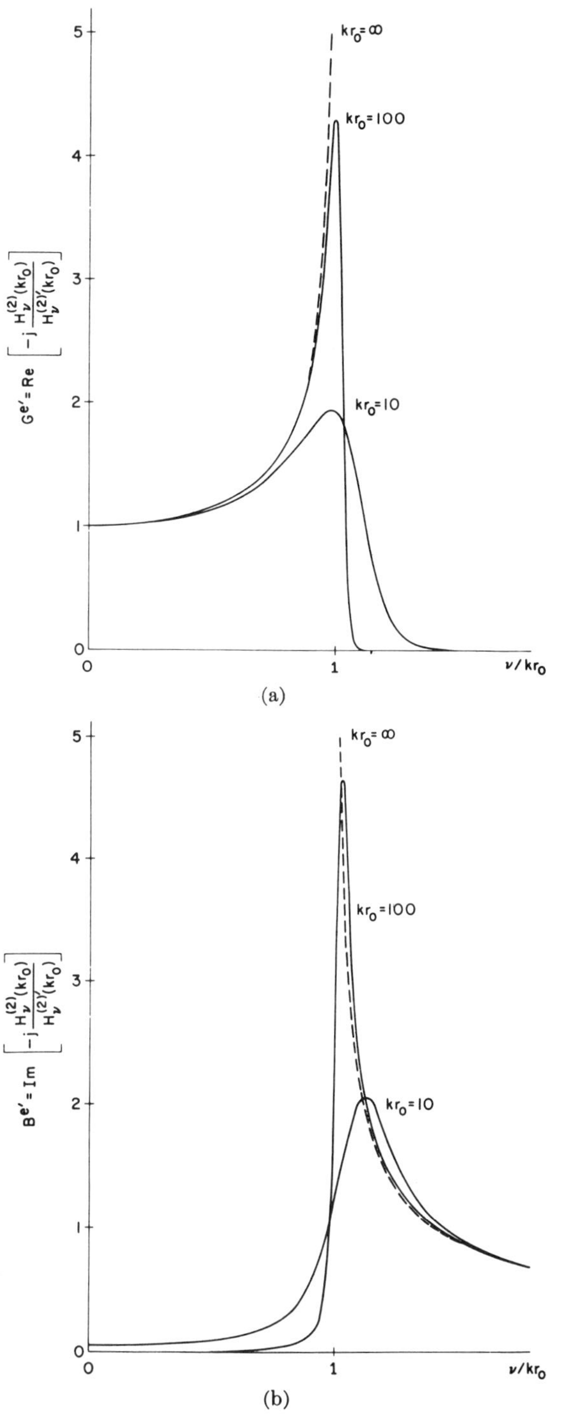


Fig. 4. Modal admittance crater for circular arrays (LSE polarization).

radiating condition ($G'^e > 1$). Such a behavior constitutes the circular array analog of the incipient grating lobe in an $E$-plane scanned planar array. The terminology "grating lobe" is actually inappropriate here since the radiation pattern of such a higher mode is omnidirectional, rather than a narrow beam.

Although the behavior of $T(\nu,\nu_0)$ on the real $\nu$ axis is sufficient for the harmonic series calculations, the knowledge of the $T(\nu,\nu_0)$ behavior in the complex $\nu$ plane is required to carry out the complete asymptotic evaluation of (17). In particular, the only singularities on the integrand are the complex periodic first-order poles of $T(\nu,\nu_0)$

$$\nu_p{}^q = \pm(\nu_p{}^0 + 2\pi q/\beta) = \pm[\nu_p{}^0 + qkr_0(\lambda/a)], \qquad q = 0,\pm1,\cdots \quad (23)$$

which are solutions of the characteristic equation

$$Y_a{}^*(\nu_0) + Y_a(\nu) = 0. \quad (24)$$

From the physics of the problem one may expect the presence of two different sets of poles. One set corresponds to a perturbation of the usual creeping wave poles on a smooth cylinder [1]. The other set, if present, is a direct consequence of the surface admittance boundary conditions in a corresponding planar structure, resulting there in a guided wave for which the curvature of the cylinder is only a perturbation [16]. It can be shown that, for narrow matched slits, the only significant set of poles[3] of $T(\nu,\nu_0)$ are those of the former type.

## III. Asymptotic Evaluation of Far Fields

The developments of the harmonic series and of the integral representation have followed parallel paths and have resulted in formal structurally similar expressions. However, the two methods depart radically in their evaluation procedures. For the harmonic series one proceeds along a manner which is strictly numerical. Although formally exact for all observation angles, the series suffers, for large $kr_0$, from the usual convergence difficulties deep in the shadow region. Moreover, it fails to provide a physical interpretation for the far-field patterns. In contrast, the integral representation may be evaluated asymptotically yielding simple analytical expressions along with illuminating physical interpretation of the results in terms of the various types of significant wave contributions, one of which does not seem to have appeared elsewhere in the literature.

The asymptotic evaluation of (17) proceeds in a manner similar to that for smooth large cylinders [17]. Different representations are obtained in the lit region, the shadow region, and the shadow boundary (Fig. 1).

### *A. Lit Region*

The basic integral in (17) to be evaluated is of the form

$$F(\Phi) = \int_{-\infty}^{+\infty} V(\nu)T(\nu)\,\frac{\exp[-j\nu(\Phi - \pi/2)]}{H_\nu{}^{(2)\prime}(kr_0)}\,d\nu \quad (25)$$

where $V(\nu) \equiv V^e(\nu)/V^i(\nu)$. Using the Debye approximation for $H_\nu{}^{(2)\prime}(kr_0)$, (25) becomes

$$F(\Phi) \sim \left(\frac{\pi kr_0}{2}\right)^{1/2} \exp[j\pi/4] \int_{-\infty}^{+\infty} V(\nu)T(\nu)\,\frac{1}{(\sin\gamma)^{1/2}} \cdot \exp[-jkr_0\Psi(\gamma)]\,d\nu \quad (26)$$

[3] An additional set of creeping waves is expected in the case of a dielectric layer covering the array face.

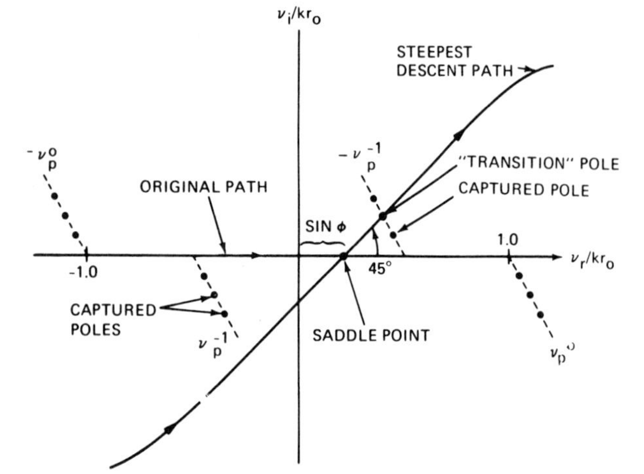


Fig. 5. Contour deformation for asymptotic evaluation in lit region.

where $\cos \gamma = \nu/kr_0$ and

$$\Psi(\gamma) = (\gamma + \Phi - \pi/2) \cos \gamma - \sin \gamma. \tag{27}$$

The saddle point is located at

$$\nu_s = kr_0 \sin \Phi. \tag{28}$$

The steepest-descent path (SDP) through the saddle point is defined by

$$\operatorname{Re} \Psi(\gamma) = \operatorname{Re} \Psi(\gamma_s) = -\cos \Phi \tag{29}$$

and is shown qualitatively in Fig. 5. By Cauchy's theorem one has

$$\int_{-\infty}^{+\infty} (\cdot)\, d\nu = \int_{\text{SDP}} (\cdot)\, d\nu + 2\pi j \sum_{p,q} [\operatorname{Res}(-\nu_p{}^q) - \operatorname{Res}(\nu_p{}^q)]. \tag{30}$$

The pole contributions in the lit region for the circular array appear because of the angular periodicity and do not have an equivalent in the case of an isolated element.



1) *Saddle Point Contribution*: Upon application of the standard formulas [18] the first-order saddle point evaluation of (17) becomes

$$E_\Phi{}^{(0)}{}_{\text{SDP}}(r,\Phi) \sim \frac{kr_0}{2} V^{\text{inc}} n_0 V(kr_0 \sin \Phi) T(kr_0 \sin \Phi) \left(\frac{2}{\pi kr}\right)^{1/2} \cdot \exp[-j(kr - \pi/4 - kr_0 \sin \Phi)]. \tag{31}$$

This contribution is observed solely in the lit region, appears to originate from the driven element, and is referred to as the direct contribution or the space wave. As in the planar array case, it is proportional to the dominant mode transmission coefficient and, from (22), may be expected to exhibit a dip at angles

$$|\Phi| = \sin^{-1}(\lambda/a - 1). \tag{32}$$

Also, the selection of the matching network may now be interpreted as maximizing this contribution at an angle $|\Phi_0| = \sin^{-1}(\nu_0/kr_0)$.

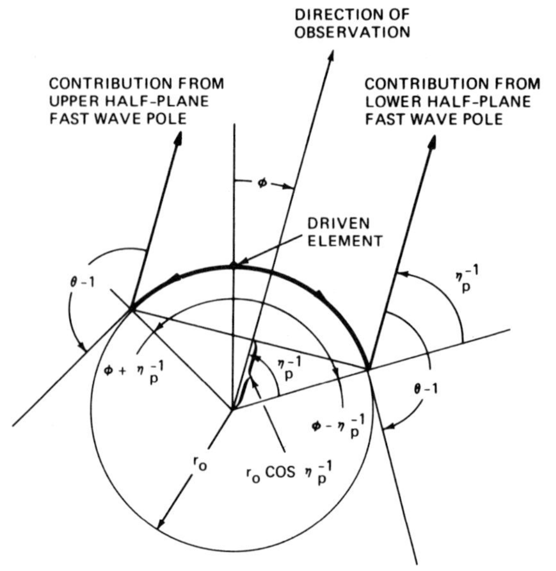


Fig. 6. Ray interpretation of pole contributions in lit region.

2) *Pole Contributions*: From (26), the residue at a typical fast wave pole $\nu_p{}^q$ of $T(\nu)$ may be written

$$\operatorname{Res}(\nu_p{}^q) = \exp[j\pi/4] V(\nu_p{}^q) \left(\frac{\pi kr_0}{2 \sin \gamma_p{}^q}\right)^{1/2} \cdot \{\lim (\nu - \nu_p{}^q) T(\nu)\} \exp[-jkr_0 \Psi(\gamma_p{}^q)]. \tag{33}$$

In (33) the pattern angle $\Phi$ exclusively enters through $\Psi$ in the exponential; the remainder of the expression depends only on the pole location and represents the creeping wave excitation coefficient.

It can be shown that the poles of $T(\nu)$ are of the form

$$\nu_p{}^q = kr_0(1 + q\lambda/a + kr_0{}^{-2/3}\tau_p), \qquad q = 0 \pm 1, \pm 2, \cdots \tag{34}$$

and, for $q \neq 0$, represent the harmonics of the fundamental $q = 0$ creeping wave.

As seen from Fig. 5, only the $q < 0$ poles are captured in the lit region. Also, for element spacings $\frac{1}{2} < a/\lambda < 1$, the dominant pole contributions in the lit region are those corresponding to $q = -1$.

Since the important poles are in general close to the real axis, first-order expressions for the exponent in (34) are found to be

$$-jkr_0\Psi(\gamma_p{}^{-1}) \simeq kr_0{}^{1/3}(\Phi - \eta_p{}^{-1}) \operatorname{Im} \tau_p + j[kr_0(\Phi - \eta_p{}^{-1}) \sin \eta_p{}^{-1} + kr_0 \cos \eta_p{}^{-1}] \tag{35}$$

with $\sin \eta_p{}^{-1} = (1 - \lambda/a)$.

This form suggests the ray interpretation shown in Fig. 6. For the case of lower half-plane poles ($\eta_p{}^{-1} < 0$, $\operatorname{Im} \tau_p < 0$), for example, the real part of (35) indicates that a typical ray travels along the surface clockwise through an angle $(\Phi - \eta_p{}^{-1})$ and decays at a rate of

$kr_0^{1/3}$ Im $\tau_p$ nepers per radian. The first term in the imaginary part of (35) represents the phase change as the ray travels from the source along the surface; its associated phase velocity is $c/v = \sin \eta_p^{-1}$, and the ray thus represents a backward fast creeping wave. The second term in the imaginary part of (35) is the phase, referenced to the cylinder center, associated with the free space path of the ray and implies that, irrespective of $\Phi$, the fast creeping wave ray emerges from the surface at a constant angle $\theta_{-1} = \pi/2 - \eta_p^{-1}$ with the local tangent (Fig. 6). Rays due to upper half-plane poles follow a path which is the mirror image of that for the lower half-plane poles. As seen from Fig. 5, the angular range in which these poles contribute is approximately defined by $\Phi \gtrless \pm\eta_p^{-1}$ for the lower and upper half-plane poles, respectively.

A typical fast wave pole contribution to the radiated electric field becomes

$$E_\Phi^{(0)}(r,\Phi)\,|_{\text{pole}} \sim V^{\text{inc}}\eta_0 V(\nu_p^{-1})\left(\frac{2\pi k}{r_0 \cos \eta_p^{-1}}\right)^{1/2} a_p \left(\frac{2}{\pi k r}\right)^{1/2} \cdot \exp\{-j[kr + kr_0\Psi(\gamma_p^{-1})]\} \quad (36)$$

where

$$a_p = \frac{G_a'(\nu_0)}{1 + 2\tau_p Y'^2 kr_0^{-2/3}} \left|\frac{1}{V(kr_0)}\right|^2 \quad (37)$$

and

$$Y' = \frac{\zeta_0 \beta r_0}{|V(kr_0)|^2} \cdot [Y_a^*(\nu_0) + \sum_{m\neq 0} |V(kr_0 + 2\pi m/\beta)|^2 Y^e(kr_0 + 2\pi m/\beta)]. \quad (38)$$

As $kr_0$ increases, the rate of change of the phase term in (36) as a function of the observation angle $\Phi$ also increases; however, the relative field magnitude of the contribution of the captured pole decreases as $kr_0^{-1/2}$. Thus, with increasing $kr_0$, the interference between this creeping ray and the space wave may be expected to produce a ripple of increasing frequency but decreasing amplitude.

3) *Fast Wave Pole Transition*: When the integral along the SDP is approximated by the saddle point contribution, the overall result experiences a discontinuity as a function of $\Phi$ everytime the path crosses a pole. A more accurate evaluation yields a continuous transition field [19]. The overall result for the asymptotic evaluation of the integral (26) is given by the sum of the saddle point contribution (31), the residue contributions from the fully captured poles (36), and the contribution from the transition poles (poles near the SDP). The latter is

$$F(\Phi)\,|_{\text{transition poles}} \sim \exp[j(kr_0 \cos \Phi + \pi/4)] \sum_p V(\nu_p^{-1}) a_p \left(\frac{\pi k r_0}{2 \sin \gamma_p^{-1}}\right)^{1/2} \cdot \left[\frac{1}{b_p}\left(\frac{\pi}{kr_0}\right)^{1/2} + j2\pi \exp[-kr_0 b_p] \frac{Q[-jb_p(kr_0)^{1/2}]}{\sqrt{\pi}}\right] \quad (39)$$

where

$$b_p = \exp[j\pi/4][\Psi(\gamma_p^{-1}) + \cos \Phi]^{1/2} \quad (40)$$

and the transition function is

$$Q(y) = \int_y^\infty \exp(-x^2)\, dx. \quad (41)$$

*B. Shadow Region $\pi/2 < |\Phi| < \pi$*

As for smooth cylinders [1], the original path of integration along the real axis can be closed at infinity in the lower (upper) half-plane for $\Phi > \pi/2 (\Phi < -\pi/2)$. The integral (25) is then expressed as a residue series at the poles of $T(\nu)$. For $\Phi > \pi/2$, for example,

$$F(\Phi) = -2\pi j \sum \text{Res}(\nu_p^q). \quad (42)$$

For element spacings $\frac{1}{2} < a/\lambda < 1$, the dominant contribution in the shadow region arises from the $q = 0$ term, the evaluation of which follows a standard procedure and is given by

$$\text{Res}(\nu_p^0) \sim \frac{1}{V^*(kr_0)} \frac{G_a'(\nu_0) \exp[j5\pi/6]}{[1 + 2\tau_p Y'^2 kr_0^{-2/3}]} \left(\frac{kr_0}{2}\right)^{2/3} \cdot \frac{\exp[-j(kr_0 + kr_0^{1/3}\tau_p)(\Phi - \pi/2)]}{\text{Ai}'(2^{1/2}\tau_p \exp[-j2\pi/3])}. \quad (43)$$

For $\Phi - \pi/2 > 0$ sufficiently large, the residue series is rapidly convergent, since contributions from successive values of $p$ decay exponentially. However, when $\Phi - \pi/2 \leq 0(kr_0^{-2/3})$, as seen from (43), the convergence slows down and an alternate technique must be used.

Each term in the $q = 0$ residue series may be given the usual interpretation of a creeping ray.

*C. Shadow Boundary*

Near the shadow boundary $||\Phi| - \pi/2| \leq 0(kr_0^{-1/3})$, the residue series is poorly convergent, and the techniques used in the lit region cannot be extended since the Debye approximation to the Hankel functions is not valid for $\nu \approx kr_0$. This difficulty has already been resolved for smooth cylinders with a surface admittance [20] and a similar procedure is used here.

By Cauchy's theorem applied to (25), we may write, for $\Phi \approx \pi/2$,

$$F(\Phi) = \int_{-\infty}^{+\infty} I(\nu;\Phi)\, d\nu = \int_\Gamma I(\nu,\Phi)\, d\nu - 2\pi j \sum_{\text{captured poles}} \text{Res}(\nu_p^q) \quad (44)$$

where the path $\Gamma$ is defined as in [1]. The integral along $\Gamma$ in (44) may be approximated by

$$F(\Phi)\,|_\Gamma \sim -\frac{j \exp[j2\pi/3]}{2^{1/3}} kr_0^{2/3} G_a'(\nu_0) \frac{1}{V^*(kr_0)} \cdot \exp[-jkr_0(\Phi - \pi/2)] g[kr_0^{1/3}(\Phi - \pi/2);\alpha] \quad (45)$$

with $\alpha = -j(kr_0/2)^{-1/3}Y' \exp[-j2\pi/3]$, where

$$g(\xi;\alpha) = \int_{\infty \exp(-j2\pi/3)}^{\infty} \frac{\exp[-j\xi\tau]}{\mathrm{Ai}(2^{1/3}\tau \exp[-j2\pi/3]) + \alpha \mathrm{Ai}'(2^{1/3}\tau \exp[-j2\pi/3])} \, d\tau. \tag{46}$$

The shadow boundary transition function $g(\xi;\alpha)$ has already been encountered [20] and is a generalization of the Fock function.

### *D. Summary of Ray Interpretation*

The various ray contributions to the far field have previously been described in [11]. The direct ray originates at the element and travels directly to the observation point; it contributes only for $|\Phi| < \pi/2$. The creeping rays corresponding to the same $p$, slow and fast, travel along the cylinder, with the same attenuation rate but different phase velocities, and continuously shed energy: the slow rays radiate along the cylinder tangent, the fast rays, away from the tangent at a constant angle $\theta_{-1}$, corresponding to a locally planar array, with element spacing $a$, excited with a phase delay of $\sim ka$ per element, and independent of the observation direction.

In the lit region, the far field is the result of the superposition of the direct ray and one ($|\Phi| > \Phi_c$) or two ($|\Phi| < \Phi_c$) fast dominant creeping rays. In the shadow region, only the slow $q = 0$ creeping rays contribute significantly. The relative attenuation of all creeping rays is determined by the distance which they travel along the surface before radiating into free space.

Contributions from creeping waves which travel several times around the cylinder may be interpreted as being due to the image sources, $s \neq 0$.

## IV. Example: Thin Slit Array

For illustration purposes a TEM parallel-plate waveguide-fed array of $0.1\lambda$ slits is considered. The aperture field is assumed to be uniform across the gap. Matching of the array under the equiphase condition $\nu_0 = 0$ ("broadside" match) was selected for all calculations. As a consequence, the results obtained are independent of the parallel-plate spacing and weakly dependent on the slit width.

In the numerical evaluation of the element patterns, the modal series for the external contribution to the active admittance was evaluated by taking 31 external modes into account. The harmonic series summation was truncated at $n_{\max} = 2kr_0$. The location of the poles of $T(\nu)$ was computed by a perturbation of the resonance equation about $\nu = kr_0$. Typical results, shown in Fig. 7, indicate that the pole locations in the $\tau$ plane are weakly dependent on $a/\lambda$ and $kr_0$ for the range of parameters considered.

Figs. 8–11 show the results of the harmonic series and of the integral representation for $kr_0 = 105$ and $a/\lambda = 0.6$. A direct comparison of the gain patterns $G(\Phi)$ normalized to the unit cell gain ($2\pi a/\lambda$) obtained by the two methods is shown in Fig. 8. Very good agreement is obtained for all angles. This provides a check of the accuracy of the analytical evaluation techniques employed. The apparent erratic behavior of the harmonic series deep in the shadow region may be attributed to truncation effects and to residual small inaccuracies in the computation of the Hankel function and its derivative. The isolated element pattern for the same radius, shown in the same figure (calculated via harmonic series), clearly shows by contrast the influence of the array environment. A detailed presentation of the results of asymptotic calculations is shown in Fig. 9. In the lit region, the total pattern is the result of the superposition of the space wave and contributions from fast wave poles. These poles were considered in pairs ($\pm\nu_p^{-1}$), and their number in a "chain" was varied from 1 ($p = 1$) to 5 ($p = 1,\cdots,5$). The location of the dip in the space wave corresponds very closely to the 41.5° predicted on the basis of (32) for $a/\lambda = 0.6$. For $0 \leq \Phi \leq 25°$ and $50° \leq \Phi < 70°$, the contribution from the fast creeping poles was taken to be their residues. In the region $25° < \Phi < 50°$ the pole transition effects were taken into account. As seen from Fig. 9, the general outline of the pattern in the lit region is determined by the space wave; the ripple in the main beam may be directly attributed to interference between the space wave and the fast creeping waves, only one clockwise and one counterclockwise of which need be taken into account to obtain good accuracy. The pattern in the shadow region was computed from a residue series of only slow wave poles ($\nu_p^0$ for $\Phi > 90°$) whose number was also varied from 1 to 5. Good accuracy is seen to be obtained closer to the penumbra by taking more slow wave poles into account. The pattern in the shadow boundary transition $70° < \Phi < 110°$ was determined by a numerical integration of the transition function given in (46). Corresponding plots of the element pattern phase are shown in Fig. 10 with the reference taken at the element. The phase variation exhibited may be directly correlated with the amplitude variation shown in Fig. 8 and may be attributed to the same effects. In contrast to the isolated element case, the array element pattern does not exhibit a well-defined phase center over the entire lit region; this variation should be taken into account in array pattern computations and synthesis.

The example shown in Fig. 11 for $kr_0 = 105$ and $a/\lambda = 0.8$ is particularly interesting: the amplitude and phase of the contributions from the clockwise and counterclockwise fast creeping waves, when combined with the space wave, are such as to make the broadside realized gain 3 dB above that determined by the element unit cell, i.e., $2\pi a/\lambda$. This is accomplished at the expense of a much reduced beamwidth. Also, in contrast to the case of smaller spacings, the effect of the clockwise fast creeping wave can be seen after the "grating lobe" angle (15°) since it has undergone only a small attenuation from its origin near $-15°$.

Fig. 12 shows a comparison of the element patterns for $kr_0 = 105$ and various spacings. Plotting these patterns

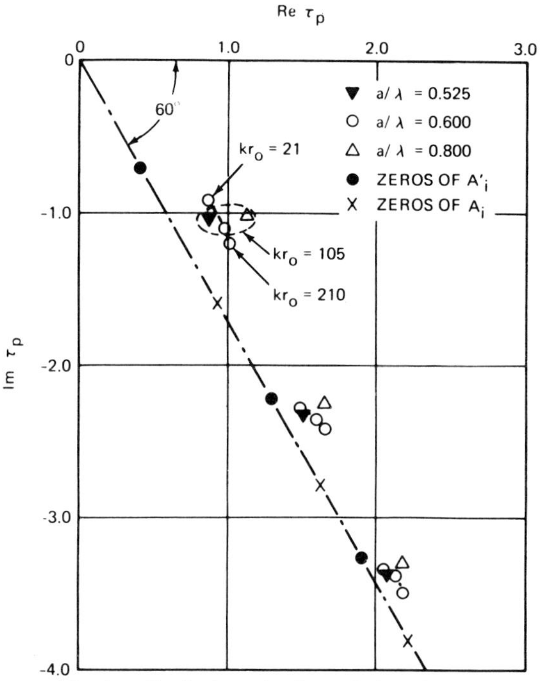

Fig. 7. Typical results for pole locations.

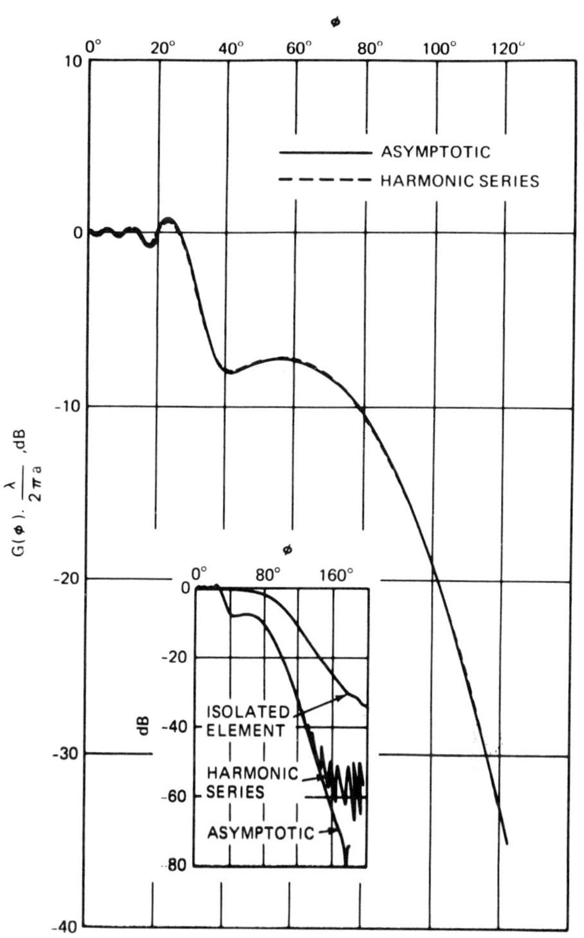

Fig. 8. Typical element pattern; $kr_0$ = 105, $a/\lambda$ = 0.6.

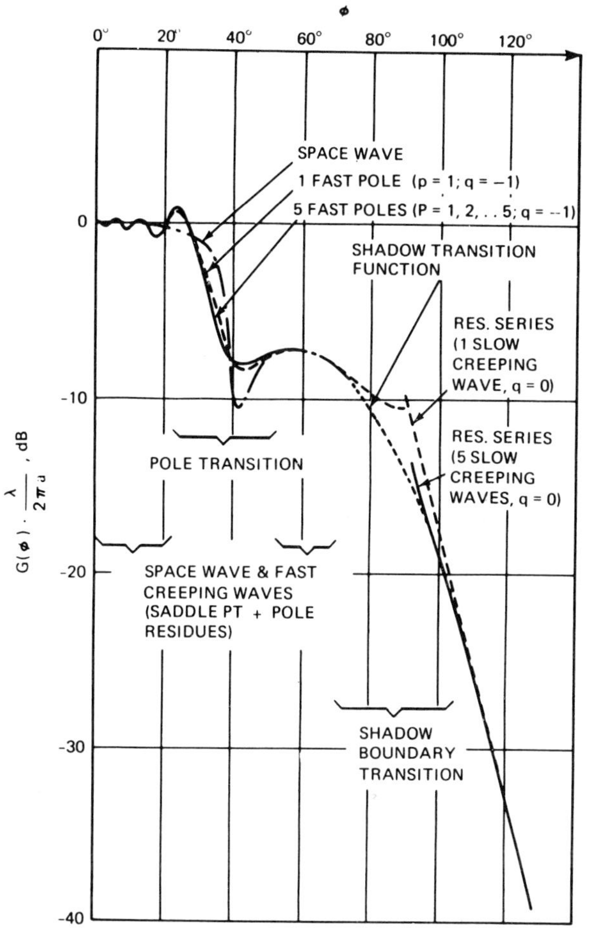

Fig. 9. Detailed breakdown of asymptotic results; $kr_0$ = 105, $a/\lambda$ = 0.6.



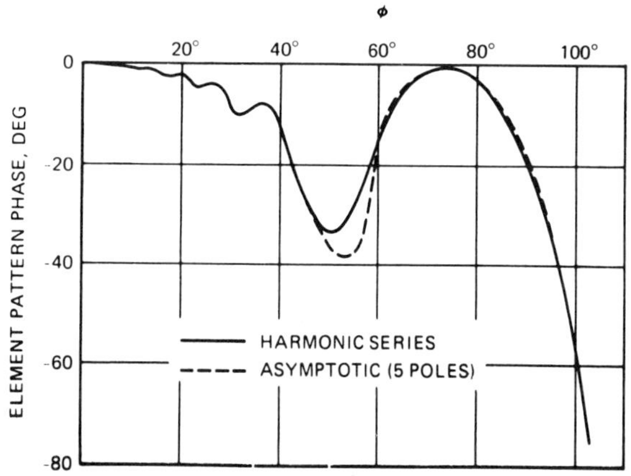

Fig. 10. Element pattern phase; $kr_0$ = 105, $a/\lambda$ = 0.6.

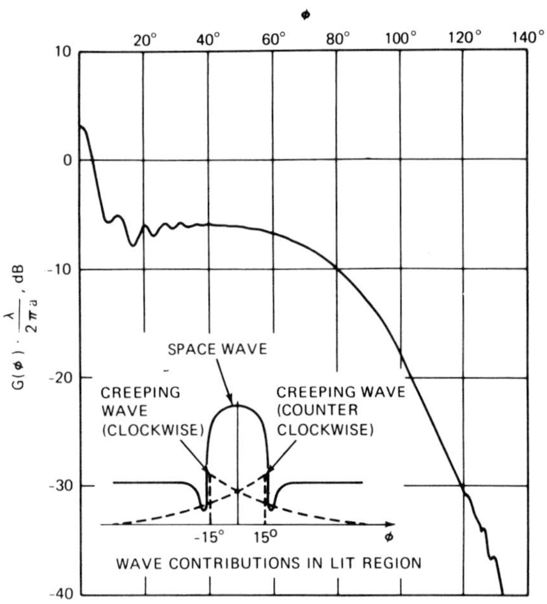


Fig. 11. Element pattern; $kr_0 = 105$, $a/\lambda = 0.8$.

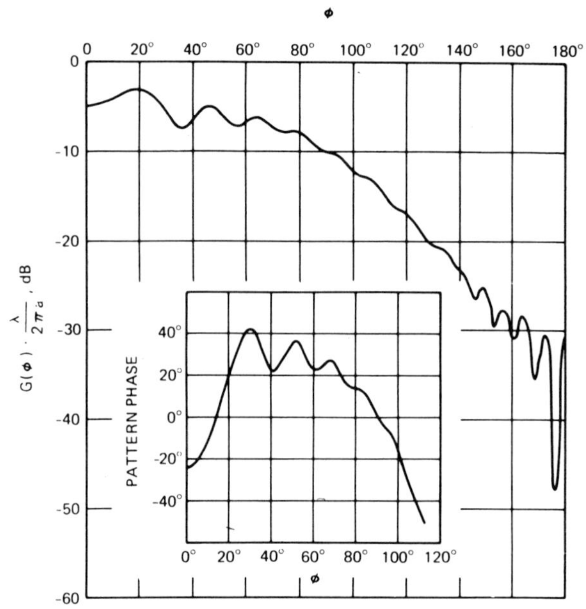


Fig. 13. Element pattern amplitude and phase; $kr_0 = 21$, $a/\lambda = 0.8$.

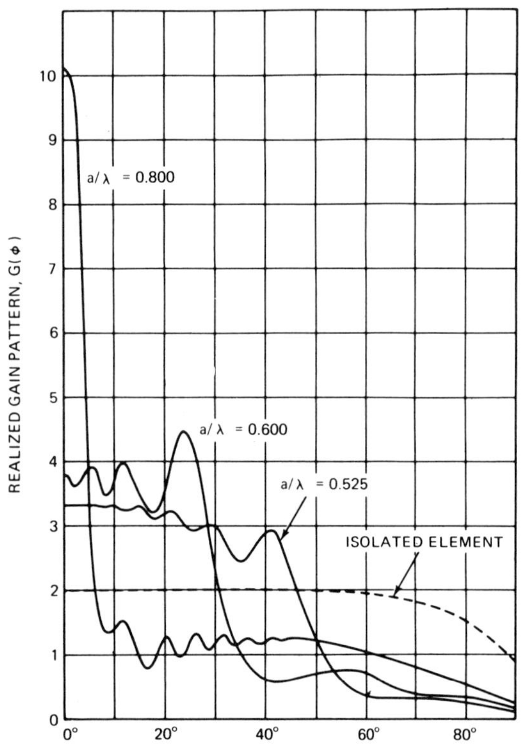


Fig. 12. Effect of spacing on element gain function; $kr_0 = 105$.

in terms of the absolute gain function provides an unusual perspective on the effect of spacing.

When the radius is changed, the above-mentioned effects are drastically modified. For the case $kr_0 = 21$ and $a/\lambda = 0.8$ shown in Fig. 13, the contribution of the fast creeping waves tends to be out of phase with the space wave resulting in a broadside realized gain approximately 4.5 dB lower than the unit cell gain. The main beam is correspondingly broader and the pattern phase varies by 65° within the main beam. In general, these unusual effects are characteristic of the larger spacings since $\Phi_c$ is small and therefore the fast creeping waves do not have to travel very much along the surface before radiating near broadside.

## V. Conclusions

Excellent agreement in numerical results has been obtained between the harmonic series and asymptotically evaluated element pattern of arrays of narrow slits on conducting cylinders, even for $kr_0$ as low as 21. The results show that the patterns in the array environment differ greatly from that of an isolated element.

In contrast to the harmonic series, the asymptotic expressions provide directly a physical interpretation of the results. In the lit region, the gross features of the pattern are determined by the space wave which appears to be emanating directly from the element. This contribution is proportional to an angle-dependent modal transmission coefficient $T(kr_0 \sin \Phi)$ and may be obtained by substituting the circular admittance crater into the appropriate planar array formulas. Near the usual grating-lobe angle, $\Phi = \sin^{-1}(\lambda/a - 1)$, $T(kr_0 \sin \Phi)$, and therefore the field pattern, exhibits a dip whose extent increases as $kr_0^{1/3}$; a rapid phase variation is also associated with this effect. The other contributors to the pattern in the lit (main lobe) region are fast creeping waves which are guided waves traveling along the impedance-matched and match-terminated array, induced by the periodicity of the structure and its curvature (hence, not present in the planar case). These creeping waves interfere with the space wave and result in a significantly rippled main beam. The appearance of such fast creeping waves in the lit region does not seem to have been previously exhibited in the literature. In most of the examples shown, the realized element gain is determined by the unit-cell size rather than the slit-aperture size. However, in contrast to large planar arrays, substantial variations from the unit cell gain can occur, due to the effects of fast creeping waves.

The element pattern in the shadow region is determined by a few (or possibly one) creeping waves, as in the case of an isolated element on a smooth cylinder. However, the rate of decay in the array case is greater, due to the additional loss mechanism introduced by the terminated elements.

It may be concluded that, in the design of a practical size circular array, neither the effect of the array environment, nor that of curvature in general can be neglected. In particular, the realized gain pattern provides a direct indication of the relative efficiency of various elements in a common direction, and the phase information shows that a correction term must be added for a true "co-phasal" operation of the array.

## References

[1] J. R. Wait, *Electromagnetic Radiation from Cylindrical Structures.* New York: Pergamon, 1959 (contains extensive bibliography).
[2] G. Hasserjian and A. Ishimaru "Currents induced on the surface of a conducting circular cylinder by a slot," *J. Res. Nat. Bur. Stand.*, sec. D., vol. 66, no. 3, pp. 33–35, 1962.
[3] ——, "Excitation of a conducting cylindrical surface of large radius of curvature," *IRE Trans. Antennas Propagat.*, vol. AP-10, pp. 264–273, May 1962.
[4] A. A. Oliner and R. G. Malech, "Mutual coupling in infinite scanning arrays," in *Array Theory and Practice*, vol. 2, (Microwave Scanning Antennas), R. C. Hansen, Ed. New York: Academic Press, 1966.
[5] J. D. Tillman, "The theory and design of circular antenna arrays," University of Tennessee Engineering Experiment Station, Knoxville, 1968.
[6] R. W. P. King, R. B. Mack, and S. S. Sandler, *Arrays of Cylindrical Dipoles.* London: Cambridge University Press, 1968.
[7] T. T. Taylor, "A synthesis method for circular and cylindrical antennas composed of discrete elements," *IRE Trans. Antennas Propagat.*, vol. AP-7, pp. 251–261, August 1952.
[8] G. Ziehm, "Optimum directional pattern synthesis of circular arrays," *Radio Electron. Eng.* (London), vol. 28, p. 341, November 1964.
[9] F. I. Tseng and D. K. Cheng, "Pattern synthesis of circular arrays with many directive elements," *IEEE Trans. Antennas Propagat.* (Communications), vol. AP-16, pp. 758–759, November 1968.
[10] G. V. Borgiotti and Q. Balzano, "Mutual coupling analysis of a conformal array of elements on a cylindrical surface," *IEEE Trans. Antennas Propagat.*, vol. AP-18, pp. 55–63, January 1970.
[11] J. C. Sureau and A. Hessel, "Element pattern for circular arrays of axial slits on large conducting cylinders," *IEEE Trans. Antennas Propagat.* (Communications), vol. AP-17, pp. 799–803, November 1969.
[12] L. B. Felsen and N. Marcuvitz, "Modal analysis and synthesis of electromagnetic fields," Microwave Research Institute, Polytechnic Institute of Brooklyn, Brooklyn, N. Y., Rep. PIBMRI-1083-62, November 1962.
[13] S. Edelberg and A. A. Oliner, "Mutual coupling effects in large antenna arrays, pt I.: slot arrays," *IRE Trans. Antennas Propagat.*, vol. AP-8, pp. 286–297, May 1960.
[14] M. Abramowitz and I. A. Stegun, Eds., *Handbook of Mathematical Functions* (Applied Mathematics Series 55). Washington, D. C.: NBS, 1964.
[15] H. A. Wheeler, "The grating-lobe series for the impedance variation in a planar phased-array antenna," *IEEE Trans. Antennas Propagat.* (Communications), vol. AP-13, pp. 825–827, September 1965.
[16] S. Barone, "On the Green's function for a circular cylinder," Tech. Res. Group, Rep. 121-SR-2, ASTIA Doc. AD 245 423, January 1960.
[17] L. L. Bailin and R. J. Spellmire, "Convergent representations for the radiation fields from slots on large circular cylinders," *IRE Trans. Antennas Propagat.*, vol. AP-5, pp. 374–383, October 1957.
[18] M. Born and E. Wolf, *Principles of Optics.* New York: Pergamon, 1959, pp. 749–750.
[19] B. Van der Waerden, "On the method of saddle points," *Appl. Sci. Res.*, vol. B2, pp. 33–45, 1951.
[20] J. R. Wait and A. M. Conda, "Pattern of an antenna on a curved lossy surface," *IRE Trans. Antennas Propagat.*, vol. AP-6, pp. 348–359, October 1958.



*Reprinted by permission from* IEEE TRANSACTIONS ON ANTENNAS AND PROPAGATION
Vol. AP-19, No. 1, January 1971
pp. 64-74

# MISCELLANEOUS SYSTEM ASPECTS VI

PROCEEDINGS OF THE IEEE, VOL. 56, NO. 11, NOVEMBER 1968

# Phased-Array Beam Steering by Multiplex Sampling

MAJOR A. JOHNSON, SENIOR MEMBER, IEEE

***Abstract*—One approach to phased-array antenna beam forming and steering is to multiplex the element signals into a single channel. Appropriate "sampling" of the resulting multiplexed signal can provide electronically steered and shaped beams. This paper describes four practical system concepts for linear and ring arrays based on this approach and discusses significant interrelationships between the various concepts.**

**For linear arrays of elements, the two alternatives are to frequency or time multiplex the element signals. Frequency multiplexing of the element signals produces time-multiplexed beam output signals, and time multiplexing the element signals produces frequency-multiplexed beam output signals. It is also shown here that appropriate correlation "sampling" may be used with either of these to produce easily one or more continuously and electronically steered signal bandwidth beam outputs.**

**Ring arrays of elements may be multiplexed and "sampled" in a somewhat analogous pair of techniques. It is also pointed out that the beam steering in this case may be visualized as the linear phase steering of a set of linear phase modes into which the signal received at the array may be resolved. Ring array beam forming and steering may thus be directly understood in terms of the previous linear array techniques.**

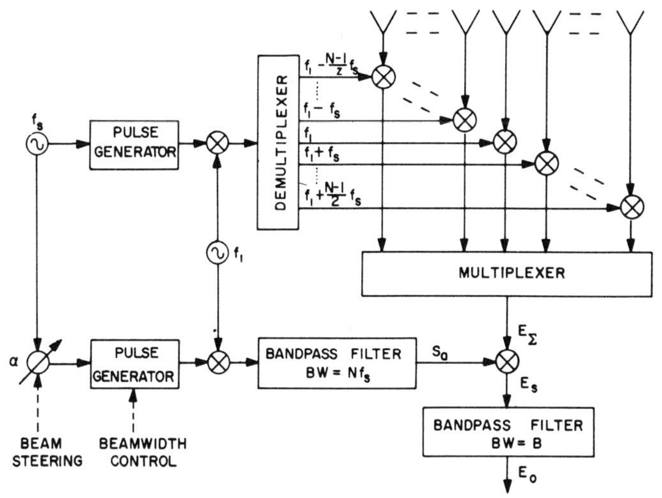


Fig. 1. Frequency-multiplexed linear array.

## Introduction

The CLASSICAL methods of beam forming and beam steering for array antennas use either elemental control elements, i.e., controllable phase shifters, or time-delay elements, or element interconnection matrices which produce a number of simultaneous beam outputs. In the latter case, beam steering becomes a matter of choosing and switching to the particular beam outputs desired.

Consideration of rapid scanning, especially for "receive only" arrays, has led to a number of beam forming and steering techniques which may be thought of as element signal multiplexing techniques. This paper will discuss four of these which are applicable to linear and ring arrays, and which can economically provide many simultaneous phase-steered beams. Extensions of these techniques can also be made to planar or surface arrays. These will not be described at this time, however.

## Frequency-Multiplexed Linear Array

Repeated rapid scanning of the beam pattern of a linear antenna array may be achieved by appropriate frequency multiplexing of the elemental signals. Although this basic scanning approach has been described in patents [1], [2] and in the literature [3], the use of sampling with this system has been described only in classified reports [4]. A review of the system operated with sampling is in order here, since it is most direct in concept and will serve as a good comparison for the other systems to be discussed.

Manuscript received March 18, 1968; revised June 12, 1968.
The author is with the Heavy Military Electronics Department, General Electric Company, Syracuse, N. Y. 13201

This basic system—which has been called MOSAR in General Electric Company work—is illustrated in Fig. 1.

The received element signals $E_n$ from an equispaced linear array are frequency multiplexed by means of a set of "scanning" local oscillator signals (SCALO's) which are spaced by a fixed frequency $f_s$. Addition of this set of SCALO frequencies, which increases linearly across the array, to the received element signals is equivalent to impressing a phase rate which increases linearly across the array, causing a repetitive scan of the array beam at a frequency $f_s$. This may be shown mathematically as follows.

We shall write the element signals as

$$E_n = We_n f(t) e^{j(\omega_0 t + nkd\eta)}$$

where $We_n$ represents applied elemental attenuation or weighting; $f(t)$ and $\omega_0$ are the modulation and carrier angular frequency, respectively, of the incoming wave $k = 2\pi/\lambda$; $d$ is the element-to-element spacing; and $\eta$ is the cosine of the angle between the source direction and the line of the array. For convenience, we shall consider that the array has an odd number $N$ of elements, with $n$ taken as zero for the center element and running therefore in integral values from $-(N-1/2)$ to $N-1/2$ across the array.

Due to the coherent method of generating the SCALO frequencies as illustrated, we may write the $n$th SCALO signal as

$$Sc_n = e^{j(\omega_1 + n\omega_s)t}.$$

Considering only the first-order sum frequency output of the mixers, the output of the $n$th mixer will be

$$We_n f(t)e^{j(\omega_0+\omega_1)t}e^{jn(kd\eta+\omega_s t)}$$

so that the multiplexer output will be[1]

$$E_\Sigma = f(t)e^{j(\omega_0+\omega_1)t}\sum We_n e^{jn(kd\eta+\omega_s t)}.$$

The waveform envelopes and spectra of these various signals are illustrated in Figs. 2 and 3 for a five-element array.

For the case of $We_n$ equal to unity we may also write

$$E_\Sigma = f(t)e^{j(\omega_0+\omega_1)t}\frac{\sin\frac{N}{2}(kd\eta+\omega_s t)}{\sin\frac{1}{2}(kd\eta+\omega_s t)}.$$

This may be seen to be a sin $Nx$/sin $x$ angular beam which repetitively scans with time at a repetition rate $f_s$ as discussed.

If this scanning process is to preserve all possible far-field information, we must make $f_s$ greater than the total RF bandwidth $B$ resulting from the modulation $f(t)$. We must also band limit the element signals to $B$ before addition to avoid noise overlap, as shown for the scanning signal section of Fig. 3. This would only be good multiplexing practice, of course, but may also be considered as satisfying the Shannon sampling criterion, where we are talking about the scanning beam "sampling" each far-field direction at a rate $f_s$. This provides sufficient samples to regenerate the $f(t)$ envelope as shown for the scanning signal section of Fig. 2.

The price we have paid for multiplexing all this information into one channel is, of course, bandwidth, since the total bandwidth of the "scanning" signal is obviously close to $Nf_s$ in total bandwidth and this must be made greater than $NB$. For modern radar application, this can be several gigahertz and this makes it obvious why the next step—beam sampling—is important. Modern phased-array radars most often interface directly with digital computers, and to date at least, several gigahertz is too much for practical computers. Several steerable beam outputs are desirable, each output having only the bandwidth $B$. Beam sampling can provide this form of output; Fig. 1 illustrates the genera-
248
tion of one beam output by sampling.

The sampling signal $S_a$ can be thought of as a train of short pulses, at a repetition rate $f_s$ which can be time phased to gate the scanning signal according to the beam direction desired. Practically speaking, these sampling pulses will be band limited to $Nf_s$ and so the mathematical analysis is more rigorous if we continue in terms of frequency components and write

$$S_a = e^{j\omega_1 t}\sum_n Ws_n e^{jn(\omega_s t+\alpha)}$$

where $\alpha$ is the beam-steering phase noted in Fig. 1, and $Ws_n$ are relative amplitudes of the sampling signal spectral lines.

We shall consider the sampling mixer to be an ideal time

[1] The limits of all summations used in this paper will be from $-(N-1)/2$ to $(N-1)/2$ corresponding to $N$ terms. We will omit these limits for simplicity.

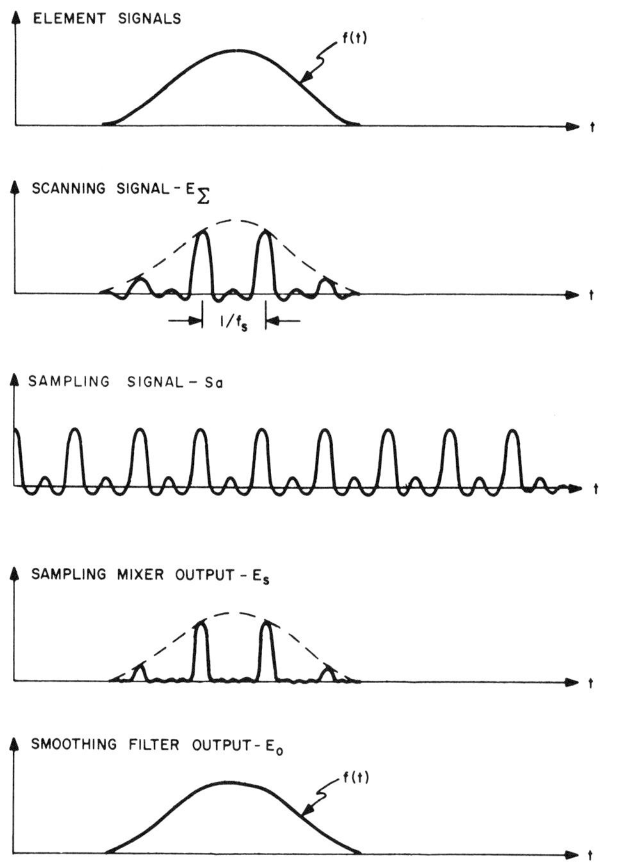


Fig. 2. MOSAR waveform envelopes.

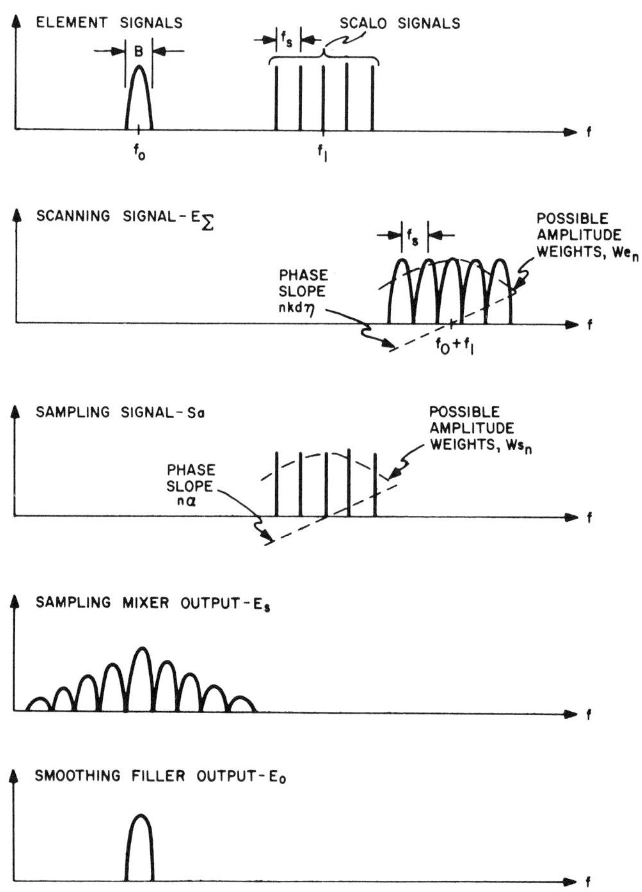


Fig. 3. MOSAR spectra.

multiplexer and we will only take the difference frequency output, which in our complex nomenclature corresponds to taking the product of one input times the complex conjugate of the other. The output of the sampling mixer will therefore be given by

$$E_s = E_\Sigma \times S_a^*$$
$$= f(t)e^{j\omega_0 t}\left[\sum_n We_n e^{jn(kd\eta + \omega_s t)}\right]\left[\sum_n Ws_n e^{-jn(\omega_s t + \alpha)}\right]$$
$$= f(t)e^{j\omega_0 t}\sum_n\sum_m We_n Ws_m e^{j[n(kd\eta + \omega_s t) - m(\omega_s t + \alpha)]}.$$

This expression is somewhat complicated, but we are going to follow the sampling mixer with a bandpass filter, of bandwidth $B$, centered about $\omega_0$. Thus, we are only interested in the terms of the double sum which corresponds to a fixed phase shift of the $\omega_0$ carrier; that is, those for which $m = n$ so that all $\omega_s t$'s in the exponents disappear. All others will be sufficiently filtered out by this final filter (which we shall call the smoothing filter) so that they may be ignored.[2] The resulting filter output thus becomes

$$E_0 = f(t)e^{j\omega_0 t}\sum_n We_n Ws_n e^{jn(kd\eta - \alpha)}.$$

This may be seen to be of the same form as the output of a simple linear phased array using phase shifters set to $-n\alpha$ and with element attenuation values which are the product $We_n Ws_n$, contributed in the MOSAR system by the element weighting and the sampling spectrum. For the simple case of $We_n Ws_n = 1$, this would give us

$$E_0 = f(t)e^{j\omega_0 t}\frac{\sin\frac{N}{2}(kd\eta - \alpha)}{\sin\frac{1}{2}(kd\eta - \alpha)},$$

i.e., a beam steered to

$$\eta_0 = \frac{\alpha}{kd}.$$

Figs. 2 and 3 illustrate the sampling mixer and smoothing filter outputs for $\alpha = kd\eta$, i.e., the main beam steered to the direction of the received signal.

In practice, one would probably let $We_n Ws_n = T_n$, a set of elemental Chebyshev or Taylor weights so as to produce a beam having low near-in sidelobes. It may also be noted that the resulting beam may be broadened by electronically broadening the sampling pulse generator. This action serves to further taper the sampling weights $Ws_n$, and produces the desired resultant broadened beam.

Generation of a number of simultaneous beams requires only a duplication of sampling pulse generators and sampling mixers and provides an economical means of building a system with multiple independent beam-steering requirements.

[2] We must note here that the output carrier frequency, which corresponds here to $\omega_0$, must be greater than $N$ times the scanning frequency to keep video frequency components in the sampling mixer output below the smoothing filter passband.

## Time-Sampled Linear Array

It has been shown that frequency multiplexing of the elements results in a time-multiplexed beam signal. Shanks [5] has shown that the reverse is also true: time multiplexing of the element signals results in a frequency-multiplexed beam signal. It is interesting to inquire whether sampling can be applied to this conceptual system in a similar manner and we shall show that this can be done.

Fig. 4 illustrates the basic components of this system. We illustrate the element sampling as done by mixers using a pulse-modulated carrier. This is not necessary—Shanks' implementation used RF switches—but mixers can do the job quite readily and will serve to maintain our analogy with the previous system. We may analyze the operation of this system as follows.

The $n$th element signal, as before, may be written as

$${}_nE_e = We_n f(t)e^{j(\omega_0 t + nkd\eta)}$$

where $We_n$ represents element weighting which could be inserted by elemental attenuators. The pulsed local oscillator (LO) signal is represented as

$$E_{LO} = \sum_m a_m e^{j(\omega_1 + m\omega_s)t}.$$

This is a form similar to the sampling signal of the previous system. If we reference time to the center of the tapped delay line, the pulsed LO signal at the $n$th tap will then be

$${}_nE_{LO} = \sum_m a_m e^{j(\omega_1 + m\omega_s)(t - n\tau)}$$

where $\tau$ is the delay between successive taps. We wish to choose the total delay equal to the period of the sampling pulse train so as to provide samples as often as possible without starting a new set of samples before the old is finished; therefore, we shall take

$$N\tau = \frac{1}{f_s} \quad \text{or} \quad \omega_s\tau = \frac{2\pi}{N}.$$

In order to avoid any phase steering effects in our analysis due to $f_1$, we shall choose $f_1$ so that $f_1\tau$ is an integral number of cycles so that $\omega_1\tau = 2\pi p$, where $p$ is any integer. In an actual system, a variation of $f_1\tau$ could be calibrated out, but the most direct solution to this may be to phase lock the $f_1$ oscillator to the delay line. With these substitutions, the pulsed LO signal at the $n$th tap may be written

$${}_nE_{LO} = e^{j\omega_1 t}\sum_m a_m e^{jm[\omega_s t - (2\pi n/N)]}.$$

The sum frequency output of the $n$th element sampling mixer will be given by

$${}_nE_1 = ({}_nE_{LO}) \times ({}_nE_e)$$
$$= We_n f(t)e^{j(\omega_0 + \omega_1)t}\sum_m a_m e^{jm\omega_s t}e^{jn[kd\eta - (2\pi m/N)]}.$$

The adder output will, therefore, be

$$E_\Sigma = \sum_n {}_nE_1$$
$$= f(t)e^{j(\omega_0 + \omega_1)t}\sum_m a_m e^{jm\omega_s t}\sum_n W_{en} e^{jn[kd\eta - (2\pi m/N)]}.$$

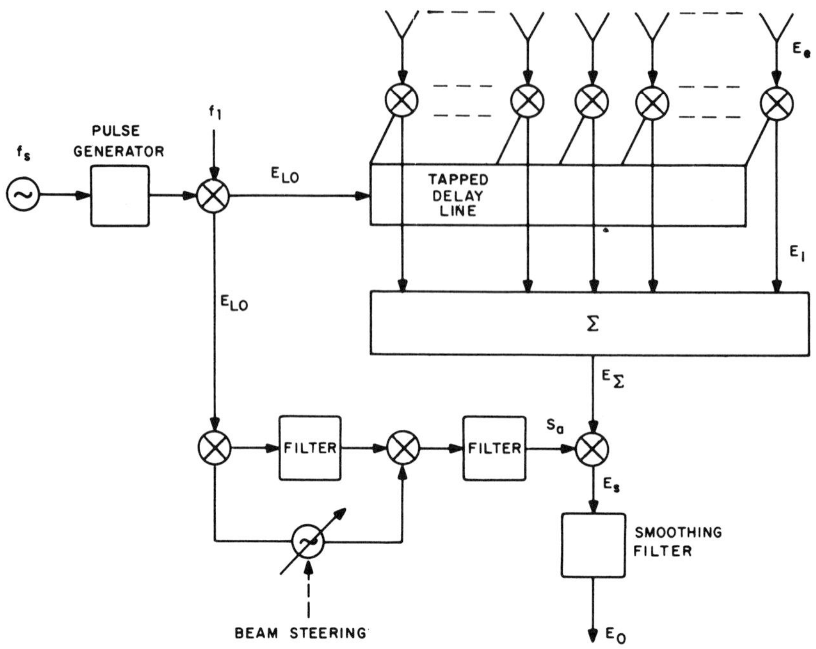


Fig. 4. Time-sampled array.

For $We_n = 1$, this may be seen to be equal to

$$E_\Sigma = f(t)e^{j(\omega_0+\omega_1)t} \sum_m a_m e^{jm\omega_s t} \frac{\sin N/2\left[kd\eta - \dfrac{2\pi m}{N}\right]}{\sin 1/2\left[kd\eta - \dfrac{2\pi m}{N}\right]}.$$

In essence, we have modulated the line spectrum of a sampled signal with the antenna pattern function. As the target direction $\eta$ increases, the frequency of maximum spectral content increases, and vice versa. Each of the spectral lines, in fact, corresponds to a particular beam direction, these directions occurring at intervals of

$$kd\Delta\eta = \frac{2\pi}{N} \quad \text{or} \quad \Delta\eta = \frac{\lambda}{Nd}$$

which is the distance to the first null of the $\sin Nx/\sin x$ pattern and corresponds to a Shannon sampling interval of the far-field pattern coordinate $\eta$. Consequently, we would expect that full gain could be realized for signals arriving from any direction $\eta$, whether it corresponds exactly to one of the spectral lines or not. We shall see further when we complete the sampling analysis that this is so.



One way of visualizing the formation of the "frequency scanning" signal we have obtained is to consider separately the effect of each of the spectral components of the pulsed LO signal. Note that each spectral component produces a phase slope across the taps of the delay line which corresponds to one of the particular beam directions we have noted; that is, it exactly compensates for the phase slope of a signal received from that direction. Consequently, the mixer output frequency component corresponding to that LO spectral line will be in phase for all mixers for a signal arriving at the array from that particular direction, and will produce a maximum output from the signal adder at that frequency. Signals arriving from other directions will produce less output at that frequency, and may produce greater output at one of the other output frequencies, which corresponds to a more favorable phase slope of one of the LO spectral lines.

Now that we have obtained the "scanning" signal, let us further consider what could be done here which would be equivalent to the MOSAR time sampling process; that is, how we would achieve a single-beam output signal which is electronically steerable in a continuous manner. Perhaps the best way to reason through this is to consider that the sampling process of that system is one of generating a sampling signal which corresponds in general format to the scanning signal for a continuous received carrier, and then correlating this with the scanning signal by the use of a mixer and narrow-band filter. Following this approach, we should use a "sampling" signal here which might be of the form.

$$S_a = e^{j\omega_1 t} \sum_m e^{jm\omega_s t} \frac{\sin N/2\left[\alpha - \dfrac{2\pi n}{N}\right]}{\sin 1/2\left[\alpha - \dfrac{2\pi n}{N}\right]}$$

or, more generally,

$$S_a = e^{j\omega_1 t} \sum_m e^{jm\omega_s t} \sum_n W_{sn} e^{jn[\alpha - (2\pi m/N)]}$$

where $\alpha$ is a desired element-to-element steering phase, and $W_{sn}$ is some impressed element-to-element weighting.

The difference frequency output of the steering mixer would then be

$$\begin{aligned} E_s &= E_\Sigma \times S_a^* \\ &= f(t)e^{j\omega_0 t} \sum_n \{We_n W s_n e^{jn(kd\eta-\alpha)} \\ &\quad \times [\sum_m e^{jm[\omega_s t - (2\pi n/N)]}][\sum_m e^{-jm[\omega_s t - (2\pi n/N)]}]\} \\ &= f(t)e^{j\omega_0 t} \sum_n \{We_n W s_n e^{jn(kd\eta-\alpha)} \sum_r \sum_s e^{j(r-s)[\omega_s t - (2\pi n/N)]}\}. \end{aligned}$$

To complete our story, if the "sampling" mixer output is passed through the smoothing filter, only those terms of the double sum for which $r$ equals $s$ will be passed, so that the smoothing filter output will be given by

$$E_0 = Nf(t)e^{j\omega_0 t} \sum_n We_n Ws_n e^{jn(kd\eta - \alpha)}$$

or for the case of $We_n Ws_n = 1$,

$$E_0 = Nf(t)e^{j\omega_0 t} \frac{\sin \frac{N}{2}(kd\eta - \alpha)}{\sin \frac{1}{2}(kd\eta - \alpha)}$$

which is of the same form as for the simple phase-shifter array. Thus, we see that our correlation approach did yield the desired result, a beam which could be continuously steered between the particular beam positions associated with the "frequency scanning signal" spectral lines.

Fig. 4 illustrates a method of generating a sampling signal of the desired type. The pulsed LO spectrum is frequency offset by mixing with a controllable frequency oscillator and is passed through a suitable bandpass filter. If, after filtering, the resulting spectrum is returned to its original frequency position by mixing with the same controllable frequency oscillator output, a signal is produced which has the desired characteristics. It has the basic spectral character of a set of fixed lines, but these are modulated (by the fixed filter) in a bandpass manner which may be continuously steered in frequency by the controllable oscillator. The bandpass filter shape provides indirectly the sampling weights $Ws_n$ for beam shaping.

Perhaps a few words about the comparative application and/or implementation of this system against the frequency-multiplexed element system are in order. In the first place, the choice of $f_s$ is governed by exactly the same considerations; that is, we must have $f_s \geq B$. If we want to scan the full angle of space we note that the number of independent beams (Shannon samples of $\eta$) for a linear array is equal to the number of elements, and therefore the pulsed LO signal for the sampled element system must have the same number of lines $N$; consequently, the LO spectrum, the scanning signal, and the sampling signal have the same bandwidth as for MOSAR. If only a reduced angular sector need be scanned, however, we need only use sufficient LO spectral lines to cover that sector, and scanning and sampling signal bandwidths may be reduced correspondingly. An advantage of the time-sampled linear array technique lies in the fact that the elements of the array need not be equally spaced, provided that the LO delay taps have delays proportional to the position of the corresponding elements, as measured from the end of the array. However, due to the phase variation of the LO carrier frequency $f_1$ with tap spacing, a set of fixed elemental phase compensators will be required.

The principal new components for this system are the tapped LO delay line, the steering control oscillator, and the beam shaping filter. These correspond to the SCALO demultiplexer, the steering phase shifter, and the controlled pulse generator of the frequency-multiplexed element system. The multiplexer of that system is here replaced by a corporate adder, but since elemental band limiting is still necessary, the complexity of the elemental components is little changed.

The tapped delay line does not appear to offer any special implementation difficulty since, for a wide-band system, it would not be very long and could be constructed of coiled coaxial line or waveguide. The steering oscillator is straightforward in principle, but the problem of accurately relating beam position to the steering control signal is somewhat more complex. The beam-shaping filter would perhaps provide the greatest difference; a more detailed analysis of the use of different filter types has not been made. Of course, an $N$-tap transverse filter would do the required job exactly, and although this may seem overly complicated for a linear array, it might be quite reasonable for a large planar array. This system does not seem to have the same beam-shaping ease as the frequency-multiplexed element system where a simple pulse width change did the trick. However, a reasonable design of a variable-width bandpass filter may be quite feasible with sufficient ingenuity applied. In summary, then, it appears that the sampled element system might find applicability in similar kinds of usage and lacks primarily the detailed investigation which has been accorded the MOSAR approach.

## Basic Ring Array Patterns

Since pattern expressions for ring arrays may be less familiar than for linear arrays, we shall spend a little time on the basic patterns for phased ring antennas. For mathematical simplicity, we shall consider the ring to be continuous rather than an array of discrete elements [6]. We shall use the basic coordinate system of Fig. 5. Note that $\phi$ represents the angular position of a point on the ring, while $\psi$ represents the angular position of some far-field point $P$ as measured in the plane of the ring.

If a signal of the form $f(t)e^{j\omega_0 t}$ is incident on the ring from the direction $(\theta, \psi)$ and we reference its phase to the center of the ring, we may note that the signal received at any point on the ring is given by

$$E_r = f(t)e^{j\omega_0 t}e^{jkR \sin\theta \cos(\psi - \phi)}$$

where $k = 2\pi/\lambda$ as before.

In order to maximize the response of this ring antenna to a signal from $(\theta_0, \psi_0)$, we should be sure that the polarization of the receiving elements is matched to that of the signal, and we should shift the phase of each element signal by $-kR \sin\theta_0 \cos(\psi_0 - \phi)$ and add the resulting signals. Representing this summation by an integral around the ring, we would have

$$E_\Sigma = f(t)e^{j\omega_0 t} \int e^{jkR[\sin\theta \cos(\psi - \phi) - \sin\theta_0 \cos(\psi_0 - \phi)]} d\phi$$

where the limits of integration are 0 to $2\pi$.

Referring to Fig. 6 for the definition of $\Delta\eta$ [7] and $\beta$, the bracketed part of the exponent in this expression may be

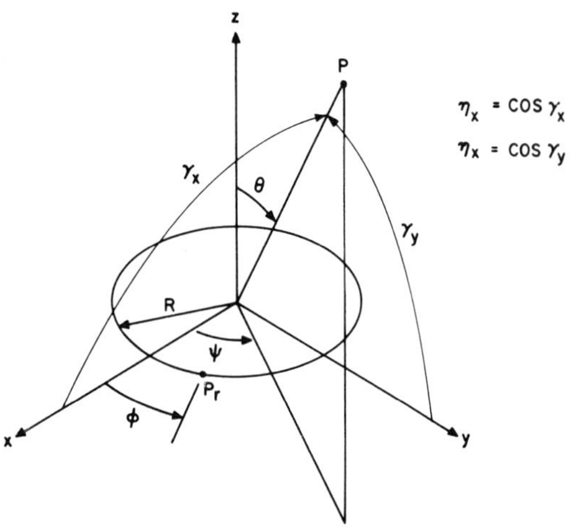


Fig. 5. Ring coordinate system.

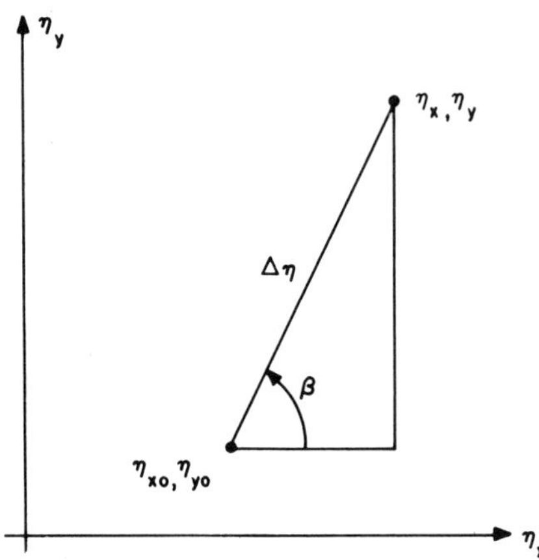


Fig. 6. $\eta$-plane relationships.

written

$$[\ ] = \Delta\eta \cos(\beta - \phi)$$

so that the array response may therefore be expressed as

$$E_\Sigma = f(t)e^{j\omega_0 t} \int e^{jkR\Delta\eta \cos(\beta-\phi)} d\phi$$



or

$$E_\Sigma = 2\pi f(t)e^{j\omega_0 t} J_0(kR\Delta\eta)$$

which is a pattern of circular symmetry in the direction cosine plane ($\eta$-plane), as would be expected.

Since ring array steering is generally in terms of the spherical coordinates $(\theta, \psi)$, we will be interested in writing this pattern as a function of those coordinates. We shall do this by noting that

$$\Delta\eta = \sqrt{\sin^2\theta + \sin^2\theta_0 - 2\sin\theta\sin\theta_0\cos(\psi - \psi_0)}$$

or

$$\Delta\eta = g(\theta, \theta_0, \psi - \psi_0)$$

so that

$$E_\Sigma = 2\pi f(t)e^{j\omega_0 t} J_0[kRg(\theta, \theta_0, \psi - \psi_0)],$$

the direct dependence on $\psi - \psi_0$ being important for some of the systems we will discuss.

The principal response pattern cut in the $\theta$-direction (a radial in the $\eta$-plane) results from letting $\psi = \psi_0$ so that

$$\Delta\eta = \sin\theta - \sin\theta_0$$

and

$${}_\theta E_\Sigma = 2\pi f(t)e^{j\omega_0 t} J_0[kR(\sin\theta - \sin\theta_0)].$$

The principal response pattern cut in the $\psi$-direction (a concentric arc in the $\eta$-plane) results from letting $\theta = \theta_0$, so that

$$\Delta\eta = 2\sin\theta_0 \sin\frac{\psi - \psi_0}{2}$$

and

$${}_\psi E_\Sigma = 2\pi f(t)e^{j\omega_0 t} J_0\left[2kR\sin\theta_0\sin\frac{\psi - \psi_0}{2}\right].$$

The $J_0$ response patterns obtained here may be noted to have rather high sidelobes, and we shall discuss later the possibilities for sidelobe reduction in ring array antenna systems. These pattern expressions will be adequate, however, for illustrating the method of operation of the various ring array steering systems.

## Gitzendanner Scan System

The basic MOSAR scanning approach was seen to result from the idea of providing a suitable set of local oscillator signals to elemental mixers of a linear array so that the summed mixer output would be a continuously scanning beam. Such an approach is also directly applicable to a ring array and was first proposed by Gitzendanner [8] of the General Electric Company's General Engineering Laboratory, and was later investigated by Applebaum [9] of the General Electric Company's Heavy Military Electronics Department.

To illustrate the Gitzendanner scanning idea mathematically, we note as before that the signal received at the elements of the ring may be written as

$$E_r = f(t)e^{j\omega_0 t}e^{jkR\sin\theta\cos(\psi-\phi)}.$$

We are interested in producing a beam which scans linearly and continuously in $\psi$, that is, around the axis of the ring. Consequently, we would expect that the LO signal should have a sinusoidal phase modulation of the form $-kR\sin\theta\cos\omega_s t$, where $\omega_s$ is the angular scanning rate. Because of the ring symmetry, the local oscillator for each element must be of the same form except for the modulation phase timing; that is, an element further around the ring in the direction of the scan must have its peak phase modulation at a later time. We can provide the required LO signals by the implementation shown in Fig. 7. The input LO signal we shall write as

$$E_{LO} = e^{j\omega_1 t}e^{-jkR\sin\theta_0\cos\omega_s t}.$$

This LO signal is distributed to the various elemental

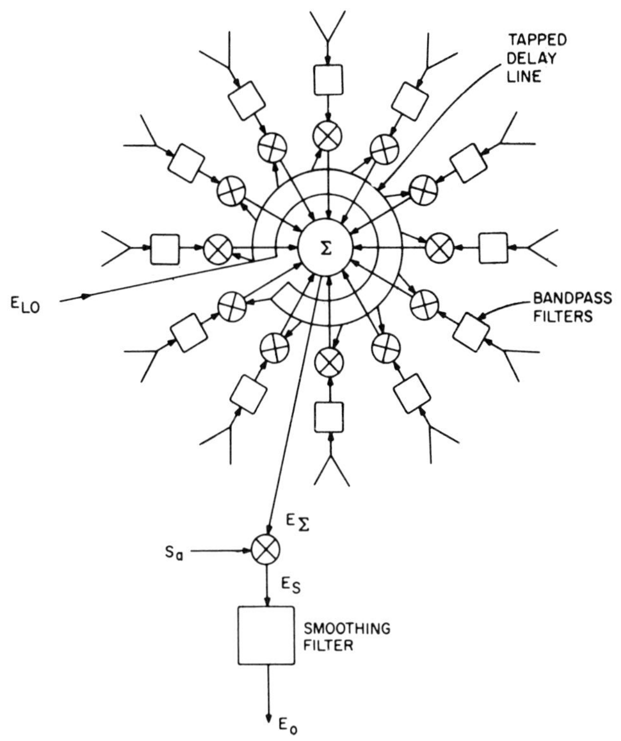


Fig. 7. Ring array beam-forming system.

mixers by a tapped delay line having a total delay equal to $1/f_s$, so that the delay to an element at $\phi$ is given by $\tau = \phi/\omega_s$, thus, the delayed LO signal at an element at position $\phi$ is given by

$$_{\tau}E_{LO} = e^{j\omega_1[t-(\phi/\omega_s)]}e^{-jkR\sin\theta_0\cos(\omega_s t-\phi)}$$

As with the sampled element system, we would like the LO carrier phase at the element taps to be equal, so we shall choose an LO carrier such that

$$\omega_1 = pN\omega_s$$

where $N$ is the number of elemental taps and $p$ is any integer. Then

$$e^{j(\omega_1/\omega_s)\phi} = e^{jpN(2\pi n/N)} = e^{j2\pi pn} = 1$$

and we shall therefore write

$$_{\tau}E_{LO} = e^{j\omega_1 t}e^{-jkR\sin\theta_0\cos(\omega_s t-\phi)}.$$

The mixer outputs will then be given by

$$\begin{aligned} E_1 &= E_r \times ({}_{\tau}E_{LO}) \\ &= f(t)e^{j(\omega_0+\omega_1)t}e^{jkR[\sin\theta\cos(\psi-\phi)-\sin\theta_0\cos(\omega_s t-\phi)]} \end{aligned}$$

Approximating the mixer output summation by an integral around the ring, we have

$$E_\Sigma = f(t)e^{j(\omega_0+\omega_1)t}\int_0^{2\pi} e^{jkR[\sin\theta\cos(\psi-\phi)-\sin\theta_0\cos(\omega_s t-\phi)]}d\phi.$$

We see that this integral is of the same form as the one we had for the phased ring with $\psi_0$ replaced by $\omega_s t$; therefore, we have a pattern at a fixed value $\theta_0$, but with $\psi_0$ scanning linearly at a rate $\omega_s$, which may be written

$$E_\Sigma = 2\pi f(t)e^{j(\omega_0+\omega_1)t}J_0[kRg(\theta, \theta_0, \psi - \omega_s t)].$$

Although the pattern is represented by a different function, the nature of this waveform is similar to that of a linear array MOSAR. In this case, however, the scan is linear in the real angle $\psi$, corresponding to a linear scan around a concentric circle in the $\eta$-plane.

By analogy with the previous systems, we would expect that the elemental signals must be band limited and that we must have $f_s \geq B$ to avoid information loss, although this expression does not make this clear. We might also expect that the scanning signal bandwidth would be too great for further processing and that sampling would be desirable. To get a better feeling for these bandwidth effects and the sampling process, let us re-derive the $E_\Sigma$ function in complex Fourier series form so as to be able to evaluate the spectral form of the various functions.

We shall start by expanding the ring element signal expression as follows:[3]

$$\begin{aligned} E_r &= f(t)e^{j\omega_0 t}e^{jkR\sin\theta\cos(\psi-\phi)} \\ &= f(t)e^{j\omega_0 t}\sum j^n J_n(kR\sin\theta)e^{jn(\psi-\phi)} \end{aligned}$$

Similarly, the delayed LO signal may be expanded to

$$_{\tau}E_{LO} = e^{j\omega_1 t}\sum j^n J_n(-kR\sin\theta_0)e^{jn(\omega_s t-\phi)}.$$

The summation of the mixer outputs may therefore be written as

$$\begin{aligned} E_\Sigma &= \int_0^{2\pi}[E_r \times ({}_{\tau}E_{LO})]d\phi \\ &= f(t)e^{j(\omega_0+\omega_1)t}\sum_m\sum_n j^{m+n}J_m(kR\sin\theta)J_n(-kR\sin\theta_0) \\ &\times e^{j(m\psi+n\omega_s t)}\int_0^{2\pi} e^{-j(m+n)\phi}d\phi. \end{aligned}$$

The integral in this expression is zero for all terms of the double sum except for $m = n$ where the value of the integral is $2\pi$. Since $J_{-n}(-x) = J_n(x)$, we have

$$E_\Sigma = 2\pi f(t)e^{j(\omega_0+\omega_1)t}\sum_n J_n(kR\sin\theta)J_n(kR\sin\theta_0)e^{jn(\omega_s t-\psi)}$$

which is a Fourier expansion of the previously obtained expression.

Let us now examine our bandwidth questions in light of these expressions. The LO signal and scanning signal have line spectra with the lines spaced $f_s$ apart, as would be expected from the basic repetitive scanning. Consequently, we must have the $f_s$ greater than the $f(t)$ bandwidth to avoid information loss by spectral overlap in the mixer outputs and scanning signal. The bandwidth of the LO and scanning signals in these expressions is seen to depend upon the magnitude of the Bessel functions involved as a function of their index and argument. For very large argument ($kR\sin\theta_0$) the number of significant spectral components approaches twice the value of the argument. This is

[3] We shall continue our convention of summation limits, that is, $-(N-1)/2$ to $(N-1)/2$, but in this case take $N$ to be infinite. We shall see later the number of terms which are necessary in a practical case.

maximum for $\theta=\theta_0=\pi/2$, that is, for scanning in the plane of the ring so that the Bessel arguments are $kR=2\pi R/\lambda$ and the number of spectral lines is approximately $\pi$ times the number of wavelengths in the diameter. This is somewhat more than for a linear array system with array length equal to the ring diameter, where a $\lambda/2$ element spacing would require a number of spectral lines equal to twice the number of wavelengths in the diameter. The increase in bandwidth is reasonable when it is considered that the ring scans a beam through a full circle while a linear array scans only the half circle in front of the array. For smaller ring diameters, a somewhat wider bandwidth is required.

It is clear that sampling may be as important to a system of this type as for the linear array systems, so let us continue our analysis by consideration of the sampling process. Since $E_\Sigma$ is seen here to be of a similar form to that of MOSAR, we shall consider a sampling signal of the same form:

$$S_a = e^{j\omega_1 t} \sum e^{jn(\omega_s t - \psi_0)}$$

where the sampling bandwidth is taken as at least equal to the significant scanning signal bandwidth as indicated above.

The sampling mixer output will then be given by

$$\begin{aligned} E_s &= E_\Sigma \times S_a^* \\ &= 2\pi f(t) e^{j\omega_0 t} \sum_m \sum_n J_m(kR \sin\theta) J_m(kR \sin\theta_0) \\ &\quad e^{jm(\omega_s t - \psi)} e^{-jn(\omega_s t - \psi_0)} \end{aligned}$$

As before, we shall choose a smoothing filter which passes significantly only the terms for which $m=n$ and the frequency offset terms are eliminated. Then

$$E_0 = 2\pi f(t) e^{j\omega_0 t} \sum_n J_n(kR \sin\theta) J_n(kR \sin\theta_0) e^{jn(\psi_0 - \psi)}.$$

By analogy with the two previous expressions for $E_\Sigma$ we note that this is equivalent to

$$E_0 = 2\pi f(t) e^{j\omega_0 t} J_0[kRg(\theta, \theta_0, \psi - \psi_0)]$$

254 which is identical with the output of the directly phased steered ring.

To summarize what the Gitzendanner system has done, we have generated a suitably phased set of LO signals to collimate the elemental received signals around the ring. By using a tapped transmission line to distribute the LO signal around the ring, we have produced a beam which scans in $\psi$ at a rate provided by the transmission line, this rate ideally being somewhat higher than the signal bandwidth, as with MOSAR. Notice that this beam can be made to scan around in the plane of the ring ($\theta_0=\pi/2$), or around a cone having a common axis with the ring ($\theta<\theta_0<\pi/2$). The LO signal must be properly generated for the value of $\theta_0$ desired, but one or more beams, steerable in $\psi$, may be generated by the use of sampling.

For the case of a system designed primarily for scanning in the plane of the ring, the poor sidelobes of the $J_0$ pattern in that plane can be improved by amplitude tapering of the ring element signals. Applebaum's patent [9] points out that this can be done by using bilinear elemental mixers and amplitude modulation of the LO signal. This would provide an element tapering which rotates around the ring along with the element phasing. The desired form of weighting would probably use only the elements on the side of the ring toward the beam direction, with maximum amplitude in the beam direction and tapered off toward both sides.

When sampling is used, we may note that spectral weighting of the sampling pulse spectrum can be used to control the pattern sidelobes in the $\psi$ direction. In this case, therefore, the sampling spectrum weights might be chosen as

$$Ws_n = \frac{T_n}{J_n^2(kR \sin\theta_0)}$$

in order to produce a Taylor weight pattern in the $\psi$ direction for $\theta=\theta_0$. This has one pitfall which would have to be avoided: $J_n$ will have zeros at certain values of $n$ depending upon the argument. If the ring radius can be adjusted to keep the zeros sufficiently separated from the required integral values of $n$, this procedure will possibly be suitable.

The implementation required for this system may be seen to combine the LO distribution system of the linear array time-sampled system with the MOSAR sampling components. The new required component is the phase-modulated (or frequency-modulated) LO signal generator. This must be a very precise sinusoidal phase modulation, especially for large rings. For a single ring, this might best be accomplished by a wide-band sinusoidally frequency-modulated oscillator, such as a reflex klystron or voltage-tuned magnetron with feedback from a very linear discriminator to maintain the required phase-modulation precision. Alternatively, a pulse-modulated LO signal might be passed through a transverse filter having the same length as the LO distribution line and at least as many taps as elements on the ring, the precise phasing being provided by precision phase shifters on the taps.

## Time-Sampled Ring

Another approach to ring array scanning has been proposed and used as a radio direction finder[4] rather than a beam former, but when beam sampling is added it is closely related in concept to the sampled Gitzendanner system. As a direction finder, the basic approach has been described by considering the output of a single omnidirectional antenna which is swung around a circle at a constant angular rate. The swinging antenna sinusoidally phase modulates the received signal; the timing of this modulation as compared with the timing of the antenna rotation depends upon the signal angle of arrival in the plane of the antenna motion. When the antenna is toward (or away) from the incident signal direction, a maximum phase advancement (or retardation) is realized, but the output frequency is equal to the incident signal frequency. Ninety degrees of rotation past these points the phase modulation is at the

[4] For example, the AN/TRD-8 direction finder built by the Heavy Military Electronics Department of the General Electric Company.

average value, but the phase change rate is maximum so as to provide a maximum negative or positive Doppler shift, respectively, of the incoming signal. A discriminator on the output can provide the timing of this Doppler modulation and thereby the angle of arrival. This type of direction finder has been known, therefore, as a Doppler direction finder.

In the implementation of this approach, because of the practically required swing circle and rotation rates, a ring of fixed elements has been used, the elements being sequentially time sampled to provide a simulation of the swinging element. This function is illustrated directly by Fig. 7, if the LO signal is considered to be pulse modulated (as for the time-sampled linear array system). By using the elemental band-limiting filters of Fig. 5, together with a rotation frequency higher than the filter bandwidths, one would expect that all incident signal information within the design frequency band had been preserved, and that we need only apply the proper kind of "sampling" of the summed output signal to generate any desired receiving beam. That this is true can be illustrated mathematically in a direct parallel to the Gitzendanner system analysis.

As before, we can write the element signals as

$$E_r = f(t)e^{j\omega_0 t}e^{jkR \sin\theta \cos(\psi-\phi)}$$

The pulsed LO signal can be written as

$$E_{LO} = e^{j\omega_1 t}\sum e^{jn\omega_s t}.$$

At the delay line taps the LO signal will be

$${}_\tau E_{LO} = e^{j\omega_1[t-(\phi/\omega_s)]}\sum e^{jn(\omega_s t-\phi)}.$$

Again choosing $\omega_1 = rN\omega_s$, this simplifies to

$${}_\tau E_{LO} = e^{j\omega_1 t}\sum e^{jn(\omega_s t-\phi)}$$

The mixer outputs will be

$$\begin{aligned} E_1 &= E_r \times ({}_\tau E_{LO}) \\ &= f(t)e^{j(\omega_0+\omega_1)t}\sum e^{jn(\omega_s t-\phi)}e^{jkR\sin\theta\cos(\psi-\phi)} \end{aligned}$$

Approximating the mixer output summation by an integral around the ring, we can write

$$\begin{aligned} E_\Sigma &= f(t)e^{j(\omega_0+\omega_1)t}\sum\int_0^{2\pi} e^{jn(\omega_s t-\phi)}e^{jkR\sin\theta\cos(\psi-\phi)}\,d\phi \\ &= 2\pi f(t)e^{j(\omega_0+\omega_1)t}\sum j^n J_n(kR\sin\theta)e^{jn(\omega_s t-\psi)} \\ &= 2\pi f(t)e^{j(\omega_0+\omega_1)t}e^{jkR\sin\theta\cos(\omega_s t+\psi)} \end{aligned}$$

This may be seen to be the incident signal with a sinusoidal phase modulation whose timing depends upon the incident signal direction $\psi$, as expected. Notice that the $\theta$-coordinate information is also included in this signal in terms of the peak value of phase modulation.

"Sampling" to produce a single controllable beam can be visualized better for this system, as with the time-sampled linear array, as a correlation process. Thus, the sampling signal we should use is one having a similar kind of sinusoidal phase modulation, such as

$$\begin{aligned} S_a &= e^{j\omega_1 t}e^{jkR\sin\theta_0\cos(\omega_s t-\psi_0)} \\ &= e^{j\omega_1 t}\sum j^n J_n(kR\sin\theta_0)e^{jn(\omega_s t-\psi_0)}. \end{aligned}$$

A phase-modulated LO signal and a pulse-modulated sampling signal were used for the Gitzendanner system, whereas in this case we are using a pulse-modulated LO signal and a phase-modulated sampling signal.

The output of the sampling mixer in this case will be given by

$$\begin{aligned} E_s &= E_\Sigma \times S_a^* \\ &= 2\pi f(t)e^{j\omega_0 t}\sum_m\sum_n j^{m+n}J_m(kR\sin\theta)J_n(-kR\sin\theta_0) \\ &\quad\times e^{-j(m\psi+n\psi_0)}e^{j(m+n)\omega_s t} \end{aligned}$$

As before, the output of the smoothing filter will include only those terms for which the frequency offsets are zero, i.e., $m=-n$:

$$\begin{aligned} E_0 &= 2\pi f(t)e^{j\omega_0 t}\sum_n J_n(kR\sin\theta)J_n(kR\sin\theta_0)e^{jn(\psi_0-\psi)} \\ &= 2\pi f(t)e^{j\omega_0 t}J_0[kRg(\theta,\theta_0,\psi-\psi_0)] \end{aligned}$$

which is again the same form as for the basic phased ring array.

Comparing this time-sampled ring system with the Gitzendanner system, we see that it uses essentially the same components, the only difference being the pulse and phase modulation interchange between LO and beam sampling signals. The time-sampled ring system has the advantage that the beam coordinate angle $\theta_0$ need not be introduced in the LO signal, which uses pulses in this case, but can be independently set for each of a number of parallel beam sampling channels.

## Ring-Linear Conversion

We noted for a linear array that frequency-multiplexed elements produced time-multiplexed beams and vice versa, this being a consequence of the Fourier transform relationship between the aperture and the far-field pattern space. For the ring antenna, we have found that a form of element time multiplexing, as in the Gitzendanner system, leads directly to time-multiplexed beam signals in the $\psi$ direction. Consideration of these facts might lead one to the conjecture that there must be some kind of a Fourier transform relationship between a ring aperture and a linear aperture; as a matter of fact there is, as we shall now illustrate.

We previously noted that we could express the ring aperture excitation as a sum of orthogonal linear phase modes; that is, we wrote

$$\begin{aligned} E_r &= f(t)e^{j\omega_0 t}e^{jkR\sin\theta\cos(\psi-\phi)} \\ &= f(t)e^{j\omega_0 t}\sum_n j^n J_n(kR\sin\theta)e^{jn(\psi-\phi)} \end{aligned}$$

Notice that the $n=0$ term has equal phase around the ring, the $n=1$ term decreases phase by $2\pi$ for each turn around the ring in the direction of increasing $\phi$, etc. Sandeman [10] investigated ring modes of this form and noted that the far-field pattern corresponding to each mode had the same form as the ring mode itself; that is, uniform amplitude and linear phase variation with angle being 0 for the 0 mode, $2\pi$ change per turn for the first, $4\pi$ for the second, etc. When a complete set of such modes is present, a pattern maximum can be obtained in the $\psi$-direction in which the far-field phases are equal. At other values of $\psi$ there will be a linear set of phases across the modes and the pattern value will be less. Steering the $\psi$-direction of the maximum can be accomplished by providing linear phase control of the various modes.

Clasen *et al.* [11] describe a system using a special antenna to excite a set of these spiral-phase modes and a linear electronic beam steering system to control the beam direction. Directly performing the desired Fourier transform (series expansion) for an array by means of a beam-forming matrix has been proposed by Chadwick and Glass [12].

The Butler [13] matrix may be noted to have the kind of characteristic we want; that is, each output corresponds to a linear phase mode across the input ports, successive modes having a difference of $2\pi$ in total phase change across the set of input ports. These matrices are also lossless when ideal components are assumed and have equal numbers of input and output ports. Although the Butler matrix, per se, is built of four-port hybrids and therefore has $2^n$ inputs (or outputs), there can be, in principle, matrices of this type having any integral number of input (or output) ports, and we shall use the term Butler matrix to include this more general case.

The type of system which is being discussed is shown in Fig. 8. Since we are using, for simplicity, a continuous-aperture expression rather than an array of discrete elements, we shall consider the outputs of the Butler matrix to be the Fourier series coefficients given by $E_n$ in the expression

$$E_r(\phi) = \sum_n E_n e^{-jn\phi}.$$

From our previous expansion, we note that

$$E_n = f(t)e^{j\omega_0 t} j^n J_n(kR \sin\theta) e^{jn\psi}.$$

If we apply amplitude weights $W_n$ and combine the signals in a linear phase-steering system as illustrated, we will have

$$E_0 = \sum_n W_n e^{-jn\alpha} E_n$$
$$= 2\pi f(t) e^{j\omega_0 t} \sum_n j^n W_n J_n(kR \sin\theta) e^{jn(\psi-\alpha)}.$$

This may be seen to be in the same form as for a linear array, where we had

$$E_\Sigma = f(t)e^{j\omega_0 t} \sum_n T_n e^{jn(kd\eta - \alpha)}.$$

In this case, we are steering directly in $\psi$ rather than $kd\eta$, but we can obtain a good pattern in $\psi$ with low sidelobes by making

$$j^n W_n J_n(kR \sin\theta) = T_n$$

or

$$W_n = \frac{T_n}{j^n J_n(kR \sin\theta)}$$

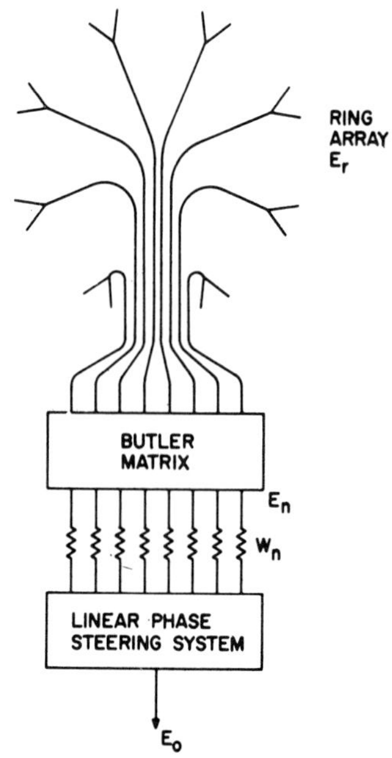


Fig. 8. Ring-linear conversion.

where $T_n$ are Taylor or Chebyshev weight values as have been developed for a linear array. Notice that $R$ and $\theta$ have to be known in order to properly choose the values of $W_n$ and that we have only obtained good sidelobes in the $\psi$-direction. This system therefore would be more appropriate for scanning, say, in the plane of the ring, rather than for an application where it is necessary to steer to various positions near the ring axis.

It may now be noted, in retrospect, that the beam sampling of the Gitzendanner system may be considered to provide beam steering and sidelobe control by means of phase steering and weighting of such a set of orthogonal spiral-phase modes. Referring to the orthogonal mode expansion of the ring elemental signals discussed in the section called "Time-Sampled Ring," we note that the individual mode amplitudes are given by $J_n(kR \sin\theta)$. The scanning process produces a signal $E_\Sigma$, whose spectral terms are seen by their $J_n(kR \sin\theta)$ amplitude factor to have a one-to-one correspondence with the ring orthogonal modes. In the sampling process, therefore, the sampling signal spectral lines provide directly linear phase steering and weighting of those modal signals. The weighting required in that case appears in slightly different form because of the correlation action of the Gitzendanner LO signal with the ring signals.

## Conclusion

The frequency-multiplexed linear array system described here has been quite thoroughly developed for specific radar applications. The time-sampled linear array, Gitzendanner scanning system, and time-sampled ring system have been implemented at least experimentally, down to the sampling point. The application of sampling to these systems has not previously been suggested, however, to the best of the writer's knowledge, and is presented here as an interesting possibility. The linear phase steering of spiral-phase modes of circular antennas has also been implemented experimentally as previously noted.

The successful application of these techniques depends upon the bringing together of the proper combination of requirement, technique concept, and component feasibility so as to meet the requirement more fully or more economically than other technique and component combinations could do at that time. The fundamental advantages of mixer array steering systems should be the economy and control flexibility permitted by mixers—which are basically simple components—and by multiplexing, which mixers can handle. In practice, the cost of systems of these types is determined, basically, not by the diodes which are the active core of mixers, but by the passive filters and interconnecting transmission lines which are required in actual array implementation. In meeting specific array parameter requirements with the various multiplexing systems, the limitation most often felt is the bandwidth limit of the various components required. This limit is currently several thousand megahertz, and is obtained at an $X$-band carrier frequency.

Widespread future application of these techniques, then, depends upon whether less expensive methods are found to fabricate the required filters and transmission lines, and upon whether further bandwidth capability of the various active and passive components can be achieved. It is the writer's firm belief that these techniques have much to offer to many applications, and that we should accept the challenge to make them even more useful through the development of low-cost, wide-band components.

## References


[1] H. M. Lewis, "System for locating a radiated-signal reflector," U. S. Patent 2 426 460, August 26, 1947.

[2] D. F. Langenwalter and K. M. Stevenson, "Receiver scanning system," U. S. Patent 3 012 244, December 5, 1961.

[3] H. V. Cottony and A. C. Wilson, "A high resolution rapid-scan antenna," *J. Research NBS*, vol. 65D, pp. 101–110, January–February 1961.

[4] P. V. Howells, "MOSAR-array multiplex beamforming technique," *Symp. Record, 9th Ann. Radar Symp.* (University of Michigan, Ann Arbor), June 1963 (classified as "secret").

[5] H. E. Shanks, "A new technique for electronic scanning," *IRE Trans. Antennas and Propagation*, vol. AP-9, pp. 162–166, March 1961.

[6] For a more exact treatment of ring arrays, see H. L. Knudsen, "Radiation from ring quasi-arrays," *IRE Trans. Antennas and Propagation*, vol. AP-4, pp. 452–472, July 1956.

[7] The basic concept of the direction cosine or $\eta$-plane, as we shall call it, is well covered by W. H. von Aulock, "Properties of phased arrays," *Proc. IRE*, vol. 48, pp. 1715–1727, October 1960.

[8] L. G. Gitzendanner, "Receiver scanning system," U. S. Patent 2 852 772, September 16, 1958.

[9] S. P. Applebaum, "Electronic scanning of circular arrays," U. S. Patent 3 076 193, January 29, 1963.

[10] E. K. Sandeman, "Spiral-phase fields," *Wireless Engr.*, vol. 26, pp. 96–105, 1949.

[11] C. P. Clasen *et al.*, "A radial antenna and multiple amplifier system for electronic scanning," *RCA Rev.*, vol. 22, pp. 543–554, 1961.

[12] G. G. Chadwick and J. C. Glass, "Investigation of a multiple beam scanning circular array," Radiation Systems, Inc., Alexandria, Va., AFCRL-63-136.

[13] J. L. Butler, in *Microwave Scanning Antennas*, vol. III, R. C. Hansen, Ed. New York: Academic Press, 1966, ch. 3.


*Reprinted from the* PROCEEDINGS OF THE IEEE
VOL. 56, NO. 11, NOVEMBER, 1968
pp. 1801-1811

*Reprinted by permission from* IEEE TRANSACTIONS ON ANTENNAS AND PROPAGATION
Vol. AP-17, No. 5, September 1969
pp. 645-647


## Reduced Sidelobes for Butler-Matrix-Fed Linear Arrays

***Abstract*—It is shown that simple transposition of the elements of a linear array fed by a Butler matrix can be used to improve the sidelobe level without significant degradation of the other characteristics of the array. The beams remain collimated with gain equal to $N$ times element gain, where $N$ is the number of elements in the array. The beam crossover level is lowered slightly but not so much as would be the case for a conventionally tapered aperture distribution. Improved sidelobes are obtained for $N \geq 14$, and data are presented for $N$ as high as 50. Specifically, sidelobe levels of 15.54, 16.54, 17.86, and 18.56 dB are obtained for $N = 18$, 31, 37, and 44, respectively. Although the improvement is modest for low $N$, it appears that the available improvement increases directly with $N$.**

### Introduction

The purpose of this communication is to consider the possibility of achieving improved sidelobes from Butler-matrix-fed multibeam arrays without significant degradation of other desirable pattern characteristics. It is assumed that, in addition to the best possible sidelobes, the desirable pattern characteristics are 1) maximum possible array gain from an $N$-element array of $N$ times the element gain, 2) beam crossover levels of about $-4$ dB relative to maximum gain, and 3) no dissipation within the feed system. The last condition eliminates the technique of introducing a lossy tapered aperture distribution on an otherwise conventional Butler array.

The basic configuration being considered is a regularly spaced linear array of $N$ elements fed by an $N$-port ($N$ inputs and $N$ outputs) Butler matrix to produce $N$ equally spaced beams, each with a gain of $N$ times the element gain.[1] The sidelobe level tends to $-13.2$ dB for large $N$, and the crossover level is approximately $-3.9$ dB.

The results presented in the following paragraphs are based on the condition, established by the author in a recent paper, that each element of a Butler array can assume not one, but an infinite number of locations [2]. This communication considers only symmetrical linear arrays, but it is felt that similar results can be obtained with planar arrays.[2] No analysis has been done for asymmetrical arrays, and it cannot be estimated whether the use of asymmetry would yield improved results.

### Theory

Fig. 1 depicts the covering condition as applied to a linear array. For a more complete discussion of the covering condition, the reader is referred to [2]. Without loss of generality, it is assumed that the matrix network produces one distribution in which all radiating elements are excited in phase. The beams can be shifted arbitrarily later by appropriate phase shifts at the elements without affecting these results. It is easily shown that the other distributions will produce a total phase progression of $2\pi n(N-1)/N$ radians across the array, where $-N/2 \leq n \leq N/2$, so that if one were to extend the array to one more element, that element would have a phase of $2\pi n(N-1)$ radians and would therefore be in phase with the one on the opposite end of the array. Thus Fig. 1 depicts the covering condition as applied to a linear array. The basic point is that those radiating elements with identical numbers can be connected to the same terminal of the Butler matrix. The power need not be equally distributed among these elements, however. For example, in Fig. 1 the elements numbered 1 are shown connected together. The feed lines are adjusted in length so that all elements numbered 1 have the same phase. By power division, one can apply a smooth taper. The basic point is that element 1, for example, can be located at any position marked 1, and a similar statement holds for each of the other array elements. By power division, one can apply a smooth taper to the array to achieve arbitrarily low sidelobes. The following aspects of this technique should be noted:

1) for an $N$-port Butler matrix, the array will have more than $N$ elements, and the array gain will be more than $N$ times the element gain;

2) as a result of 1), the beamwidth will be narrower than for the conventional arrangement, and the crossover level will be reduced below $-4$ dB.

Therefore, the application of the tapering technique will involve a tradeoff between degraded crossover level and improved sidelobes and array gain. Fig. 2 illustrates the results of imposing a limited taper on a continuous uniform aperture distribution, which provides an indication of the characteristics of an array with a large number of elements. The taper yields an aperture distribution of the form

$$e(\chi) = \begin{cases} \sqrt{p(\chi)} = 1, & |\chi| \leq 1 - a \\ \sqrt{1 + \cos\left[\dfrac{\pi(\chi+1)}{2a}\right]}, & -1 - a < \chi < -1 + a \\ \sqrt{1 + \cos\left[\dfrac{\pi(\chi-1)}{2a}\right]}, & 1 - a < \chi < 1 + a. \end{cases}$$

The amplitude of the radiation pattern is

$$E(\psi) = \frac{\psi \sin((1-a)\psi) + b\cos((1+a)\psi)}{b^2 - \psi^2} + \frac{\sin((1-a)\psi)}{\psi}$$

where $b = \pi/4a$ and $\psi = (2\pi s/\lambda)\sin\theta$, with $s$ the element spacing and $\theta$ the angle from the array normal. It is seen from Fig. 2 that the absolute sidelobes improve from 13.2 dB for $a = 0$ to 21.4 dB for $a = 1$, but the absolute crossover level degrades from 3.9 dB for $a = 0$ to 7.4 dB for $a = 1$. The gain relative to the untapered array, on the other hand, ranges from 0 dB for $a = 0$ to 2.1 dB for $a = 1$ so that the relative crossover level degrades to 9.5 dB.

The remainder of this communication is concerned with symmetrical linear arrays satisfying the previously cited covering condition. These arrays are formed as shown in Fig. 3, by transposing elements from each end to positions near the opposite end corresponding to the same element position in an unfilled adjacent array. Because of symmetry, the element arrangement at each end of the array has complementary mirror symmetry, i.e., if

Manuscript received October 31, 1968; revised February 19, 1969. This work was supported in part under a Central Research Project of the Institute for Defense Analyses, Arlington, Va.

[1] Although Butler matrices are generally assumed to have $2^n$ elements, this communication assumes the availability of multibeam forming matrices with arbitrary $N$. Synthesis of such matrices is not commonly known, but it was derived by Nolen [1].

[2] In fact, unpublished work on phased array subarrays, done without regard for application to multibeam arrays but nevertheless satisfying the covering condition in [1], has also achieved sidelobe reduction.

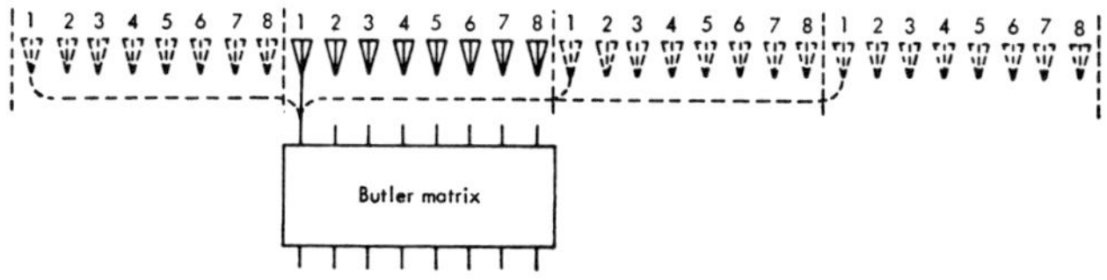


Fig. 1. Covering condition for linear array.

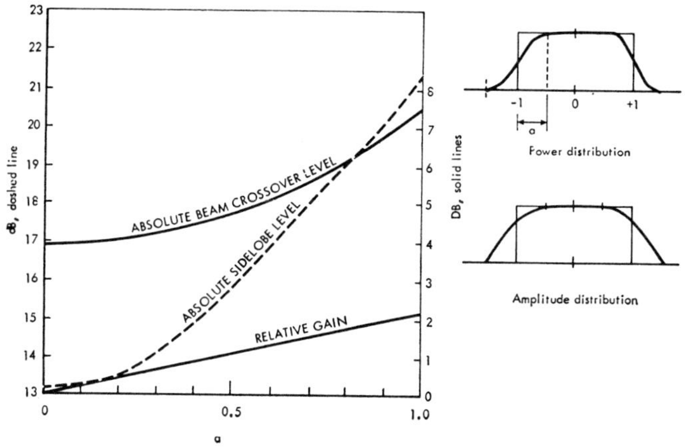


Fig. 2. Characteristics of continuous aperture with variable tapered region.

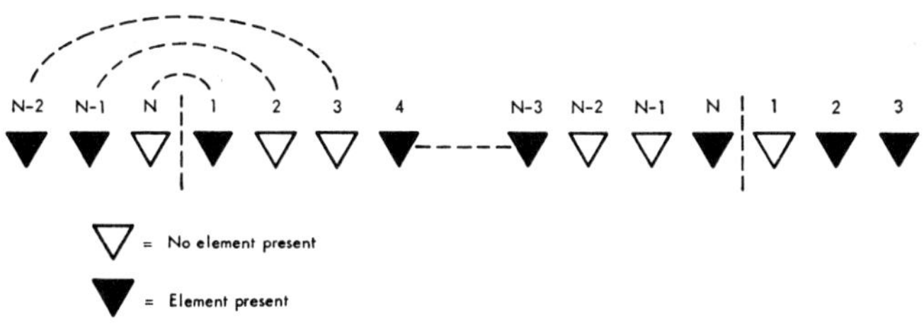


Fig. 3. Array being considered.

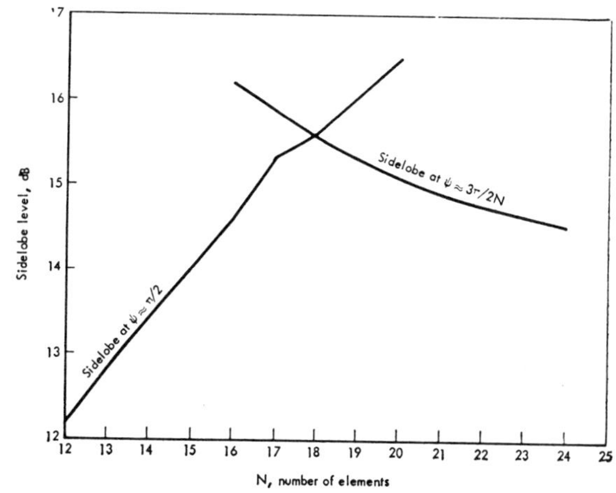


Fig. 4. Characteristics of array of form $N$,1.

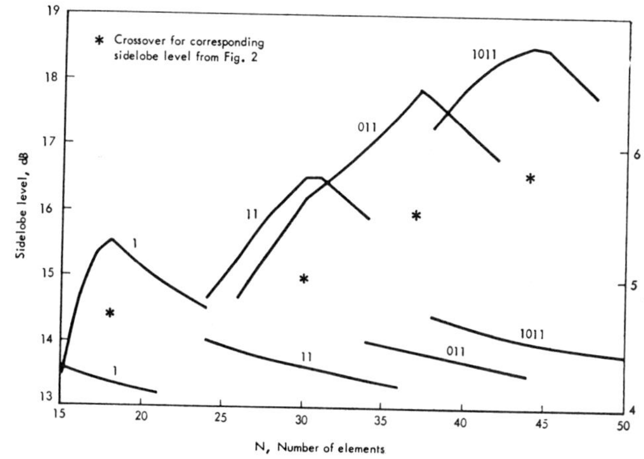


Fig. 5. Sidelobe and crossover levels for up to 4 elements displaced.

the elements outside the original end of the array are folded over the boundary, they fill up gaps on the inside of the boundary. Thus it is possible to completely define the array by the number of elements $N$ and the element arrangement outside one end of the original array.

The radiation pattern of the array is now considered. For an $N$-element array, the pattern of the original unmodified array is

$$E_0(\psi) = \frac{\sin N(\psi/2)}{\sin \psi/2}$$

with $s$ the element spacing and $\theta$ the angle from the normal to the array. If the array is modified by moving elements to the $k$th position from each end of the array, the pattern becomes

$$E_m(\psi) = \frac{\sin N(\psi/2)}{\sin \psi/2} + 2 \cos \frac{(N + 2k - 1)\psi}{2} - 2 \cos \frac{(N/2 - 2k + 1)\psi}{2}$$

$$= \frac{\sin N(\psi/2)}{\sin \psi/2} \left[1 - 4 \sin \frac{\psi}{2} \sin \frac{(2k - 1)\psi}{2}\right] \qquad (1)$$

If several elements are moved to new positions, it is possible to extend (1) to account for the entire modified array

$$E_m(\psi) = \frac{\sin N(\psi/2)}{\sin \psi/2} \left[1 - 4a_k \sin \frac{\psi}{2} \sum_{k=1}^{n} \sin \frac{(2k - 1)\psi}{2}\right] \qquad (2)$$

where $a_k$ is the coefficient of the $k$th outside element position, i.e., $a_k = 0$ if the position is unoccupied and $a_k = 1$ if the position is occupied. Therefore, the array distribution is defined by the parameters $N, a, \cdots, a_k$.

Equation (2) indicates that the modified array pattern is the product of the unmodified pattern and a relatively simple factor. Thus the modified pattern has the same zeros as the unmodified plus new zeros determined by the second factor. The overall expression is so easily and quickly computed that it was considered preferable to determine the best distributions by analysis rather than synthesis. A program was written in BASIC language for use on a time-share computer, evaluating (2). The program samples $E_m(\psi)$ in search of extrema and reads them out in decibels. A glance at the printout determines the maximum sidelobe level.



## RESULTS

The procedure has been to test distributions of the form $N, a_1, a_2, \cdots, a_n$, keeping all the $a_k$ fixed and varying $N$. It is clear that for large $N$ the sidelobe level will approach the 13.2-dB value of the unmodified array. For low $N$, in general, the sidelobe level will be degraded below 13.2 dB. For intermediate values of $N$ and judicious choice of the $a_k$, the sidelobes will be improved. The function of sidelobe level versus $N$, although defined only for integer $N$, approximates a curve. Nevertheless, the approximate curve is not smooth because the highest sidelobe changes abruptly from one portion of the pattern to another as $N$ is varied.

Fig. 4 illustrates the results obtained for arrays of the form $N$,1, the simplest modification that can be made. For values of $N$ less than 18, sidelobes in the neighborhood of $\psi = \pi$ are highest, but for $N$ greater than 18, the near-in sidelobe is dominant. At $N = 18$, the two sidelobes are approximately equal.

Fig. 5 is a plot of results obtained by displacing blocks of up to four elements from each end of the array. It is seen that with increasing $N$, it is possible to obtain greater sidelobe reduction. In addition, the beam crossover levels are plotted, and they are only slightly less than the 3.9-dB characteristic of uniform distributions. The absolute crossover levels for the corresponding absolute sidelobe levels of the continuous tapered distribution of Fig. 3 are plotted for comparison.

## Conclusion

A simple method has been described for improving the sidelobe level of Butler-matrix-fed linear arrays with a number of elements greater than 14. The improvement is modest for small $N$ and becomes more significant for larger $N$. The beam crossover level has been maintained at a high level, and gain is unchanged. This investigation has analyzed arrays of up to approximately 45 elements. It is suggested that the following areas represent reasonable extensions of this work: investigate planar arrays, analyze larger linear arrays, consider asymmetrical arrays.

J. Paul Shelton
Delex Systems, Inc.
Arlington, Va. 22209

## References

[1] J. C. Nolen, "Synthesis of multiple beam networks for arbitrary illuminations," Ph.D. dissertation, The Johns Hopkins University, Baltimore, Md., April 21, 1965.
[2] J. P. Shelton, "Multibeam planar arrays," *Proc. IEEE*, vol. 56, pp. 1818–1821, November 1968.

# Granularity of Beam Positions in Digital Phased Arrays

BURRELL R. HATCHER, MEMBER, IEEE

***Abstract*—Many phased arrays are designed to use digital or digitalized phase shifters in which the phase shift varies in discrete steps rather than continuously. In such cases, the beam can be steered only in discrete steps.**

**Granularity, defined as the finest realizable increment between adjacent beam positions, is an important parameter in the description of the performance of a phased array antenna using digital phase shifters. Two approaches to the problem of determining granularity are developed in this paper. The first is based upon the expansion by its moments of complex aperture distribution functions; the second is based upon approximating the actual discrete phase distribution by a linear continuous one. Results of the two approaches are shown to be in excellent agreement, with a variation of less than 3.6 percent being obtained.**

**Arrays employing binary digital phase shifters are explicitly considered, but the techniques can be generalized to other types of digital, or digitalized, phase shifters.**

**For other than the first beam position, there are alternate phase distributions which might be used to obtain a given beam position. The selection of a specific phase distribution depends, at least in part, upon the computer logic and round-off procedures. It is shown that these alternate phase distributions do not produce large differences in the beam position. The spread in beam position is between 2 and 3 percent for the alternate distributions.**

**The effect of an amplitude taper on granularity is also considered. It is found that when the edge elements are less intensely illuminated than those at the center, the first beam position off-broadside is nearer broadside than for an array having the same phase distribution but uniform amplitude.**

**The average granularity of a phased array has been determined, and is plotted against the number of elements with the number of phase bits as a parameter.**

## I. Introduction

FIG. 1 SHOWS two continuous, linear phase distributions. The first, Fig. 1(a), is antisymmetrical and may offer advantages when the phase shifter drivers can be used to drive more than one phase shifter. Note, however, that for a given maximum absolute value of phase shift at the end(s) of the array, the second distribution [Fig. 1(b)] gives a smaller angle of scan. However, for any given scan angle, the differential phase shift from one end of the array to the other is the same.

In the case of a continuous, linear phase distribution, the peak of the main lobe is normal to the phase front.[1] With respect to the array normal, the beam occurs at an angle

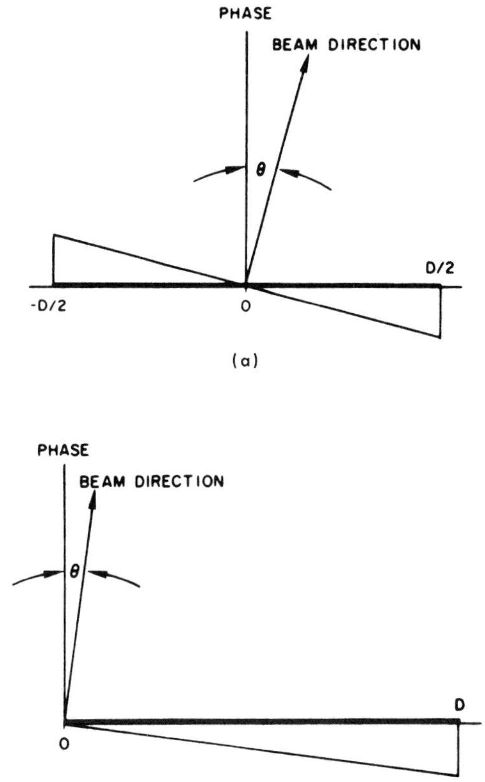


Fig. 1. Continuous phase distributions. (a) Antisymmetrical phase distribution. (b) Asymmetrical phase distribution.

$$\theta_0 = -\sin^{-1}\left[\frac{\phi(D/2)}{kD/2}\right] \qquad \text{(Case A)} \qquad (1a)$$

$$\theta_0 = -\sin^{-1}\left[\frac{\phi(D)}{kD}\right] \qquad \text{(Case B)} \qquad (1b)$$

where $\phi(x)$ is the phase distribution along the array, and $k = 2\pi/\lambda$.

It should be noted that although these two expressions give the same result if the phase distribution is linear and if $2\phi(D/2)$ for the antisymmetrical distribution equals $\phi(D)$ for the asymetrical distribution, such is not the case if the phase in quantized, as illustrated in Fig. 2.

When binary digital phase shifters and discrete elements are used, the determination of the beam position is not so simple. The phase can vary only in discrete and finite steps. Thus, as one attempts to steer the beam off-broadside, the phase at an element does not change until some predetermined incremental value is exceeded, at which time the

Manuscript received March 29, 1968; revised June 20, 1968. This work was presented at the 1966 International Antenna and Propagation Symposium, Palo Alto, Calif.

The author is with the Missile Systems Division, Raytheon Company, Bedford, Mass.

[1] It is assumed throughout that the element factor is broad enough with respect to the array factor that the latter dominates the beam shape and the position near broadside, i.e., the radiators can be considered to be isotropic in the plane of scan.

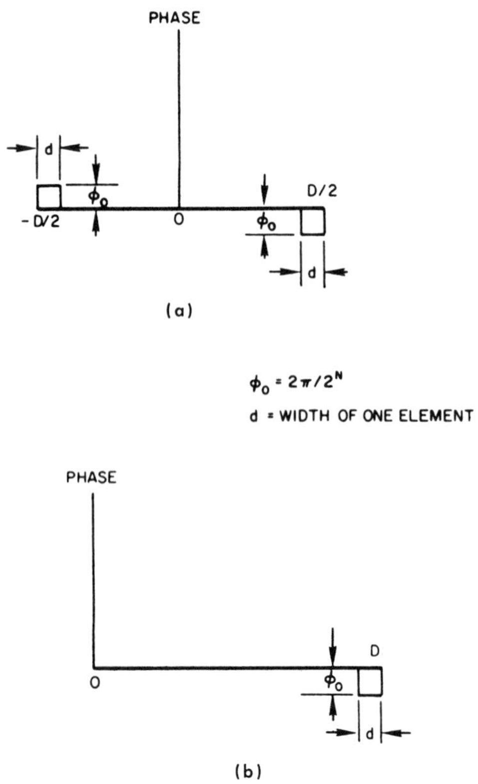


Fig. 2. Phase distribution for first beam position off-broadside. (a) Antisymmetrical distribution. (b) Asymmetrical distribution. $\phi_0 = 2\pi/2^N$; $d$ = width of one element.

phase suddenly increases by one phase bit. As a result, the beam can be steered only in discrete steps.

For the remainder of the paper, it will be assumed that the phase shifters are binary digital, so that the smallest phase incremental is $2\pi/2^N$ radians for an $N$-bit phase shifter. An analogous treatment can be applied to arrays using other types of digital phase shifters, however.

Fig. 2 illustrates the quantized phase distribution for the first beam position off-broadside for the two origins of phase reference; Fig. 2(a) is the antisymmetrical phase distribution and Fig. 2(b) is the symmetrical distribution.



For the antisymmetrical distribution, the radiation pattern of a linear aperture is given by

$$E(u) = \frac{D}{2} \sum_{m=0}^{\infty} \frac{j^m \mu_m}{m!} u^m \tag{2}$$

where $u = kD/2 \sin \theta$ and $\mu_m$ is a moment used to express the aperture distribution [1].

To determine the position of the main beam, the resulting expression is differentiated and equated to zero. Truncation of the resulting series results in an error. However, since higher-order terms are multiplied by powers of $u$ and the latter is proportional to $\sin \theta$, the neglected terms do not contribute significantly to the result. Retaining only the first three terms of the result, the maximum of $E(u)$ is found to occur at

$$u_1 \simeq \frac{\mu_2 - \sqrt{\mu_2^2 - 2\mu_1\mu_3}}{-j\mu_3} \tag{3}$$

If $2\mu_1\mu_3 \ll \mu_2^2$, (tantamount to retaining only the first two terms of the differentiated series) a further approximation can be made to obtain[2]

$$u_1 = \frac{kD}{2} \sin \theta_1 \simeq j \frac{\mu_1}{\mu_2}. \tag{4}$$

For a large array, the beamwidth is approximately $\lambda/D$ radians, and for small angles, $\sin \theta \simeq \theta$. In terms of a fraction of a beamwidth, the beam position off-broadside is approximately

$$\frac{\theta_1}{BW} \simeq \frac{u_1}{\pi} = \frac{j\mu_1}{\pi\mu_2}. \tag{5}$$

If the aperture distribution function were not symmetric and if the phase were not antisymmetric, $E(u)$ would be a complex function. The method of moments might still be used, but in this case, the beam direction must be computed from the power pattern, $P(u) = E(u) \cdot E(u)^*$, and the moments are complex quantities. In this case, the result obtained upon retaining only the first three terms in $P(u)$ is

$$u_1 = j \frac{\mu_0^*\mu_1 - \mu_0\mu_1^*}{\mu_0\mu_2^* + \mu_0^*\mu_2 - 2\mu_1\mu_1^*}. \tag{6}$$

## II. Method of the Fitted Straight Line

If the phase distribution could be adequately approximated by a straight line, the angular position of the main beam could be found by using (1). This approach has the advantage that asymmetrical phase distribution, such as that illustrated in Fig. 2(b), may be analyzed much more easily than by using the method of moments.

Spencer [3] has shown that such an approximation to the wavefront can be made by a weighted least-mean-square (LMS) error criterion, where the weighting function is the aperture amplitude function. That is, the approximating phase distribution is $\phi_A = -ax$, where the coefficient of $x$ is determined by minimizing the quantity

$$\int_{-D/2}^{D/2} f(x)|ax - \phi(x)|^2 dx \quad \text{or} \quad \int_0^D f(x)|ax - \phi(x)|^2 dx \tag{7}$$

depending upon whether the phase distribution is antisymmetrical or asymmetrical.

The beam position is then given by

$$\theta = \sin^{-1}\left(\frac{aD/2}{kD/2}\right) = \sin^{-1}\left(\frac{aD}{kD}\right) = \frac{a\lambda}{2\pi} + \frac{1}{6}\left(\frac{a\lambda}{2\pi}\right)^3 \cdots \tag{8}$$

regardless of whether the phase distribution is antisymmetrical or asymmetrical. In terms of beamwidths, the

[2] The magnitude of the error resulting from this approximation at a four-bit quantization level may be determined from Table I in Section VI.

equations reduce to a simple formula for small values of $\theta$, and for beamwidth equal to $\lambda/D$:

$$\frac{\theta}{BW} = \frac{aD}{2\pi}.$$

For the case of a uniform amplitude distribution,[3] $f(x)=1$, and the antisymmetrical phase distribution illustrated in Fig. 2(a), the slope (a) is determined from

$$\frac{d}{da}\left[\int_0^{D/2-d}(ax)^2dx + \int_{D/2-d}^{D/2}(\phi_0 - ax)^2dx\right] = 0 \qquad (9)$$

where again, $\phi_0 = 2\pi/2^N$. From this, one obtains

$$aD = 12\phi_0\left(\frac{d}{D}\right)\left(1 - \frac{d}{D}\right). \qquad (10)$$

Upon substituting this result into (9), the first position off-broadside for the antisymmetrical distribution [Fig. 2(a)] is found to be

$$\frac{\theta_1}{BW} = \frac{12d}{2^N D}\left(1 - \frac{d}{D}\right) \qquad (11)$$

while for the "single-ended" or asymmetrical phase distribution [Fig. 2(b)] a similar analysis gives the result

$$\frac{\theta_1}{BW} = \frac{3d}{2^N D}\left(1 - \frac{d}{2D}\right). \qquad (12)$$

## III. Successive Beam Positions

Not only is the LMS fitted straight line the easiest to calculate for the asymmetrical case in which the phase is referenced to the end of the array [Fig. 2(a)], but it is also more convenient in the determination of the second, and subsequent, beam positions.

Fig. 3 illustrates two candidate phase distributions for the second beam position. Exactly which one will be used in an array will depend upon the logic used to compute and distribute the beam-steering command. Likewise, Fig. 4 illustrates the alternative distributions for the third beam position, and Fig. 5, those for the fourth position. It will be shown later that the alternative phase distributions for a given beam position produce very slight differences in the beam position.[4] These figures are all shown for the case in which the phase is referenced to and steered from one end of the array.

Fig. 6 is a generalized two-step phase distribution. The slope of the LMS fitted straight line is given by

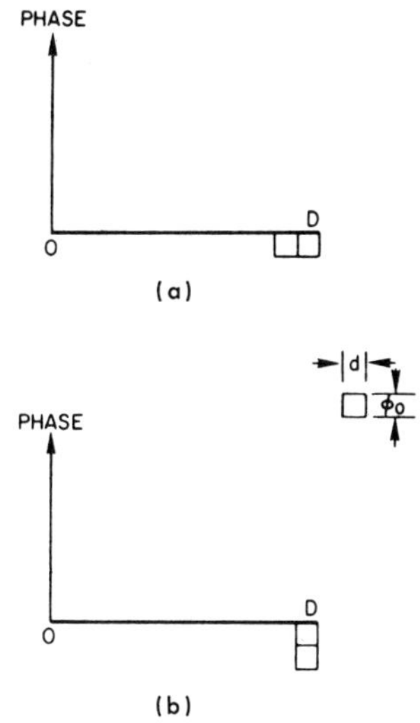


Fig. 3. Phase distribution for second beam position. (a) First candidate. (b) Second candidate.

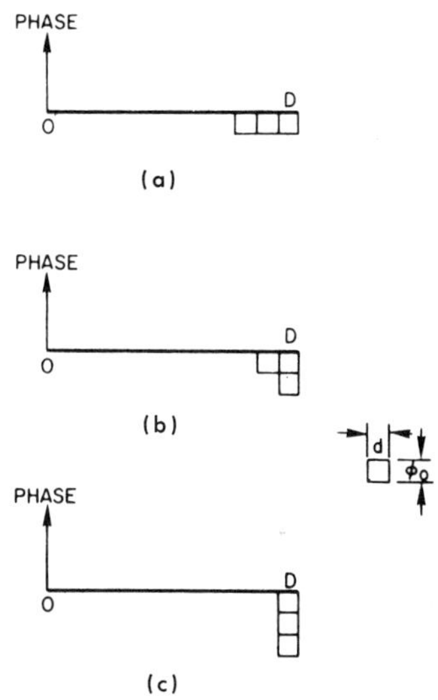


Fig. 4. Phase distribution for third beam position. (a) First candidate. (b) Second candidate. (c) Third candidate.

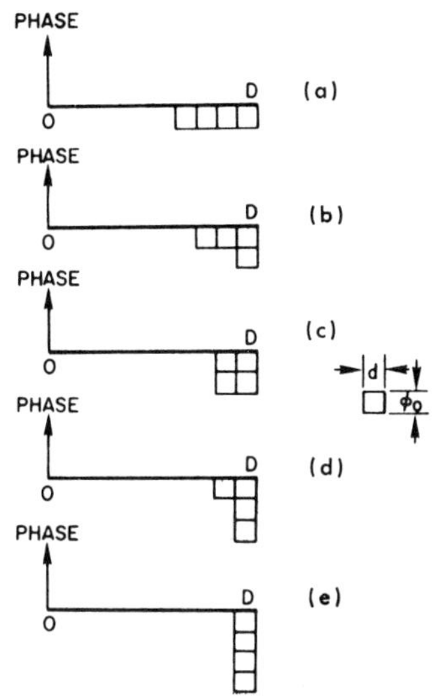


Fig. 5. Phase distributions for fourth beam position. (a) First candidate. (b) Second candidate. (c) Third candidate. (d) Fourth candidate. (e) Fifth candidate.

[3] The uniform amplitude distribution is of special interest since, other things being equal, its granularity is greater than for the usual tapered distributions in which the ends of the array are less intensely illuminated than the center. Therefore, this represents a limiting case.

[4] Exactly which of the candidate phase distributions is actually used for a particular beam position may have serious ramifications on sidelobe levels and monopulse null depths, or alternatively, the latter considerations may dictate computing logic. However, these aspects are beyond the scope of the present paper.

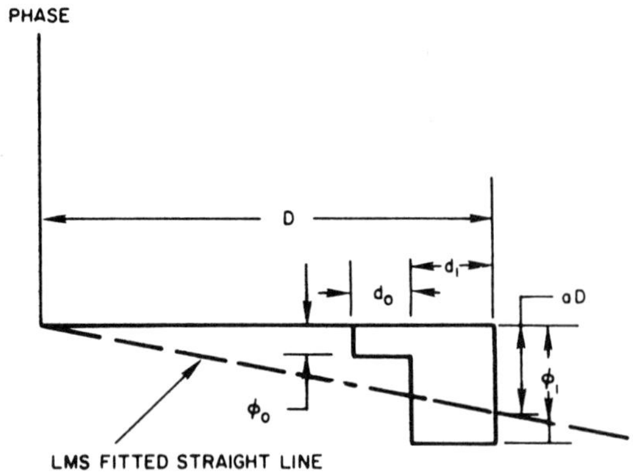

Fig. 6. Two-step phase distribution.

$$\frac{d}{da}\left[\int_0^D |ax - \phi(x)|^2 dx\right]$$
$$= \frac{d}{da}\left[\int_0^{D-d_1-d_0} a^2x^2dx + \int_{D-d_1-d_0}^{D-d_1} (a^2x^2 - 2\phi_0 ax + \phi_0^2)dx + \int_{D-d_1}^{D} (ax^2 - 2\phi_1 ax + \phi_1^2)dx\right] = 0. \quad (13)$$

From (13), one obtains

$$aD = 3\phi_1\left[\frac{d_1}{D}\left(1 - \frac{d_1}{2D}\right) + \frac{\phi_0 d_0}{\phi_1 D}\left(1 - \frac{d_0}{2D}\right)\right]. \quad (14)$$

This result permits the easy evaluation of the beam position for each of the phase distributions illustrated in Figs. 2(b), 3, 4, and 5 through the use of (9).

## IV. Example with a Nonuniform Amplitude Distribution

When the aperture amplitude distribution is other than uniform, the LMS straight line method, as given here, cannot be expected to give accurate results, since it does not weigh the fact that the edge elements, where the phase change first occurs, are illuminated less intensely than those nearer the center. However, there is no such limitation on the method using moments.

The following example illustrates the results obtained for an amplitude distribution given by

$$E(x) = \cos\left(\frac{\pi x}{D}\right) \quad (15)$$

for the first beam position off-broadside, i.e., for

$$\phi(x) = \phi_0[H(x + D) - H(x + D - d) - H(x - D + d) + H(x - D)] \quad (16)$$

where $H(x)$ is the unit step function occurring at $x=0$, and $\phi_0$ is used to denote $2\pi/2^N$.

The moments used in the first approximation to the beam position, i.e., the first and second moments, are given by

$$\mu_1 = 2j\left(\frac{2}{\pi}\right)^2\left[\sin\frac{\alpha\pi}{2} + (1-\alpha)\frac{\pi}{2}\cos\frac{\alpha\pi}{2} - \frac{\pi}{2}\right]\sin\phi_0 \quad (17)$$
$$\mu_2 = 2\left(\frac{2}{\pi}\right)^3\left\{(1-\alpha)\pi\sin\frac{\alpha\pi}{2} + \left[(1-\alpha)^2\left(\frac{\pi}{2}\right)^2 - 2\right]\cos\frac{\alpha\pi}{2}\right\} + 2\left(\frac{2}{\pi}\right)^3\cos\phi_0\left\{\left(\frac{\pi}{2}\right)^2 - 2 - (1-\alpha)\pi\sin\frac{\alpha\pi}{2} - \left[(1-\alpha)^2\left(\frac{\pi}{2}\right)^2 - 2\right]\cos\frac{\alpha\pi}{2}\right\}$$

where $\alpha = 2d/D$.

## V. Average Granularity

When the phase distribution is such that each and every element differs in phase by one phase bit, from its neighbor (Fig. 7), the resulting "staircase" distribution approximates a continuous linear phase front as well as is possible with discrete elements.

The phase at the last element in the $M$-element array is $(M-1)2\pi/2^N$, and the resulting beam position is displaced from broadside by approximately

$$\frac{\theta}{BW} = \frac{M-1}{2^N}\text{ beamwidths.} \quad (18)$$

Starting from broadside, the number of phase bits which must be added to the array to obtain this distribution is $\frac{1}{2}M(M-1)$. Since each of these phase bits corresponds to a beam position, there are also $\frac{1}{2}M(M-1)$ beam positions between broadside and the first "exact" position off-broadside. Therefore, the average beam displacement between beam positions for the asymmetrical phase distribution is

$$\left\langle\frac{\Delta\theta}{BW}\right\rangle = \frac{M-1}{2^N}\,\frac{2}{M(M-1)} = \frac{2}{2^N M} \quad (19)$$

or, since $M = D/d$, the average granularity may be expressed as

$$\left\langle\frac{\Delta\theta}{BW}\right\rangle = \frac{d}{2^{N-1}D}. \quad (20)$$

Taking the ratio of this result to that of (12),

$$\frac{\langle\Delta\theta\rangle}{\theta_1} = \frac{d/2^{N-1}D}{3d(1-d/2D)/2^N D} = \frac{2}{3(1-d/2D)} \simeq \frac{2}{3}. \quad (21)$$

In other words, the average granularity is approximately two-thirds the first beam position. This lends some credence to the common assumption that the first beam position represents the greatest granularity [4].

## VI. Results and Conclusions

The results obtained by the two methods for a uniform amplitude distribution and the antisymmetrical phase distribution of Fig. 2(a) are compared in Table I, in which the

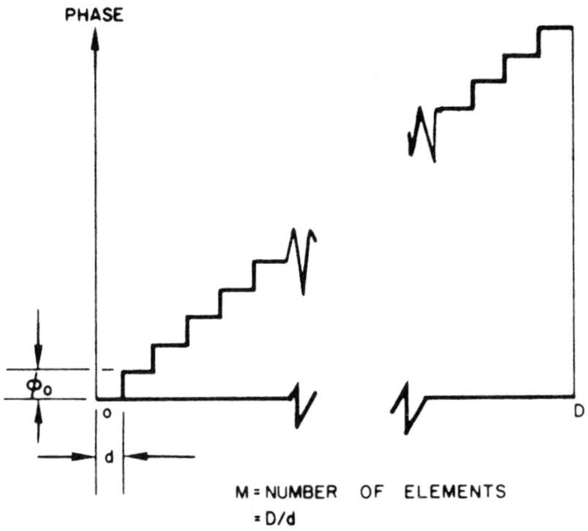


Fig. 7. Staircase phase distribution.

TABLE I

FIRST BEAM POSITION OFF-BROADSIDE (IN BEAMWIDTHS) FOR FOUR-BIT QUANTIZATION-UNIFORM AMPLITUDE

| $D/d=2/d$ | Moments (Quadratic Approximation) | Moments (Linear Approximation) | LMS Fitted Straight Line |
|---|---|---|---|
| 8 | 0.0830 | 0.0836 | 0.0821 |
| 16 | 0.0442 | 0.0444 | 0.0440 |
| 32 | 0.0222 | 0.0224 | 0.0227 |
| 64 | 0.0112 | 0.0113 | 0.0115 |
| 128 | 0.0056 | 0.0057 | 0.0058 |

first beam positions are tabulated for a number of values of $D/d$. The three columns give, from left to right, respectively, the results for the method of moments using the quadratic approximation of (3), the method of moments using the linear approximation of (4), and the LMS fitted straight line (11). A four-bit quantization level (i.e., $N=4$) was assumed.

The results of the three approximations are in remarkable agreement, indicating that any of these approximations may be used to roughly the same accuracy, providing that the limiting conditions of validity are not approached too closely. These conditions are that $d \ll D$, and that $N$ is large enough that $\theta \ll$ beamwidth.

Fig. 8 gives the first beam position (expressed in beamwidths) for a uniform amplitude distribution as calculated by the LMS straight line method, as a function of the ratio $D/d$ for a range of quantization levels ($N$), for the antisymmetrical phase distribution of Fig. 2(a). Fig. 9 is a similar plot for the asymmetrical or "single-ended" distribution of Fig. 2(b).

Table II is a tabulation of the beam position for each of the uniform amplitude candidate phase distributions for the first four beam positions (Figs. 3–5). The positions were calculated for $D=32d$, and for an incremental phase of four bits, i.e., $\pi/8$ radians. As can be seen, the different candidate distributions for a given beam position yield very nearly the same results. The smallness of the difference is, in fact, surprising.

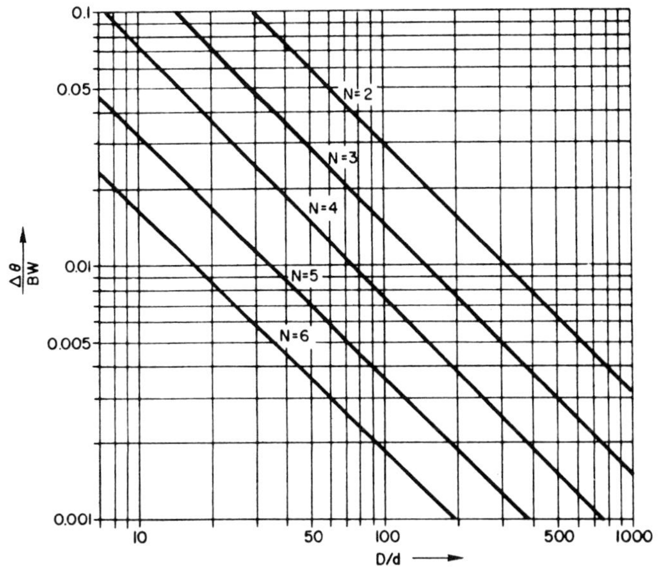


Fig. 8. First beam position, uniform amplitude, antisymmetrical phase distribution.

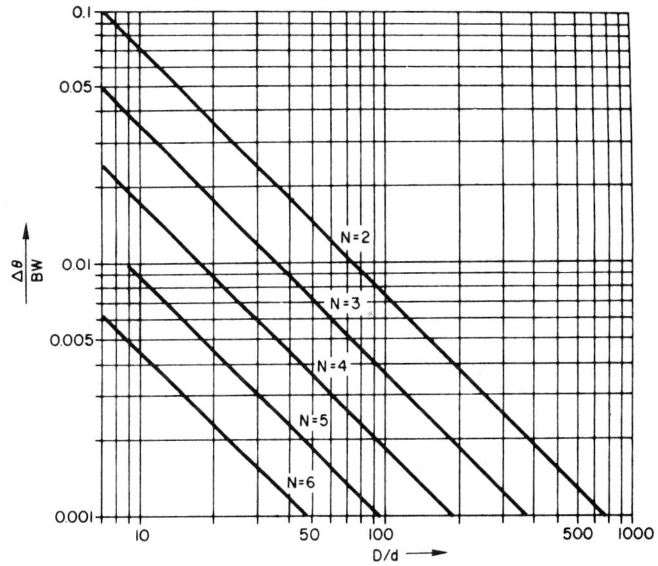


Fig. 9. First beam position, uniform amplitude, asymmetrical phase distribution.

TABLE II

BEAM POSITION OFF-BROADSIDE (IN BEAMWIDTHS) FOR FOUR-BIT QUANTIZATION, $D=32d$, ASYMMETRICAL PHASE DISTRIBUTION

| Beam Position | $\theta/BW$ | Fig. No. |
|---|---|---|
| 1 | 0.00576 | 2(b) |
| 2A | 0.01135 | 3(a) |
| 2B | 0.01155 | 3(b) |
| 3A | 0.01676 | 4(a) |
| 3B | 0.01731 | 4(b) |
| 3C | 0.01731 | 4(c) |
| 4A | 0.02303 | 5(a) |
| 4B | 0.02290 | 5(b) |
| 4C | 0.02265 | 5(c) |
| 4D | 0.02310 | 5(d) |
| 4E | 0.02303 | 5(e) |

TABLE III

FIRST BEAM POSITION OFF-BROADSIDE (IN BEAMWIDTHS) FOR FOUR-BIT QUANTIZATION-COSINE AMPLITUDE TAPER, ANTISYMMETRICAL PHASE DISTRIBUTION

| $D/d=2/\alpha$ | $\theta_1/BW$ | Percent of Position for Uniform Amplitude |
|---|---|---|
| 8 | 0.04180 | 51.0 |
| 16 | 0.01065 | 24.2 |
| 32 | 0.00255 | 11.2 |
| 64 | 0.000861 | 7.5 |
| 128 | 0.000328 | 5.7 |

Table II affords a measure of the granularity by observing the difference between adjacent beam positions. This granularity depends very strongly upon the number of elements in the array, or the number that are steered together.

The first beam position for an antisymmetrical phase, cosine amplitude distribution was computed following the procedure described in Section IV. The results for a four-bit quantization level are given in Table III for direct comparison with those of the uniform amplitude distribution given in Table I.

As would be expected intuitively, the first beam position of a tapered amplitude array is a fraction of the first beam position of a uniform amplitude array. Furthermore, as would also be expected, the fraction decreases as the number of elements in the array ($D/d$) increases for a given quantization level.

The average granularity between broadside and the first "exact" position was derived in Section V. The result is plotted in Fig. 10 for the asymmetrical case as a function of the number of elements (or array size to element size ratio), with the number of phase bits a parameter.

As the results for the average granularity depend only upon the phase and not the amplitude distribution, they are obviously independent of the latter. Since the granularity near broadside of a tapered amplitude distribution is less than that of a uniform amplitude distribution, the mean

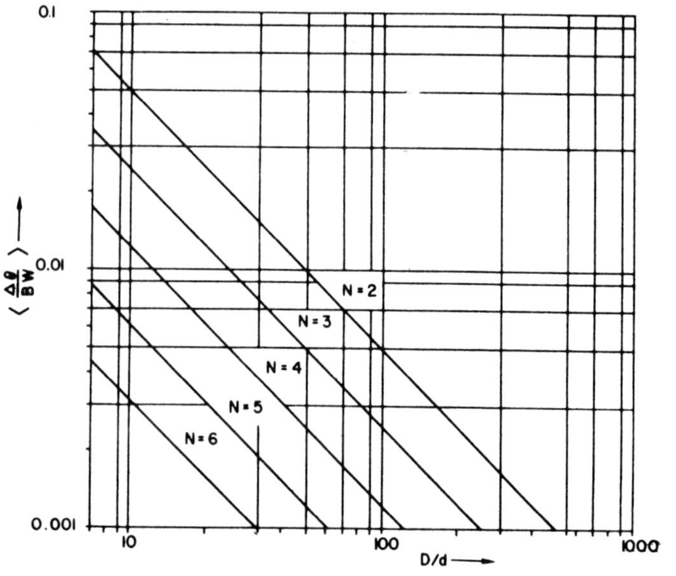


Fig. 10. Average separation between adjacent beam positions, asymmetrical phase distribution.

value theorem would imply that somewhere in the range between broadside and the first "exact" position, the granularity of a tapered distribution is greater than that of a uniform distribution. This conclusion casts doubt upon the validity of the hypothesis that the first beam position off-broadside is the case of greatest granularity for tapered amplitude distributions.

## REFERENCES

[1] S. Silver, *Microwave Antenna Theory and Design*. New York: McGraw-Hill, 1947, p. 184.

[2] T. C. Cheston and H. M. Grady, "A phased array using subarray techniques," *Digest, 1965 Internat'l Symp. IEEE Group on Antennas and Propagation*, pp. 98–101.

[3] R. C. Spencer, "A least square analysis of the effect of phase errors on antenna gain," Cambridge Field Station, Air Materiel Command, Cambridge, Mass., CFS Rept. E5025, January 1949.

[4] J. L. Allen *et al.*, "Phased array radar studies, July 1960 to July 1961," M.I.T. Lincoln Lab., Lexington, Mass., Tech. Rept. 236, p. 125, November 13, 1961.

# An Analytical Approach to the Coverage of a Hemisphere by N Planar Phased Arrays

JOHN L. KMETZO, MEMBER, IEEE

***Abstract*—The coverage of a hemisphere by $N$ planar phased arrays arranged in pyramids or pyramidal frustra is discussed. Transformations between the spherical coordinates of individual arrays and of the hemisphere of coverage are derived. These results are then used to determine the minimum value of the maximum scan angle required of each array and the positioning of the arrays. Numerical results are given.**

**These transformations are also used to minimize the number of elements needed in a given array aperture. This is shown for an example with both equilateral triangular and rectangular element spacing. It is further demonstrated how the use of an isosceles triangular element spacing affords additional reduction of the number of elements needed.**

## INTRODUCTION

IN ESTABLISHING the number of planar arrays necessary to provide hemispherical coverage above a plane, two general classes of geometry may be considered. The $N$ arrays may be arranged in a pyramid of $N$ sides, or in a pyramidal frustrum of $(N-1)$ sides with the top of the frustrum accommodating the $N$th array. A recent paper[1] has solved the problem of determining the minimum value of the maximum scan angle required of each array by means of a graphical technique.

In this paper a set of transformations is derived which leads to an analytical solution of this problem. Furthermore, it is shown that the transformations are a powerful tool for use in minimizing the number of elements needed in an array while still avoiding the formation of end-fire grating lobes.

## COORDINATE TRANSFORMATIONS

Figure 1 defines the standard right-handed spherical coordinate system that will be used. The relationships between the Cartesian coordinates $(x, y, z)$ and the spherical coordinates $(r, \theta, \phi)$ are given by the following transformations:

$$\begin{vmatrix} x \\ y \\ z \end{vmatrix} = \begin{vmatrix} r(\sin\theta)(\cos\phi) \\ r(\sin\theta)(\sin\phi) \\ r(\cos\theta) \end{vmatrix} \qquad (1)$$

$$\begin{vmatrix} r \\ \theta \\ \phi \end{vmatrix} = \begin{vmatrix} (x^2+y^2+z^2)^{1/2} \\ \cos^{-1}(z/r) \\ \tan^{-1}(y/x) \end{vmatrix} \qquad (2)$$

Manuscript received October 26, 1966. The work reported in this paper was sponsored by the Air Force Avionics Laboratory, Wright-Patterson AFB, Ohio, under Contract AF33(615)-2986.

The author is with the Radio Division, The Bendix Corporation, Baltimore, Md.

[1] G. H. Knittel, "Choosing the number of faces of a phased-array antenna for hemisphere scan coverage," *IEEE Trans. on Antennas and Propagation*, vol. AP-13, pp. 878–882, November 1965.

Two sets of variables are defined: $(x, y, z; r, \theta, \phi)$ are hereafter referred to as "coverage coördinates" while $(x_0, y_0, z_0; r_0, \theta_0, \phi_0)$ are referred to as "array coordinates." The $(x\text{-}y)$ plane of the coverage coordinate system lies in the flat side of the hemisphere of coverage, while the $(x_0\text{-}y_0)$ plane of the array coordinate system lies in the plane of an array face. Furthermore, since only the direction angles at a far-field point are to be considered, it is assumed that the array faces are located at a point corresponding to the common origin of the coverage and array coordinate systems.

The above discussion is summarized in Fig. 2, which shows the orientation of one of the side faces of a four-sided pyramid. The plane of the face is tilted back from the normal to the coverage plane at a "tilt angle" $\alpha$. This corresponds to a CCW rotation about the $y$-axis by $(90°-\alpha)$. Thus the transformation $T$ of the coverage coordinates into the array coordinates is given by:[2]

$$T = \begin{vmatrix} \cos(90°-\alpha) & 0 & -\sin(90°-\alpha) \\ 0 & 1 & 0 \\ \sin(90°-\alpha) & 0 & \cos(90°-\alpha) \end{vmatrix} = \begin{vmatrix} \sin\alpha & 0 & -\cos\alpha \\ 0 & 1 & 0 \\ \cos\alpha & 0 & \sin\alpha \end{vmatrix}.$$

Then:

$$\begin{vmatrix} x_0 \\ y_0 \\ z_0 \end{vmatrix} = \begin{vmatrix} \sin\alpha & 0 & -\cos\alpha \\ 0 & 1 & 0 \\ \cos\alpha & 0 & \sin\alpha \end{vmatrix} \begin{vmatrix} x \\ y \\ z \end{vmatrix}. \qquad (3)$$

Since the inverse of an orthogonal transformation matrix is identical to its transpose, it follows that

$$\begin{vmatrix} x \\ y \\ z \end{vmatrix} = \begin{vmatrix} \sin\alpha & 0 & \cos\alpha \\ 0 & 1 & 0 \\ -\cos\alpha & 0 & \sin\alpha \end{vmatrix} \begin{vmatrix} x_0 \\ y_0 \\ z_0 \end{vmatrix}. \qquad (4)$$

Substituting (1) into (3) gives

$$\begin{vmatrix} x_0 \\ y_0 \\ z_0 \end{vmatrix} = \begin{vmatrix} r(\sin\theta)(\cos\phi)\sin\alpha - r(\cos\theta)\cos\alpha \\ r(\sin\theta)\sin\phi \\ r(\cos\theta)\sin\alpha + r(\sin\theta)(\cos\phi)\cos\alpha \end{vmatrix},$$

and substituting the above results into (2) gives:

[2] Herbert Goldstein, *Classical Mechanics*. Reading, Mass.: Addison-Wesley, 1952, pp. 99–101.

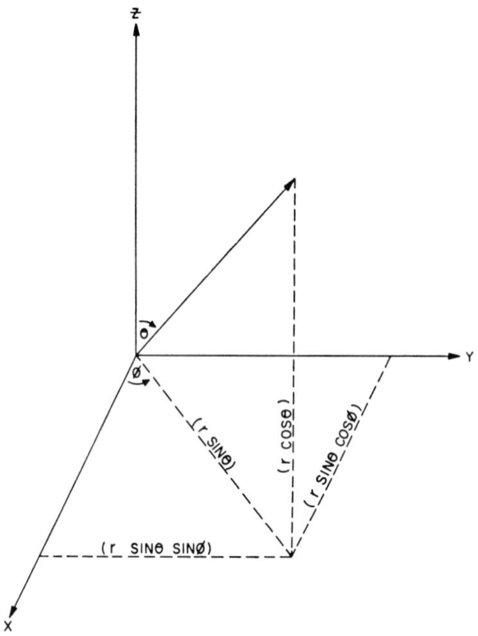


Fig. 1. Definition of spherical coordinate systems used.

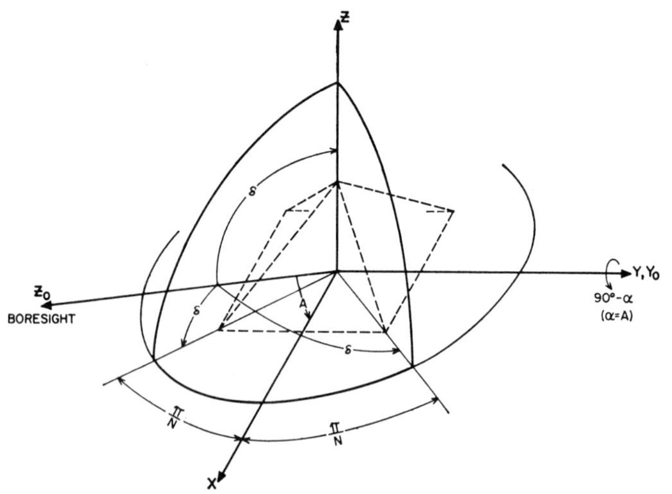


Fig. 2. Geometry of coverage by a pyramid.



$$\theta_0 = \cos^{-1}[\cos\theta(\sin\alpha) + \sin\theta(\cos\phi)\cos\alpha] \tag{5}$$

$$\phi_0 = \tan^{-1}\left[\frac{\sin\theta(\sin\phi)}{\sin\theta(\cos\phi)\sin\alpha - \cos\theta(\cos\alpha)}\right]. \tag{6}$$

Equations (1), (4), and (2) may be combined in a similar manner to give:

$$\theta = \cos^{-1}[\cos\theta_0 \sin\alpha - \sin\theta_0 \cos\phi_0 \cos\alpha] \tag{7}$$

$$\phi = \tan^{-1}\left[\frac{\sin\theta_0 \sin\phi_0}{\sin\theta_0 \cos\phi_0 \sin\alpha + \cos\theta_0 \cos\alpha}\right]. \tag{8}$$

Equations (5)–(8) now allow the beam steering coordinates of the array to be transformed into coordinates of the hemisphere of coverage and vice versa. We may now turn our attention to the problem of determining the minimum value $\delta$ of the scan angle and the corresponding value $A$ of the tilt angle required to satisfy the conditions of hemispherical coverage.

## Determination of Scan and Tilt Angles

Consider first the simple pyramid. In Fig. 2, it may be seen that the boundary condition for coverage of the zenith by the conical scan requires

$$A = 90° - \delta. \tag{9}$$

At the intersection in the horizon plane of the conical scan of two adjacent sides the following boundary conditions exist:

$$\theta = \pi/2$$
$$\phi = \pi/N$$
$$\theta_0 = \delta.$$

These boundary conditions substituted into (5) and (9) result in

$$\cos\delta = \cos(\pi/2)\sin(90° - \delta) + \sin(\pi/2)\cos(\pi/N)\cos(90° - \delta)$$

$$\cos\delta = 0 + \cos(\pi/N)\sin\delta$$

$$\delta = \tan^{-1}\left[\frac{1}{\cos(\pi/N)}\right]. \tag{10}$$

The geometry for the case of the $N$-array pyramidal frustrum is shown in Fig. 3. As in the case of the pyramid, there are a set of boundary conditions at the horizon:

$$\theta = \pi/2$$
$$\phi = \pi/(N-1)$$
$$\theta_0 = \delta.$$

Substituting into (5)

$$\cos\delta = \cos(\pi/2)\sin A + \sin(\pi/2)\cos(\pi/N - 1)\cos A$$

$$\cos\delta = \cos(\pi/(N-1))\cos A. \tag{11}$$

A second set of boundary conditions exists at the intersection of the cone of scan of a side array with the cone of scan of a top array. As before,

$$\theta_0 = \delta$$
$$\phi = \pi/(N-1).$$

Since the coordinates of the top array ($\alpha = 90°$) correspond to coverage coordinates, (5) or (7) shows that

$$\theta = \delta.$$

Once again substituting into (5) gives

$$\cos\delta = \cos\delta\sin A + \sin\delta\cos(\pi/(N-1))\cos A. \tag{12}$$

The two unknowns, $A$ and $\delta$, are now determined from (11) and (12)

$$1 - 2\sin\delta\cos^2(\pi/(N-1)) - \sin^2\delta\sin^2(\pi/(N-1)) = 0 \tag{13}$$

$$A = \sin^{-1}[1 - \sin\delta]. \tag{14}$$

Equation pairs (9)-(10) and (13)-(14) have been solved for various numbers of arrays and the results are summarized in Table I.

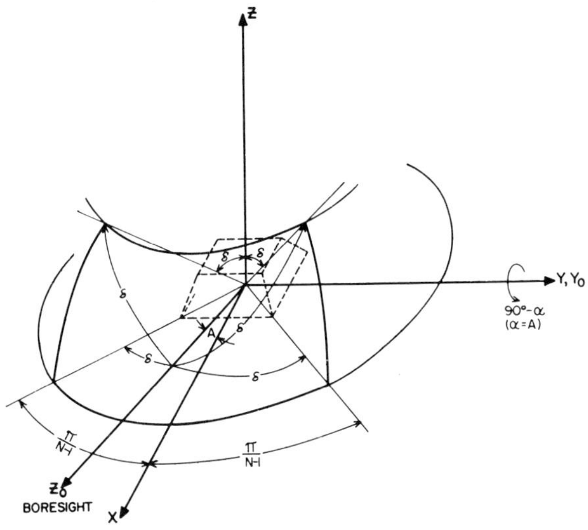


Fig. 3. Geometry of coverage by a pyramidal frustrum.

TABLE I

SCAN AND TILT ANGLES FOR HEMISPHERICAL COVERAGE

| Number of Arrays | Pyramid | | Pyramidal Frustrum | |
|---|---|---|---|---|
| $N$ | $\delta$ (deg) | $A$ (deg) | $\delta$ (deg) | $A$ (deg) |
| 3 | 63.435 | 26.565 | — | — |
| 4 | 54.736 | 35.264 | 60.287 | 7.555 |
| 5 | 51.027 | 38.973 | 47.059 | 15.542 |
| 6 | 49.107 | 40.893 | 40.678 | 20.377 |
| 7 | 47.982 | 42.018 | 37.268 | 23.232 |
| 8 | 47.266 | 42.734 | 35.262 | 25.004 |
| 9 | 46.781 | 43.219 | 33.985 | 26.169 |
| 10 | 46.437 | 43.563 | 33.124 | 26.971 |
| . | | | | |
| . | | | | |
| . | | | | |
| . | | | | |
| $\infty$ | 45.000 | 45.000 | 30.000 | 30.000 |

## GRATING LOBE FORMATION—TRIANGULAR SPACING

The usefulness of (5) and (6) in designing an array with maximum element spacing on the basis of avoidance of end-fire grating lobes will now be shown. A five-array pyramidal frustrum is taken as an example. Figure 4 is a polar representation in coverage coordinates of the scan region required of each array. It may be seen that the scan region of a side array is bordered by lines of constant $\phi$ or constant $\theta$. The scan region of one of these faces may be transformed into the array coordinates of that face by means of (5) and (6). Figures 6, 7, and 9 show this scan region in the side array coordinate system.

The elements of the array are assumed to be arranged in a triangular spacing pattern[3] as shown in Fig. 5. If the location of a grating lobe in array coordinates is indicated by $(\theta', \phi')$, it may be shown that the coordinates of the beam giving rise to this grating lobe are given by

$$\theta_0 = \sin^{-1}[(\sin\theta' \sin\phi' - j\lambda/2d)^2 + (\sin\theta' \cos\phi' - \lambda(2k+j)/D)^2]^{1/2} \quad (15)$$

$$\phi_0 = \tan^{-1}\left[\frac{\sin\theta' \sin\phi' - j\lambda/2d}{\sin\theta' \cos\phi' - \lambda(2k+j)/D}\right]. \quad (16)$$

These equations result in six grating lobe regions corresponding to values of $(k,j)=(\pm 1, 0)$, $(0, \pm 1)$, and $(\pm 1, \mp 1)$. Actually, there are in fact an infinite number of grating lobe regions corresponding to combinations of values of $k$, $j = \pm 1, \pm 2, \cdots, \pm N$. However, these "higher-order" grating lobes would always be contained within the regions of the primary grating lobes (especially for element spacings between $\lambda/2$ and $\lambda$) and so are not of any practical significance.

The grating lobes may be located at end-fire by setting $\sin\theta' = 1$. Equations (15) and (16) then simplify to the following expressions:

$$\theta_0 = \sin^{-1}[(\sin\phi' - j\lambda/2d)^2 + (\cos\phi' - \lambda(2k+j)/D)^2]^{1/2} \quad (17)$$

$$\phi_0 = \tan^{-1}\left[\frac{\sin\phi' - j\lambda/2d}{\cos\phi' - \lambda(2k+j)/D}\right]. \quad (18)$$

Equations (17) and (18) contain an independent variable $\phi'$ which varies from 0° to 360° around the array plane. The locus of points $(\theta_0, \phi_0)$ denotes the border of the end-fire grating lobe regions for each set of values $(k, j)$. These regions are indicated in Figs. 6 and 7. If the beam is steered at any position $(\theta_0, \phi_0)$ outside of these regions, no grating lobe will appear. If the beam is steered into these regions, a corresponding grating lobe will appear above the array plane. The object of design, then, is to increase the element spacing until the grating lobe regions border the region of desired scan. This has been shown in Fig. 6 for the case of equilateral triangular spacing, where $d = (\sqrt{3}/2)D$.

In Fig. 7 the more general case of isosceles triangular spacing, where the dimensions $d$ and $D$ are considered to be independent, is shown. This allows the interelement spacing to be increased so that the grating lobe regions more closely border the scan region of the array. This may be clearly seen from a comparison of Figs. 6 and 7.

The aperture area associated with each element is

$$S = d \cdot D.$$

For the equilateral case we have:

$$S_{eqi} = (\sqrt{3}/2)D^2 = 0.398\lambda^2$$

and for the isosceles case we have:

$$S_{iso} = 0.685\lambda \times 0.644\lambda = 0.441\lambda^2.$$

[3] E. D. Sharp, "A triangular arrangement of planar-array elements that reduces the number needed," *IEEE Trans. on Antennas and Propagation*, vol. AP-9, pp. 126–129, March 1961.

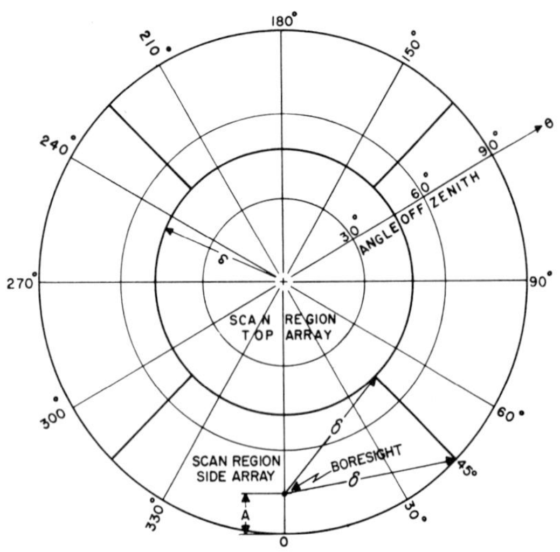


Fig. 4. Coverage regions of a 5-array pyramidal frustrum.

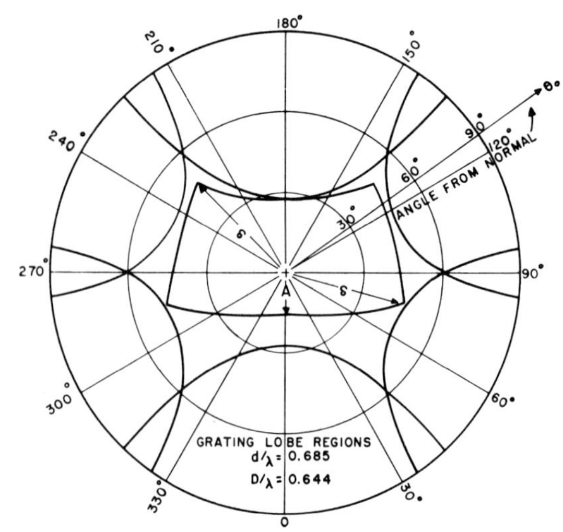


Fig. 7. Scan and grating lobe regions for an isosceles spaced triangular array on the side of a 5-array pyramidal frustrum.

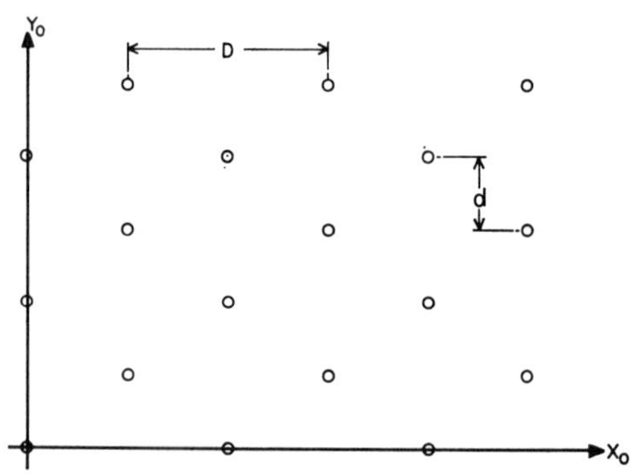


Fig. 5. Element arrangement—triangular spacing.

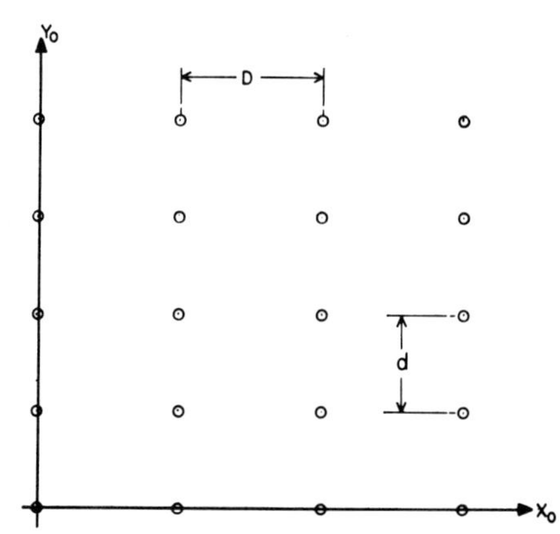


Fig. 8. Element arrangement—rectangular spacing.

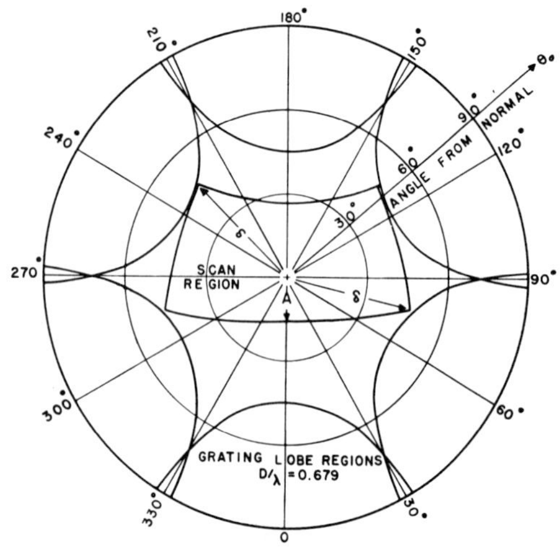


Fig. 6. Scan and grating lobe regions for an equilateral spaced triangular array on the side of a 5-array pyramidal frustrum.

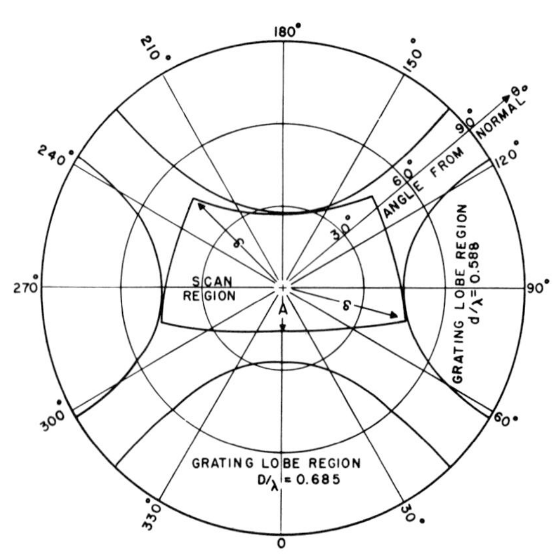


Fig. 9. Scan and grating lobe regions for a rectangular spaced array on the side of a 5-array pyramidal frustrum.

Thus, for a given aperture area the isosceles spacing will employ fewer elements that the equilateral spacing:

$$N_{iso}S_{iso} = N_{eqi}S_{eqi}$$

$$N_{iso} = \frac{S_{eqi}}{S_{iso}} N_{eqi} = \frac{0.398}{0.441} N_{eqi}$$

$$N_{iso} = 0.904\, N_{eqi}.$$

This indicates that an additional 9.9 percent savings in elements would be obtained in this case by using isosceles rather than equilateral triangular spacing.

## Rectangular Spacing

The same methods as have been outlined above may also be applied to a rectangular element grid. If a spacing grid as shown in Fig. 8 is assumed, then the coordinates of the beam as a function of grating lobe location are:

$$\theta_0 = \sin^{-1}\left[(\sin\theta' \sin\phi' - j\lambda/d)^2 + (\sin\theta' \cos\phi' - k\lambda/D)^2\right]^{1/2} \quad (19)$$

$$\phi_0 = \tan^{-1}\left[\frac{\sin\theta' \sin\phi' - j\lambda/d}{\sin\theta' \cos\phi' - k\lambda/D}\right] \quad (20)$$

where there are now four primary grating lobe regions corresponding to values of $k=\pm 1, j=\pm 1; |k| \neq |j|$. Figure 9 shows maximized element spacing for a rectangular spaced array on the side face of the five-array pyramidal frustrum.

The borders of the grating lobe regions were found by setting $\sin\theta'=1$ and varying $0 \leq \phi' \leq 360°$ as before and evaluating (19) and (20).

## Conclusion

It should be pointed out that the coverage of the side face of a pyramidal frustrum has been discussed in the preceding examples. As may be seen from Fig. 3, the scan region of a top array is conical and therefore requires a different element arrangement. For a given conical scan angle $\delta$ the element spacing required is

$$d = \frac{\lambda}{1 + \sin\delta}$$

and

$D=(2/\sqrt{3})d$, Triangular (Equilateral) Spacing
$D=d$, Rectangular (Square) Spacing.

This paper has presented a method of determining the coverage of a hemisphere provided by one array of an arrangement of $N$ arrays. The transformations involved allow the positioning of the array to be determined and its maximum required angle of scan to be minimized. This also results in a ready method of maximizing the element spacing while still avoiding the formation of end-fire grating lobes. Other uses for these transformations could undoubtedly be found.

## Acknowledgment

The author is especially indebted to W. E. Schneider of the Bendix Radio Division for many interesting and informative discussions on these topics. Thanks are also due to H. S. Gruenberger of the Bendix Radio Division for his encouragement and cooperation in the preparation and review of this paper, to Mrs. Shirley Brown for her help with the drawings, and to June Horton for typing the manuscript.